D1236287

COLLEGE GEOMETRY
A UNIFIED DEVELOPMENT

TEXTBOOKS in MATHEMATICS

Series Editor: Denny Gulick

PUBLISHED TITLES

ABSTRACT ALGEBRA: AN INTERACTIVE APPROACH
William Paulsen

COLLEGE GEOMETRY: A UNIFIED DEVELOPMENT
David C. Kay

COMPLEX VARIABLES: A PHYSICAL APPROACH WITH APPLICATIONS AND MATLAB®
Steven G. Krantz

ESSENTIALS OF TOPOLOGY WITH APPLICATIONS
Steven G. Krantz

INTRODUCTION TO ABSTRACT ALGEBRA
Jonathan D. H. Smith

INTRODUCTION TO MATHEMATICAL PROOFS: A TRANSITION
Charles E. Roberts, Jr.

INTRODUCTION TO PROBABILITY WITH MATHEMATICA®, SECOND EDITION
Kevin J. Hastings

LINEAR ALBEBRA: A FIRST COURSE WITH APPLICATIONS
Larry E. Knop

LINEAR AND NONLINEAR PROGRAMMING WITH MAPLE™: AN INTERACTIVE, APPLICATIONS-BASED
APPROACH
Paul E. Fishback

MATHEMATICAL AND EXPERIMENTAL MODELING OF PHYSICAL AND BIOLOGICAL PROCESSES
H. T. Banks and H. T. Tran

ORDINARY DIFFERENTIAL EQUATIONS: APPLICATIONS, MODELS, AND COMPUTING
Charles E. Roberts, Jr.

TEXTBOOKS in MATHEMATICS

COLLEGE GEOMETRY
A UNIFIED DEVELOPMENT

David C. Kay

University of North Carolina

Asheville, North Carolina, USA (retired)

CRC Press
Taylor & Francis Group
Boca Raton   London   New York

CRC Press is an imprint of the
Taylor & Francis Group an **informa** business
A CHAPMAN & HALL BOOK

Acknowledgment is due for certain items in this book including Historical Notes, which appear at different points throughout this book. Some of these notes were taken from Kay, David C., *College Geometry: A Discovery Approach*, pp. 19, 58, 79, 128, 188, 256–257, and 424–425, © 2001 Addison Wesley Longman Inc. Reproduced by permission of Pearson Education, Inc.

CRC Press
Taylor & Francis Group
6000 Broken Sound Parkway NW, Suite 300
Boca Raton, FL 33487-2742

© 2012 by Taylor & Francis Group, LLC
CRC Press is an imprint of Taylor & Francis Group, an Informa business

No claim to original U.S. Government works

Printed in the United States of America on acid-free paper
Version Date: 20110518

International Standard Book Number: 978-1-4398-1911-1 (Hardback)

This book contains information obtained from authentic and highly regarded sources. Reasonable efforts have been made to publish reliable data and information, but the author and publisher cannot assume responsibility for the validity of all materials or the consequences of their use. The authors and publishers have attempted to trace the copyright holders of all material reproduced in this publication and apologize to copyright holders if permission to publish in this form has not been obtained. If any copyright material has not been acknowledged please write and let us know so we may rectify in any future reprint.

Except as permitted under U.S. Copyright Law, no part of this book may be reprinted, reproduced, transmitted, or utilized in any form by any electronic, mechanical, or other means, now known or hereafter invented, including photocopying, microfilming, and recording, or in any information storage or retrieval system, without written permission from the publishers.

For permission to photocopy or use material electronically from this work, please access www.copyright.com (http://www.copyright.com/) or contact the Copyright Clearance Center, Inc. (CCC), 222 Rosewood Drive, Danvers, MA 01923, 978-750-8400. CCC is a not-for-profit organization that provides licenses and registration for a variety of users. For organizations that have been granted a photocopy license by the CCC, a separate system of payment has been arranged.

Trademark Notice: Product or corporate names may be trademarks or registered trademarks, and are used only for identification and explanation without intent to infringe.

Library of Congress Cataloging-in-Publication Data

Kay, David C., 1933-
   College geometry : a unified development / David C. Kay.
      p. cm. -- (Textbooks in mathematics)
   Includes bibliographical references and index.
   ISBN 978-1-4398-1911-1 (alk. paper)
   1. Geometry--Textbooks. I. Title. II. Series.

QA445.K388 2011
516.0076--dc22
                                                              2011001821

Visit the Taylor & Francis Web site at
http://www.taylorandfrancis.com

and the CRC Press Web site at
http://www.crcpress.com

To Katie

Contents

Preface . . . . **xi**
Author . . . . **xv**

1   Lines, Distance, Segments, and Rays. 1
   1.1   Intended Goals   1
   1.2   Axioms of Alignment   2
   1.3   A Glimpse at Finite Geometry   8
   1.4   Metric Geometry   18
   1.5   Eves' 25-Point Affine Geometry: A Model for Axioms 0–4   21
   1.6   Distance and Alignment   25
   1.7   Properties of Betweenness: Segments and Rays   27
   1.8   Coordinates for Rays   35
   1.9   Geometry and the Continuum   45
   1.10 Segment Construction Theorems   47

2   Angles, Angle Measure, and Plane Separation. 53
   2.1   Angles and Angle Measure   53
   2.2   Plane Separation   59
   2.3   Consequences of Plane Separation: The Postulate of Pasch   63
   2.4   The Interior of an Angle: The Angle Addition Postulate   69
   2.5   Angle Construction Theorems   82
   2.6   Consequences of a Finite Metric   86

3   Unified Geometry: Triangles and Congruence 95
   3.1   Congruent Triangles: SAS Hypothesis   95
   3.2   A Metric for City Centers   100
   3.3   The SAS Postulate and the ASA and SSS Theorems   110
   3.4   Euclid's Superposition Proof: An Alternative to Axiom 12   113
   3.5   Locus, Perpendicular Bisectors, and Symmetry   116
   3.6   The Exterior Angle Inequality   128
   3.7   Inequalities for Triangles   132
   3.8   Further Congruence Criteria   144
   3.9   Special Segments Associated with Triangles   153

4   Quadrilaterals, Polygons, and Circles 161
   4.1   Quadrilaterals   161
   4.2   Congruence Theorems for Convex Quadrilaterals   165
   4.3   The Quadrilaterals of Saccheri and Lambert   166
   4.4   Polygons   176
   4.5   Circles in Unified Geometry   186

5   Three Geometries **203**
    5.1   Parallelism in Unified Geometry and the Influence of α 203
    5.2   Elliptic Geometry: Angle-Sum Theorem   210
    5.3   Pole-Polar Theory for Elliptic Geometry   214
    5.4   Angle Measure and Distance Related: Archimedes' Method   217
    5.5   Hyperbolic Geometry: Angle-Sum Theorem   224
    5.6   A Concept for Area: AAA Congruence   228
    5.7   Parallelism in Hyperbolic Geometry   231
    5.8   Asymptotic Triangles in Hyperbolic Geometry   236
    5.9   Euclidean Geometry: Angle-Sum Theorem   251
    5.10   Median of a Trapezoid in Euclidean Geometry   256
    5.11   Similar Triangles in Euclidean Geometry   260
    5.12   Pythagorean Theorem   266

6   Inequalities for Quadrilaterals: Unified Trigonometry **281**
    6.1   An Inequality Concept for Unified Geometry   281
    6.2   Ratio Inequalities for Trapezoids   284
    6.3   Ratio Inequalities for Right Triangles   295
    6.4   Orthogonal Projection and "Similar" Triangles   297
    6.5   Unified Trigonometry: The Functions $c(\theta)$ and $s(\theta)$ 306
    6.6   Trigonometric Identities   311
    6.7   Classical Forms for $c(\theta)$ and $s(\theta)$ 315
    6.8   Lambert Quadrilaterals and the Function $C(u)$ 316
    6.9   Identities for $C(u)$ 320
    6.10   Classical Forms for $C(u)$ 322
    6.11   The Pythagorean Relation for Unified Geometry   325
    6.12   Classical Unified Trigonometry   328

7   Beyond Euclid: Modern Geometry **341**
    7.1   Directed Distance: Stewart's Theorem and the Cevian Formula   341
    7.2   Formulas for Special Cevians   347
    7.3   Circles: Power Theorems and Inscribed Angles   353
    7.4   Using Circles in Geometry   359
    7.5   Cross Ratio and Harmonic Conjugates   374
    7.6   The Theorems of Ceva and Menelaus   381
    7.7   Families of Mutually Orthogonal Circles   399

8   Transformations in Modern Geometry **411**
    8.1   Projective Transformations   411
    8.2   Affine Transformations   425
    8.3   Similitudes and Isometries   438
    8.4   Line Reflections: Building Blocks for Isometries and Similitudes   448
    8.5   Translations and Rotations   464
    8.6   Circular Inversion   477

9   Non-Euclidean Geometry: Analytical Approach **489**

    9.1   Law of Sines and Cosines for Unified Geometry 490

    9.2   Unifying Identities for Unified Trigonometry 496

    9.3   Half-Angle Identities for Unified Geometry 498

    9.4   The Shape of a Triangle in Unified Geometry: Cosine Inequality 501

    9.5   The Equations of Gauss: Area of a Triangle 507

    9.6   Directed Distance: Theorems of Menelaus and Ceva 515

    9.7   Poincarè's Model for Hyperbolic Geometry 525

    9.8   Other Models: Surface Theory 539

    9.9   Hyperbolic Parallelism and Bolyai's Ideal Points 545

Appendix A: Sketchpad Experiments **557**

Appendix B: Intuitive Spherical Geometry **569**

Appendix C: Proof in Geometry **573**

Appendix D: The Real Numbers and Least Upper Bound **577**

Appendix E: Floating Triangles/Quadrilaterals **581**

Appendix F: Axiom Systems for Geometry **587**

Solutions to Selected Problems **591**

Bibliography **621**

Index **623**

Preface

The famous book by Nathan A. Court, *College Geometry* (1952), used for many years for courses by that same title for teachers and mathematics majors, was the original inspiration for this book. When the author accepted a position at Oklahoma in 1965, Dr. Court had retired from teaching but was still lecturing. The content of those lectures, and his book, covered many elegant concepts that, in the author's opinion, should be taught in such a course today. The title of this book thus pays homage to Court and this wonderful area of geometry that has somehow been lost on generations of mathematics students. The author's original book by that title (published in 1969 by Holt, Rinehart, and Winston, out of print for many years) attempted to embrace much of the material traditionally taught in geometry courses at that time, with a slightly new twist. It included the study of all three classical geometries and was designed to reflect other goals that students of geometry should achieve, including an appreciation of just how hyperbolic geometry comes about. (It is, after all, a very natural occurrence in an axiomatic setting similar to that found in Euclid's *Elements*.) Thus, the three classical geometries were given a common axiomatic basis in just 11 axioms in the author's 1969 book. An updated version of that system (with 12 axioms) is developed here under the name *unified geometry*, also reflected in the title.

The author believes in gradually developing a central theme in any course in mathematics, including geometry, providing a backdrop for various other topics that can be introduced at the appropriate time, rather than covering several themes all at once, with a potpourri of topics. A concerted effort was made to approach every topic as a fresh, new concept and to carefully define and explain the concept where appropriate. Thus, ideas obvious to the teacher are not treated as such in this book. (In fact, such words as "obvious" and "clear" have been purged.) An important, often neglected aspect of mathematics education is that each generation of students needs the same core basics as previous generations, in spite of the temptation to move on to new ideas and to skip such basics.

This book was written to communicate directly with a beginning student in geometry (although with some background as normally encountered in the first semester of calculus—a recommended prerequisite for this material). Limits and the concept of least upper bound occur frequently. In spite of this, the book starts out on a very elementary level. Chapter 1 begins with common ideas about points, lines, and distance—material that evolves to include the concept of a line and its coordinate system (for both Euclidean and spherical geometry). Chapter 2 continues the development to include angles and half-planes. The "plot" becomes more involved in Chapter 3 where much of the basic geometry normally encountered in a high school course is covered, including a new congruence criterion for triangles, and elementary inequalities crucial for later topics. Chapter 4 shows how to develop quadrilaterals and circles on the sphere (as well as in the Euclidean plane), and then Chapter 5 explores in some detail the idea that unified geometry is actually the study of *three* distinctly different geometries.

Having laid a proper foundation, the topics now build to a climax in Chapter 6, where the trigonometry of the right triangle, including the Pythagorean theorem (along with the classical formulas for spherical and hyperbolic geometry), is finally established. Here is where extensive use of limits occurs; the sine and cosine functions are defined in terms of limits of certain ratios. Once the domain of graduate level mathematics, this material is explained in a manner understandable by the ambitious undergraduate student. This answers the natural question of whether it is possible to develop trigonometry for non-Euclidean geometry without the use of models.

The rest of the book is then devoted to applications of Euclidean geometry in solving problems, some of which are actually practical (Chapter 7), including transformation theory in Chapter 8, and further details on non-Euclidean geometry in Chapter 9.

The axioms are thus formulated to include both Euclidean and elliptic geometry, and, as an unintended consequence (to the reader), hyperbolic geometry. This makes the study largely observable in the real world, which also has far-reaching implications. The reader will at first believe that the purpose is to study the foundations of the geography of the world we live in (the great circle routes jetliners travel and shortest distance on a sphere are discussed frequently). In Chapter 5 it is discovered that all the while, a third (and quite famous) geometry was also included in that study. Such material is the first step on the road toward a possible study of Riemannian geometry, Einstein's relativity, and theories of cosmology.

Admittedly, a course based on this book is somewhat more sophisticated than the ordinary course in geometry for teachers (for which the author's book, *College Geometry: A Discovery Approach*, 2nd Edn. [2001. Addison, Wesley, Longman], was designed), and is intended more generally for mathematics majors as well as more advanced mathematics education students. The teaching of high school geometry is addressed, with material found in high school textbooks currently used often discussed and/or criticized. Some problems are similar to, or compared with, those found in high school textbooks.

The book, however, was mainly written for a student who simply wants to learn more about geometry and about the foundations that explain where certain mathematical concepts originate. For example, one encounters metric spaces in Chapter 1—the concept of the continuum is discussed later in that chapter as it applies to lines; Chapter 2 provides the proof of the crossbar principle, an indispensable part of geometry; Chapter 3 deals with two versions of the SAS postulate and equivalence relations. Also, numerous other important concepts are covered in the rest of the chapters (the general linear group $GL(2)$ is introduced in a problem in Chapter 8). The spirit of this book is elementary geometry having implications in advanced geometry.

Several passages of explanation are sometimes included, providing the reader assistance and encouragement. Frequently a sequence of figures appears to illustrate various steps in a proof (where one is normally considered adequate). Sometimes, a figure is used to illustrate the statement of a theorem, and a separate figure is used to illustrate its proof. The Moments for Reflection challenge students to think for themselves and to participate in the development by self-discovery. Other features of the book include the following:

- Carefully crafted problems, graded according to (perceived) difficulty
- Instructions for specific experiments using *Geometer's Sketchpad* software
- Topological shapes of geometric objects frequently discussed
- Historical notes to point out the human aspects of geometric discovery

Parts of this book may be used for existing (or new) courses in geometry, and that pursuit will be left to the instructor to decide what to use and what to omit. We believe that the instructor knows best how to use the book, so no attempt is made to anticipate the various teaching applications this book may have.

A few statistics will further enlighten the features of this book:

- More than 500 drawings
- More than 700 problems, all tested for feasibility
- Solutions for most odd- and a few even-numbered problems
- Eighty examples to illustrate concepts and problem solving
- Six undergraduate projects (suggested in the opening chapters)

Note: For the convenience of the instructor in selecting possible test problems or other purposes, solutions are given for selected odd-numbered problems. With rare exceptions when deemed helpful by the author even-numbered problems are noted by an asterisk following the problem number. The asterisk for even-numbered problems indicate answers that appear at the back. Also, for greater clarification, the symbol ⬊ appears at the end of each formal proof, and at the end of each note.

Author

David C. Kay received his doctorate at Michigan State University in 1963 in the area of metric spaces and has taught mathematics at the University of Wyoming, the University of Oklahoma, the University of North Carolina at Asheville (UNCA), and, after retirement, at North Georgia College and State University. He was chairman of the department at UNCA from 1983 to 1992 and retired from teaching there in 1997. He is the author of 20 research papers in refereed journals in the areas of metric geometry and convexity theory, and has written three books on geometry and its applications: *College Geometry* (Holt, Rinehart, and Winston 1969); Schaum's Outline Series, *Tensor Calculus* (McGraw-Hill 1988); and *College Geometry: A Discovery Approach*, 2nd edition (Addison, Wesley and Longman 2000).

1

Lines, Distance, Segments, and Rays

1.1 Intended Goals

It is appropriate to take a moment to address the goals and style of this book. It is written primarily with you, the student, in mind. The attempt is made as much as possible to explain what is being done (defining terms or proving theorems) and why (the historical background). The appendices are to fill in the gaps where necessary. Being a book about geometry, naturally the objects of study are sets of points in the plane or in space, either realized in actuality in coordinate geometry (models and examples) or as imaginary entities in the axiomatic development. There are many ways to approach the subject of geometry, but virtually all secondary school books on geometry remain traditional in nature. There is a very good reason for this; Euclid's axiomatic approach has been found to be the best way to learn the subject of geometry, in spite of various attempts at innovation. As Euclid himself is supposed to have once said, in so many words, that there is no "quick and easy" way to learn geometry. But it is perhaps the best way. It is certainly the most elegant.

Euclid began with a set of 10 principles he called *common notions* and *postulates*. The five common notions were algebraic in nature ("if equals are added to equals, the wholes are equal"), and the five postulates were those dealing with geometric objects (e.g., points, lines, circles, angles). For example, one of the postulates reads "a finite straight line [a *segment* in our development] can be produced continuously in a line." The famous Postulate Five basically states: *If two lines do not make either equal or supplementary angles with a transversal, they intersect*. This fifth postulate of parallels started a controversy in mathematics about which much will be said later. Despite the many shortcomings found in the *Elements*, this work serves as the beginners' handbook. The presentation is concise, and the objects of study can be easily drawn and visualized through diagrams. This method of development has come to be known as **synthetic geometry** (geometry without numbers).

In order to avoid falling into the trap of presuming Euclid's Postulate 5 to be true, and thus missing out on one of the most interesting and prevailing pursuits in mathematics, we purposely set out to study a geometry that everyone would agree cannot make use of the parallel postulate of Euclid in any way. Our development will include *spherical geometry*.

The reader will at first be engaged in formulating axioms that will include in a natural and familiar way the geometry of the sphere, as well as ordinary (Euclidean) geometry *simultaneously*. Only the most basic knowledge about spherical geometry—evident from the geography of the earth's surface—will be used in the process; Appendix **B** provides some details for those who need it.

This book is intended for students who have had at least a semester of calculus, including a study of limits. This prerequisite is highly recommended, both from the standpoint of mathematical maturity and because later on we plan to use a basic knowledge of limits and continuity. The book can be used for the traditional course in geometry required for teachers and mathematics majors at a slightly advanced level—the same course for which the standard text for many years was the epic book *Elementary Geometry from an Advanced Standpoint*, by Edwin Moise (Moise 1990). It is interesting that only minor changes in the axioms found in that book will provide a natural (familiar) foundation for a geometry here, which will become known as *unified geometry*. Thus, as in Moise, the traditional and very efficient proofs of Euclid can be used. Not only that, in Chapter 6, inequalities for quadrilaterals are developed in the spirit of Euclid, which will enable us to define the trigonometric functions for non-Euclidean geometry and to develop a *non-Euclidean Pythagorean theorem*. This latter part is where the use of limits becomes necessary and requires you to have the knowledge normally achieved in one or two semesters of calculus. But from the standpoint of axiomatics, this is rather appealing since past treatments of this are virtually intractable to the average (undergraduate) student of geometry or mathematics major in general. The last part of the book (the last three chapters) presents the classical theory.

1.2 Axioms of Alignment

The most primitive concepts in geometry are points and lines and their properties. We often think of a point as a particle or a small dot on a piece of paper, but the ideal concept of a point is really more like a location rather than a physical object. Euclid's description of *point*—"that which has no parts"—seems inadequate, to say the least. And his description of *line*—"that which has length but no breadth"—is just as vague. What does Euclid mean by "no parts" and "no breadth"? Such esoteric objects would certainly be invisible, and cannot have an accurate physical representation.

In fact, we should not expect an accurate physical representation or self-contained, clear definitions in an axiomatic treatment like that found in the *Elements*. Such ideas and images can best be handled by merely taking them as *undefined terms* and letting the axioms set forth the properties we want them to have. The axioms can be arbitrary, as long as they are consistent, and they do not have to reflect properties in the "real world." The great philosopher and mathematician Bertrand Russell once said "Mathematics may be defined as the subject in which we never know what we are talking about, nor whether what we are saying is true."

Historical Note

Very little is known concerning the man Euclid, except that around 300 BC he was a professor of mathematics at the University of Alexandria in Greece, and he was the author of the celebrated *Elements*. The stories told of Euclid show that he was a serious, uncompromising scholar. When asked by King Ptolemy whether there were an easier way to learn geometry (besides studying the *Elements*), Euclid is said to have responded, "There is no royal road to geometry!" Euclid's great contribution to mathematics was the elegant, perfectly logical arrangement of topics in the *Elements*, much of it from earlier ancient sources. This monumental work consists of 13 books containing 465 propositions with detailed proofs for each, all based on only 10 basic axioms. The work was such a success that all other works of mathematics in antiquity disappeared. The logical arrangement and concise arguments for this material was so effective it became the standard for secondary school students throughout the world, and in earlier times, for those seeking the doctorate in mathematics at leading universities. Needless to say, the *Elements* have had a great and lasting impact on science, its influence extending even to modern times.

The initial axioms presented here will characterize the interaction between points and lines. For example, points *belong to* or *lie on* lines (not the other way around), and lines are certain *collections* (or *sets*) of points. Thus set theory emerges as a natural way to handle such concepts. Recall that **set membership** is denoted by $x \in S$ (the element x **belongs to** the set S), the **union** of two sets S and T, denoted $S \cup T$, is the set of all elements belonging to either S, T, or both, and the **intersection** of S and T, denoted $S \cap T$ is the set of elements belonging only to both sets. The **empty set** is that set having no elements, denoted by $\emptyset$. A set S is said to be a **subset** of T, denoted $S \subseteq T$, whenever the elements of S belong to T, with $S \subset T$ denoting the **proper subset** relation, when $S \neq \emptyset$ and $S \neq T$. The **universal set**, or **universe**, is the set of all elements (points) under consideration—assumed to be nonempty. Since our object is to study primarily *plane geometry*, the universe will be denoted by $\mathcal{P}$ (the set of all points lying in some "plane").

Thus, taking *point* and *line* as the undefined terms, (with $\mathcal{P}$ as the nonempty set of all points—the *universe*), the first three postulates for our geometry may be stated as follows (the first will eventually be eliminated, later to be proven as a theorem).

> ***Axiom 0***   *A line is a proper subset of the universal set.*
> ***Axiom 1***   *Each line is a set of points having at least two members.*
> ***Axiom 2***   *There exist two points, and each two points belong to some line.*

These axioms guarantee that there exist at least three points, since there exists a line having two points (**Axiom 2**), and there exists an

Figure 1.1 A model for axioms 0–2.

additional point not on it (**Axiom 0**). Thus, $\mathcal{P}$ consists of at least *three noncollinear points*. The smallest possible geometry therefore consists of just three points and three lines, as depicted in Figure 1.1. We could also have four points and three lines, where two of the lines intersect in the same two points (analogous to a situation possible in spherical geometry), as illustrated in Figure 1.2. Note that we are not requiring that "two points determine a line" (a phrase meaning that *each two points lie on a unique line*).

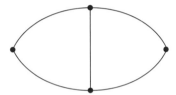

Figure 1.2 Another model.

Such drawings as in Figures 1.1 and 1.2 are called **models** for the axioms; a **model** is merely a *realization* or *representation* of the axiomatic system—just an *example*, if you wish. In such drawings, we assume that the dots represent *points* and that the thin lines or arcs are the *lines*, and that the only points belonging to the lines are those dots exhibited in the drawing, and shown on the particular line or arc that contains it.

Example 1

Verify that the following system of points and lines satisfies **Axioms 0**, **1**, and **2**, and draw a figure that represents a model for this geometry, labeling the points appropriately.

Points: $\mathcal{P} = \{A, B, C, D\}$
Lines: $l_1 = \{A, B\}$, $l_2 = \{A, C\}$, $l_3 = \{A, D\}$, $l_4 = \{B, C\}$, $l_5 = \{B, D\}$, $l_6 = \{C, D\}$

Solution

Axiom 0 is valid since not all points lie on the same line, and **Axiom 1** is true by observation (each line, six in all, contains exactly two points). **Axiom 2** is satisfied because each two points are contained by some line. A model is shown in Figure 1.3.

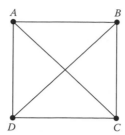

Figure 1.3 Example 1.

Remember that the line segments in the diagram of Figure 1.3 represent sets of only two elements each (the dots) and not the entire geometric segment shown in the figure. No further relationships are to be inferred from this diagram. Thus, the intersection of line segments in a model, such as *AC* and *BD* in Figure 1.3, does not necessarily represent a point in the geometry being depicted. However, it is sometimes possible to represent nonintersecting lines by nonintersecting arcs in the diagram. Such a model is called **geometrically faithful**. This can be done in two different ways for the geometry of Example 1, as shown in Figure 1.4. Thus, we seem to have three models (at least the diagrams differ) for this geometry. But we might regard these models as equivalent, or *isomorphic* in some way, since they have the same number of points and lines, and the points are joined by the lines in the same manner in both models. Thus, three significant problems in axiomatic geometry emerge:

1. To formulate a definition for the term *isomorphic*.
2. To determine whether two models are isomorphic.
3. To prove whether all models of a geometry are isomorphic. Such a geometry is called *categorical*. (The geometries we eventually develop are known to be categorical.)

Note: The main purpose of models in axiomatic mathematics is to show that the axioms are *consistent* (not self-contradictory) and *independent* (cannot be proved as theorems). A model can show consistency, often by direct observation of its features, because a contradiction in the axiomatic system would show up in the model. A model can also show independence, since if a model exists which illustrates, say, axioms A, B, and C, but shows that axiom D is not valid, then it would be impossible to prove axiom D from axioms A, B, and C. ◣

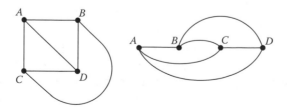

Figure 1.4 Geometrically faithful models.

Example 2

Find whether each of the following geometries obeys **Axioms 0, 1**, and **2**; explain.

(a) Points: $\mathcal{P} = \{A, B, C, D\}$
Lines: $l_1 = \{A, B, C\}$, $l_2 = \{A, B, D\}$, $l_3 = \{B, C\}$, $l_4 = \{C, D\}$
(b) Points: $\mathcal{P} = \{A, B, C, D, E\}$
Lines: $l_1 = \{A, B, C\}$, $l_2 = \{B, C, D\}$, $l_3 = \{C, D, E\}$, $l_4 = \{D, E, A\}$

Solution

(a) Although somewhat bizarre (one line is a proper subset of another—in stark contrast to Euclidean geometry), this geometry does satisfy the axioms since each line contains two or more points, and each pair of points, that is, the pairs (A, B), (A, C), (A, D), (B, C), (B, D), and (C, D), belong to one or more lines.

(b) This geometry satisfies **Axioms 0** and **1**, but does not satisfy **Axiom 2**: There is no line containing the points B and E.

Example 3

Figure 1.5 depicts a certain geometry satisfying **Axioms 0, 1, and 2**. Name the set of all points ($\mathcal{P}$) and the subsets that are the lines from the figure (model).

Solution

Points: $\mathcal{P} = \{A, B, C, D, E\}$
Lines: $l_1 = \{A, B, C\}$, $l_2 = \{A, D, C\}$, $l_3 = \{A, E, C\}$, $l_4 = \{A, C\}$, $l_5 = \{B, D, E\}$.

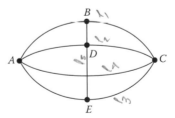

Figure 1.5 Example 3.

So far, the examples given of geometries satisfying **Axioms 0–2** have all been finite—having a *finite number of points*. It is easy to come up with examples of infinite geometries satisfying **Axioms 0–2**. In fact, one may make a given infinite set satisfy these axioms. For example, if $\{A_1, A_2, \ldots, A_n, \ldots\}$ is any sequence of real numbers (taken as points), simply take as lines all possible two-element subsets $\{A_1, A_2\}$, $\{A_1, A_3\}, \ldots, \{A_i, A_j\}, \ldots$.

Problems (Section 1.2)

Group A

1. Decide whether the following system of points and lines satisfies **Axioms 0**, **1**, and **2**. If so, construct a diagram to represent this system.
    Points: $\mathcal{P} = \{A, B, C, D\}$
    Lines: $l_1 = \{A, B, C, D\}$ $l_2 = \{A, D\}$

2. Decide whether the following system of points and lines satisfies **Axioms 0**, **1**, and **2**. If so, construct a diagram to represent this system.
    Points: $\mathcal{P} = \{A, B, C, D, E\}$
    Lines: $l_1 = \{A, B\}$, $l_2 = \{A, C\}$, $l_3 = \{A, D\}$, $l_4 = \{A, E\}$, $l_5 = \{B, C\}$, $l_6 = \{B, D\}$, $l_7 = \{B, E\}$, $l_8 = \{C, D\}$, $l_9 = (C, E\}$

3. In Problem 2, you should have found that the given geometry does not satisfy **Axiom 2**. Make a simple correction that makes it valid. In general, if every line contains exactly two points in an n-point geometry satisfying **Axioms 0–2**, how many lines must it have?

4. The diagram for this problem is a model for a certain geometry (Figure P.4). Name the set of all points and the subsets that are the lines. (Assume A, B, and C are collinear, as indicated.)

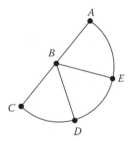

Figure P.4

5. The diagram for this problem is a model for a certain geometry (Figure P.5). Name the set of all points and the subsets that are the lines.

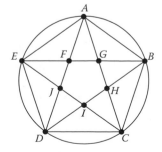

Figure P.5

Group B

6. Consider the Euclidean plane, and instead of ordinary lines, let the line *segments* $\overline{PQ}$ for all pairs of distinct points P, Q be taken as the "lines." Show that this system of points and lines satisfies **Axioms 0–2**. What seems unnatural to you about this geometry?

7. As in Problem 6, let the lines be segments $\overline{PQ}$ for all ordinary points $P \ne Q$, but (a) consider only those points on a circle, or (b) those points on or interior to a circle.
    (a) Show that if the points lie on a circle, the principle "two points determine a unique line" is valid.
    (b) Show that if the points lie on or inside a circle, some pairs of points determine a unique line, but all the rest are contained by infinitely many lines.

8.* Consider the following assumptions that constitute an axiomatic system:
    1. There exist five points and two lines.
    2. Each line is a subset of points having at least two elements.
    3. The two lines have no points in common.
    Show that there is only one model. If Axiom 1 reads: *There exist six points and two lines*, there exist two different (non-isomorphic) models. Show this.

9. Show that the following axiomatic system has essentially only one model, even though there are many diagrams for it (thus a definition for isomorphic diagrams suggests itself; see also Problem 8 in this regard):
    1. There exist three points and three lines.
    2. Each line contains exactly two points.

Group C

10. Propose a definition for "isomorphic models" and test it on previous examples to see if it is a workable definition (see Problems 8 and 9).

1.3 A Glimpse at Finite Geometry

Finite geometry is closely related to the important field of *combinatorial mathematics*. Our interest in this area is simply to obtain a variety of models for **Axioms 0–2** above. While a finite geometry is patently non-Euclidean, many finite geometries possess intricate Euclidean-like properties. This phenomenon is quite intriguing and we think you will find this interesting.

Example 1: Seven-Point Triangle Geometry

Consider the following properties for a geometry on seven points:

1. There exist seven points and six lines.
2. Every line contains exactly three points.
3. If two lines intersect, they do so in only one point.

Create a drawing using dots for points and segments for lines that accurately represents these axioms. That is, produce a geometrically faithful model for the geometry.

Solution

As before, we can represent a line with three points by simply drawing a line segment or curve with three dots on it, as in the drawings below.

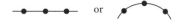

Now we are supposed to have seven points in our "geometry". So we start out with a sketch of seven dots, and, experimentally, add a few line segments for the six lines, like this:

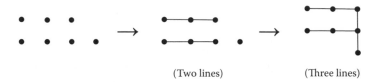

(Two lines)　　　　(Three lines)

However, if we continue with this choice of lines, even if we allow curved arcs, it turns out that it is impossible to draw in the remaining three lines, each having three points, without violating property (3). So we start over, as in the following sequence of diagrams (note that we are free to move the dots, representing the points, at will):

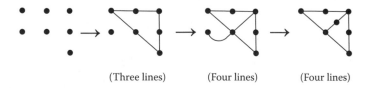

(Three lines)　　　(Four lines)　　　(Four lines)

At this point (luck having been on our side), we almost immediately see how to complete the diagram that illustrates the given axiomatic system, as shown in Figure 1.6a. (The points have been assigned labels in this diagram for later reference.) The final result is a triangle with the vertices, midpoints of sides, and centroid for the seven *points*, and the sides and medians of the triangle for the six *lines*.

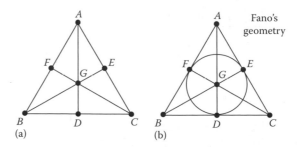

Figure 1.6 Two 7-point geometries.

The seven-point geometry of Example 1 can also be specified (exactly) by listing the subsets of points that define the six lines. This is the actual definition of the geometry. For example, two of the lines would be $l_1 = \{A, F, B\}$ and $l_2 = \{A, G, D\}$. The remaining lines are indicated in the table below, which gives all the lines (indicated as the columns).

l_1	l_2	l_3	l_4	l_5	l_6
A	A	A	B	B	C
F	G	E	G	D	G
B	D	C	E	C	F

However, the seven-point geometry we have constructed is very crude, and does not satisfy **Axiom 2**. Can you see why? The remedy is to add one more line, namely

$$l_7 = \{D, E, F\}$$

and to use a circle drawn through the points D, E, and F (the incircle of $\triangle ABC$) to represent this additional line, as in Figure 1.6b. The new model now satisfies not only **Axioms 0–2**, but the following stronger set of conditions, those that define a much-studied area of geometry.

Postulate System for Projective Planes

Postulate 1 There exist at least three noncollinear points.
Postulate 2 Every line contains at least three distinct points.
Postulate 3 Each two points belong to exactly one line.
Postulate 4 Each two lines intersect in exactly one point.

Any geometry satisfying these four properties is called a **projective plane**. The smallest possible projective plane is the famous **Fano's geometry**—the modified seven-point triangle geometry of Figure 1.6b. In general, a projective plane can have almost any number of points. Note that in terms of the axioms adopted for our study of geometry, Postulate 1 is our **Axiom 0**, Postulate 2 implies our **Axiom 1**, and

Postulate 3 implies our **Axiom 2**. Postulate 4 disallows parallel lines, so ordinary Euclidean geometry is not a projective plane. However, the *modified* Euclidean plane with the addition of its so-called points at infinity for each family of parallel lines is a projective plane, and it provides the most important example of an infinite projective plane. We skip the details for now.

Finite projective geometry abounds with many interesting ideas and unsolved problems. The above four postulates imply, for example, that if a single line has $n + 1$ points ($n \geq 2$), then *every* line has $n + 1$ points, that every point lies on exactly $n + 1$ lines, and that there are altogether $n^2 + n + 1$ points and $n^2 + n + 1$ lines in the geometry. The number n is called the **order** of the projective plane. Leonhard Euler (1707–1783) showed that there does not exist a projective plane of order 6. But a projective plane exists for any order n which is either a prime number (like 2, 3, 5, and 7) or the power of a prime. A major unsolved problem for many years was whether one exists having order 10—that is, one having 111 points. A computer proof has allegedly answered this question in the negative. The 13-point projective plane (order = 3) is introduced in Problem 9 below.

A few additional examples of finite geometries (not necessarily projective planes) will now be considered.

Example 2: Star Geometry on 12 Points

Consider the following relations between points and lines:

1. There are 12 points and 6 lines.
2. Each line contains exactly four points.
3. Each point lies on exactly two lines.
4. If two lines intersect, they do so in only one point.

(a) Find a geometrically faithful model for this axiomatic system.

(b) Is this geometry a projective plane?

Solution

(a) We might start out with the drawing below (left), which depicts two intersecting lines and four points on each line; we can add another line and two additional points as shown in the next diagram.

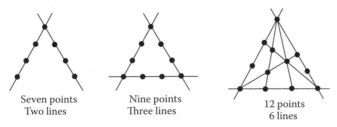

Seven points
Two lines

Nine points
Three lines

12 points
6 lines

This yields a triangular figure, where axioms (2) and (4) are valid. We still need to come up with three additional points and three more lines to complete the model.

With a little thought, and ruling out a diagram like the one at the far right, which does not work, we find that another triangle (inverted) duplicating the one above will exactly fit into the points already located. This forms the six-pointed star, as shown in Figure 1.7.

(b) No, this geometry is not a projective plane since Postulate 3 is violated.

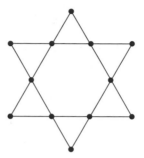

Figure 1.7 A star geometry.

Example 3: Square Geometry on Nine Points

Consider the axioms listed below for the *square geometry* on nine points.

1. There exist nine points and nine lines.
2. Each line contains exactly three points.
3. Each point lies on exactly three lines.
4. If two lines intersect, they do so in only one point.

(a) Find a geometrically faithful model for this axiomatic system, and label the points.

(b) In this model, find two parallel lines, and then for those two lines and for each point on one of those lines, test the Euclidean property: *If a line intersects one of two parallel lines, it intersects the other also.*

Solution

(a) One can start by experimenting with nine points arranged in a matrix of rows and columns, as shown in the diagram below, at left. By merely drawing the horizontal and vertical lines, we come up with a model that makes axiom (2) valid, and the first half of axiom (1). This gives us six lines (l_1 through l_6), so

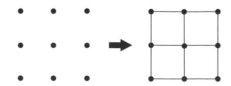

we must add one line passing through each point to make axiom (3) valid. We can add a diagonal (l_7), but if we add both diagonals this would violate axiom (3). It becomes clear that in addition to the diagonal, we need to add two more curved lines, l_8 and l_9, as shown in Figure 1.8.

(b) We leave this to the reader as a simple exercise on working with models.

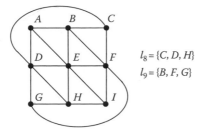

$$l_8 = \{C, D, H\}$$
$$l_9 = \{B, F, G\}$$

Figure 1.8 A square geometry.

Example 4

Consider the axioms listed below, which are the defining properties of an **affine geometry**.

Postulate System for Affine Planes

Postulate 1 There exist at least three noncollinear points.
Postulate 2 Every line contains at least two points.
Postulate 3 Each two points belong to exactly one line.
Postulate 4 [*Parallel Postulate*] Given a line and a point not on it, there exists exactly one line containing that point that does not intersect the given line.

Find a geometric model for this axiomatic system having nine points, with three points on each line, and verify the property for this particular affine plane: *Each point lies on exactly four lines*.

Solution

We observe that the model for the nine-point square geometry of the previous example lacks only three lines for a model that validates postulates (1), (2), and (3). Only lines passing through C and G, through B and D, and through H and F need be added, and this can be done while preserving the features we originally had. Thus, one need only draw the remaining diagonal for line *CEG* and two more curved lines to represent lines *AFH* and *BDI* (see Figure 1.9).

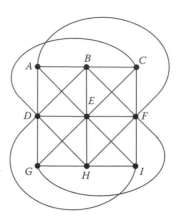

Figure 1.9 Model for affine plane on 9 points.

In general, an affine plane will have n points on every line for some integer $n \geq 2$ (called its **order**), every point is contained by $n + 1$ lines, and there are altogether n^2 points and $n^2 + n$ lines. The geometry in Example 4 is an affine plane of order 3, with 9 points and 12 lines. The ordinary Euclidean plane is also regarded as an affine plane (having infinitely many points) because these same axioms are valid. It will be left to the reader to show that an affine plane satisfies our **Axioms 0–2**.

Before going on, it should be mentioned that there is a unifying link between projective and affine planes of the same order arising from the following construction: If a line and all its points be removed from any projective plane, keeping the original collinearity relationships among the remaining points, the result is an affine plane of the same order. In this connection, the language "point at infinity" (or "ideal point") and "line at infinity" (or "ideal line") is often applied to those deleted points and the deleted line. One often asserts, for example, that *two lines are parallel if and only if they meet at a point at infinity*. Conversely, certain points may be *added* to a given affine plane that makes it a projective plane of the same order (see Problem 15 below).

Problems (Section 1.3)

Group A

1. Draw a geometric model for the following two axioms, using straight line segments to represent the lines if possible:
    (1) There are three points and three lines.
    (2) Each line contains exactly two points.

2. Draw a geometric model for the following two axioms, using straight line segments to represent the lines if possible:

(1) There are four points and four lines.
(2) Each line contains exactly two points.

3. Find a geometric model for the following axioms, and determine whether or not it is a projective plane:
(1) There are 5 points and 10 lines.
(2) Each line contains exactly two points.
(3) Each point lies on exactly four lines.

4. Find a geometric model for the following axioms, and determine whether or not it is a projective plane (see Example 2 for ideas; you can also "dualize" the geometry of Problem 3):
(1) There are 10 points and 5 lines.
(2) Each line contains four points.
(3) Each point lies on exactly two lines.
(4) If two lines intersect, they do so in only one point.

5. A well-known puzzle involves a landscape designer who wants to plant 10 fruit trees in an orchard so that there are 4 trees in each row and the outermost trees lie on a circle. How should the trees be planted? (See Problem 4.)

6. Verify that the diagram in Figure P.6 provides a geometric model for the nine-point square geometry of Example 3 by verifying each of the axioms stated in Example 3. (This diagram makes use of a famous theorem of Euclidean geometry known as Pappus' theorem, proven later, in Chapter 7. Thus, this geometry possesses not only a geometrically faithful model, but one in which the lines are represented by *straight line segments*.)

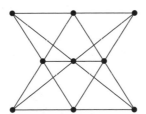

Figure P.6

7. A geometry is defined by letting the points be labeled A, B, C, ..., I (the first nine letters of the alphabet) and taking as lines those sets of points whose labels occur as a horizontal row or vertical column in either of the following matrices:

A	B	C		A	H	F
E	F	D		D	B	I
I	G	H		G	E	C

Show that this geometry is identical to one of those introduced in this section.

Group B

8. Why must the following proposition be true in any affine plane: *If a line intersects one of two parallel lines, it must intersect the other also?*

9. The 13-point projective plane can be defined by letting the points be labeled A, B,..., M (the first 13 letters of the alphabet) and taking as lines those points whose labels lie in any one of the columns of the following matrix:

A	A	A	A	B	B	B	C	C	D	D	D
B	E	H	K	E	F	G	E	F	G	E	F	G
C	F	I	L	H	I	J	I	J	H	J	H	I
D	G	J	M	K	L	M	M	K	L	L	M	K

Carry out the following operations based on this matrix.
(a) Find the lines (identify their points) determined by B and E, and by C and F.
(b) Find the point of intersection of lines BE and CF.
(c) Determine whether the lines BC, EF, and HI are concurrent. If so, name the point of concurrency.

10. **Desargues' theorem**   For the specific triangles BEI and CFH in the 13-point projective plane of Problem 9, verify Desargues' theorem: *If corresponding pairs of vertices of two triangles lie on concurrent lines* (in this case, lines BC, EF, and IH), *then the three pairs of corresponding sides intersect in three collinear points* (in this case, BE and CF, EI and FH, and BI and CH, as shown in Figure P.10). Identify the line of collinearity for this example. See *Sketchpad Experiment 1* in Appendix **A** demonstrating Desargues' theorem in the Euclidean plane.

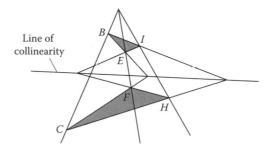

Figure P.10

11. Repeat the instructions of Problem 10 for the specific triangles EFM and DBK.

Group C

12. As an experiment, consider the points of the 13-point projective plane of Problem 9, and delete the four points of line $AKLM$ throughout, resulting in the matrix:

B	E	H	B	B	B	C	C	C	D	D	D
C	F	I	E	F	G	E	F	G	E	F	G
D	G	J	H	I	J	I	J	H	J	H	I

where everything else is left unchanged. Do you observe anything of interest? Do parallel lines exist? Can you find a geometric model for this geometry considered previously? If so, use as labels for the points of your model those of the above (the letters *B*, *C*,..., *J*) in such a manner that the lines in your model agree with those indicated in the above matrix.

13. **The Harmonic Construction**   If three points *A*, *B*, and *C* are given on any line, a fourth point *D* on that line, called the *harmonic conjugate of B with respect to A and C*, may be constructed as follows: Draw (or consider) any three concurrent lines through *A*, *B*, and *C*, meeting at point *P* and locate any point *Q* on line *BP* (see Figure P.13). Let *R* be the intersection of lines *AQ* and *PC*, and *S* the intersection of lines *BR* and *PA*. Then draw line *QS*. The intersection of line *QS* and the given line is the desired harmonic conjugate *D*.

> **Theorem**   The construction given above is independent of the points *P* and *Q*. That is, the same point *D* results regardless of the choice of points *P* and *Q* in the construction.

Verify this theorem for the 13-point projective plane and your own choice of points *A*, *B*, and *C*, choosing two distinctly different pairs of points *P*, *P'* and *Q*, *Q'* for the construction.

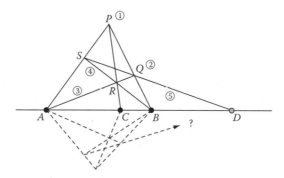

Figure P.13

Note: It can be shown that in any projective plane the theorem quoted above is a corollary of Desargues' theorem. For an experiment in the Euclidean plane involving this theorem, see *Sketchpad Experiment 14* in Appendix **A**. ◣

14. Prove that deleting a line and all its points in any projective plane, and retaining all other points, lines, and original collinearity relations, will produce an affine plane.

15.   How could one add points (and a line) to an affine plane in order to yield a projective plane? (See Problem 12 for ideas.)

16.   In connection with Problem 15, suppose the ordinary Euclidean plane with which you are familiar is augmented by the addition of certain points to make it a projective plane. Investigate some of its properties. (This yields an *infinite* projective plane, whose construction otherwise appears to be difficult or impossible. In this connection, see Problem 24, Section 1.7.)

17.   If you are familiar with the terms, an affine plane (and thus, a projective plane) can be constructed algebraically from an arbitrary field F in the following manner: Let the *points* be ordered pairs (x, y) where $x \in F$ and $y \in F$, and the *lines* sets of points (x, y) such that either $x = a$ for $a \in F$ (a *vertical line*), or $y = mx + b$ for $m, b \in F$ (a *non-vertical line*). Verify the four axioms for affine geometry given in Section 1.3. The finite field Z_3 (integers mod 3) will produce the 9-point geometry of Example 4. (With care, the algebra required to prove these results can be carried out without using the commutative law of multiplication, a system known as a *division ring*.)

Note: An example of a division ring that is not a field is provided by the *quaternions*, the set of numbers represented by $a + bi + cj + dk$, where a, b, c, and d are real numbers, i, j, and k are pure imaginary ($i^2 = j^2 = k^2 = -1$), and $ij = -ji = k$, $jk = -kj = i$, and $ki = -ik = j$, keeping all other algebraic laws intact. It is well known that only infinite division rings can fail to satisfy the field axioms. ◣

1.4 Metric Geometry

The next axiom to be added to our system will provide the important concept of distance. One might think there is very little to be said about distance that is not already obvious or trivial. Aside from the fact that the distance concept must be introduced axiomatically, the practice of measuring the distance between two points with a ruler or tape measure indeed involves the assumption that some underlying concept of distance already exists and is reliable. When we measure first a chair, couch, or desk, and then the space available for it, we assume that our measurements, if correct, will faithfully predict whether or not the chair, couch, or desk will fit into that space. On a sphere, or any other "smooth" surface in Euclidean space, we assume that "distance" is already defined as the length of the shortest arc on the surface joining two points.

The axiom on distance which follows will simply be added to those we already have, consisting, so far, of the *axioms of alignment* (**Axioms 0–2**). At this point we begin to blend the strictly synthetic approach with that of real numbers, which will greatly reduce the amount of labor in "proving the obvious." In this process, Euclid's basic presentation will still be intact. However, it should be pointed out that there is, after all, an underlying concept of measurement in the Elements, although not explicitly stated (note the concept of the "addition of segments," and the idea that one segment can be "greater" than another). The concept of distance can be avoided only at considerable effort.

Axiom 3 [Metric Axiom]   *To each pair of points (A, B), distinct or not, there corresponds a real number AB, called **distance**, or the distance from A **to** B, which satisfies the properties:*

   (a) *$AB \geq 0$ with equality if and only if $A = B$.*
   (b) *$AB = BA$.*

The first property **(a)** simply states the trivial fact that the distance between two distinct points is always positive—a property sometimes referred to as **positive definiteness**. The property in **(b)**, often referred to as **symmetry**, guarantees that the order of measurement makes no difference. For example, if you switch ends of the ruler you will still get the same measurement, or if you travel from A to B and then return from B to A, the distance—but not necessarily the energy expended—is the same both ways.

A third property is usually included in the metric concept, called the **triangle inequality**, which will be proved later: If A, B, and C are any three points, distinct or not,

   (c) $AB + BC \geq AC$

(see Figures 1.10 and 1.11). If a set of points satisfy **Axiom 3**, together with the triangle inequality, it is called a **metric space**. A significant area of

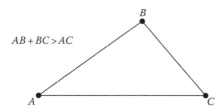

Figure 1.10 Triangle inequality.

research in mathematics is devoted exclusively to the study of metric spaces. In topology, for example, a large body of literature involves various topological properties that will guarantee the existence of a consistent metric.

Now if it so happens that the equality occurs in the triangle inequality, that is, if

$$AB + BC = AC$$

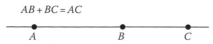

Figure 1.11 Triangle equality.

as illustrated in Figure 1.11, then it is only natural to assume that A, B, and C are collinear, and that B is somehow "between" A and C—that the ordering of the points on the line is either A–B–C (first A, then B, and finally C), or the reverse. This idea is actually quite important—almost no other concept plays a more critical role in geometry. An entire section will be devoted to it later (Section 1.7).

Example 1

Any set of points can be made into a metric space, and also satisfying **Axioms 0–3**. To construct a metric from an arbitrary set of points, define for any two points A and B, $AB = 0$ if $A = B$, and $AB = 1$ if $A \neq B$. (Such a metric is called the **discrete metric**.) Take a line to be *any pair of distinct points*. For instance, suppose the universe is the set of all apple trees in an orchard (Figure 1.12). Each line would consist of just two trees, taken pairwise, and all pairs would yield the lines for the geometry. The metric would be as indicated in the figure—the "distance" between any two trees = 1.

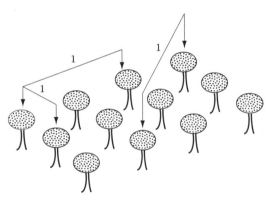

Figure 1.12 Example 1.

Moment for Reflection

Is the triangle inequality valid for the discrete metric as defined in Example 1? Consider any three points A, B, and C, and the sum $AB + BC$. What must be true?

Essentially arbitrary sets of nonnegative real numbers can be realized as the distance set for a metric geometry that includes the triangle inequality. A geometric structure is not necessarily implied by this process however, unless one wants the points to be embedded in Euclidean space, for instance. As an example, suppose $\boldsymbol{D} = \{0, 8, 11\}$ is a given set of distances. One could take as points A, B, C, and D, and as lines, pairs of points, and then assign distance in the following manner: $AB = 8$, and $AC = AD = BC = BD = CD = 11$. **Axioms 0–3** and the triangle inequality can be easily verified.

1.5 Eves' 25-Point Affine Geometry: A Model for Axioms 0–4

One example of a finite metric geometry satisfying Postulates 1–4 for an affine plane (previous section) is really amazing, which appears in Howard Eves' *Survey of Geometry,* Volume I, p. 432 (Eves 1963). It consists of an affine plane of order 5, having 25 points and $5^2 + 5 = 30$ lines. Not only are such basic concepts as Desargues' theorem and the harmonic relation valid (because the order is a prime number), a host of other significant properties hold as well. Some examples are: *The diagonals of a parallelogram intersect at the midpoint on each diagonal*, and *the medians and altitudes of a triangle are concurrent.* These are but a few valid properties in this system that seem very familiar from Euclidean geometry.

Let the *points* be designated by $A, B, C, \ldots, Y$ (all the letters of the English alphabet except Z), and let the *lines* be defined as the set of points whose labels occur as a *horizontal row* or *vertical column* in any one of the following matrix blocks.

	(1)					(2)					(3)			
A	B	C	D	E	A	I	L	T	W	A	X	Q	O	H
F	G	H	I	J	S	V	E	H	K	R	K	I	B	Y
K	L	M	N	O	G	O	R	U	D	J	C	U	S	L
P	Q	R	S	T	Y	C	F	N	Q	V	T	M	F	D
U	V	W	X	Y	M	P	X	B	J	N	G	E	W	P

For example, one of the 30 lines is IVOCP, occurring as the second column in matrix (2); another is KLMNO occurring as a row in matrix (1). Since each two distinct points determine a unique line, we shall designate the line passing through P and Q in general by $\overleftrightarrow{PQ}$. Two lines are called *parallel* if they do not intersect and *perpendicular* if they occur as a *row and column in the same matrix*. Using the usual geometric symbols, in matrix (3) for example, we find XKCTG ∥ HYLDP and XKCTG ⊥ RKIBY at K. The line XKCTG is also the *perpendicular bisector* of "segment" (R, I) at K since K lies midway between points R and I. This concept is used cyclically as if the points of a line were placed on an imaginary circle; thus XKCTG is also the perpendicular bisector of (B, Y) at K because K is considered midway between points B and Y on the line RKIBY. To be more precise, a point M in general is said to be the *midpoint* of the pair (P, Q) provided that the points of line $\overleftrightarrow{PQ}$ are modeled as a circle in the order in which they occur in the matrix (as in Figure 1.13), and point M is exactly half-way between P and Q. Then define a *bisector* of the pair (P, Q) as any line passing through the midpoint of (P, Q). A line l is then the *perpendicular bisector* of a pair of points if it is both perpendicular to the line determined by that pair and bisects the pair.

Although the concepts defined above might seem artificial or arbitrary, they lead to some results that resemble very familiar theorems in Euclidean geometry.

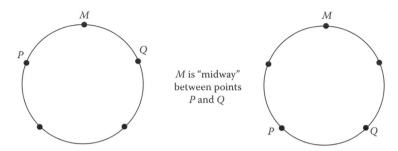

Figure 1.13 "Midpoints" in Eves' geometry.

> **Theorem 1**   For any point P and any line l, there exists exactly one line passing through P that is perpendicular to l.

Suppose we define *median* and *altitude* the same way as in ordinary geometry. A *triangle* is taken to be any *noncollinear triple of points*. A *parallelogram* is taken to be any set of *four consecutive points* (the *vertices*) such that *opposite sides* (that is, opposite pairs of vertices) lie on parallel lines. Then additional theorems are:

> **Theorem 2**   The medians of a triangle are concurrent.
> **Theorem 3**   The altitudes of a triangle are concurrent.
> **Theorem 4**   The diagonals of a parallelogram bisect each other.

One can even define a concept for distance, which leads to further Euclidean-like theorems. This is taken up in Problems 7–13 below for you to pursue on your own. We now give several examples to show how to verify theorems (or conjectures) in the 25-point geometry.

Example 1

Perform the following operations in the 25-point geometry.

(a) Find the line (i.e., identify its points) which passes through U and N.
(b) Find the perpendicular to line UN passing through point V.
(c) Find the midpoint of the pair (F, N).

Solution

(a) This line occurs as the fourth column in matrix (2): THUNB.
(b) We must search among the points of the same matrix (2) for the desired perpendicular. It is found to be line SVEHK.
(c) We must first identify line FN; it is found in matrix (2) having points Y, C, F, N, and Q. Thinking of these points arranged uniformly on a circle, we find that Y is the desired midpoint.

Example 2

Verify Theorem 3 (*the medians of a triangle are concurrent*) for triangle FUN. Find the point of concurrency (the "centroid").

Solution

(See Figure 1.14.) We first find the midpoints of the "sides" of ΔFUN. The midpoint of (F, N) was already determined as Y in Example 1. For (U, N), we find line UN = THUNB, so the midpoint of (U, N) is T. For (F, U), line FU = AFKPU and the midpoint of (F, U) is A. Next, we determine the points on the three medians: UY = UVWXY, FT = VTMFD, and NA = ARJVN. We find that the unique point which these lines have in common is V.

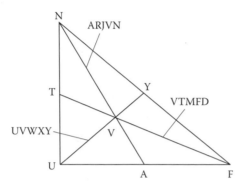

Figure 1.14 Example 2.

Problems (Section 1.5)

Group A

1. In Eves' 25-point geometry carry out the following operations.
    (a) Find the perpendicular to line PV passing through N.
    (b) Find the midpoint of (R, S).
    (c) Find the perpendicular bisector of the pair (S, W).

2. Perform the following operations for Eves' 25-point geometry.
    (a) Find the perpendicular to line JU passing through W.
    (b) Find the midpoint of (G, V).
    (c) Find the perpendicular bisector of the pair (E, V).

3. Find the medians of ΔCAT, and verify Theorem 2 for this triangle.

4. Find the altitudes of ΔCAT, and verify Theorem 3 for this triangle.

5. Find the perpendicular bisectors of $\triangle$CAT, and verify that they are concurrent, as in Euclidean geometry.

6. Find the fourth vertex of the parallelogram three of whose vertices are A, Q, and N, in consecutive order, and verify Theorem 4 for this parallelogram.

Group B

7. **Distance in Eves' 25-point geometry**   A concept for distance for Eves' 25-point geometry can be formulated as follows: Given any two points $P \neq Q$ in general, let the *distance PQ* from P to Q be defined as follows: $PQ = 1$ iff P and Q are adjacent points on the circular line model for PQ as in Figure 1.13, and $PQ = 2$ otherwise. Thus, all distances between distinct points are either 1 or 2. Nevertheless, many familiar Euclidean propositions involving distance are valid. One example is the familiar *equidistant locus theorem*:

    Each point on the perpendicular bisector of a pair of points A and B is, in general, equidistant from A and B.

    Find at least two examples in the 25-point geometry that demonstrate this proposition.

8. Find at least two examples in the 25-point geometry of each of the following propositions, taking a *rectangle* to be any parallelogram having adjacent sides perpendicular:
    (a) The opposite sides of a parallelogram are congruent (have equal lengths).
    (b) The diagonals of a rectangle are congruent.

Group C

9. Using the equidistant locus theorem stated in Problem 7, give a pseudo-proof of the theorem in Eves' 25-point geometry that the perpendicular bisectors of the sides of any triangle are concurrent. (See the warning in Problem 12; in spite of this, the theorem mentioned here still holds true.)

10. The point of concurrency of the perpendicular bisectors of the sides of a triangle (see Problem 8) is obviously the *circumcenter* of the triangle, the center of a circle (called the *circumcircle*) passing through the vertices of the given triangle. Find the remaining points of the circumcircle of triangle ABR (besides A, B, and R).

11. See if you can verify or prove that every circle in Eves' 25-point geometry has the same number of points? (To get started, work Problem 10 first to see how many points should lie on a circle.)

12.   It is interesting to discover Euclidean-like propositions for the 25-point geometry on your own. For example, what theorem might be true involving points at proportional distances on two sides of a triangle? There is also a familiar theorem concerning the tangents to a circle. In your exploration, however, you should be fully aware of the following:

> **WARNING**: *A point that is equidistant from two other points need not lie on the perpendicular bisector of those two points.*

13.   Taking the usual definition for a circle and its center, show that all 25 points of Eves' geometry can be arranged so that they lie on two concentric circles, one of radius 1, the other of radius 2, plus the common center, by drawing and labeling an appropriate figure. Can a line and a circle intersect in 3 or more points? Does this result have any bearing on the warning given in Problem 12?

1.6 Distance and Alignment

As we have seen, the property "two points determine a line" or, equivalently, "two distinct points lie on exactly one line" is not a valid proposition in general for a geometry satisfying **Axioms 0–2**. Since the concept of distance is independent of these two axioms, the addition of **Axiom 3** does not help. In spherical geometry, two lines can sometimes intersect in two distinct points (e.g., two meridian lines on the earth's surface meet at the North and South poles). So if we want to include spherical geometry in our study, we must somehow allow this behavior of lines. That is, we want to allow certain pairs of points to contain more than one line.

The key is the distance concept in Euclidean and spherical geometry. In Euclidean geometry distance is unbounded, while distance on a sphere is bounded. On the earth's surface, for example, the earth's circumference is approximately 24,800 mi, and two points can be at most 12,400 mi apart, as measured along the shortest great circle arc. And two points at this distance apart must be *polar points* (like the North and South poles), and there are an infinite number of lines through those two points. All other pairs of points determine a unique line.

If a set S of real numbers is bounded, then its least upper bound exists, denoted by LUB (S), as discussed in some detail in Appendix D. Thus, on the earth's surface, we can assert that the least upper bound of the set of all distances is 12,400 (if miles are used for the unit of measure). But if a set S is unbounded, the least upper bound does not exist, and we customarily write LUB (S) = ∞ for this case. So our geometry must take into account these two cases.

First, one defines the *least upper bound for distance*:

(1) $\alpha = \text{LUB } \{XY \colon X,\ Y \text{ arbitrary points}\}$

Notes:

1. By the property of α as an upper bound, for all points A and B we have $AB \leq \alpha$.
2. For later reference, we take $\alpha < \infty$ to mean simply that α is a (bounded) real number. The two cases for α are then $\alpha = \infty$ (unbounded distance), and $\alpha < \infty$ (bounded distance).
3. On a sphere of radius r, the circumference is $2\pi r$ and the maximum shortest distance measurable along a great circle is therefore $\alpha = \pi r$. (On the earth's surface, with $r = 3{,}960$ mi roughly, the value for α is approximately 12,400, as mentioned above.)
4. On a sphere of radius r, if $AB < \alpha$ then A and B cannot be poles (endpoints of a diameter), and in this case there is only one great circle passing through A and B. ◣

The appropriate axiom can now be stated.

> **Axiom 4**   *If the distance between two distinct points is less than α, they lie on exactly one line.*

Note that in Euclidean geometry, where $\alpha = \infty$, $AB < \alpha$ for all points A and B, and Axiom 4 yields the stronger statement.

> **Axiom 4′**   *Two distinct points determine a unique line.*

Axiom 4 allows the use of the familiar symbol for the line containing A and B (provided $0 < AB < \alpha$):

$$\overleftrightarrow{AB}$$

In view of Axiom 4 we can then assert that if C and D are any two points on line $\overleftrightarrow{AB}$ and $0 < CD < \alpha$, then lines $\overleftrightarrow{AB}$ and $\overleftrightarrow{CD}$ both contain the distinct points C and D, hence must be the same line. Thus, $\overleftrightarrow{AB} = \overleftrightarrow{CD}$. Note that a slight rewording of **Axiom 4** also tells us that if two distinct lines l and m intersect at distinct points A and B, then $AB = \alpha$. Thus:

> **Theorem**   If C and D lie on line $\overleftrightarrow{AB}$ and $0 < CD < \alpha$, then $\overleftrightarrow{AB} = \overleftrightarrow{CD}$. If lines l and m intersect at distinct points A and B, then $AB = \alpha$.

When $\alpha = \infty$, the preceding properties of lines become the familiar ones: C and $D \in \overleftrightarrow{AB}$ implies $\overleftrightarrow{AB} = \overleftrightarrow{CD}$ and two lines can intersect at only one point.

1.7 Properties of Betweenness: Segments and Rays

In the foundations of geometry it is often assumed that real numbers are forbidden and cannot be used in any way—after Euclid's characteristic style. Thus, order concepts must be introduced axiomatically. In such treatments, the relation of order, or *betweenness* as we shall call it, is an *undefined relation*, and its properties must be introduced as axioms. For example, one finds postulates like "if A–B–C then C–B–A" or "if A–B–C and B–C–D then A–B–D and A–C–D". In such treatments, the real numbers are ultimately developed geometrically. One such development is found in the book by David Hilbert, *The Foundations of Geometry* (Hilbert 1902), originally written in 1898. See Appendix F, where Hilbert's axioms are cited explicitly.

Such studies of geometry, although good mathematics and true to the cause of preserving the synthetic method, are very tedious, and progress takes place at a snail's pace. We prefer to reverse the order of topics and to use distance and real numbers to *define* betweenness—the approach taken by Moise, as mentioned at the outset. This makes the task much easier, without sacrificing too much from the traditional. While retaining a majority of ideas from synthetic (axiomatic) geometry as in the *Elements*, this approach places the study of geometry squarely in the area of metric spaces, a well-established modern concept, where the metric axiom, **Axiom 3**, is the basic concept.

Definition 1   A point B is said to lie **between** points A and C, and we denote this relationship by (ABC), provided that

   (a) A, B, and C are distinct, collinear points, and
   (b) $AB + BC = AC$.

An immediate consequence of this definition is that if (ABC), then $AB + BC = AC$ or, by symmetry, $BA + CB = CA$ (or $CB + BA = CA$), and thus (CBA). This result reveals the order aspect of betweenness: If one encounters A, B, and C on some line, in that order (first A, then B, and finally C), then the reverse order is also true: first C, then B, and finally A. A good thought exercise for you (Problem 18 below) is to show that if (ABC), then neither (BAC) nor (ACB) can be true.

The betweenness relation of three points on a line leads to a simple definition for the order of *four* points on a line.

Definition 2   If A, B, C, and D are four collinear points such that the four betweenness relations (ABC), (ABD), (ACD), and (BCD) hold, then the aggregate of these relations will be denoted by $(ABCD)$.

Historical Note

David Hilbert (1862–1943) is widely regarded as the greatest mathematician of the twentieth century. Born in the town of Königsberg, East Prussia (made famous for its seven bridges and the well-known problem in graph theory), he was the son of a district judge who wanted him to become a lawyer. Having found that his greatest interests were in mathematics, Hilbert eventually earned his doctorate from Königsberg University in May 1885, and arrived at Göttingen University to teach mathematics in 1895, exactly 100 years after Gauss had been a student there. His research was so far-reaching that it has been estimated that every significant mathematical discovery made in the twentieth century was involved in some way with Hilbert's work. His famous list of 23 problems in many different fields of mathematics became a bench-mark for distinguished mathematical achievement. After having solved a major problem in number theory that beautifully extended Gauss's reciprocity theorem, he turned to geometry and wrote his *Foundations of Geometry* in 1898. He adopted the familiar style of Euclid, and its logical perfection (analogous to that found in the *Elements*), quickly went through 7 editions. Hilbert's favorite comment was: "Instead of points, lines, and planes, one must be able to say at all times tables, chairs, and beer mugs," to emphasize the abstract nature of strictly axiomatic geometry.

You need to be aware that certain natural properties, such as the notion that (ABC) and (BCD) imply the order $(ABCD)$, cannot be proven at this time, and in fact, in spherical geometry such ideas are not always valid. At this point it cannot even be asserted that there exists a point between two given points. This desirable feature requires a further axiom, and will be introduced later. However, we do have the following basic result that can be proven.

Theorem 1   If (ABC), (ACD) and (ABD), then (BCD) and therefore, $(ABCD)$.

Proof   By hypothesis and the theorem of Section **1.6**, (ACD) implies $0 < CD = AD - AC < AD \leq \alpha$ and $\overleftrightarrow{CD} = \overleftrightarrow{AB}$ so that A, B, C, and D are collinear. We also obtain directly from (ABD)

$$BD = AD - AB$$

Substituting $AC + CD$ for AD (using (ACD)) and $AC - BC$ for AB (using (ABC)), this becomes

$$BD = (AC + CD) - (AC - BC) = BC + CD$$

proving the relation (BCD). The relation $(ABCD)$ then follows by definition. ◣

It is significant that in a metric space we can prove the third betweenness relation (ABD) in the hypothesis of Theorem 1 from the first two. Owing to concepts to be introduced later this turns out to be unnecessary, but it is interesting to prove as much as we can at this early stage, so we include the result and its proof.

Theorem 2   If the triangle inequality is valid, then (ABC) and (ACD) imply $(ABCD)$.

Proof   In view of Theorem 1, it suffices to prove (ABD). As in the proof of Theorem 1, points A, B, C, and D are collinear. By the triangle inequality,

$$AD \leq AB + BD$$

If we can prove that $AD \geq AB + BD$, then equality would follow and (ABD) would result. But observe that, also by the triangle inequality

$$BD \leq BC + CD$$

Now simply add AB to both sides and use the given betweenness relations (ABC) and (ACD):

$$AB + BD \leq AB + BC + CD = AC + CD = AD$$

That is, $AB + BD \leq AD$, as desired. Thus, (ABD) and $(ABCD)$ follow. ◣

The familiar concept of segments and rays clearly depends on betweenness properties, which can now be defined.

Definition of Segment   Let A and B be any two distinct points, and l any line containing them. A **segment** joining A and B is the set of all points P on l such that either $P = A$, $P = B$, or (APB). The points A and B are called **endpoints**, while all other points are **interior points**.

Definition of Ray   A **ray** from A through B is the set of all points P on a particular line passing through A and B such that either $P = A$, $P = B$, (APB), or (ABP). Point A is called the **endpoint** or **origin** of the ray, and any other point at a distance less than α from A is an **interior point**.

The set-theoretic version of the above two definitions is as follows (using the traditional notation for segments and rays):

(1) $\overline{AB} = \{P : P \in l, P = A, P = B, \text{ or } (APB)\}$

(2) $\overrightarrow{AB} = \{P : P \in l, P = A, P = B, (APB), \text{ or } (ABP)\}$

Note that at this stage of development it can only be guaranteed that segments and rays contain the two points A and B in the definition; we have not yet proven or assumed the existence of a point P on $\overline{AB}$ such that (APB). (This deficiency, along with others, will be corrected in the next section.) Also note that more than one segment can join A and B (if $AB = \alpha$), and more than one ray having endpoint A and containing point B can exist. However, if $AB < \alpha$, the line l of the definition of segment is unique and there is only one segment having endpoints A and B. A similar statement follows for the ray from A through B.

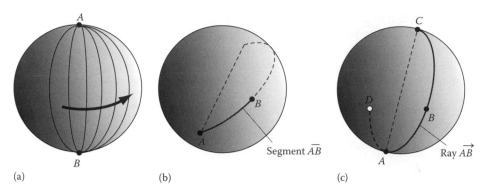

(a)   (b)   (c)

Figure 1.15 Spherical segments and rays.

Although these sets are familiar objects in Euclidean geometry, it might be well to examine what segments and rays look like on a sphere. First, if $AB = \alpha$ (the distance half-way around the sphere), then evidently A and B are the endpoints of a diameter, and the points on the sphere lying between A and B cover the *entire sphere* (see Figure 1.15a). Thus, in this case, the segment joining A and B is not defined (there exist many such segments and rays). If $AB < \alpha$, then segment $\overline{AB}$ consists of all points on the minor arc $\overset{\frown}{AB}$ of the unique great circle passing through A and B (as shown in Figure 1.15b). Finally, the ray $\overrightarrow{AB}$ consists of all points on the semicircle $\overset{\frown}{AC}$ passing through A and B, as shown in Figure 1.15c. Note that in spherical geometry, a ray can be said to have *two* endpoints, since, evidently, the ray $\overrightarrow{AB}$ is the same as the ray $\overrightarrow{CB}$. (This can be proven axiomatically.) Note also the non-Euclidean behavior of collinear points A, B, C, and D, as shown in Figure 1.15c: here, (DAB) and (ABC) hold, but the anticipated relation $(DABC)$ does not. Note also that if A, B, and D were equally spaced on the great circle of A, B, and D, then neither (ABD), (BAD), nor (ADB) would be valid, another non-Euclidean behavior.

Moment for Reflection

There are two models for our axioms so far: the Euclidean plane as represented by coordinates, and spherical geometry taking place on the surface of any sphere. In the former, if (ABC) holds, we know that there exists a point D such that $(ABCD)$ holds. Do you think this is true for spherical geometry? Examine a variety of examples where (ABC) holds for three points as you consider this question. The previous discussion gives you a strong hint.

Up to now, all concepts and results were based on the simple definition of betweenness presented above—there were no additional assumptions made involving betweenness. The axiom to be introduced next is an important concept of betweenness, one that implies the familiar property that a line can be expressed as the set of points on two "opposing" rays. That is, the points of any line $\overleftrightarrow{AB}$ consist of those points lying on the two rays $\overrightarrow{BA}$ and $\overrightarrow{BC}$ where (ABC).

Figure 1.16 Axiom 5.

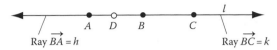

Figure 1.17 Theorem 3: $l = \overrightarrow{BA} \cup \overrightarrow{BC}$.

To motivate and understand how this axiom works, imagine point D on line l as it moves from left to right, assuming various positions on that line (Figure 1.16).

> **Axiom 5** If A, B, C, and D are any four distinct collinear points such that (ABC), then either (DAB), (ADB), (BDC), or (BCD).

The first application of this axiom establishes the property of lines mentioned above. Can you see how to prove it from the definition of ray using **Axiom 5**? Figure 1.17 indicates a "proof by picture." Providing the details makes a good initial exercise for you in the use of the axioms and definitions assumed so far.

> **Theorem 3** If the relation (ABC) holds and l is the line of collinearity of A, B, and C, then l is the union of the two rays $\overrightarrow{BA}$ and $\overrightarrow{BC}$.

> **Definition 3** The rays $\overrightarrow{BA}$ and $\overrightarrow{BC}$ of Theorem 5 are called **opposite rays**. In general, if two rays h and k have a common endpoint and their union is a line, they are said to be **opposite rays**.

Problems (Section 1.7)

Group A

1. If (ABC), $AB = 3$, and $BC = 4$, find AC.

2. If $(ABCD)$, $AB = 3$, $BC = 4$, and $CD = 5$, find AD.

3. If $(ABCD)$, prove that $AD = AB + BC + CD$.

4. Suppose that in a certain metric geometry satisfying **Axioms 0–5** points A, B, C, and D are collinear, and that

    $$AB = 4 = AC, \quad AD = 6, \quad BC = 8, \quad BD = 9, \quad CD = 1$$

    (a) What betweenness relations follow among these points, by definition?
    (b) Show that the triangle inequality is not satisfied.
    (c) Show that the conclusion of Theorem 2 is not true.

5. Suppose that in a certain metric geometry satisfying **Axioms 0–5** points A, B, C, and D are collinear and that

$$AB = 3, \quad AC = 5, \quad AD = 4, \quad BC = 2, \quad BD = 5, \quad CD = 3$$

   (a) What betweenness relations follow among these points, by definition?
   (b) Is the triangle inequality valid for these points?
   (c) Is the conclusion of Theorem 2 valid for these points?

6. In Problem 5, show that the given distances can be realized on a spherical great circle having circumference 12 ($\alpha = 6$).

7. Using the proper notation and the betweenness relations as indicated in Figure P.7, give one representation for each segment and ray determined by the points A, B, C, and D that contains point E in its interior.

Figure P.7

8. As in Problem 7, name each segment and ray determined by the points A, B, C, and D exhibited in Figure P.7 that does *not* contain point E.

9. Two toothpicks having equal lengths are laid side-by-side, overlapping by 3 cm (Figure P.9). The distance between the outer ends of the toothpicks is 16 cm.
   (a) How long is each toothpick?
   (b) If the ends of the two toothpicks are as indicated, what betweenness relations were needed to reach your conclusion in **(a)**?

Figure P.9

10. Two toothpicks are each of length 5 cm laid side by side (as in Figure P.10 for the previous problem) such that the total length is only 7 cm. How much do they overlap?

11. A single mark is made on a stick 12 ft long 7 in. from one end (Figure P.11). Show that all 12 integer distances from 1 ft to 12 ft can be measured using this stick with just this one mark.*

Figure P.11

---

* This problem is adapted from one found in the book by H. Jacobs, *Geometry* (Jacobs 2003, p. 90).

12. In reference to Problem 11, find all other positions on the 12 in. stick where a mark at x would enable one to measure all the distances from 1 to 12 in. (assuming x is a positive integer). For an extra challenge, allow x to be any positive real number less than 12 (Figure P.12).

Figure P.12

13. Prove that $\overline{AB} \subset \overrightarrow{AB} \subset \overleftrightarrow{AB}$ if A and B are any two points such that $0 < AB < \alpha$.

14. Suppose you are told that (ABC), (ACD), and (BCD), and that $AB = 1$, $BC = 3$, and $BD = 9$. Deduce the other remaining distances AC, AD, and CD.

15. If (ABC) and $AB = p/q\, BC$, show that

$$AB = \frac{p}{p+q}\, AC \quad \text{and} \quad BC = \frac{q}{p+q}\, AC$$

16. Suppose (ABC) and that $AC/AB = AB/BC$ (AB is thus the geometric mean of AC and BC).
  (a) Show that AC/AB is the **Golden Ratio** $\tau = \frac{1}{2}(1 + \sqrt{5})$.
  (b) Using everything you know about Euclidean geometry, suppose $\overline{AP}$ is constructed $\perp \overline{AB}$ with $AP = AB$, and M is the midpoint of $\overline{AB}$. If C is constructed on ray $\overrightarrow{AB}$ such that $MC = MP$, show that $AC/AB = \tau = AB/BC$.

Group B

17. Using everything you know about Euclidean geometry, show that if A, B, and C are points on a line satisfying the relation in Problem 16, and, if D is constructed such that $BD = CD = AB$, then $m\angle DBC = 72°$. Thus emerges a compass straight-edge construction of a regular pentagon. [**Hint:** Prove $\triangle DBC \sim \triangle ADC$, hence $\angle BDC \cong \angle DAC \cong \angle ADB$.]

18. Prove the following:

> **Theorem** If (ABC), then (CBA) and neither (BAC) nor (ACB) can hold.

19. Suppose you are working in a geometry satisfying **Axioms 0–5** and you come upon four collinear points A, B, C, and D such that

$$AB = 3, \quad BC = 4, \quad AC = 7, \quad CD = 8, \quad AD = 15.$$

Can you deduce the value for BD from just this much information?

20. In Problem 19, suppose that the geometry you are working in satisfies the triangle inequality. Now can you deduce the value for *BD*?

21. Show that Eves' geometry on 25 points satisfies **Axioms 1–5**.

Group C

22. Defining (*ABCDE*) in the obvious manner, what between-ness relations involving three points would guarantee this relation in a metric space?

23. **The Roundup Metric**   For all points on the *x*-axis in the coordinate plane, represent all points by their *x*-coordinate. A concept for distance is defined as follows:

$$d(x,y) = |x - y| \quad \text{if } |x - y| \text{ is an integer}$$
$$= \text{least integer} > |x - y| \text{ otherwise}$$

[Examples: $d(2, 5) = 3$, $d(2, 5.5) = 4$, and $d(-2, 5.5) = 8$.]
  (a) Identify the segment $\overline{25}$ [**Hint:** The answer is {2, 3, 4, 5}. Explain.]
  (b) Identify the segment $\overline{2a}$ where $a = 5.5$.
  (c) Identify the segment $\overline{ab}$ where $a = 2.5$ and $b = 5.5$.
  (d) Prove the triangle inequality for this metric.

24. A quite general kind of geometry may be constructed satisfying **Axioms 0–5**, using the coordinate plane, with points as ordinary points $P(x, y)$ in the plane, and lines virtually any class *C* of curves $y = f(x)$. For example, consider the class of cubic curves of the form $y = (ax + b)^3$ and vertical lines $x = a$, where a and b are constants. For the distance concept, for any two points A and B determine the unique curve in *C* passing through A and B, and take AB to be the arc-length along that curve from A to B. (For each such pair of points, one must determine a and b for which $y = (ax + b)^3$ passes through those two points.) You are to answer these questions: How many lines pass through the points $A(0, 0)$ and $B(2, 8)$? How many pass through $C(1, 8)$ and $D(5, 64)$? (Show that $a = 1/2$ and $b = 3/2$ for this second case.) Examine **Axioms 0–5** in detail for this geometry.

25. Consider the *distance set* $D = \{0, 1, 2, 3, 5\}$ for a metric space. An abstract set of points and lines can be defined such that **Axioms 0–5** and the triangle inequality are satisfied, with the given set as the distance set (set of all distances). Determine such a set of points. Can these points be placed in the plane so that the distance assigned to them will be Euclidean?

26.   **Undergraduate Research Project**   Generalize the idea in Problem 25 to allow for arbitrary distance sets, leading to either finite or infinite geometries satisfying **Axioms 0–5** and **3(c)**. Sample problems: How many points would you need if D were a set of n positive integers and zero? If D were closed under addition (example: $D = \{0, 3, 5, 8, 11, 13, 14, 16, 17, 18, \ldots\}$) how could one construct a geometry having precisely these distances? (1) If $D = \{0 = a_0, a_1, a_2, \ldots, a_n\}$ is a finite set of nonnegative reals, show that a geometry with $2n$ points can be constructed having D as distance set. (2) If $D = \{0 = a_0, a_1, a_2, \ldots, a_n, \ldots\}$ is a set of nonnegative integers closed under addition, show that for points P_i ($i = 0, 1, 2, \ldots$) the assignment $P_i P_j = a_i + a_j$ will suffice ($i \neq j$). (3) Finite version of (2), where D is the sum of all pairs of members of a subset of D.

1.8 Coordinates for Rays

In Euclidean geometry a familiar concept is the so-called *ruler postulate*, a prominent feature in high school geometry texts. It asserts simply that there is a one-to-one correspondence between the points of a line and the real numbers such that the distance between any two points is the difference between the numbers assigned to them, or their *coordinates*—illustrated in Figure 1.18 (where $AB = 10 - 4 = 6$). In general, if $A[a]$ and $B[b]$ denote any two points with their respective coordinates, a and b, then $AB = |a - b|$. This assignment of points on a line to the real numbers leads naturally to what is called the *real number line*. It might seem appropriate to make this assumption here, but if we want to continue to include spherical geometry in our study, the appropriate postulate applies to rays instead of entire lines.

> **Axiom 6 [Ruler Postulate for Rays]**   *There is a one-to-one correspondence between the points of any ray h and the set of nonnegative real numbers x, $0 \leq x \leq \alpha$, called **coordinates**, such that:*
>
> (a) *The coordinate of the endpoint O of h is zero (termed the **origin** of the coordinate system), and all other points have positive coordinates.*
> (b) *If A[a] and B[b] are any two points of h as indicated by their coordinates, then*
>
> $$AB = |a - b|$$

One immediately sees that if $A[a]$, $B[b]$, and $C[c]$ are arbitrary points of a ray such that $a < b < c$, then

$$AB = b - a, \quad BC = c - b, \quad \text{and} \quad AC = c - a$$

Figure 1.18 Ruler postulate.

so that

$$AB + BC = (b-a) + (c-b) = c - a = AC.$$

Thus, (ABC) holds. This same result holds when $c < b < a$. For the converse, suppose that (ABC) holds, and consider the six permutations of the coordinates a, b, and c under inequality. These can be grouped into three pairs of orders which imply, under the previous comment, either (ABC), (BAC), or (ACB), the latter two prohibited by the assumption (ABC). The conclusion is either $a < b < c$ or $c < b < a$. This proves:

Theorem 1 If $A[b]$, $B[b]$, and $C[c]$ are any three points of a ray, with their coordinates, then (ABC) iff either $a < b < c$ or $c < b < a$.

This result makes it clear that the order of points on a ray under metric betweenness corresponds precisely to their order under coordinate betweenness. This results in many natural properties on betweenness, all true in both Euclidean and spherical geometry. These may be listed as follows:

Betweenness Properties for Points A, B, C, and D on a Ray

(1)   Exactly one of the betweenness relations (ABC), (ACB), or (BAC) must hold.
(2)   If (ABC) and (ACD), then $(ABCD)$; similarly, if (BCD) and (ABD), then $(ABCD)$.
(3)   If (ABC) and (BCD) then $(ABCD)$.
(4)   If (ABC) and (ADC) then either $(ABDC)$ or $(ADBC)$.
(5)   If (ABC) and (ABD) then either $(ABCD)$ or $(ABDC)$.
(6)   If $AB < \alpha$, then there exists a point E on h such that (ABE).
(7)   If the origin of h is O and $OA = OB$, then $A = B$

Since a ray can be passed through any two points, the above imply two further properties that answer questions raised earlier.

(8)   For any two distinct points A and B, there exists C such that (ACB).
(9)   For any four distinct points A, B, C, and D, if (ABC) and (ACD), then $(ABCD)$.

The next result is the strongest version possible of (1) for collinear points that is valid in spherical geometry. We have already observed that if A, B, and C are distinct points lying on a line, we cannot make the claim that one of the relations (ABC), (BAC), or (ACB) must hold (as we can in Euclidean geometry). But under certain restrictions, this conclusion is valid, as we shall see.

A useful device will be introduced at this point in order to facilitate these results in the case $\alpha < \infty$. Given a point A on line l, we prove the existence of a unique point A^* on l such that $AA^* = \alpha$. Such a point A^* certainly exists by taking ray $\overrightarrow{AW}$ on l and the point A^* on that ray whose coordinate is α. Suppose there were a second point P on l such that $AP = \alpha$ and $P^* \neq A$. By **Axiom 6** there exists Q on ray $PA^* \subset l$ such that (PQA^*), $Q \neq A$. Consider the distinct points A, P, Q, and A^* on line l. By **Axiom 5**, either (APQ), (PAQ), (QAA^*) or (QA^*A), but each of these imply either $AP < \alpha$ or $AA^* < \alpha$, a contradiction. Therefore, no such point P exists and A^* is the unique point on line l such that $AA^* = \alpha$. (Note that A and A^* correspond to poles on a sphere, such as the North and South Poles, where the distance between is half-way around the sphere—the maximal distance possible.)

Definition   The point A^* constructed above is called the **polar opposite** of A on line l.

The uniqueness of A^* allows us to show that for any point B on l such that $0 < AB < \alpha$, $A^* \in \overrightarrow{AB}$. For, let P be that point on $\overrightarrow{AB}$ whose coordinate is α. Then $AP = \alpha$ and therefore $A^* = P \in \overrightarrow{AB}$. This, in turn, implies:

(10)   If B is any point on line l distinct from A and A^*, then (ABA^*).

To prove this, consider ray $\overrightarrow{AB}$ on l and the polar opposite A^* of A on l. Since $A^* \in \overrightarrow{AB}$ and $A^* \neq B$, suppose the coordinates of A, B, and A^* are, respectively, 0, b, and α. Then $0 < b < \alpha$ and by Theorem 1 (ABA^*).

Theorem 2: Betweenness for Three Collinear Points

If A, B, and C arc any three distinct points lying on line l, then either (ABC), (BAC), (ACB), or (BA^*C), where A^* is the polar opposite of A on line l (if $\alpha < \infty$). The cases are mutually exclusive unless $B = C^*$.

Proof   A trivial case occurs when any pair of the points are polar opposites, for then by **(10)** one of the pairs (ABC), (BAC), or (ACB) holds. Hence assume that AB, AC, and BC are each less than α. If $\alpha = \infty$ let A^* be chosen on line l such that (ABA^*) and $A^*B > BC$, and if $\alpha < \infty$ let A^* represent the polar opposite of A; by **(10)** (ABA^*) and in either case, (ABA^*). By **Axiom 5** applied to point C and the relation (A^*BA), either (CA^*B), $C = A^*$, (A^*CB), (BCA), or (BAC). Since the first, and the last two possibilities finish the proof, it remains to show that the cases $C = A^*$ or (A^*CB) imply one of the desired cases. But since (A^*BA), $A^* = C$ implies (CBA), and (A^*CB) implies (A^*CBA) by **(9)**, or (CBA), as desired.

For the second part of the theorem, suppose (BA^*C) (thus $\alpha < \infty$) and either (ABC), (BAC), or (ACB) are true simultaneously. If (ABC) then (ABA^*C) and (AA^*C) implying $AA^* < \alpha$, a contradiction, and similarly for (ACB). It follows that (BAC) is the only possibility. By simple algebra [using (ABA^*) and (ACA^*)], one obtains $BC = \alpha$, hence $B = C^*$. ◊

Corollary A   If A, B, and C are distinct, collinear points such that $AB + BC \le \alpha$, then exactly one of the betweenness relations (ABC), (BAC), or (ACB) must hold.

Proof   In view of the previous theorem, it suffices to show that $(CA\!*\!B)$ does not hold in the case $\alpha < \infty$. If $(CA\!*\!B)$, then $CA\!* = BC - A\!*\!B = BC - (\alpha - AB) = AB + BC - \alpha$. Hence, by hypothesis, $CA\!* \le 0$, a contradiction, unless $C = A\!*$, which would imply (ABC). ◢

An obvious consequence (from a direct application of Corollary A) is

Corollary B   If $\alpha = \infty$ and A, B, and C are distinct, collinear points, then either (ABC), (BAC), or (ACB), the cases being mutually exclusive.

A property of rays is often taken for granted and overlooked in geometry. The angle indicated by the numeral 1 in Figure 1.19 is routinely expressed by either $\angle DAB$ or $\angle DAC$. The mathematical reason for this is that the rays $\overrightarrow{AB}$ and $\overrightarrow{AC}$ are the *same set of points*, so that by definition of angle (given formally in the next chapter),

$$\angle DAB = \overrightarrow{AD} \cup \overrightarrow{AB} = \overrightarrow{AD} \cup \overrightarrow{AC} = \angle DAC.$$

This property is a direct result of properties **(2)**, **(4)**, and **(5)** above, but must be proven. The argument can be much simplified by a notational trick, eliminating a host of logical cases. One first proves two preliminary results

(11)   If (ACB) then $\overrightarrow{AB} \subseteq \overrightarrow{AC}$.

(12)   If (ABC) then $\overrightarrow{AB} \subseteq \overrightarrow{AC}$.

The proofs being similar, consider **(11)**: Let $P \in \overrightarrow{AB}$ (to prove that $P \in \overrightarrow{AC}$). Excluding trivial cases, by definition of ray either (APB) or (ABP). Since (ACB) holds, then by **(4)**, either $(APCB)$ or $(ACPB)$, or by **(2)**, $(ACBP)$. Therefore, either (APC) or (ACP) and $P \in$ ray $\overrightarrow{AC}$ as desired. (Property **(5)** is used to prove **(12)**.)

The result of these two propositions is that (ACB) implies $\overrightarrow{AB} \subseteq \overrightarrow{AC}$ and, interchanging B and C in **(12)**, $\overrightarrow{AC} \subseteq \overrightarrow{AB}$, or $\overrightarrow{AB} = \overrightarrow{AC}$. Similarly, changing notation in **(11)**, (ABC) implies $\overrightarrow{AC} \subseteq \overrightarrow{AB}$ and $\overrightarrow{AB} \subseteq \overrightarrow{AC}$ by **(12)**; again $\overrightarrow{AB} = \overrightarrow{AC}$. Thus, if either (ACB) or (ABC), $\overrightarrow{AB} = \overrightarrow{AC}$. This proves

Theorem 3   If C is any interior point of ray $\overrightarrow{AB}$ then $\overrightarrow{AC} = \overrightarrow{AB}$.

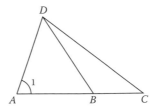

Figure 1.19 A familiar assumption.

Our final goal for this section is to establish a coordinate system for entire lines that is valid for both Euclidean and spherical geometry. An example will illustrate the basic problem.

Example 1

Suppose the space shuttle orbits the earth at an altitude of 100 mi and passes directly over Cairo, Mandalay, and Tokyo, which lie approximately on a great circle passing through Cape Canaveral, as illustrated in Figure 1.20. The astronauts aboard clock their mileage, using points on the ground as reference. Using Cape Canaveral (*O*) as the starting point and assigning zero to it, the number 8 is assigned to Cairo (*C*) since Cairo is approximately 8000 mi from Cape Canaveral (using 1000 mi as unit). Since Mandalay (*M*) lies at the approximate opposite pole from Cape Canaveral—12,400 mi away—they would assign it the number 12.4. Finally, Tokyo (*T*) would be assigned the number 15 since the distance from Mandalay to Tokyo is approximately 2600 mi. (Since the earth's circumference is 24,800 mi, the astronauts would assign points close to Cape Canaveral numbers approaching 24.8 on their first pass.

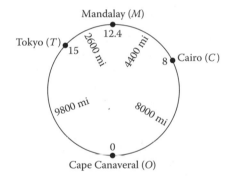

Figure 1.20 Example 1.

The distance from *T* to *O* is not the difference of the coordinates which were assigned to them (i.e., 15 − 0 = 15), but is given instead by the formula

$$OT = 2\alpha - |x - y|$$

which in this case is computed as 2(12.4) − 15 = 24.8 − 15 = 9.8 → 9800 mi (see Figure 1.20).

Establishing coordinates on a line is simple in theory. As suggested by the coordinate *x*-axis in Euclidean geometry, simply take the coordinates for all points on the ray *h'* opposite ray *h* in **Axiom 6** to be the *negative* of the coordinates as guaranteed by **Axiom 6**. Then if $\alpha = \infty$, one obtains for any two points $A[a]$ and $B[b]$ on opposite sides of point *O*, as usual,

(13) $AB = |a - b|$

For the case $\alpha < \infty$ however, if $a - b > \alpha$ the distance cannot be simply $|a - b|$.

Example 2

Suppose in Example 1, coordinates are taken on the great circle in Figure 1.20 so that those on ray $\overrightarrow{OC}$ are positive (counting from O) and those on ray $\overrightarrow{OT}$ are negative (counting from O). Thus T now has the coordinate -9.8. Show that CT (distance from Cairo to Tokyo) is given by the same formula as introduced above:

$$CT = 2\alpha - |x - y|$$

where $\alpha = 12.4$, and x and y are the coordinates for C and T.

Solution

$$CT = 2(12.4) - |8 - (-9.8)| = 24.8 - 17.8 = 7 \rightarrow 7000 \text{ mi}$$

(The distance from Cairo to Tokyo using Figure 1.20 is given by $4400 + 2600$.)

Using Example 2 as a model, since (AOB) and $OA = \alpha - a$ and $OB = \alpha - (-b)$, then $AB = (\alpha - a) + (\alpha + b)$, or, evidently,

(14) $AB = 2\alpha - |a - b|$ (when $|a - b| > \alpha$)

It is interesting that the relation **(14)** can be established axiomatically (using the axioms and propositions established so far). A key theorem used in this endeavor validates the general features of Examples 1 and 2 needed to prove the coordinatization theorem, including the result **(14)**. The coordinatization theorem itself will be left to the reader (see Problem 18 below).

Theorem 4   Suppose that (PAQ) and $l = \overrightarrow{AP} \cup \overrightarrow{AQ}$, with $PQ < \alpha$. For any two interior points $B \in \overrightarrow{AP}$ and $C \in \overrightarrow{AQ}$, then

(a) if $AB + AC \le \alpha$, (BAC) holds, or
(b) if $AB + AC > \alpha$, then (BA^*C) holds, where A^* is the polar opposite of A on l.

Proof   By Theorem 3, $\overrightarrow{AB} = \overrightarrow{AP}$ and $\overrightarrow{AC} = \overrightarrow{AQ}$. It follows that neither (ABC) nor (ACB) can hold, for otherwise, $\overrightarrow{AB} = \overrightarrow{AC}$ and $\overrightarrow{AP} = \overrightarrow{AQ}$ contradicting (PAQ). Also, note that A, B, and C are distinct points and $AB, AC < \alpha$. Now take the two cases stated in the theorem.

(a) By Corollary A above, either (ABC), (BAC), or (ACB), of which only (BAC) is valid. [If (ABC), for example, then $\overrightarrow{AP} = \overrightarrow{AC} = \overrightarrow{AB} = \overrightarrow{AQ}$ or $Q \in \overrightarrow{AP}$, contradicting (PAQ).]
(b) By **(10)** it follows that (ACA^*), and by **Axiom 5**, either (BAC), (ABC), (CBA^*), or (CA^*B)—the desired conclusion. If (BAC) then $\alpha > BC = BA + AC$, contradicting the hypothesis, and we showed above that (ABC) cannot hold. If (CBA^*), then with (ACA^*), it follows that $(ACBA^*)$ or (ACB), which cannot hold. This leaves (BA^*C), as desired. ◣

Theorem 5: Coordinatization of Lines

The points of any line $\overrightarrow{AB}$ may be placed into one-to-one correspondence with the set of real numbers x, $-\alpha < x \leq \alpha$, called *coordinates*, such that

(a) the coordinate of A is zero (called the **origin** of the coordinates system), and that of B is the distance AB,

(b) if $P[x]$ and $Q[y]$ are any two points of l, indicated with their coordinates, then

$$PQ = |x - y|, \qquad \text{if } |x - y| \leq \alpha$$

$$= 2\alpha - |x - y|, \quad \text{if } |x - y| > \alpha.$$

Note: Observe the following interpretation of Theorem 5 when $\alpha = \infty$. Here the case $|x - y| > \alpha$ never occurs, leaving simply $PQ = |x - y|$, the familiar ruler postulate for a line in Euclidean geometry. ◣

Since the coordinatization theorem for finite α is likely to be unfamiliar to most readers, we include a few examples to show how it works.

Example 3

Suppose the points A, B, and C on some line have coordinates indicated by $A[20]$, $B[-20]$, and $C[12]$. Find AB, BC, and AC and determine any betweenness relations that may exist if

(a) $\alpha = 30$ (see Figure 1.21),
(b) $\alpha = 35$,
(c) $\alpha = \infty$.

Solution

(a) Observe that by Theorem 5,

$$|20 - (-20)| = 40 > 30 \quad \rightarrow \quad AB = 2\alpha - 40 = 60 - 40 = 20$$

Similarly,

$$|-20 - 12| = 32 > 30 \quad \rightarrow \quad BC = 60 - 32 = 28$$

$$|20 - 12| = 8 < 30 \quad \rightarrow \quad AC = 8$$

Since $AB + AC = 20 + 8 = 28 = BC$, then we conclude that (BAC) holds.

(b) We have here

$$|20 - (-20)| = 40 > 35 \quad \rightarrow \quad AB = 2\alpha - 40 = 70 - 40 = 30$$

$$|-20 - 12| = 32 < 35 \quad \rightarrow \quad BC = 32$$

$$|20 - 12| = 8 < 35 \quad \rightarrow \quad AC = 8$$

In this case no betweenness relations hold.

(c) All three absolute values are less than α, so we obtain $AB = 40$, $BC = 32$, and $AC = 8$. Thus (ACB) holds.

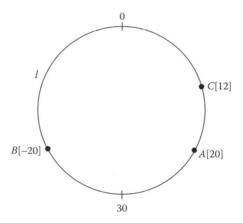

Figure 1.21 Example 3.

Example 4

Suppose $\alpha = 10$. Consider the points $A[-9]$, $B[2]$, and $C[5]$ on line l, as shown in Figure 1.22.

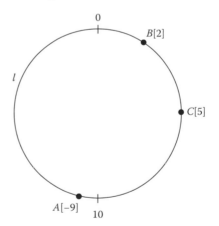

Figure 1.22 Example 4.

Since $-9 < 2 < 5$, we might conclude from this that (ABC) holds. However, this is not true. Let the reader calculate the distances AB, BC, and AC (by Theorem 5) showing that (ACB) instead.

Example 5

Suppose points $C[c]$, $D[-c]$, and $E[\frac{1}{2}c]$ are points on some line, for real c, and that $3\alpha/4 < c < \alpha$. Prove that (DCE) holds.

Solution

$$\left|c - (-c)\right| = 2c > \alpha \quad \rightarrow \quad CD = 2\alpha - 2c$$

$$\left|(-c) - \tfrac{1}{2}c\right| = \tfrac{3}{2}c > \tfrac{9}{8}\alpha > \alpha \quad \rightarrow \quad DE = 2\alpha - \tfrac{3}{2}c$$

$$\left|c - \tfrac{1}{2}c\right| = \tfrac{1}{2}c < \tfrac{1}{2}\alpha < \alpha \quad \rightarrow \quad CE = \tfrac{1}{2}c$$

It then follows that

$$DC + CE = 2\alpha - 2c + \tfrac{1}{2}c = 2\alpha - \tfrac{3}{2}c = DE$$

Since C, D, and E are distinct collinear points, (DCE) follows by definition.

Problems (Section 1.8)

Group A

1.   Suppose points on ray h are, with their coordinates, $A[10]$, $B[2]$, and $C[9]$. Find AB, BC, and AC.

2.   In Problem 1, what betweenness relation holds? What theorem in this section directly predicts this relation?

3.   Points A and B lying on line l have the respective coordinates 3 and −4. Find the distance AB if
    (a) $\alpha = \infty$
    (b) $\alpha = 7$
    (c) $\alpha = 5$; in this case, make a sketch of a circle indicating the origin $P[0]$ and polar opposite point $Q[5]$, properly locating the points A and B.

4.   Points A and B lying on line l have the respective coordinates 3 and −4. Find the distance AB if
    (a) $\alpha = \infty$
    (b) $\alpha = 50$
    (c) $\alpha = 21$; make a sketch, as in Problem 3(c).

5.   Suppose $\alpha = 12$ and points on line l are, with their coordinates, $A[-10]$, $B[1]$, and $C[9]$ (Figure P.5).
        Find AB, BC, and AC. Since $-10 < 1 < 9$ why do we not obtain (ABC) as one might expect?

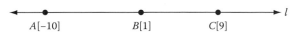

$A[-10]$ $\qquad\qquad$ $B[1]$ $\qquad\qquad$ $C[9]$

Figure P.5

6. Work Problem 5 if $\alpha = 100$, showing that in this case, (ABC).

7. Solve the toothpick problem (Problem 9, Section **1.7**) using coordinates.

8. If $AB = \alpha$ and C is any point collinear with A and B, prove directly that either $A = C$, $B = C$, or (ACB).

9. If $AB = \alpha$ then prove there exists no point C such that (ABC).

10. In Example 2 (using Figure 1.20), suppose the Hawaiian islands (H) lie on the same great circle as before, at a distance of 3,000 mi from Tokyo.
  (a) Find the correct coordinate to be assigned to H.
  (b) Using the coordinate found in **(a)**, use **(14)** to calculate the distance from the Hawaiian islands to Cairo, and **(13)** for that from the Hawaiian islands to Cape Canaveral.

Group B

11. If on some line l we plot all points $P[x]$ possible such that $-3 \leq x \leq 20$, what geometric object best describes the given set if
  (a) $\alpha = \infty$
  (b) $\alpha = 10$

12. Suppose two points on line l are $A[10]$ and $B[x]$, and that $AB = 8$. Assuming coordinates for a line, what are the possible values for x if
  (a) $\alpha = \infty$
  (b) $\alpha = 15$

13. Suppose two points on line l are $A[-12]$ and $B[x]$, and that $AB = 4$. Assuming coordinates for a line, what are the possible values for x if
  (a) $\alpha = \infty$
  (b) $\alpha = 15$

14. Prove that if $AB < \alpha$, there exists C on line $\overleftrightarrow{AB}$ such that (ABC).

15. Suppose you are working in a geometry satisfying **Axioms 0–6** and you come upon four collinear points A, B, C, and D such that

$$AB = 3, \quad BC = 4, \quad AC = 7, \quad CD = 8, \quad AD = 15.$$

Can you deduce the value for BD from just this much information? (Compare with Problem 19, Section **1.7**.)

Group C

16.   What conditions would make the cases in **Axiom 5** mutually exclusive when $\alpha < \infty$?

17.   The following problem appeared recently in a high school geometry text: On the line indicated in Figure P.17, if $WY = 17$, $WZ = 23$, and $XZ = 21$, find XY. It is tacitly assumed that the student will use the betweenness relations evident from Figure P.17. With $\alpha = \infty$, work this problem without using Figure P.17. (There is more than one answer.) [**Hint:** Use coordinates. For ideas, see Problems 12 and 13 above.]

Figure P.17

18.   Use Theorem 4 to prove the coordinatization theorem for lines, Theorem 5.

1.9 Geometry and the Continuum

The correspondence between the points of a line and the real numbers on the interval $(-\alpha, \alpha]$ has a direct bearing on the geometric characteristics of lines, rays, and segments. At first, we could only say that a line contained at least *two points* (**Axiom 2**). Even after the introduction of **Axioms 3–5** a line might contain only finitely many points, or an infinite number of isolated points, like the integers in the real number system. A set is said to be **countable** iff its members are either finite in number, or they can be put into one-to-one correspondence with the natural numbers 1, 2, 3,...; if an infinite set does not possess such a correspondence, it is called **uncountable**. It is interesting that the rational numbers are countable (like the integers themselves), but the entire set of reals is uncountable. The rationals are **dense** on the real number line, in the sense that between any two rationals, regardless of how close they are, there exists another rational. This property also carries over to arbitrary real numbers.

An infinite model for **Axioms 0–5** where every line has precisely a countably infinite number of points is provided by taking the *points* to be all *lattice points* in the coordinate xy-plane (points having *integer coordinates*), and as *lines* those sets of lattice points lying on any ordinary line whose equation is $y = ax + b$ (a and b constant and necessarily rational), or on any vertical line, as illustrated in Figure 1.23. Then take as distance the ordinary Euclidean distance as provided for by the distance formula in the coordinate plane. Moreover, it is easy to see that *each pair of points lies on a unique line*. Hence **Axioms 0–5** are satisfied. The interesting thing about this geometry is that the points on any line are equally spaced at a constant positive distance apart (always greater than or equal to 1) on each line, as illustrated in Figure 1.23, and thus form a *discrete set* on that line.

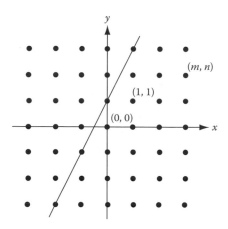

Figure 1.23 A discrete geometry.

The addition of **Axiom 6** produces a radically different character for lines. In set theory one speaks of a **continuum** of points if the set is uncountable and in one-to-one correspondence with the real numbers. A **continuum in a metric space**, however, requires certain topological properties. The simplest way to express this is to say that the space "has no holes" or "missing points." On the real number line, for example, the set of rational numbers is "dense" but also has missing points, or "holes," everywhere on the line. One such occurs where the irrational $\sqrt{2}$ is supposed to be, somewhere between 1.4 and 1.5, which is not rational. However, all these missing points have sequences of rational points converging to them. The term *continuum* itself as associated with a line suggests that a line consists of a *continuous stream of points*, with every bounded, increasing or decreasing sequence of points converging to a point on the line. The technical requirements for a **continuum on a line** l are then as follows:

(1)   The points of l are **dense** in l; that is, between each two points of l there exists a third point.

(2)   Line l is **finitely compact**, that is, each bounded sequence of points on l has a subsequence converging to some point on l.

We can immediately see that, due to the coordinatization theorem, a line in our geometry satisfies these two properties since their real number coordinates satisfy them. So every line consists of a *continuum of points* and is *uncountable*. Each segment is also a continuum, as well as each ray. This endorses ones intuitive notion regarding such geometric objects (e.g., as a "continuous stream of points").

When the pioneers of modern geometry formulated its foundations from a strictly axiomatic approach, notably Hilbert and Pasch, the two properties (1) and (2) were established using consequences of the betweenness axioms, together with the additional two axioms known as *axioms of continuity*. But the axioms of continuity are merely substitutes for analogous properties that characterize the real number system analytically (see Appendix D). Instead of algebra and analysis, Hilbert used the properties of segments and rays. This does achieve a synthetic treatment devoid of real numbers, but it also duplicates well-known developments of the real number system in modern mathematics. So one may as well assume the coordinatization axiom to begin with, as we have done (along with Moise) and forego all the work involved in developing the real numbers geometrically. In so doing, the subject of geometry is not as pure synthetically, but it moves along at a faster pace, while preserving the essential synthetic spirit.

It is interesting to carry these ideas further and apply them to the entire (Euclidean) plane (and, for that matter, to dimension 3 and higher). Property (1) above is already true: between any two points of the plane, there exists a point between them. As for property (2), suppose we have a sequence of points P_1, P_2, P_3,..., P_n,.... In order to find a point P in the plane and a subsequence of $\{P_n\}$ converging to it, one projects the sequence orthogonally onto the coordinate axes, and uses the finite compactness property for lines in order to extend this to the given sequence. For further details, see any book on real analysis.

1.10 Segment Construction Theorems

Frequently we need to reconstruct a given segment at a point along an arbitrary line, thus, in effect, "moving" the given segment from one place to another. Although not specifically defined, we prefer to use the traditional term *construct* in order to convey this concept. (One defines the **length** of a segment $\overline{AB}$ to be merely the distance AB.) Further, at this point we begin using the traditional term **congruent segments** to designate *segments having the same length*, and the symbol $\cong$ for congruence. This concept will be discussed in greater detail in Chapter 3.

Segment-Construction Theorem If $\overline{AB}$ and $\overline{CD}$ are any two segments and $AB < CD$, a unique segment $\overline{CE}$ may be constructed on $\overline{CD}$ congruent to $\overline{AB}$, that is, there exists a unique point E on segment CD such that $CE = AB$ and (CED) (see Figure 1.24).

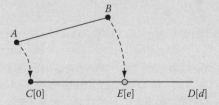

Figure 1.24 Duplicating a segment.

Proof Choose coordinates on ray $\overrightarrow{CD}$ with C as origin and D having coordinate $d \equiv CD$. Take point E as the unique point on that ray having coordinate $e \equiv AB$. Since $0 < e < d$, we have (CED) and $CE = |0 - e| = e = AB$, or $\overline{CE} \cong \overline{AB}$ as desired. ⬟

Definition 1   If point M lies on segment $\overline{AB}$ such that $AM = MB$, then M is called the **midpoint** of $\overline{AB}$. If a line, ray, or segment passes through M, it is said to **bisect** given segment $\overline{AB}$.

One then obtains these corollaries of the segment construction theorem.

Midpoint Construction Theorem   The midpoint of any segment exists and is unique.

Segment-Doubling Theorem   If $\overline{AB}$ is a segment whose length is less than $\alpha/2$, there exists a point C on ray $\overrightarrow{AB}$ such that B is the midpoint of segment $\overline{AC}$ (see Figure 1.25).

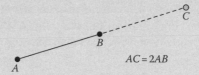

$AC = 2AB$

Figure 1.25 Doubling a segment.

Frequently, the term *extended segment*, including the concept of *extending a segment*, often used in geometry, is defined as follows:

Definition 2   If (ABC), segment $\overline{AC}$ is called an **extension** of segment $\overline{AB}$ in the direction B (or an extension of segment $\overline{BC}$ in the direction A).

Segment-Extension Theorem   Any segment can be extended in either direction to any desired length less than or equal to α. That is, given a segment $\overline{AB}$ and real number c, $AB < c \leq \alpha$, there exists a unique point C on ray $\overrightarrow{AB}$ such that (ABC), and $AC = c$ (see Figure 1.26).

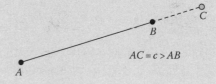

$AC = c > AB$

Figure 1.26 Extending a segment.

It would seem a trivial matter that two extensions of the same segment cannot produce two different rays, as illustrated in Figure 1.27. Although visually obvious, it is interesting to justify this axiomatically. The Moment for Reflection explores this with you.

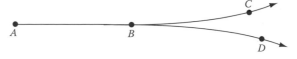

Figure 1.27 Branching possible?

Moment for Reflection

How could it be proven that the situation pictured in Figure 1.27 cannot occur? Would this come in direct conflict with any previous results that you know of?

A related property, which has been more or less taken for granted up to this point, is not difficult to prove once one recalls Theorem 2 following **(11)** and **(12)** in Section **1.8**.

Uniqueness of Opposite Rays   There exists exactly one ray that is opposite any given ray.

Proof   (See Figure 1.28.) Suppose $h = \overrightarrow{AB}$ ($AB < \alpha$). Then there exists C on $\overleftrightarrow{AB}$ such that (CAB). Thus, $\overleftrightarrow{AB} = \overrightarrow{AB} \cup \overrightarrow{AC}$ and $h' \equiv \overrightarrow{AC}$ is opposite h. (Thus, an opposite ray h' of h exists.) If ray $\overrightarrow{AD}$ also lies opposite h, then $\overleftrightarrow{AB} = \overrightarrow{AB} \cup \overrightarrow{AD}$ and $D \in \overleftrightarrow{AB}$. If $D \in h$, then by Theorem 2 of Section **1.8**, $\overrightarrow{AD} = h$, a contradiction that $\overrightarrow{AD}$ was opposite ray h. Otherwise, $D \in h'$ and $\overrightarrow{AD} = h'$, as desired. ◳

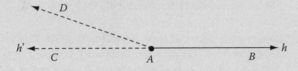

Figure 1.28 Existence of unique opposite ray.

Problems (Section 1.10)

Group A

1. Suppose that points A, B, C, and D are given such that (ABC) and (ABD) hold. Must D lie on line $\overleftrightarrow{BC}$? Discuss, or prove your answer.

2. Prove that if $A \neq B$ and $\overline{AB} \subset \overline{AC}$ (thus $\overline{AC} \neq \overline{AB}$ and $B \neq C$), then (ABC).

3. Prove that if (ABC) then $\overline{AB} \subset \overline{AC}$. (Showing that there exists a point in $\overline{AC}$ not in $\overline{AB}$ is part of the problem.)

4. Extend the segment construction theorem to include the case $AB > CD$. Prove there exists a unique point E on ray $\overrightarrow{CD}$ such that $CE = AB$ and (CDE). Use this to prove the segment-doubling theorem.

5. Prove that any segment can be *trisected*, that is, a given segment $\overline{AB}$ contains points C and D such that $(ACDB)$ and $AC = CD = DB$.

6. Suppose that $(ABCD)$ and $AB = CD$. Show that the midpoints of segments $\overline{AD}$ and $\overline{BC}$ coincide.

[**Hint:** Use coordinates.]

Group B

7. On a sphere the relation $\overrightarrow{AB} \cup \overrightarrow{BA} = \overleftrightarrow{AB}$ is not true. In an example of your own, identify the points on line $\overleftrightarrow{AB}$ that do not belong to the set $\overrightarrow{AB} \cup \overrightarrow{BA}$. (Use meridian lines running from the North Pole to the South Pole for examples.)

8. In Euclidean geometry, if (ABC) and (BCD) hold, then $(ABCD)$. Find a counterexample to this in spherical geometry.

9. Prove that if a pair of nontrivial segments S and T coincide, and one pair of endpoints coincide, then the other pair must coincide. [**Hint:** See Problem 2.]

10. Prove that if $\overline{AB} = \overline{CD}$ then either $A = C$ and $B = D$, or $A = D$ and $B = C$. [**Hint:** Show first that if $A = B$ or $C = D$ then $\overline{AB} = \{A\} = \overline{CD} = \{C\}$ and $A = C = B = D$. Otherwise, $A \neq B$ and $C \neq D$; then prove that either $A = C$, $B = D$, $A = D$, or $B = C$; if all four were false, then A, B, C, and D are four distinct points and the hypothesis implies that (ACB), (ADB), (CAD), and (CBD). Show this leads to a contradiction. Use Problem 9 to complete.]

11. Median Inequality in a Metric Space Referring to Figure P.11, prove that if d is the length of the *median* $\overline{CM}$ to a side of $\triangle ABC$ (where M is the midpoint of side $\overline{AB}$), and $a = AC$, $b = BC$, then

(1) $$d \leq a + b$$

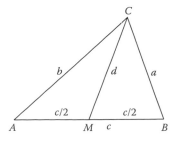

Figure P.11

12. **Cevian Inequality in a Metric Space**   Prove that if d is the length of a *cevian* $\overline{CD}$ of $\triangle ABC$ (where D is any point of side $\overline{AB}$), and $a = BC$, $b = AC$, then

(1′) $d \leq a + b$

[**Hint:** Use the triangle inequality three times; see Figure P.12.]

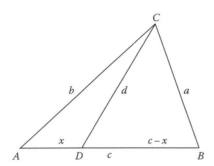

Figure P.12

13. **Median Inequality in Euclidean Geometry**   In Euclidean geometry, the median inequality of Problem 11 can be strengthened to

(2) $d < \frac{1}{2}(a + b)$

by using a simple argument in elementary geometry, using congruent triangles. Fill in all missing details in the proof given below (see Figure P.13).

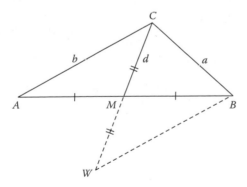

Figure P.13

Given $AM = MB$, construct segment $\overline{CW}$ such that M is the midpoint of $\overline{CW}$ by the segment-doubling theorem, making $CM = MW$. By SAS $\triangle BMW \cong \triangle AMC$ and $WB = AC$. Triangle inequality implies $CW < a + BW$.

See *Sketchpad Experiment 2* in Appendix **A** for an experiment that will demonstrate this inequality and lead you to explore a few related ideas.

Group C

14. **Dyadic Rationals**   A **dyadic rational** is any real number of the form $n/2^k$ where k and n are nonnegative integers and that $0 \le n \le 2^k$. The value of k in $n/2^k$ is called the **order**. For example, the dyadic rationals of orders 1, 2, and 3 are:

$$0, \quad \tfrac{1}{2} \quad 1 \qquad\qquad \text{(order 1)}$$
$$0, \quad \tfrac{1}{4}, \ \tfrac{1}{2}, \ \tfrac{3}{4}, \ 1 \qquad \text{(order 2)}$$
$$0, \quad \tfrac{1}{8}, \ \tfrac{1}{4}, \ \tfrac{3}{8}, \ \tfrac{1}{2}, \ \tfrac{5}{8}, \ \tfrac{3}{4}, \ \tfrac{7}{8}, \ 1 \quad \text{(order 3)}$$

By increasing the order, the dyadic rationals begin to "fill up" the interval [0, 1] on the real number line. Prove that the dyadic rationals are *dense* on [0, 1]. That is, prove that given any real number a, $0 < a < 1$, and for any $\varepsilon > 0$, there exists a dyadic rational p within ε of a (i.e., $|a - p| < \varepsilon$). [**Hint:** Choose k such that $1/2^k < \varepsilon$, and let n be the largest integer such that $0 \le n \le 2^k a$.]

15. **Dyadic Rationals in Geometry**   Show that the following geometric construction will produce a point D_p on the segment $\overline{AB}$ such that $AD_p = pAB$, where p is a dyadic rational of order k, and $d_p < pa + qb$, where $d_p = CD_p$ and $p = AD_p /AB$, $q = 1 - p$.

STEP 1 Let M be the midpoint of segment $\overline{AB}$; use **(2)** for order $k = 1$ ($d_{1/2} < 1/2a + 1/2b$).

STEP 2 Let K be the midpoint of segment $\overline{AM}$, L that of segment $\overline{MB}$, and set $D_{1/4} = K$, $D_{1/2} = M$, and $D_{3/4} = L$. Show that $d_{1/4} < 1/4a + 3/4b$ and $d_{3/4} < 3/4a + 1/4b$ (order $k = 2$).

STEP 3 Continue the process inductively, for larger orders k.

(If $d_p < pa + (1 - p)b$ has been proven for all dyadic rationals of order k, the construction in the $(k + 1)$st step indicated above will produce points D_r such that $AD_r = rAB$, and $d_r < ra + (1 - r)b$ follows for all dyadic rationals of order $k + 1$. Carry this out, completing the proof.)

16. **Cevian Inequality for Euclidean Geometry**   From Problem 15 and mathematical induction, it has thus been proven that the inequality $d_n < p_n a + q_n b$ holds for any dyadic rational p_n of arbitrary order n, where $d_n = CD_n$, $p_n = AD_n/AB$, and $q_n = D_n B/AB \equiv 1 - p_n$. To extend this to arbitrary p and thus for arbitrary $D \in AB$, define $d = CD$, $p = AD/AB$, and $q = 1 - p$. There exists a sequence of dyadic rationals p_n converging to p, thus $d_n < p_n a + q_n b$ for all n. Now by the triangle inequality, $|d - d_n| = |CD - CD_n| \le DD_n = |AD - AD_n|$. Use this to show that $\lim (d_n - d) = 0$ and $d_n \to d$. Thus $d = \lim d_n \le \lim (p_n a + q_n b) = pa + qb$, which proves the general cevian inequality (in Euclidean geometry)

(3) $d \le pa + qb$

(One can obtain the strict inequality by locating $D' \in \overline{AB}$ such that D' is the midpoint of $\overline{AB}$ and applying (2) and (3) above.)

2

Angles, Angle Measure, and Plane Separation

The list of axioms will now be expanded to include the familiar ideas about angles. An apparently trivial fact that each line separates the plane into two "sides" or "half planes" will also be introduced, which is the basis for further properties of angles. First, we introduce the concept of angles and their measure. At this point in our development, we discontinue the consistent use of arrow symbols and bars for lines, rays, and segments. The convention of letting the identity of the objects of the discussion be defined by context will be followed from now on, except when greater clarification is desired.

2.1 Angles and Angle Measure

The angle concept to be used is the obvious one, as stated in the following definition.

> **Definition 1**   An **angle** is the set of points belonging to each of two rays having a common endpoint. The two rays are called the **sides**, and the common endpoint, the **vertex**. An angle is said to be **degenerate** if its sides coincide and **straight** if the sides are opposite rays.

Thus, Figure 2.1 shows a nondegenerate, nonstraight angle, consisting of the two rays h and k having point A as vertex. Several acceptable notations for an angle are possible, unlike the situation for a segment where the two endpoints are sufficient. If h = ray AB and k = ray AC, as in Figure 2.1, the angle thus formed is the set $h \cup k$. It would therefore make sense to use the notation $\overline{hk}$ (resembling segment notation) or $\angle hk$. Then, when angle measure is introduced, hk can be used for the measure of angle hk, consistent with the precedent already set for segments. This is often convenient, so we use this frequently, even though it is unconventional. Alternatively, the same angle can be denoted by the traditional $\angle BAC$,

53

where A is the vertex (with AB and $AC < \alpha$), and its measure, $m\angle BAC$, or, if no ambiguity is involved, by $\angle A$ and $m\angle A$, respectively.

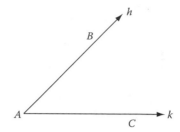

Figure 2.1 The angle *BAC*.

As in the case of distance, an axiom is required to introduce the concept of angle measure into our geometry. Since antiquity, the most common way to measure angles is to divide a circle into 360 equal parts and to define one of those parts as "an angle of one degree." Arbitrary angles can then be measured accordingly, using fractional parts of a degree when necessary. With this approach, it would seem that an angle of measure 250° might make sense, for example. But that same angle, as the *union of two rays*, could also have measure 110° using the same criterion (do you see why?). Obviously, this ambiguity is to be avoided in a coherent concept for angle measure. Keep in mind that an angle, by definition, is simply the union of two rays, so its measure can only depend on the position of those two rays *relative to each other*.

The axiom that introduces the measure of angles is observed to be logically equivalent to the axiom that introduced distance (**Axiom 3**), where one substitutes the notion of the sides of an angle for that of the endpoints of a segment. Both axioms consist of the positive definite property and symmetry.

> **Axiom 7**   *To each angle $\overline{hk}$, there corresponds a real number hk, called its **measure**, satisfying the properties*
>
> **(a)** $hk \geq 0$, with equality if and only if $h = k$
> **(b)** $hk = kh$

It would be quite fanciful to pursue a geometry in which angle measure is unbounded. Indeed, in trigonometry, one presumably considers "angles" whose measures are arbitrary real numbers. However, we take as another axiom one that establishes the boundedness of angle measure.

> **Axiom 8**   *The measure of an angle is less than or equal to the measure of any straight angle.*

Analogous to the least upper bound for distance, we define β to be the least upper bound for angle measure, which, by **Axiom 8**, exists as a (bounded) real number.

(1) $\beta = \text{LUB } \{hk: h \text{ and } k \text{ are the sides of an angle}\}$

Moment for Reflection

Can you explain the apparent discrepancy between bounded angle measure here and that assumed in trigonometry, where an angle of 1000° is possible? What concepts are different in the two theories?

It seems obvious that β is the measure of a straight angle. Indeed, suppose $\angle hh'$ is a given straight angle and x its measure. Then, by **(1)**, $x \le \beta$. If $x < \beta$, then, by **Axiom 8**, the least upper bound of angle measure would be strictly less than β, a contradiction. Thus

> **Theorem** The measure of a straight angle equals the least upper bound β defined above.

There are several logical choices for the number β. The value $\beta = 1/2$ corresponds to the use of revolutions, with one revolution corresponding to 360°, $\beta = \pi$ corresponds to the use of radian measure, while $\beta = 180$ corresponds to degree measure. We shall keep with the traditional for elementary geometry and choose $\beta = 180$, at least for now. Note that angle measure, being a real number, does not require the special degree symbol. This will be used for clarification only, particularly when we want to distinguish degree measure from radian measure.

One can define betweenness for concurrent rays (i.e., rays having the same endpoint) analogous to that of collinear points. Indeed, there is a duality between *collinear points* and *concurrent rays* that frequently proves useful. Thus, we state

> **Definition 2** A ray u is said to lie *between* rays h and k, and we denote this relationship by (huk), provided that
>
> **(a)** h, k, and u are distinct *concurrent* rays (having a common endpoint)
> **(b)** $hu + uk = hk$

> **Definition 3** If h, k, u, and v are four concurrent rays such that the four betweenness relations (hku), (hkv), (huv), and (kuv) are all valid, then the aggregate of these relations will be denoted by $(hkuv)$.

The logical consequences of these two definitions are the direct analogues of previously proven properties of point betweenness. For example, one can show that (huk) implies (kuh) and that if (hku), (hkv), and (huv), then (*hkuv*) (the "dual" of Theorem 1, Section **1.7**). We shall omit these results since they are all rather obvious—one can easily use previous arguments to prove them. One might toy with the idea of dualizing previous axioms (such as **Axiom 6**) to give us the needed properties for angle measure, but we take a more traditional approach.

Problems (Section 2.1)

Group A

1. As discussed above, a logical value for the measure of ∠*AOB* in Figure P.1 for this problem might be 270°. Since **Axiom 8** prohibits angle measure greater than 180, what other logical choice for *m*∠*AOB* could be made from Figure P.1 figure? Discuss.

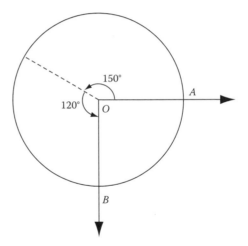

Figure P.1

2. It appears that lines *OA* and *OB* of Problem 1 are *perpendicular* (not yet defined formally). From this standpoint, we might want *m*∠*AOB* = 90. Formulate a geometric criterion for *perpendicular lines* without using the concept of right angle directly, and then see where that leads (intuitively).

3. Figure P.3 shows three rays *h*, *k*, and *u* from *O*, where the positive (counterclockwise) degree rotation from *h* to *k* is 105 and that from *h* to *u* in the opposite (negative) direction is 90. Find the positive degree rotation from ray *k* to ray *u*.

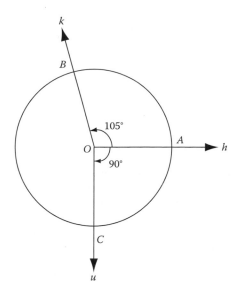

Figure P.3

4. In Problem 3, name the logical choices for $m\angle AOB$, $m\angle AOC$, and $m\angle BOC$ that satisfy **Axioms 7–8**. Does (uhk) hold? (huk)? (hku)?

5. Suppose that $hk = 40$ and $ku = 30$. Under what circumstances would (huk) hold?

6. Suppose that $hk = 30$ and $ku = 40$. Under what circumstances would (huk) hold?

7. If rays OA and OA' are opposite rays and $m\angle AOB = 75$, what is a logical choice for $m\angle BOA'$?

Group B

8. How far apart (in degrees) are the minute hand and the hour hand of a clock at
    (a) 2:30? [**Hint:** The answer is *not* 120°.]
    (b) 6:45?

9. **The Defective Protractor Problem**   A protractor was used to measure the angle sum of a triangle in a Euclidean plane. However, it was later found that the protractor was defective: the angle at the bottom edge was found to be off by 1°, and instead of a straight angle at the center, the angle measured exactly 179°, as shown in Figure P.9. Find the error made in the angle sum of the triangle, accurate to four decimals.

10.* If in Problem 9 the angle sum of the triangle measured 180.3°, find the actual angle measure at the bottom edge of the defective protractor (masquerading as 180°).

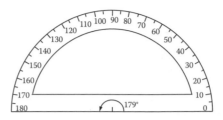

Figure P.9

11. **Using the Circular Protractor**   A circular protractor is read in the positive direction and, as shown in Figure P.11, indicates all positive rotations from 0° to 360°. This creates a coordinate system for the rays from O, the center of the protractor. Rotations can then be calculated mathematically from just the coordinates using the formula $|x - y|$ for the rotation from ray $h[x]$ to the ray $k[y]$. Observe the particular rays OA, OB, and OC in Figure P.11.

  (a) Find the positive degree rotation from ray OA to ray OB and from ray OB to ray OC.

  (b) Show that the sum of these two rotations gives the positive rotation from OA to OC.

  (c) Find $m\angle AOB$, $m\angle BOC$, and $m\angle AOC$. Is $m\angle AOC$ the sum of $m\angle AOB$ and $m\angle BOC$?

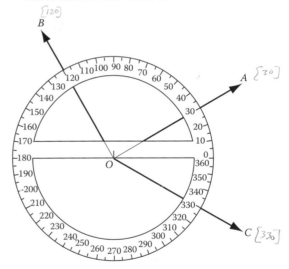

Figure P.11

12. **Coordinates and Angle Measure**   Using the circular protractor to coordinate the rays from O using the real numbers $0 \le x < 360$, show that your answers in Problem 11(c) for $m\angle AOB$, $m\angle BOC$, and $m\angle AOC$ follow the general formula for the angle measure hk for any two rays $h[x]$ and $k[y]$ in terms of their coordinates x and y:

$$hk = |x - y| \qquad \text{if } |x - y| \le 180$$
$$= 360 - |x - y| \qquad \text{if } |x - y| > 180$$

13. In Problem 11, if ray OC' is the opposite ray of OC, show that $(\overrightarrow{OA}\ \overrightarrow{OB}\ \overrightarrow{OC'})$ holds.

14. If (huk), prove (kuh) but neither (hku) nor (khu).

15. Prove: If (huk), (hkv), and (huv), then $(hukv)$.

Group C

16. **Vertical Angles Are Congruent** If rays h and k are concurrent at O and h' and k' are the rays opposite h and k, respectively, the angles hk and $h'k'$ are called *vertical angles*. Using coordinates (as in Problem 11), prove that $hk = h'k'$. [**Hint:** Rotate the protractor about its center so that h has coordinate 0 and k or k' has coordinate $x < 180$. Use the formula for angle measure as given by Problem 12.]

2.2 Plane Separation

If a cruise ship were to sail the ocean along a great circle from point A to point B, both points lying in the northern hemisphere, it is not altogether trivial that the ship will never cross the equator, regardless of how close A and B might be to it. But one could reason as follows: Assume that arc AB is the shortest path on the sphere from A to B. Suppose the ship crosses the equator, and suppose that the first crossing occurs at P, as in Figure 2.2. At this point, the ship passes into the southern hemisphere. Then, in order to get back to point B in the northern hemisphere, the ship must cross the equator again at, say, point Q. But now we have two great circles, the ship's path and the equator, intersecting at two distinct points P and Q, and we must conclude that P and Q are opposite poles, with $PQ = \alpha$, and, consequently, $AB > \alpha$—a contradiction. Therefore, the ship remains in the northern hemisphere.

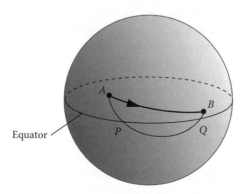

Figure 2.2 A great circle path from A to B.

This property, which hemispheres evidently have, is of great significance. It may be formulated in the following manner:

Definition   A set S is called *convex* if, for each pair of points A and B belonging to S such that $AB < \alpha$, all the points of segment $\overline{AB}$ lie in S. (See Figure 2.3 for a few examples.)

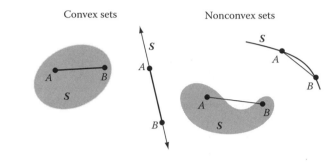

Figure 2.3 Convex and nonconvex sets.

Convexity is a relatively modern concept, and, although simply stated, it leads to many interesting and significant results. Indeed, the theory of convexity is a major field in mathematics, rich in elegant propositions, and unsolved problems. A fundamental theorem in convexity that we shall use here is the following:

Theorem   The intersection of two convex sets is convex.

Proof   Suppose S and T are convex. To show that $S \cap T$ is convex, let $AB < \alpha$, with $A \in S \cap T$ and $B \in S \cap T$. Then $A, B \in S$ and $A, B \in T$. If P is any point of segment $\overline{AB}$, then by convexity of S and T, $P \in S$ and $P \in T$. This means that $P \in S \cap T$ and $\overline{AB} \subseteq S \cap T$. Therefore, by definition, $S \cap T$ is convex. ◣

The apparent fact that each hemisphere in spherical geometry is convex and that every great circle divides the sphere into two hemispheres (similarly for a line in Euclidean geometry that divides the plane into two half planes) leads us to the next axiom, **Axiom 9**, that will be crucial to future development. First, a model will be constructed showing that **Axiom 9** cannot be proven from the previous axioms (**Axioms 0–8**).

Suppose we start with an ordinary sphere S of radius r (the maximal distance is therefore $\alpha = \pi r$). So far, our axioms are evidently valid for the geometry taking place on S, with the great circles as lines and, for any two points A and B, $AB = $ shortest arc length on the great circle passing through A and B. Thus, **Axioms 0–4** are certainly satisfied for S. The next two axioms we added in Chapter 1 can be easily verified (you should consider this a brief thought-exercise as you review **Axioms 5** and **6**). What about the axioms on angle measure, that is, **Axioms 7** and **8**? We simply adopt the calculus notion of angle measure between two *intersecting curves* (in this case, great circles) and take the measure of the smallest angle between the tangents to the curves at the point of intersection. Figure 2.4 illustrates this idea explicitly: the measure of $\angle BAC$ on S is taken as θ, the measure of the smaller angle between the two tangents at A. Since at any point

A on *S* the tangents at *A* lie in a (Euclidean) plane, angle measure for *S* coincides with that of the Euclidean plane at *A*, and thus **Axioms 7–8** are valid. Spherical geometry, therefore, satisfies **Axioms 0–8**.

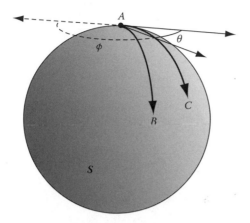

Figure 2.4 Spherical angle measure.

To obtain a model where **Axiom 9** is denied, we are going to do something really unusual. Suppose we agree to *identify* points on *S* that are at *opposite poles* (when $AB = \alpha$). With some thought, the resulting geometry can be seen to satisfy **Axioms 0–4** and the maximal distance becomes $\alpha' = \alpha/2$. To see this, consider a point *P* as it moves along the great circle *C* of *S* starting at point *A*, as shown in Figure 2.5. As *P* approaches point *B*, the point that is one-fourth the way around *C*, the distance *AP* approaches $\alpha/2$. But as *P* goes beyond *B* to point *Q*, the distance from *A* begins to decrease (since $AQ = AQ' < AB' = AB$, where *Q'* and *B'* are the opposite poles of *Q* and *B*). A *ray* is thus observed to be a one-quarter arc of a great circle, while a *segment* is a proper subarc thereof. **Axiom 6** on coordinates for a ray, which in this context implies **Axiom 5**, will be left to the reader, and the axioms on angle measure at each point are self-evident, as in the case of the ordinary sphere. This modified spherical geometry is then another model for **Axioms 0–8**. Thus, at this point, we have two different models for **Axioms 0–8**, for which $\alpha < \infty$.

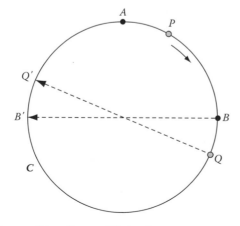

Figure 2.5 Distance *AP* on the modified sphere.

But consider a line *l* and a point *A* not on it, as shown in Figure 2.6. Apparently, we cannot decide on which side of *l* point *A* lies, for its opposite pole *A'* would seem to lie on the opposite side, and yet *A* = *A'*. This evidently implies that in this geometry, *every line has only one "side"*!

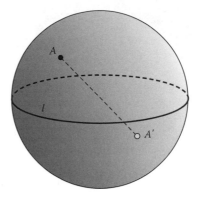

Figure 2.6 On which side of *l* is point *A*?

Moment for Reflection

A more convincing argument showing that **Axiom 9** is independent from **Axioms 0–8** involves a concept of implied equivalence in such phrases as *same as* or *same side as*. Consider how the natural transitivity of the relation *same side as* works in geometry: Suppose that *A* lies on the same side of a line as *B*, and *B* lies on the same side as *C*. Then, the question is whether *A* and *C* lie on the same side, which, as will be proven later for axiomatic spherical geometry, is a consequence of **Axiom 9**. [To be precise, define "*A* lies on the same side of line *l* as *B*" to mean "there does not exist a point $X \in l$ such that (*AXB*)."] Can you determine whether this is true for the "pseudo-spherical" geometry considered above? Is it true for ordinary spherical geometry?

Note: The geometry we just considered is known as **single elliptic geometry** (one pole for each line), as opposed to ordinary spherical geometry, known as **double elliptic geometry** (two poles for each line). In single elliptic geometry, one can observe the desirable feature in an otherwise undesirable geometry: *two points determine a unique line*. Single elliptic geometry also provides a model for an *infinite projective plane*—one which *possesses a finite metric*. ⬔

In order to eliminate the pathological behavior of a line with only one side, we assume

Axiom 9 [Plane-Separation Postulate]   *There corresponds to each line l in the plane two regions (sets) H_1 and H_2 having the following properties*:

(a) *Each point in the plane belongs to one and only one of the three sets l, H_1, and H_2.*
(b) *H_1 and H_2 are each nonempty convex sets.*
(c) *If $A \in H_1$, $B \in H_2$, then l intersects any segment AB joining A and B at an interior point on that segment.*

The sets H_1 and H_2 mentioned in **Axiom 9** are called the **half planes determined by** line *l*. Each half plane is said to be **opposite** the other. We also assert that each line *l* has two **sides**, namely the half planes determined by *l*. Since **(b)** implies that not all points of the plane can lie on *l*, we can now eliminate **Axiom 0.** (We might note that everything stated previously could apply to just a single *line* in the coordinate plane, up until we assume **Axiom 9.**) It is often convenient to use the terminology **closed** half plane to mean a half plane plus the line that determines it. (The **open** half plane is the half plane itself.)

A useful notation for half planes is the following: If A, B, and C are three noncollinear points such that $BC < \alpha$, let

(1) $$H(A, \overleftrightarrow{BC})$$

represent the half plane H_1 or H_2 determined by $\overleftrightarrow{BC}$ that contains point A, also referred to as the **A-side** of line BC. This notation will be particularly useful in defining the interior of an angle in the next section.

2.3 Consequences of Plane Separation: The Postulate of Pasch

At this point in the development, you should be warned that the arguments to be used from now on will sometimes involve logic that is a little more sophisticated than that previously used. Up to now virtually, all the arguments were relatively short, involving only the simplest of direct reasoning (e.g., if A implies B and B implies C, then A implies C). Indirect arguments will be the staple from now on. A routine example of an indirect argument involves a concept with which most readers are no doubt familiar, the **trichotomy** property: given any two real numbers a and b, either $a > b$, $a = b$, or $a < b$. Thus, if we want to prove that $x = y$, it suffices to show that if we assume either $x > y$ or $x < y$, then a contradiction results. Closely related to this is the so-called **contrapositive** argument: In order to prove that A implies B, one assumes that B is not true and shows from this that A is not true, which contradicts the original hypothesis (A).

The first instance of indirect proof occurs now for one of several consequences of **Axiom 9**. Suppose that $A \in l$ and $B \in H_1$. Then, it will be shown that any point C such that (ACB) must also lie in H_1. By the plane separation postulate, C must lie in either H_1, line l, or H_2. If C lies on line l, then since $0 < AC < \alpha$ by hypothesis, $\overleftrightarrow{AC} = l$ and $B \in l$, a contradiction. If $C \in H_2$, then l intersects segment BC at some point D, as shown in Figure 2.7, which means that $(ACDB)$ holds hence (ADB), so that $l = \overleftrightarrow{AD}$ and $B \in \overleftrightarrow{AD} = l$, a contradiction. The only case left is $C \in H_1$, as desired. This argument proves

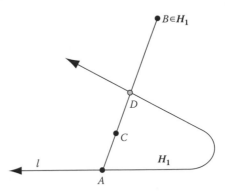

Figure 2.7 Segment AB lies in H_1.

Theorem 1   If A lies on line l and B lies in one of the half planes H determined by l, then every interior point of any segment AB also lies in H.

Corollary   If A lies on l, B lies in H_1 and (ABC) where $AC < \alpha$, then C lies in H_1.

Proof   (Argument similar to that of Theorem 1.)

Theorem 2   If A lies on line l and B lies in one of the half planes H determined by l, then every interior point of any ray AB also lies in H. (See Figure 2.8.)

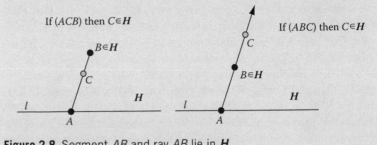

Figure 2.8 Segment AB and ray AB lie in H.

The next result will be the centerpiece of the proof of the much-used crossbar theorem in geometry, which will be proved in the next section. In order for you to understand the proof of that theorem, you need to have a clear understanding of the result which follows.

Theorem 3   Let B and F lie on opposite sides of a line l and let A and G be any two distinct points on l such that $AG < \alpha$. Then, any segment GB and any ray $\overrightarrow{AF}$ are disjoint sets (have no points in common). (See Figure 2.9.)

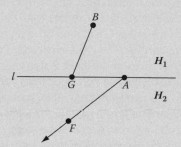

Figure 2.9 Segment GB and ray AF do not meet.

Proof   Suppose $B \in \boldsymbol{H_1}$ and $F \in \boldsymbol{H_2}$ where $\boldsymbol{H_1}$ and $\boldsymbol{H_2}$ are the half planes determined by l.

By Theorem 1, the interior points of any segment having endpoints B and G lie in $\boldsymbol{H_1}$, and by Theorem 2 those of any ray with origin A passing through F lie in $\boldsymbol{H_2}$, hence are disjoint. This leaves the boundary points $\{G, B\}$ and $\{A, C\}$ where $AC = \alpha$. By hypothesis, $G \neq A$ or C, and $B \neq A \in l$. This leaves the *logical* possibility $B = C$. Suppose $B = C$, to gain a contradiction. If m is the line through A and F defining ray AF, where $F \in \boldsymbol{H_2}$, then $m \neq l$. But since $AC = \alpha$ and $C \in \overrightarrow{AF} \subset m$, (AFC) follows (recall **(10)** Section **1.8**). Therefore, (AFB) and by Theorem 1 F lies in $\boldsymbol{H_1}$, a contradiction. ◣

We leave the proof of the following consequence of the plane separation postulate to the reader.

Theorem 4   Two points A and B not on line l lie on the same side of l iff there exists no point X on l such that (AXB).

Another consequence of the plane separation postulate is the famous postulate of Pasch—apparently a statement of the obvious— the principle that if a line intersects one side of a triangle and does not pass through any of its vertices, then it must intersect a second side, as indicated in Figure 2.10. The geometer M. Pasch was among the first to notice that this property had been tacitly assumed by Euclid. This principle will be stated without reference to triangles, however, since it is valid whether or not the points A, B, and C are collinear. Keeping with tradition, it is referred to here as a *postulate*, but it is certainly not a postulate here and will be proven as a theorem. It is logically equivalent to the separation postulate in the sense that, in the context of **Axioms 0–8**, each can be proven from the other. (Problem 8 below shows how to use the postulate of Pasch in order to establish the plane-separation postulate.)

Historical Note

Moritz Pasch (1843–1930) was a German mathematician who had a profound influence on scholarly thinking in axiomatic geometry. He was the first to realize the importance of undefined terms and their use in basic assumptions. His triangle postulate (our Theorem 4), in addition to many other developments in axiomatic systems, appeared in his 1882 work *Vorlesungen Über Neuere Geometrie*. Pasch's philosophy was to allow empirical ideas in the choice of certain undefined terms and axioms (the so-called nuclear concepts), but every deduction from these axioms should proceed from pure logic, without empirical influence. The undefined term *point* was allowed, for example, but not *line*, since no one had ever observed a complete straight line. Pasch was also the first to postulate the concept of betweenness, which was used in Hilbert's foundations of geometry in 1898. Pasch's groundbreaking ideas appeared in writings by a flourishing Italian school of geometers during the period 1860–1900, led by Peano, Veronese, Pieri, Padoa, Burali-Forti, and Levi-Civita (Burton 1998).

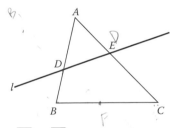

Figure 2.10 Line l cuts $\overline{AC}$ or $\overline{BC}$.

Theorem 5: Postulate of Pasch

If A, B, and C are any three points in the plane, collinear or not, and (ADB) holds for any other point D, then any line l passing through D but not passing through A, B, or C either passes through a point E such that (AEC), or through a point F such that (BFC), the cases being mutually exclusive.

Proof   Let H_1 and H_2 be the two half planes determined by l with $A \in H_1$; by the Theorem 4, $B \in H_2$, as in Figure 2.11. (Here we are using the hypothesis that l does not pass through A or B.) Then by **Axiom 9**, since C does not lie on line l, C must either lie in H_1, or in H_2, but not both. If $C \in H_2$ (as in Figure 2.11), then A and C lie on opposite sides of line l by definition and by Theorem 4, there is a point $E \in l$ such that (AEC). Now if l also passes through a point F such that (BFC), then by Theorem 4, $C \notin H_2$, a contradiction. The argument is similar if one assumes that $C \in H_1$. ◥

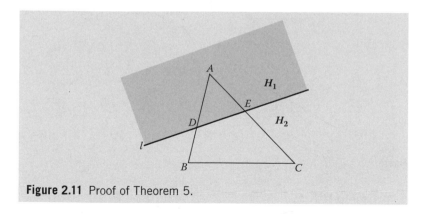

Figure 2.11 Proof of Theorem 5.

Problems (Section 2.3)

Group A

1. If solid lines, solid dots, and shaded regions indicate points that belong to S, while dashed lines and open dots those that are not, which of the following diagrams depict convex sets (Figure P.1)?

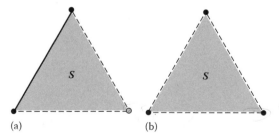

(a)   (b)

Figure P.1

2. If solid lines, solid dots, and shaded regions indicate points that belong to S, while dashed lines and open dots those that are not, which of the following diagrams depict convex sets (Figure P.2)?

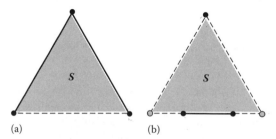

(a)   (b)

Figure P.2

3. Using the same conventions as in Problems 1 and 2, which of the sets in Figure P.3 are convex?

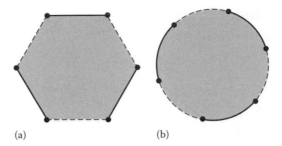

(a)   (b)

Figure P.3

Group B

4. Can the complement of a convex set ever be convex? List, without proof, the types of sets possible. (The complement of a set S in the plane $\mathcal{P}$ is the set of all points not belonging to that set, denoted $\mathcal{P} \setminus S$.)

5. Prove that an angle is not a convex set unless it is either degenerate or straight.

6. Write an explicit proof for the corollary of Theorem 1.

7. Show that if A and B are on the same side of line l, and if B and C are on the same side, then A and C are on the same side.

Group C

8. Using the analysis that follows, prove the plane-separation postulate from the postulate of Pasch using **Axioms 0–8**. Let line l be any line and suppose that A, B, and C are any three points such that (ABC), with $B \in l$.

    Define:

    $$H_1 = \{P \notin l: \text{there exists no point } X \in l \text{ such that } (AXP)\}$$

    $$H_2 = \{P \notin l: \text{there exists no point } X \in l \text{ such that } (CXP)\}$$

    Show that the sets H_1 and H_2 are convex and that they have the other properties required in **Axiom 9**. (For a more elegant proof, see the next problem.)

9. If you are familiar with the terms used here, prove the plane-separation postulate from the postulate of Pasch by defining for line l the relation for all points not on line l: $A \approx B$ iff there exists no point X on l such that (AXB). Show that $\approx$ is an equivalence relation having exactly two equivalence classes, H_1 and H_2. (**Axiom 0** is needed here.) Then show that these two equivalence classes satisfy the conditions of **Axiom 9**.

2.4 The Interior of an Angle: The Angle Addition Postulate

In any representative drawing of an angle, such as in Figure 2.1, it is not difficult to picture what the "inside" or "interior" of that angle should be. But defining the concept without recourse to figures could lead to some frustration—were it not for the concept of half planes. A singularly elegant way to handle it appears in the following definition.

Definition 1   The **interior** of any nondegenerate, nonstraight angle, $\angle ABC$, denoted by Int$\angle ABC$, is the set of all points in the plane that simultaneously lie on the A-side of line BC and on the C-side of line AB.

Making use of the half-plane notation introduced earlier, the definition for angle interior can be concisely written as

(1) $$\text{Int } \angle ABC = H(A, \overleftrightarrow{BC}) \cap H(C, \overleftrightarrow{BA})$$

(see Figure 2.12 for illustration). This formulation has the advantage of enabling one to accurately refer to the interior of an angle without relying on a figure or drawing, quite useful in proofs.

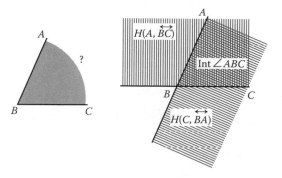

Figure 2.12 Interior of an angle.

Since it is mathematically desirable to have a definition that applies to all angles, we extend the concept of angle interior to cover degenerate and straight angles. This makes it possible to eliminate an additional axiom that would otherwise be needed. Define

(2) $$\text{Int}\angle ABC = \varnothing \text{ (the empty set) if } \overrightarrow{BA} = \overrightarrow{BC}$$

(3) $\text{Int}\angle ABC = \mathcal{P}$ (the whole plane) if $\overrightarrow{BA}$ and $\overrightarrow{BC}$ are opposite rays

In order to illustrate the concept of angle interior and to clarify its application, a sequence of diagrams is presented in Figure 2.13, where side k of angle $hk \equiv h \cup k$ gradually rotates in the positive direction, starting with $k = h$, to eventually form a straight angle (when $k = h'$, the ray opposite h), and then continues to rotate. As k passes through h', the angle interior "switches sides," in accordance with the definition. (The inset illustrates a common misconception concerning these ideas.) *Sketchpad Experiment 3* in Appendix **A** provides a realization in the Euclidean plane of the diagrams in Figure 2.13 (Problem 3).

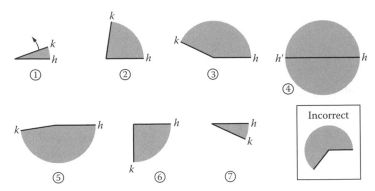

Figure 2.13 Visualizing the interior of an angle.

In spherical geometry, the interior of an angle is obtained by intersecting two *hemispheres*, which together with the sides of the angle, produces a region called a **lune**. (See Figure 2.14 for the spherical analog of Figure 2.13 where the lune grows in size, becomes the entire sphere, and then begins to decrease in size as the rotation of *k* continues.)

Figure 2.14 Interior of a spherical angle.

Now we come to another fundamental property that appears to be just a simple observation. It concerns an additive property involving the interior of an angle. Consider ∠*ABC* and point *D* in its interior (Figure 2.15). One would naturally assume that ray $\overrightarrow{BD}$ lies "between" rays $\overrightarrow{BA}$ and $\overrightarrow{BC}$ and that

$$m\angle ABD + m\angle DBC = m\angle ABC$$

The question then emerges: If ray *u* lies in the interior of angle $\overline{hk}$, then does this not imply that ray *u* lies between rays *h* and *k*? But the two ideas, a ray lying *in the interior* of the angle formed by two rays, and a ray lying *between* those two rays, are not necessarily equivalent at this point, and we find the language itself a hindrance. In fact, Problem 25 shows how to construct a counterexample—a geometry in which **Axioms 1–9** are valid, but for which the above two concepts are not equivalent. Fortunately, it will not be necessary to make this distinction, but it requires an additional axiom. One can easily see the plausibility of this property in spherical geometry.

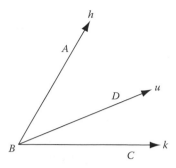

Figure 2.15 Axiom 10: (*huk*) if $D \in \text{Int}\angle hk$.

Axiom 10 [Angle Addition Postulate] *If point D lies in the interior of $\angle ABC$ or is an interior point of one of its sides, then $m\angle ABD + m\angle DBC = m\angle ABC$.*

Thus, in Figure 2.15, if u passes through an interior point of angle hk, then (*huk*), and ray u lies *between* rays h and k in both the prior meaning of the term (Section 2.1) and in the new intuitive meaning derived from the concept of angle interiors. Note that if we define the **closed interior** of an angle to be its interior plus its sides (the intersection of the closed half planes determined by its sides), denoted by $\text{INT}\angle ABC$, we can say in the case $\alpha = \infty$, if $D \in \text{INT}\angle ABC$ then $m\angle ABD + m\angle DBC = m\angle ABC$ (not true if $\alpha < \infty$).

The converse of Axiom 10 is also true, so the addition property is characteristic of angle interiors; it requires two preliminary results before it can be proven. The first is usually taken as an additional axiom, but this can be avoided with the proper handling of the interior of angles (as above).

Theorem 1: Linear Pair Theorem

If rays h and h' are opposite rays and k is any other ray concurrent with h and h', then (*hkh'*) and

$$hk + kh' = 180$$

Proof Since the interior of a straight angle $h \cup h' \equiv l$ is the entire plane, k passes through an interior point of angle hh'. Therefore, by **Axiom 10**, (*hkh'*) and $hk + kh' = hh' = 180$. ◣

At this point, we introduce several common terms associated with the above result. The angles hk and kh' of Theorem 1 are called **supplementary angles** or a **supplementary pair**—angles such that the sum of their measures is the measure of a straight angle. They are also **adjacent angles**—angles having a common side, with disjoint, nonempty interiors. Two angles are said to be **complementary** or a **complementary pair**, iff the sum of their measures is the measure of a **right angle**, which, in turn, is defined as any angle having measure one-half that of a straight angle. A line which essentially *bisects* a straight angle is evidently **perpendicular** to the line of the straight angle. (A formal definition for this concept will be given momentarily.) Finally, an **acute angle** is an angle whose measure is less than that of a right angle (90), and an **obtuse angle** is one whose measure is greater than that of a right angle.

Note: Although of frequent occurrence, supplementary and complementary angles need not be adjacent to one another. In Euclidean geometry, the acute angles of a right triangle are not adjacent, but they are complementary, and, certainly, the measures of two separate angles can have a sum of 180 without being adjacent. ◗

A standard elementary theorem can therefore be established (proof to be supplied by the reader). As in the case for segments, we use the term **congruent angles** to mean *angles having the same measure* (more on this in Chapter 3). Thus, if, for example, $\angle 1$ is complementary to $\angle 2$, $\angle 3$ is complementary to $\angle 4$, and $\angle 2$ is congruent to $\angle 4$ (written $\angle 2 \cong \angle 4$) then:

$$m\angle 1 + m\angle 2 = 90 = m\angle 3 + m\angle 4 \quad \text{or} \quad m\angle 1 = m\angle 3$$

so that $\angle 1 \cong \angle 3$ by definition. This argument can be extended to supplementary angles, proving the following elementary principle:

Two angles that are supplementary (or complementary) to the same angle, or to congruent angles, are congruent to each other.

A concept that occurs repeatedly in geometry is defined next.

> **Definition 2**   If two angles have the sides of one opposite the sides of the other, they are called **vertical angles** or a **vertical pair**.

Thus, Figure 2.16 shows a pair of vertical angles hk and $h'k'$. Alternatively, if (ABD) and (CBE) hold, as shown, then $\angle ABC$ and $\angle DBE$ are vertical angles. Reciprocally, there is another pair of vertical angles in Figure 2.16, namely, $\angle ABE$ and $\angle DBC$. Using the linear pair theorem, one can prove:

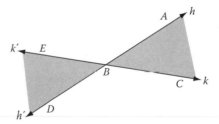

Figure 2.16 Vertical angles.

Theorem 2: Vertical Pair Theorem

Vertical angles are congruent.

Proof   If $\angle hk$ and $\angle h'k'$ are the given vertical angles, then $\angle h'k'$ is supplementary to $\angle h'k$, which, in turn, is supplementary to $\angle hk$. Therefore, $\angle hk \cong \angle h'k'$. ◗

The converse of **Axiom 10** mentioned earlier can now be established, using the linear and vertical pair theorems. These two results were consequences of **Axiom 10**, so it follows that **Axiom 10** implies its own converse, a property of geometry repeated many times in other situations. Details in its proof will be left as Problem 12.

> ***Converse of Axiom 10***   If $m\angle ABD + m\angle DBC = m\angle ABC$ then either $D \in \text{Int}\angle ABC$ or D is an interior point of one of the sides of $\angle ABC$.
>
> *Proof (outline)*   Let $h = \overrightarrow{BA}$, $k = \overrightarrow{BC}$, and $u = \overrightarrow{BD}$. In the nontrivial case, when $m\angle ABC \neq 0$ and D does not lie on the sides of $\angle ABC$, then (huk) holds. If $D \notin \text{Int}\angle hk$ then $k \neq h'$. Since $D \in \mathcal{P}$, consider the cases that follow from the half-plane postulate and **Axiom 10**: either $D \in h'$, $D \in k'$, (kuh'), $(h'uk')$, or $(k'uh)$. Using the hypothesis (huk), each of these cases leads to a contradiction, hence $D \in \text{Int}\angle hk$. (If $m\angle ABC = 0$, then $D \in \text{Int } h$.) ◨

A quite useful result shows a certain betweenness relation for rays lying on one side of a line:

> **Theorem 3**   If rays u and v are concurrent with ray h and lie on the same side of the line containing h, then either (hvu) or (huv).
>
> *Proof*   Let H be the half plane determined by h that contains u and v (except for the endpoint) let H_1 and H_2 be those determined by the line $u \cup u'$, with h in H_1 and h' in H_2, and select an interior point W of v. By the plane separation postulate, since $W \in H$ by hypothesis, either $W \in H \cap H_1 \equiv \text{Int}\angle hu$, which immediately implies (by Axiom 10) the relation (hvu), or $W \in H \cap H_2 \equiv \text{Int}\angle h'u$, which implies $(h'vu)$. We then have $h'v + vu = h'u$, or, by the linear pair theorem,
>
> $$180 - hv + vu = 180 - hu$$
>
> That is, $hu + vu = hv$ and (huv). ◨

The concept of perpendicular lines is next. Note that any two intersecting lines (as in Figure 2.17) form four angles at the point of intersection: any one of these angles is either supplementary to, or congruent to, the other three. Thus, if one of these four angles has measure 90, as in Figure 2.17, then they all have measure 90.

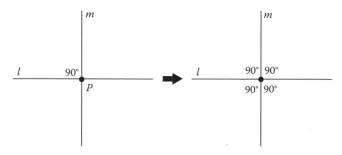

Figure 2.17 Perpendicular lines.

> **Definition 3**   Two lines l and m are said to be **perpendicular** ($\perp$) iff they intersect and form four right angles at the point of intersection. This definition extends to segments and rays: Two **rays** or **segments** are **perpendicular** iff the lines containing them are perpendicular, similarly for a segment and ray, a segment and line, etc.

Although obvious, we state a theorem that will save us from the incessant repetition of a routine argument.

Theorem 4: Criterion for Perpendicularity

If (ABC) and D is any point not on line AB, then $\overleftrightarrow{DB} \perp \overleftrightarrow{AC}$ iff $m\angle DBA = m\angle DBC$.

Proof Suppose that $m\angle DBA = m\angle DBC$, as shown in Figure 2.18. Since the two angles are supplementary by the linear pair theorem, then

$$2m\angle DBA = m\angle DBA + m\angle DBC = 180 \quad \text{or} \quad m\angle DBA = 90$$

Hence all four angles at B are right angles and $\overrightarrow{AB} \perp \overrightarrow{BD}$ by definition. Conversely, suppose that $\overrightarrow{AB} \perp \overrightarrow{BD}$. Then, $\angle DBA$ and $m\angle DBC$ are right angles and $m\angle DBA = m\angle DBC = 90$.◢

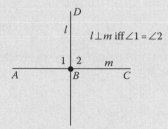

Figure 2.18 Criterion for perpendicularity.

This section will be concluded with another result involving angles and provides an alternate way to define the interior of an angle. More importantly, this result, known as the *crossbar principle*, is used repeatedly throughout our development in various ways. It enables us to conclude, for example, that an angle bisector of a triangle intersects the side opposite that angle (another obvious result no doubt). The two major results needed to prove this theorem should be reviewed before going further, namely Theorem 3, Section **2.3**, and the Postulate of Pasch.

Theorem 5: The Crossbar Theorem

Let $\angle BAC$ be a nondegenerate, nonstraight angle, where B and C are interior points of its sides. If D is an interior point of $\angle BAC$ with $AD < \alpha$, then ray $\overrightarrow{AD}$, meets some segment having endpoints at B and C at an interior point E of that segment. (See Figure 2.19.)

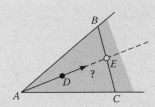

Figure 2.19 Crossbar hypothesis.

The figure reveals a triangle ABC, and we are essentially to prove that line AD meets side BC. This suggests the use of the Postulate of Pasch in some way. But in order to use that result, we must arrange things so that line AD passes through an interior point of the side of some triangle. Thus, we must add some points to the diagram in Figure 2.19. Suppose we locate, by the ruler postulate, points F and G such that (FAD) and (CAG), with $FD < \alpha$ and $CG < \alpha$, then add segment BG and ray AF to the objects we have so far. Thus, we obtain a diagram like that shown in Figure 2.20. Now it looks promising: Observing the three points B, C, and G, line AD, which is the set $\overrightarrow{AD} \cup \overrightarrow{AF}$, must meet either a segment $\overline{BC}$ having endpoints B and C, or a segment $\overline{BG}$ having endpoints B and G (since line AD does not pass through B, C, or G). The rest of the argument depends on Theorem 3, Section 2.3, and simple logical cases that can be faithfully represented by the four diagrams in Figure 2.21. These diagrams are just different ways of viewing the original figure—a *proof by picture*, if you will. Note that they depict B and F lying on opposite sides of line AC, and D and G lying on opposite sides of line AB, both of which follow by hypothesis and construction.

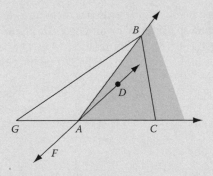

Figure 2.20 Crossbar hypothesis augmented.

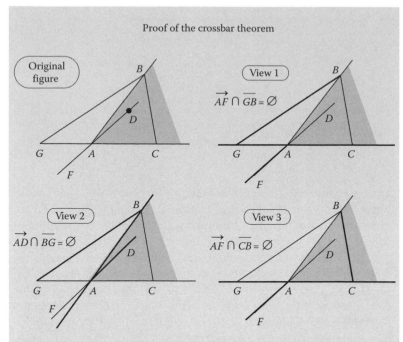

Figure 2.21 Proof of the crossbar theorem.

Proof of the Crossbar Theorem   Referring to Figure 2.21, the Postulate of Pasch applied to the points B, C, and G implies that line $\overrightarrow{AD} \cup \overrightarrow{AF}$ either meets $\overline{BG}$ at some interior point J or $\overline{BC}$ at some interior point E. But because of **View 1** in Figure 2.21 and Theorem 3 in Section **2.3**, ray $\overrightarrow{AF}$ and segment $\overline{BG}$ are disjoint sets, and **View 2** shows that $\overrightarrow{AD}$ and $\overline{BG}$ are disjoint. Hence $\overrightarrow{AD} \cup \overrightarrow{AF}$ meets $\overline{BC}$ at E. **View 3** shows that, again by Theorem 3, $\overrightarrow{AF}$ cannot meet $\overline{BC}$. Therefore, ray $\overrightarrow{AD}$ meets segment $\overline{BC}$ at an interior point E, as desired. ◣

A good exercise involving the crossbar theorem is to explore an alternative way to define the interior of an angle. In Figure 2.22, point X is shown satisfying the betweenness relation (AXC), where A and C are points on the sides of the angle. Certainly, X is an interior point of $\angle ABC$. Is the converse true?

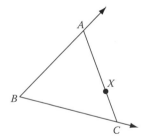

Figure 2.22 Transversal through point X.

Moment for Reflection

Try showing that, if $X \in$ Int$\angle ABC$, there exists A', C' on rays BA and BC, respectively, and Y on ray BX, such that $(A'YC')$ holds, as shown in Figure 2.23. **WARNING**: Your proof is not valid if you assume there exists a line passing through X and intersecting both sides of the angle (such a line is called a **transversal**). Although it might seem clear enough, transversals do not always exist in our geometric system. A valid identity for angle interiors is the following:

Int $\angle ABC = \{X: BX < \alpha$ and for some $Y \in \overline{BX}$, (AYC') for some $C' \in \overline{BC}\}$

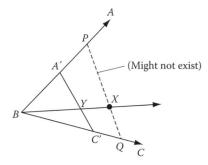

Figure 2.23 Transversals not guaranteed.

A more important corollary of the crossbar principle provides a link between betweenness for points and betweenness for rays, closing the gap on dual concepts. (Refer to Figure 2.24.)

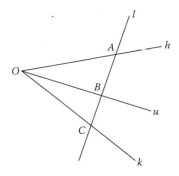

Figure 2.24 Duality of betweenness.

Theorem 6: Ray-Point Duality Theorem

Suppose that concurrent rays h, u, and k intersect line l at the respective interior points A, B, and C, on each ray, where l does not pass through O, the point of concurrency. Then (huk) iff (ABC).

Proof

1. Suppose (*huk*) holds. If $AC = \alpha < \infty$, then C is the polar opposite of A on line l, hence, as defined and shown in Section 1.8, (*ABC*) (since $u \neq h$, k and $B \neq A$, C). Otherwise, $\overleftrightarrow{AC} = l$ and it follows that $\angle AOC$ is nondegenerate and nonstraight. By the converse of **Axiom 10**, u passes through an interior point of $\angle AOC$, and by the crossbar principle, u meets segment AC at some interior point D, and it follows that $D = B$. (If not, then B and D both lie on line l, so that $BD = \alpha < \infty$ and, as points on ray u, either B or D must coincide with O, contrary to hypothesis.) Therefore, (*ABC*) holds.
2. Conversely, if (*ABC*) then it follows that $B \in \text{Int}\angle AOC$ by routine applications of Theorem 1 in Section **2.3**, and (*huk*) holds by **Axiom 10**. ⟍

Problems (Section 2.4)

Group A

1. Reproduce Figure P.1 on your paper and make a sketch showing the half plane $H(P, \overleftrightarrow{QR})$. Use shading ▥.

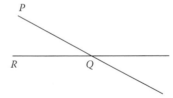

Figure P.1

2. Repeat Problem 1, showing the half plane $H(R, \overleftrightarrow{QP})$, using the shading ▤. Using the resulting cross-hatch shading, illustrate the set $\text{Int}\angle PQR$ in your sketch.

3. Carry out *Sketchpad Experiment 3* (Appendix **A**), illustrating the interior of an angle in Euclidean geometry.

4. What kinds of sets are possible in the Euclidean plane by taking the nonempty intersections of three distinct half planes? Make a sketch. *Sketchpad Experiment 4* may help you to solve this problem.

5. An old TV antenna was to be sold at an auction. The angle between the two signal-receiving arms was measured to be 72° (Figure P.5). Find the angle between each of the two arms and the main support if
   (a) The angles between the arms and the main supporting upright have equal measure
   (b) One of the angles between the arms and supporting upright has measure 6° greater than that of the other

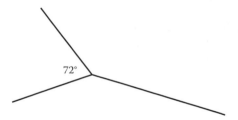

Figure P.5

6.   Using the information given in Figure P.6, find $m\angle PQR$
     and $m\angle PQS$.

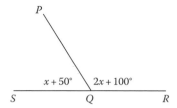

Figure P.6

7.   If $m\angle UOV = m\angle VOW$, and the measurements of the
     angles are as indicated in Figure P.7, find $m\angle UOW$.

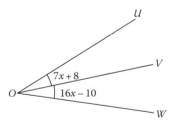

Figure P.7

8.   If $l = \overleftrightarrow{BC}$, $m\angle ABC = 72$, and ray BD makes an angle of
     measure 21 with line l (Figure P.8), what are the possible
     values of $m\angle ABD$ if
     (a) $\overrightarrow{BD}$ lies on the A-side of l?
     (b) $\overrightarrow{BD}$ lies on the opposite side of l as point A?

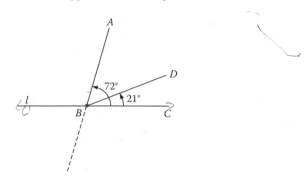

Figure P.8

9.   Ray *OD* is the opposite of ray *OB*. If $m\angle BOA = 53$ and $m\angle DOE = 1$, find $m\angle EOA$ if

(a) $(\overrightarrow{OA}\ \overrightarrow{OE}\ \overrightarrow{OD})$, as in Figure P.9.

(b) $(\overrightarrow{OA}\ \overrightarrow{OD}\ \overrightarrow{OE})$.

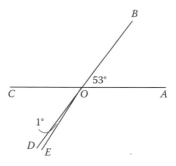

Figure P.9

10.   Use the information from Figure P.10 to find the value of *y* if $x = 35$. Assume (EAB).

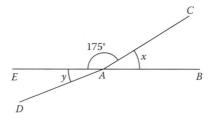

Figure P.10

Group B

11.   Show that if $(h'uk)$ and hk is a nondegenerate angle then (ukh). [**Hint:** $v'w = 180 - vw$ for any two rays *v* and *w* (why)?].

12.   Provide the details in the proof of the converse of **Axiom 10**, obtaining the contradictions mentioned.

13.   Prove the following:

> ***Theorem 7***   For any three distinct, concurrent rays *h*, *k*, and *u*, either (huk), (hku), (khu), or $(hu'k)$, where u' is the opposite ray of *u*.

[**Hint:** Show by applications of the plane separation postulate and definition of angle interior that a point *W* on ray *u* lies in the interior of either $\angle hk$, $\angle hk'$, $\angle h'k'$, or $\angle h'k$, then apply **Axiom 10** and the result of Problem 11.]

14.*   **Triangle Inequality for Rays**   Prove that if *h*, *u*, and *k* are any three distinct concurrent rays, then

$$hu + uk \geq hk$$

[**Hint:** Consider the cases in the theorem of Problem 13.]

15. Using the result of Problem 14, prove the following dual to Theorem **2**, Section **1.7**.

> **Theorem**   If (*huv*) and (*hvk*), then (*huvk*).

[**Hint:** Dualize the proof of Theorem **2**, Section **1.7**, carefully adapting it to the one needed here. A coordinate proof can also be given using coordinates via the circular protractor theorem below.]

16. Prove explicitly that if (*ADC*) holds, then $D \in \text{Int}\angle ABC$ ($0 < BA < \alpha, 0 < BC < \alpha$).

17. Angle A is supplementary to angle B and complementary to angle C, and angle B is supplementary to angle C. Find the measures of the three angles.

18. Angle A is supplementary to both angles B and C, and angle B is complementary to angle C. Find the measures of all the three angles.

19. Angle A is supplementary to angle B and complementary to angle C, and angle B has measure four times that of angle C. Find the measures of all three angles.

20. Suppose that ($\overrightarrow{AB}\ \overrightarrow{AC}\ \overrightarrow{AD}$) and ($\overrightarrow{AC}\ \overrightarrow{AD}\ \overrightarrow{AE}$), are both true betweenness relations. It might seem that ($\overrightarrow{AB}\ \overrightarrow{AC}\ \overrightarrow{AE}\ \overrightarrow{AD}$) should hold. Find a counterexample in Euclidean geometry.

21. Find a counterexample in Euclidean geometry for the conjecture: Given $\angle ABC$, there exists a ray BD such that ($\overrightarrow{BA}\ \overrightarrow{BC}\ \overrightarrow{BD}$).

22. For self-discovery: If ($\angle PML$, $\angle PMN$) is a linear pair, and rays MT and MW bisect these two angles, with $m\angle PMN = 20$ (Figure P.22), find
  (a) $m\angle PMT$
  (b) $m\angle PMW$
  (c) Rework the problem with $m\angle PMN = 50$, and then again with $m\angle PMN = 72$. Did you notice anything?

Note: *Sketchpad Experiment 5* will lead to a conclusive answer to the above in a quite effective manner. ◤

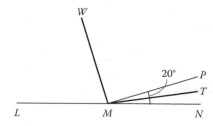

Figure P.22

Group C

23. Suppose points A, B, C, D, and E are such that A, B, and C are noncollinear, (BDC), and (AEC). Prove there exists a point F such that (AFD) and (BFE) (Figure P.23). [**Hint:** Theorem 6 will save work.]

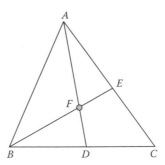

Figure P.23

24. Suppose points A, B, C, D, and E are such that A, B, and C are noncollinear, (BCD), and (AEC). Prove there exists a point F such that (AFB) and (FED) (Figure P.24). [**Hint:** Theorem 6 will save work.]

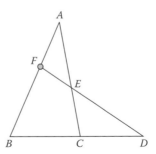

Figure P.24

25. Consider the coordinate plane, with points, lines, and distance as in Euclidean geometry. Instead of the usual angle measure, let the measure of any nondegenerate angle be 90 and that of a straight angle 180. Thus, there are only two measures for all angles in the plane. Show that **Axioms 1–9** are valid, but **Axiom 10** is not.

2.5 Angle Construction Theorems

The ability to construct one angle congruent to another at some given point in the plane is essential to the further development of geometry. As in the case of segments, this allows us to essentially "move" a given angle from one place to another. So far, there is no property that guarantees this; it requires one more axiom, **Axiom 11**—the last to be taken up in this chapter (only one more axiom is needed in order to complete the axiomatic system for the geometry we intend to study). See Figure 2.25 for an illustration of the axiom.

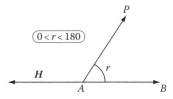

Figure 2.25 Constructing an angle with a given measure.

Axiom 11 [Angle Construction Postulate]   *Given any line AB and half plane **H** determined by that line, for every real number r between 0 and 180 there is exactly one ray AP in **H** such that m∠PAB = r.*

Taking $r = 90$, an immediate result is

Theorem 1   Given a point A on a given line l, there exists one and only one line m perpendicular to line l at A.

An important angle construction theorem (dual to the segment construction theorem in Section 1.10) is the following.

Theorem 2: Angle Construction Theorem

Let two nondegenerate, nonstraight angles ABC and DEF be given, such that $m\angle ABC < m\angle DEF$. There exists a unique ray EG lying between rays ED and EF such that angle $\angle GEF$ is *congruent* to $\angle ABC$. (See Figure 2.26.)

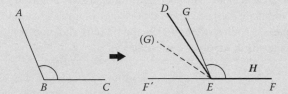

Figure 2.26 Proof of Theorem 2.

Proof   Let **H** be the half plane determined by line EF containing point D. By **Axiom 11**, there exists exactly one ray EG in **H** such that $\angle GEF \cong \angle ABC$. Now G cannot lie on lines ED or EF, hence G lies interior to either $\angle DEF$ or $\angle DEF'$, where (FEF'). If $G \in \text{Int}\angle DEF'$, then by the angle addition postulate (**Axiom 10**),

$$m\angle GEF' + m\angle GED = m\angle DEF'$$

and by the linear-pair theorem

$$180 - m\angle GEF + m\angle GED = 180 - m\angle DEF$$

or

$$m\angle DEF + m\angle GED = m\angle GEF = m\angle ABC$$

Hence $m\angle DEF < m\angle ABC$, contradicting hypothesis. Therefore, $G \in \text{Int}\angle DEF$ and ($\overrightarrow{ED}\ \overrightarrow{EG}\ \overrightarrow{EF}$) by **Axiom 10**. ◣

Now we apply these ideas to another familiar concept in geometry, that of the *bisector* of an angle.

Definition   If ray *BD* lies between rays *BA* and *BC*, and $m\angle ABD = m\angle DBC$, then ray *BD* is called the **bisector** of $\angle ABC$. (See Figure 2.27.)

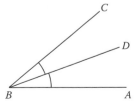

Figure 2.27 Angle bisector.

Angle Bisection Theorem   Each nondegenerate, nonstraight angle has a unique bisector.

Moment for Reflection

It is interesting to consider a theorem analogous to the above in the case of a straight angle. Is the bisector of a straight angle defined and, if so, would there be more than one bisector of a straight angle, violating uniqueness? What, then, makes the bisector of a nonstraight angle unique in terms of axiomatic geometry?

Example 1

Consider two intersecting lines *l* and *m* that form two pairs of vertical angles. Prove that the lines *p* and *q* that bisect the two pairs of vertical angles are perpendicular. (See Figure 2.28.)

Solution

Let the two pairs of vertical angles be (*hk*, *h'k'*) and (*h'k*, *hk'*), as in Figure 2.28. Lines *p* and *q* then contain the bisectors *u* of angle $\overline{hk}$ and *v* of angle $\overline{kh'}$. By the linear pair theorem, $hk + kh' = 180$, and by the definition of angle bisector, $uk = \frac{1}{2}hk$ and $vk = \frac{1}{2}kh'$. Since (*huk*) and (*hkh'*) hold, then (*hukh'*) (the result of Problem 15, Section **2.4** below). Therefore, (*h'ku*). Also, (*h'vk*), so (*h'vku*) or (*vku*). Thus,

$$uv = uk + kv = \tfrac{1}{2}hk + \tfrac{1}{2}kh' = \tfrac{1}{2}(hk + kh') = \tfrac{1}{2}(180) = 90.$$

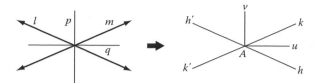

Figure 2.28 Example 1.

It is interesting that we can obtain a coordinate system for the rays from a given point O—dual to the coordinatization theorem for a line (Theorem **5**, Section **1.8**). Since this is not used much in our development, we skip the proof and ask you to supply it, according to your interests. (See Problem 16 below.)

> ***Circular Protractor Theorem***   There exists a one-to-one correspondence between the set of all rays that are concurrent at a given point O and the set of real numbers θ, $-180 < \theta \le 180$, called their *coordinates*, such that
>
> **(a)** A given ray in the set can be taken as the origin and assigned the coordinate zero, and any other ray can be assigned a positive coordinate
> **(b)** For any two rays h and k having coordinates θ and φ,
>
> $$hk = |\theta - \varphi| \qquad \text{if } |\theta - \varphi| \le 180$$
>
> $$= 360 - |\theta - \varphi| \quad \text{if } |\theta - \varphi| > 180$$

It might be noted that for the geometry on the unit sphere, distance along a great circle *coincides with angle measure at the center of the sphere* (recall the arc length formula $s = r\theta$ if θ is measured in radians). This shows that this theorem and the coordinatization theorem for lines in the case $\alpha < \infty$ are perfect duals of each other for a finite metric. In coordinate geometry, the positive and negative x-axis (rays u and u' in Figure 2.29) may be assigned the coordinates 0 and 180, all rays in the upper half plane assigned positive coordinates and, those in the lower half plane, negative coordinates, where the coordinates *measure the angles which u or u' make with h*. More on this occurs in Chapter 5 (last problem section) where details are presented for an axiomatic basis for the trigonometric functions.

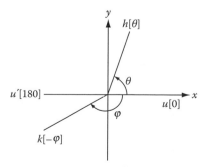

Figure 2.29 Coordinates for concurrent rays.

Example 2

Figure 2.30 depicts five concurrent rays with their respective coordinates, where $\angle COA$ is a right angle (having measure 90). Find the measures of $\angle BOE$ and $\angle COE$ in two different ways, thus showing that ray OB does not lie between rays OC and OE:

(a) By appealing to the figure and using familiar ideas of angles

(b) By the direct application of the circular protractor theorem

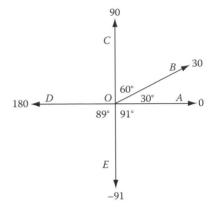

Figure 2.30 Example 2.

Solution

(a) One observes from the figure

$$m\angle BOE = 30 + 91 = 121 \quad \text{and} \quad m\angle COE = 90 + 89 = 179$$

Hence $m\angle COB + m\angle BOE = 181 \neq m\angle COE$.

(b) Using coordinates and the circular protractor postulate we have: For $\angle BOE$ whose sides have coordinates 30 and -91, observe that

$$|30 - (-91)| = 121 < 180, \quad \text{so} \quad m\angle BOE = 121$$

And for $\angle COE$, whose sides have coordinates 90 and -91,

$$|90 - (-91)| = 181 > 180, \quad \text{so} \quad m\angle COE = 360 - 181 = 179.$$

Note that these results agree with (a).

2.6 Consequences of a Finite Metric

We have developed axioms for geometry that made allowances for spherical geometry, some of them rather powerful, such as the ruler postulate and plane-separation property. But the question is: If we impose on this axiomatic system the hypothesis $\alpha < \infty$, do we then obtain properties that are characteristic of spherical geometry? For example, given any point A on a sphere, we know that we can find a second point A^* on that sphere such that AA^* is the maximal distance (on the sphere). (The North and South poles on the earth's surface is an example; in general, A^* is the point diametrically opposite point A.) Also, we know that every line (great circle) passing through A also passes through A^*, as illustrated by lines l and m in Figure 2.31. Furthermore, every other point B on the sphere lies between A and A^*. Finally, no other point besides A^* has this property relative to point A. (Point A^* is usually referred to as the *opposite pole of A*.)

It is interesting to obtain these same features axiomatically. Throughout this section, α will be a finite, positive real number. The geometry thereby obtained from **Axioms 0–11** will be called **axiomatic spherical geometry**, or later, **elliptic geometry**. (The term *spherical geometry* refers to the geometry of the sphere, a *model* for the latter.)

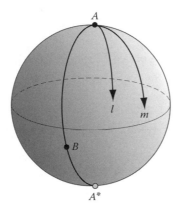

Figure 2.31 Opposite poles on the sphere.

The first step is to obtain point A^* such that $AA^* = \alpha$, given A. But this is a direct result of the ruler postulate (take any point B and ray from A through B, then locate A^* having coordinate α). Furthermore, if A and A^* lie on line l, A^* is the only point on l, such that $AA^* = \alpha$. (See the discussion concerning uniqueness preceding **(10)**, Section **1.8**.) Next, locate points P and Q on l such that (PAQ), with $C \in \overrightarrow{AP}$ and $D \in \overrightarrow{AQ}$ such that $AC = AD = 3/4\alpha$, as shown in Figure 2.32. Then, $AC + AD = 3/2\alpha > \alpha$, so by Theorem **4(b)**, Section **1.8**, (CA^*D) holds. Hence $CD = CA^* + A^*D = (\alpha - CA) + (\alpha - AD) = 2\alpha - 3/2\alpha = 1/2\alpha$. Thus, $CD < \alpha$ and segment CD is well defined. If H_1 and H_2 are the two half planes determined by line m and $P \in H_1$, then by Theorem **4**, Section **2.3**, $Q \in H_2$. By Theorem **2**, Section **2.3**, C and D lie on opposite sides of line m, and by **Axiom 9** m passes through some point X on segment CD. Thus, X lies on both lines l and m; since $A \in m$ and $X \neq A$, $XA = \alpha$ (**Axiom 4**). But by the uniqueness of the polar opposite A^* of A on line l, $X = A^*$. Thus, m passes through point A^*. This proves

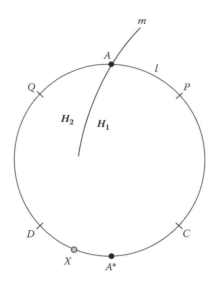

Figure 2.32 Obtaining opposite poles axiomatically.

Theorem 1   Every line passing through A also passes through A^*.

Corollary A   Given point A, there is exactly one point A^* in the plane such that $AA^* = \alpha$.

Proof   Suppose $AB = \alpha$. If l is a line passing through A, by Theorem **1** l passes through both A^* and B, with $AA^* = AB = \alpha$. Hence $B = A^*$.◣

Definition   If A and A^* are two points such that $AA^* = \alpha$, they are called **extreme points**, and each is called the **extremal opposite** of the other.

Thus, the result of Corollary A is that *every point has an associated unique extremal point*, and polar opposites on lines and extremal opposites in the plane become one and the same.

We have thus succeeded in establishing from the axioms properties that are characteristic of spherical geometry. A few more such results follow from essentially one-line proofs.

Corollary B   If two lines intersect at point A, they also intersect at the extremal opposite point A^*.

Corollary C   If A and A^* are any two extreme points, and X is any other point in the plane, then (AXA^*).

Corollary D   If A and B both lie in an open half plane, then $AB < \alpha$.

(The proofs of the preceding two corollaries will be left as Problems 10 and 11.)

An additional result will be essential to the main topic of the next chapter (the triangle). If we are given three noncollinear points A, B, and C, it follows that unique segments $\overline{AB}$, $\overline{BC}$, and $\overline{AC}$ exist (the *sides* of $\triangle ABC$), as the next result shows.

Existence of Triangles   If A, B, and C are noncollinear points, then AB, BC, and AC are each less than α. Consequently, segments $\overline{AB}$, $\overline{BC}$, and $\overline{AC}$ exist and $\angle ABC$, $\angle BAC$, and $\angle ACB$ are nondegenerate, nonstraight angles. (The set $\overline{AB} \cup \overline{BC} \cup \overline{AC}$ will be defined as the *triangle ABC* in the next chapter.)

Proof   If $AB = \alpha$, then since A and C belong to some line (**Axiom 2**) that line must pass through point B, and A, B, and C are collinear, contrary to hypothesis. Hence $AB < \alpha$. Similarly, $BC < \alpha$ and $AC < \alpha$. ◣

Problems (Section 2.6)

Group A

1. In Figure P.1 for this problem, suppose $\angle GEF$ has been constructed congruent to $\angle ABC$, as in the angle construction theorem. Find $m\angle DEG$.

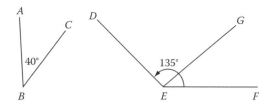

Figure P.1

2. State and prove a theorem generalizing the result of Problem 1 (with $\theta = m\angle ABC$).

3. Adjacent angles $\angle RST$ and $\angle RSW$ have measures as indicated in Figure P.3. Find (prove) the value of $m\angle WST$ in two ways:
    (a) Use the fact that the sum of the three angle measures equals 360 (see Problems 17 and 18).
    (b) Use established results without assuming (a).

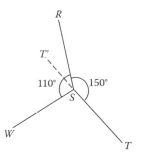

Figure P.3

4. Coordinates are as indicated for the rays shown in Figure P.4.

For (a), (b), and (c) find:
   (a) $m\angle KPW$
   (b) $m\angle KPT$
   (c) $m\angle TPW$
   (d) Show that ray PT is the bisector of $\angle KPW$

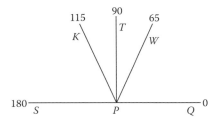

Figure P.4

5. As in Problem 4, from Figure P.5 for this problem, find
   (a) $m\angle BOD$
   (b) $m\angle COE$
   (c) The coordinate of the bisector of $\angle DOB$

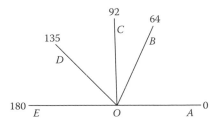

Figure P.5

6. Using the circular protractor theorem, find the measures of each of the six angles formed by pairs of concurrent rays $h[-179]$, $k[-61]$, $u[5]$, and $v[120]$, as indicated by their coordinates.

Group B

7. Suppose C and D lie on opposite sides of line AB and that $\angle ABC$ and $\angle ABD$ are supplementary. Use **Axiom 11** to show that ray BD coincides with the opposite ray BC' of BC (see Figure P.7 for a hint.)

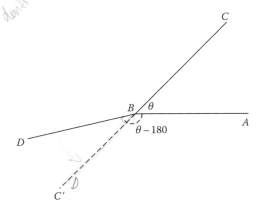

Figure P.7

8. **Converse of Vertical Pair Theorem**   Suppose C and D lie on opposite sides of line AB and that $\angle ABC \cong \angle DBE$ where (ABE) holds. Use the result of Problem 7 to prove (CBD).

9. Suppose two triangles, $\triangle ABC$ and $\triangle DEF$, have $m\angle ABC < m\angle DEF$. Show that there exists a point G on segment DF such that $\angle GEF \cong \angle ABC$. (See Figure P.9.)

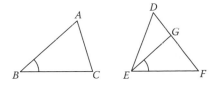

Figure P.9

10. Prove that if $AB = \alpha$ and C is any other point in the plane, then (ACB) (Corollary C above).

11. Prove Corollary D.

12. Define angle trisectors and prove that any angle can be trisected.

Note: An ancient problem was to construct the trisectors for an arbitrary angle using only a compass and straight-edge. This was ultimately proven to be impossible using the methods of abstract algebra. One way to verify this is to use a theorem of Gauss on the construction of certain regular polygons stated in Chapter 4. In particular, see Problem 8, Section **4.4**. ✎

13.   The Angle-Doubling Theorem   Given an acute angle $\angle ABC$, their exists a unique ray BD, such that $(\overrightarrow{BD}\ \overrightarrow{BC}\ \overrightarrow{BA})$ and $m\angle DBA = 2m\angle ABC$. See Figure P.13 for illustration. Prove, using the angle construction postulate. (Since this result is the dual of the segment-doubling theorem, Section **1.9**, it suffices to merely "dualize" its proof.)

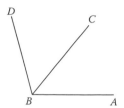

Figure P.13

14.   The Interior of a Triangle   The interior of $\triangle ABC$ may be defined as the intersection of the interiors of its angles $(\text{Int}\triangle ABC = \text{Int}\angle A \cap \text{Int}\angle B \cap \text{Int}\angle C)$. Prove that the interior of a triangle is convex and that it is the intersection of the interiors of *any two* of its angles (but for notation, $\text{Int}\triangle ABC = \text{Int}\angle A \cap \text{Int}\angle B$). (We already have by definition, $\text{Int}\triangle ABC \subseteq \text{Int}\angle A \cap \text{Int}\angle B$; it suffices to prove that if $X \in \text{Int}\angle A \cap \text{Int}\angle B$ then $X \in \text{Int}\angle C$, proving that $\text{Int}\angle A \cap \text{Int}\angle B \subseteq \text{Int}\triangle ABC$.)

15.   Prove: If $P \in \text{Int}\angle ABC$ and l is any line passing through P, then l must intersect the sides of $\triangle ABC$ at exactly two points (Figure P.15).

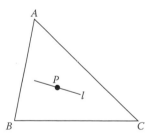

Figure P.15

Note: This result is analogous to one involving the interior of a circle: Any line passing through an interior point P of a circle must intersect the circle in exactly two points. The theorem is known as the *secant theorem* in geometry, which is much harder to prove. It will be established in Chapter 4 using least upper bound theory and properties of circles. ◣

Group C

16. **Proving the Circular Protractor Theorem** Furnish the details and reasoning to complete the proof outlined in the following steps.

(a) Begin with noncollinear rays u and v from point O and assign the coordinate 0 to u and the measure of angle $\overline{uv}$ to v. Assign 180 to u', the opposite ray of u.

(b) Given a positive real θ, $0 < \theta < 180$, by **Axiom 11**, there exists a unique ray h from O on the same side of line $u \cup v$ as v such that $hu = \theta$. Assign the coordinate θ to h. Similarly, there exists a unique ray k from O on the opposite side of line $u \cup v$ such that $ku = \varphi$, where $0 < \varphi < 180$. Assign the coordinate $-\varphi$ to k. Prove that, at this point, a one-to-one correspondence has been established between the rays from O and the real numbers θ for $-180 < \theta \le 180$.

(c) Consider any two rays $h[\theta]$, $k[\varphi]$. With $\theta > \varphi$, the proof for the case $\theta < \varphi$ being similar, suppose first that $|\theta - \varphi| = \theta - \varphi \le 180$. Then, either θ and φ have like signs, or $\theta > 0 > \varphi$. In the first case, h and k lie on the same side of line $u \cup v$ and, except for trivial cases, either (hku) or (khu) (why?). Hence $hk = |\theta - \varphi|$. If $\theta > 0 > \varphi$, let h' be the ray opposite h. It follows that either $(kh'u)$ or (huk), with the results either $-\varphi = ku > h'u = 180 - \theta$, a contradiction, or $hk = hu + uk = \theta + (-\varphi) = |\theta - \varphi|$.

(d) The last case is $|\theta - \varphi| > 180$. Unless $h = 180$, $(hu'k)$ follows (prove) and therefore $hk = hu' + u'k = (180 - hu) + (180 - uk) = 360 - \theta + \varphi = 360 - |\theta - \varphi|$.

17. If h, k and u are distinct, concurrent rays and the three angles defined by these rays are pairwise adjacent, show that $hk + ku + uh = 360$.

18. **Sum of Measures of Angles about a Point** The sum of the measures of three or more nondegenerate pairwise adjacent angles, taken consecutively, equals 360. Prove.

Unified Geometry: Triangles and Congruence

That part of Euclidean geometry that deletes any reference to parallel lines or notions directly provable from them is known as *absolute geometry*. Although perhaps puzzling, there is a perfectly logical reason for this term. One of the early pioneers in geometry, János Bolyai (1802–1860), set out to discover what must be true about three-dimensional space without introducing the concept of parallelism. Any logical consequence thereof would then be an *absolute truth* about the universe, thus the term.

Our development of geometry so far includes (axiomatic) spherical as well as Euclidean geometry, and, since there are no parallel lines in spherical geometry, the concept of parallelism cannot play a role in this endeavor. Thus, the term absolute geometry might also be used to describe our study as well. However, we prefer to use the term **unified geometry** instead, to distinguish it from Bolyai's work. This geometry then includes (at least) the two classical geometries: spherical geometry and Euclidean geometry.

3.1 Congruent Triangles: SAS Hypothesis

The terminology normally associated with triangles will now be introduced formally. First, if A, B, and C are any three noncollinear points, note that by the results in Section 2.6 one can assume that $AB < \alpha$, $BC < \alpha$, and $AC < \alpha$. Thus, the segments AB, BC, and AC are uniquely determined. Moreover, the three angles determined by these three points, such as $\angle ABC$, cannot be a degenerate or a straight angle.

Definition 1   If A, B, and C are any three noncollinear points, we define the **triangle** determined by these points, denoted by the symbol $\triangle ABC$, as the set of points

$$\triangle ABC = \overline{AB} \cup \overline{BC} \cup \overline{AC}$$

Terms commonly associated with a triangle are as follows: The segments $\overline{AB}$, $\overline{BC}$ and $\overline{AC}$ are called the **sides** of the triangle, while the original three points A, B, and C are its **vertices**. The **angles** of the triangle are $\angle A \equiv \angle BAC$, $\angle B \equiv \angle ABC$, and $\angle C \equiv \angle ACB$. The **angle sum** of a triangle is the sum of the measures of its angles. For further terms, a pictorial glossary in Figure 3.1 is provided to convey their meaning instead of formal definitions (which can be readily composed if need be).

The concept of *congruent triangles* is of fundamental importance. and it leads to all the usual geometric features you would expect to find. In his *Elements*, Euclid used the term "equal" to mean what we today call "congruent" and, unfortunately, he used the same term to describe "equal in area." For example, two polygons were said to be *equal* if their areas were equal. On the other hand, two triangles were *equal* if one triangle could be placed directly on top of the other for a perfect match, a process known as **superposition** (a term often attributed to Euclid, but never actually used by him).

In modern mathematics, the standard meaning for equality (=) is *identically equal to*. In set theory, we write $S = T$ only when the sets S and T have precisely the same elements. In geometry, point A = point B only when A coincides with B. And for two segments, $\overline{AB} = \overline{CD}$ only when the set of points $\overline{AB}$ is the exact same set of points $\overline{CD}$. But, a more general type of equality is useful, and necessary, for geometry. It is a special case of the most general kind of equality useful in mathematics, called an **equivalence relation**. By definition, this concept satisfies the following three characteristic properties of ordinary equality: For all objects x, y, and z under consideration:

(1)   **Reflexive Law.** For any x, x *equals* x (any object equals itself).
(2)   **Law of Symmetry.** If x *equals* y, then y *equals* x.
(3)   **Transitive Law.** If x *equals* y and y *equals* z, then x *equals* z.

This concept is universal in nature, applying not only to mathematics. For example, two persons could be considered "equal" if they have they same height, the same zip code, etc. In any application of the term *equal*, the concept being considered must be substituted for the italic equals in the above laws and must then obey those laws. For example, consider the relation among lines in the plane "parallel to" (meaning: having no points in common) as a proposed equivalence relation. One runs into trouble with the reflexive and transitive laws unless it is agreed that a line can be considered parallel to itself.

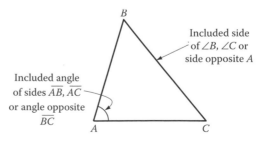

Figure 3.1 A triangle and associated terms.

The first nontrivial use of equality in geometry involving sets that are not identical occurs with respect to segments and angles. We note, for example, that the *lengths* of two different segments can be the same without the segments themselves being identically the same set of points. It is useful to assign a special name to this concept.

Definition 2   Two segments S_1 and S_2 are said to be **congruent**, denoted by $S_1 \cong S_2$, iff they have the same length. Two angles A_1 and A_2 are said to be **congruent**, denoted by $A_1 \cong A_2$, iff they have the same measure.

The concept for congruent triangles is more complicated, involving the three sides and the three angles of two triangles in some way. One must decide what must, or should, be required of these six parts in order to produce the proper concept for congruence. It presumably must mean "triangles having the same shape and size." All future work depends on this decision.

Moment for Reflection

What criteria properly describe congruent triangles? Requiring equal areas is clearly not strong enough. Perhaps, both area and perimeter might work. What do you think? Would the equality of both the area and the perimeter of two triangles make them congruent in the usual sense? (See Problem 6.)

The conventional definition for congruent triangles fortunately does not require any new concepts or axioms governing them. One does not need to worry about properties involving area and perimeter, for example, or motions in the plane that transport geometric figures from one place to another or any other imaginative concept.

Definition 3   Two triangles are **congruent** iff, under some correspondence between their vertices, corresponding sides and corresponding angles are congruent. Thus, $\triangle ABC$ and $\triangle XYZ$ are congruent under the correspondence A to X, B to Y, and C to Z, provided that

$$\overline{AB} \cong \overline{XY}, \quad \overline{BC} \cong \overline{YZ}, \quad \overline{AC} \cong \overline{XZ},$$

$$\angle A \cong \angle X, \quad \angle B \cong \angle Y, \quad \text{and} \quad \angle C \cong \angle Z$$

This congruence will be denoted in the usual manner by

$$\triangle ABC \cong \triangle XYZ$$

(It is routine to show that the concept of congruence for triangles is an equivalence relation, which will be left to the reader.)

The underlying correspondence represented by the notation for congruent triangles introduced above may be designated by

$$ABC \leftrightarrow XYZ$$

Since each triangle has three vertices, it is clear that there are six distinctly different ways in which a correspondence between two triangles can be prescribed, and a congruence under one correspondence need not be a congruence under a different one. The reflexive property of congruence is directly involved here: while a triangle is clearly congruent to itself, it is necessary to specify the correspondence as well. Thus, while it is true that, *as sets of points,*

$$\triangle ABC = \triangle ACB$$

it does not necessarily follow that

$$\triangle ABC \cong \triangle ACB$$

unless, among other things, $\angle B \cong \angle C$, for example. Such a congruence may indeed be true for a certain type of triangle (which ones?), but it is not so in general.

At this point, we introduce an abbreviation that proves to be quite convenient and has become a convention in elementary geometry. We use the acronym CPCPC* to stand for the statement: *corresponding parts of congruent polygons are congruent.* This is merely a restatement of the definition of congruent polygons (given later for polygons in general), that applies to triangles as well.

Although the definition of congruent triangles ostensibly requires all six parts of triangles to be pairwise congruent, one can often get by with less. The magic number is *three*: all the usual congruence criteria for triangles require three sets of congruent parts, and all of them require at least *one pair of congruent sides*. But not even five parts are enough if three of the five are the three pairs of angles! (See Problem 7.)

Euclid presents a superposition argument that two triangles are congruent, provided the following conditions are met, as illustrated in Figure 3.2:

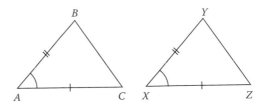

Figure 3.2 The SAS hypothesis.

---

* CPCTC is also used commonly used ("corresponding parts of congruent *triangles* are congruent"); CPCF appears in at least one high school text ("corresponding parts of congruent *figures* are congruent").

SAS Hypothesis   Suppose that under some correspondence between their vertices, two sides and the included angle of one triangle are congruent, respectively, to the corresponding sides and included angle of the other triangle.

Euclid's familiar theorem (often abbreviated by SAS and known as the *SAS congruence criterion*) is that *two triangles are congruent, provided the SAS hypothesis holds*. But this proposition cannot be proven with the results we have established so far. Showing that **Axioms 1–11** are not sufficient is the object of the next section.

Problems (Section 3.1)

Group A

1.   Name the congruent pairs of (distinct) sides and angles if $\triangle RST \cong \triangle STR$. What kind of triangle would this have to be in Euclidean geometry?

2.   Which pairs of triangles in Figure P.2 (with measures as indicated) directly satisfy the SAS hypothesis? If the familiar SAS congruence criterion is assumed, state the correct congruence(s) for these triangles.

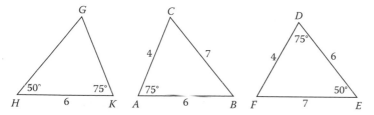

Figure P.2

3.   **Overlapping Triangles**   Suppose it is given in Figure P.3 for this problem that $\overline{AB} \cong \overline{AC}$ and $\overline{AD} \cong \overline{AE}$. Show that $\triangle ABE$ and $\triangle ADC$ satisfy the SAS hypothesis under the correspondence $ABE \leftrightarrow ACD$.

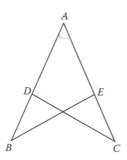

Figure P.3

4. In Figure P.4, consider $\triangle ABC$ and suppose that $AB = BC$. Regarding $\triangle ABC$ and $\triangle CBA$ as overlapping triangles, show that these "two" triangles satisfy the SAS hypothesis. If the SAS congruence criterion is assumed, what triangle congruence would be true? What conclusion(s) could be drawn?

Figure P.4

5. Prove that congruence in the set of all triangles is an equivalence relation.

6. The triangles in Figure P.6 have the dimensions as indicated. Show by direct calculation that both areas and perimeters of the two triangles are the same. [**Hint:** For the area, use Heron's formula $K = \sqrt{s(s-a)(s-b)(s-c)}$ where $s = \frac{1}{2}(a + b + c)$].

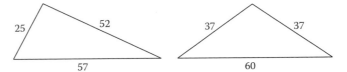

Figure P.6

7. Show that if two triangles in Euclidean geometry have respective sides of length 8, 12, and 18, and 12, 18, and 27, they are similar and therefore have five pairs of congruent parts between them, using ordinary geometry. Clearly, the triangles are not congruent. Such pairs of triangles are called *5-con*. Find another example of a pair of 5-con triangles.

8. Show that if a relation $\equiv$ (*equals*) satisfies the symmetric and transitive laws and has the additional property that each element equals another element in the set, then the reflexive law holds.

3.2 A Metric for City Centers

Consider the usual coordinate plane in Euclidean geometry, with the usual xy coordinate system. Suppose that, instead of the familiar formula for Euclidean distance, we take as metric the following: For any two points $P(x_1, y_1)$ and $Q(x_2, y_2)$,

(1)
$$PQ = |x_1 - x_2| + |y_1 - y_2|$$

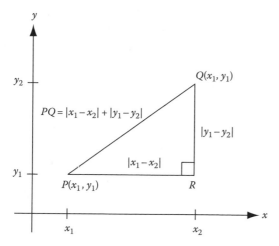

Figure 3.3 The taxicab metric for *PQ*.

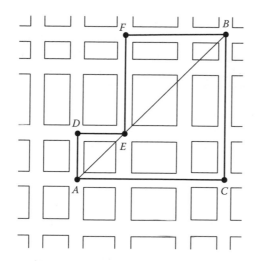

Figure 3.4 Distance in a city center.

Thus, we see that, as illustrated in Figure 3.3,

$$PQ = PR + RQ$$

Thus the distance formula **(1)** has the effect of measuring distance "around corners" instead of "as a crow flies." This would be particularly suitable in sections of a city for which the streets are parallel or perpendicular, where, in order to get from point *A* to point *B*, one must travel along the streets instead of cutting across city blocks, as indicated by Figure 3.4. For this reason, this metric is called the **taxicab**, or **Manhattan**, metric. It is a special case of the Minkowskian metric in mathematics (see Historical Note).

This concept for distance is clearly positive-definite since $PQ = 0$ iff $x_1 = x_2$ and $y_1 = y_2$, or $P = Q$. If we take lines to be ordinary Euclidean lines, we have a geometry satisfying **Axioms 1–4**, with $\alpha = \infty$. The betweenness concept depends on the metric, so the definitions for segments, rays, and angles are affected, and the remaining **Axioms 5–11** are not clear at this point.

Historical Note

The taxicab metric is a special case of a more general metric invented by Hermann Minkowski (1864–1909), taking the boundary of any bounded convex set symmetric about the origin as the "unit circle." The metric turns out to be Euclidean if that boundary is an ordinary ellipse or Euclidean circle, and non-Euclidean otherwise—like the taxicab metric. Minkowski, like Hilbert, grew up in Königsberg. When he won a prize for a competition in mathematics at the age of 15, he met Hilbert for the first time (who was two years older), and they became close friends. Some 20 years later, at Hilbert's urging, Minkowski accepted a teaching position at Göttingen in 1902, and they were colleagues there until Minkowski's untimely death of appendicitis at the age of 45. Besides Minkowski's outstanding work in the theory of numbers (using his metric), he is also known for his mathematical foundation for the Theory of Relativity, which Einstein—a student of Minkowski—used extensively in his work. The idea of the four-dimensional space/time framework commonly used for relativity, is due to Minkowski. The physicist Max Born, who was also a student of Minkowski, once said: "Minkowski laid out the whole arsenal of relativity mathematics… as it has been used ever since by every theoretical physicist."

Using a little trigonometry, observe that in Figure 3.5 we have for any line l making an angle θ with the x-axis ($\theta \neq 90$) and with ΔPQR as shown, where a, b, and c represent the Euclidea lengths of the sides of the triangle:

$$PQ = a + b = c\cos\theta + c\sin\theta = c(\cos\theta + \sin\theta)$$

That is, if the Euclidean distance from P to Q is denoted PQ^*, then

(2) $$PQ = \lambda c = \lambda PQ^*$$

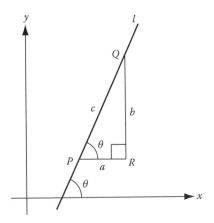

Figure 3.5 Taxicab distance $PQ = \lambda \cdot$ Euclidean distance.

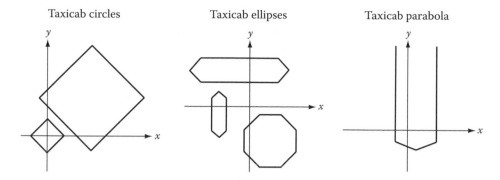

Figure 3.6 Familiar objects in taxicab geometry.

where $\lambda = \cos\theta + \sin\theta$, and we find that the taxicab distances between points are constant multiples of the Euclidean distances on each fixed line l. (If $\theta = 90$, the two metrics agree, so there is no problem when l is a vertical line.)

Note that **(2)** shows that *taxicab betweenness coincides with Euclidean betweenness*. This is in spite of the apparent contradiction in Figure 3.4 where more than one path from point A to point B exists having the same length: $AB = AE + EB = AD + DE + EF + FB = AD + DB$, for example. Here, D is **metrically between** A and B, but D is not collinear with A and B, as required by the definition of betweenness. So there is actually no discrepancy. Thus, segments, rays, and angles are the same for taxicab and Euclidean geometry. If we adopt the ordinary Euclidean angle measure for taxicab geometry, we now have a geometry satisfying **Axioms 5–11** (you should go over each of these axioms and try them out for the taxicab geometry as a thought exercise). Thus, although the metric is unusual, this new geometry satisfies all our axioms up to this point.

Taxicab geometry has many interesting and unusual features, some of which we now explore. Taxicab *circles* (defined as sets of points equidistant from a fixed point) turn out to be squares, with sides making angles of 45° and 135° with the coordinate axes, as in Figure 3.6. Taxicab *ellipses* are crystal-like hexagons or octagons, and *parabolas* are boundaries of pencil-shaped regions as in (Figure 3.6).

Here are some examples to show you how to navigate in this new, no doubt unfamiliar, territory.

Example 1

Show that the coordinate equation of the taxicab circle centered at $C(h, k)$ having radius r is

$$|x - h| + |y - k| = r.$$

Solution

The taxicab distance from C to any point $P(x, y)$ on the circle is given by **(1)**. Hence $PC = |x - h| + |y - k|$. By definition, PC has the constant value r. Therefore,

(3) $$|x - h| + |y - k| = r$$

Example 2

Find the equation of the part of the taxicab circle centered at the origin having radius r that lies in the second quadrant and then sketch its graph.

Solution

Using **(3)**, with $h = k = 0$, we obtain the equation $|x| + |y| = r$. In the second quadrant, $P(x, y)$ has $x < 0$ and $y > 0$ so that $|x| = -x$ and $|y| = y$. Hence, we obtain the equation of the line

$$-x + y = r \quad \text{or} \quad y = x + r$$

But only that part of the line lying in the second quadrant is part of the taxicab circle, as shown in the graph (the bold line in Figure 3.7); the rest of the taxicab circle is shown by dashed lines.

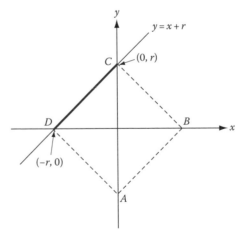

Figure 3.7 Example 2.

Example 3

Show that the triangle whose vertices are $A(2, 2)$, $B(-1, 1)$, and $C(1, -1)$, as shown in Figure 3.8, is equilateral under the taxicab metric.

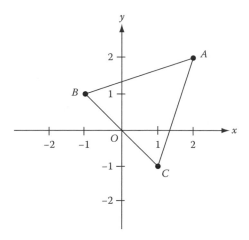

Figure 3.8 Example 3.

Solution

We calculate the lengths of the sides of $\triangle ABC$ as follows:

$$AB = \left|2-(-1)\right| + \left|2-1\right| = 3 + 1 = 4$$

$$BC = \left|-1-1\right| + \left|1-(-1)\right| = 2 + 2 = 4$$

$$AC = \left|2-1\right| + \left|2-(-1)\right| = 1 + 3 = 4$$

Thus, $AB = BC = AC$ and $\triangle ABC$ is equilateral.

Example 4

In taxicab geometry, consider the hexagonal border of the crystal-shaped figure shown in Figure 3.9, where $\overleftrightarrow{AB} \perp \overleftrightarrow{AG}$ and $\overleftrightarrow{CD} \perp \overleftrightarrow{DE}$. Define the two points F_1 and F_2 to be the centers of the squares formed by the two ends of the hexagon, as shown. Taking F_1 and F_2 as foci, show that the hexagon is an ellipse under the taxicab metric using the characteristic property for an ellipse in Euclidean geometry, $PF_1 + PF_2 = $ (constant).

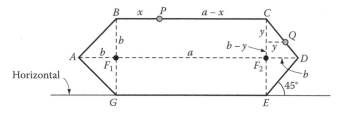

Figure 3.9 Example 4.

Solution

First, if P lies on either the top or bottom side, we have, with $BP = x$ (variable),

$$PF_1 + PF_2 = (x + b) + (a - x + b) = a + 2b = AD$$

If Q is a point on segment CD, with y the length of the leg of the right triangle shown (variable),

$$QF_1 + QF_2 = (b - y + y + a) + (b - y + y) = a + 2b = AD$$

By symmetry in the figure involved, we can deduce the same result for any point P on the hexagon. Therefore, in all cases,

$$PF_1 + PF_2 = AD$$

Finally, we can conclude that although all the axioms so far introduced are valid for the taxicab geometry, the SAS congruence criterion mentioned in the preceding section is not.

Moment for Reflection

The two triangles in Figure 3.10 have certain angles congruent and two pairs of congruent sides. Identify these parts. Do you find that the SAS hypothesis is valid? Are the triangles congruent? What do you think? Is this a counterexample?

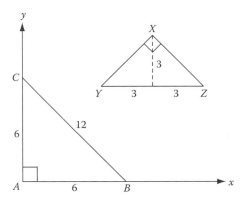

Figure 3.10 SAS congruence tested.

Problems (Section 3.2)

Group A

1.   A triangle in the coordinate plane has vertices $A(0, 0)$, $B(2, 0)$, and $C(1, \sqrt{3})$. This triangle is equilateral under the Euclidean metric (verify). For the taxicab metric, find the following:

(a) The lengths of the sides of $\triangle ABC$.
(b) The angle measures.
(c) What type of triangle is $\triangle ABC$ under the taxicab metric?

2. Three points R, S, and T are placed on a unit grid (each square a unit), as shown. Verify that $\triangle RST$ is equilateral under the taxicab metric. Is $\triangle RST$ equiangular? (This example also shows that taxicab geometry violates Euclid's SAS proposition, since that theorem, as we shall see, leads to the result that an equilateral triangle is always equiangular.) (Figure P.2)

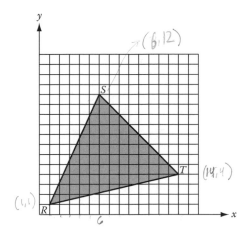

Figure P.2

3. Consider the taxicab triangle ABC with its vertices as indicated in the coordinate plane (Figure P.3).
(a) Show that $\triangle ABC$ is a right triangle.
(b) Find the taxicab lengths $a = BC$, $b = AC$, and $c = AB$.
(c) Show that the Pythagorean theorem is valid for this triangle.

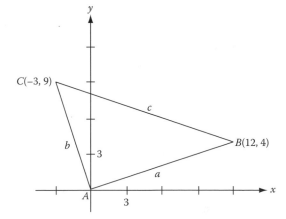

Figure P.3

4. Find a counterexample to the Pythagorean theorem in taxicab geometry.

Group B

5. A man has a newspaper stand at $W(3, 0)$, eats regularly at a cafeteria located at $E(8, 3)$, and does his laundry at a laundro-mat at $L(7, 2)$. (See Figure P.5.) He locates the desirable point R for a room that is equidistant from W, E, and L under the taxicab metric (suitable for the city blocks in this location).

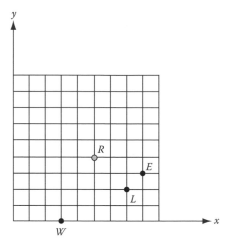

Figure P.5

(a) Find the correct coordinates of R and verify that $RW = RL = RE$.

(b) How many blocks must he walk from R to each of the three points W, E, and L?

6. Refer to Problem 5: Where should R be located if Euclidean distance is used, assuming that this part of town consists of vacant lots? Find $WR = ER = LR$ for this case. (Note that the distance is approximately 9.2 blocks, compared to 5 blocks for Problem 5!)

7. Do three noncollinear points determine a unique taxicab circle?

8. Show that a regular octagon in Euclidean geometry with two horizontal sides in the coordinate plane is a taxicab ellipse. [**Hint:** Take as foci points F_1 and F_2 located on the perpendicular bisector of two opposite nonhorizontal, nonvertical sides at a taxicab distance s from those sides, where $2s$ is the taxicab length of each of those two sides.]

9. **Equidistant Loci in Taxicab Geometry** Using the taxicab circles shown in Figure P.9, prove that all points on the jagged line z are equidistant from points A and B in the figure.

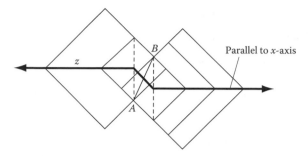

Figure P.9

Group C

10. Generalize the result (and figure) of Problem 9 to arbitrary points A and B. (You should cover the case when line AB makes a 45° angle with the x-axis—with some unusual results.)

11. **Non-Euclidean models for Axioms 1–11** A variety of models for **Axioms 1–11** may be constructed by starting with any differentiable function $f(x)$ having as domain some open interval of real numbers and range all real numbers, and with $f'(x) > 0$ for all x. (This construction was alluded to earlier in Problem 22, Section **1.7**; as an example, take $f(x) = \log x$ for $x > 0$.) Then one takes as "points" all points in the coordinate plane, and as "lines" the *vertical Euclidean lines* together with *all curves of the form*

$$y = f(ax + b), \quad a \text{ and } b \text{ constant}$$

where x is such that $ax + b$ belongs to the domain of f.
  (a) Show that f^{-1} exists and that each pair of points $P(x_1, y_1)$, $Q(x_2, y_2)$ lie on one and only one "line." The distance PQ^* is then defined as the Euclidean arc length along this "line" from P to Q. Examine intuitively the metric axioms for their validity.
  (b) If angle measure is defined as the Euclidean measure of the angles between lines or curves, show that the axioms regarding angle measure are valid.
  (c) Show that the remaining **Axioms 1–11** are valid and find a specific example (exhibit a definite choice for $f(x)$), where Euclid's SAS proposition fails.
  (d) What geometry is obtained in the special case $f(x) = x$?

12. **Undergraduate Research Project** There is an entire class of right triangles in taxicab geometry satisfying the (Euclidean) Pythagorean theorem, as in Problem 3 above. Make a full investigation, and attempt a geometric characterization of such triangles. [**Hint:** To get started, consider $A(a, b)$, $B(-ub, ua)$, and $C(0, 0)$, and find a necessary condition for u.]

13. **Undergraduate Research Project** Investigate what taxicab hyperbolas look like, defined by the set of points P such that $PF_1 - PF_2 = $ constant.

3.3 The SAS Postulate and the ASA and SSS Theorems

In view of the preceding section, the following axiom is independent of the preceding axioms and cannot be proven as a theorem. It is valid in Euclidean geometry, át least plausible for spherical geometry, and very much needed in further developments.

> ***Axiom 12 [SAS Postulate]*** *If under some correspondence between their vertices, two sides and the included angle of one triangle are congruent, respectively, to the corresponding two sides and included angle of another, the triangles are congruent under that correspondence.*

Example 1

Suppose we are given △*KLM*, as shown in Figure 3.11, with *LK = KM* and such that ray *KW* bisects ∠*LKM*, with *W* on line *LM*. Prove that *W* is the midpoint of segment *LM* and that ∠*L* ≅ ∠*M*.

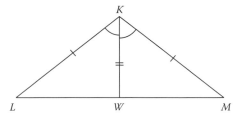

Figure 3.11 Example 1.

Solution

By definition of angle bisector, ∠*LKW* ≅ ∠*MKW* and ($\overrightarrow{KL}$ $\overrightarrow{KW}$ $\overrightarrow{KM}$). Thus, by the ray/point duality theorem (*LWM*). Also, by the reflexive property, $\overline{KW} \cong \overline{KW}$ and by definition $\overline{KL} \cong \overline{KM}$ from the given. Therefore, by SAS (**Axiom 11**), △*LKW* ≅ △*MKW*, and by CPCPC, ∠*L* ≅ ∠*M* and *LW = WM*. Since (*LWM*), *W* is the midpoint of segment *LM*, by definition.

Note: The preceding example seems trivial on first glance, and it appears that we have made it overly complicated because of betweenness considerations, an essential part of geometry. Such details will sometimes be omitted when they appear to be obvious, but they should not be overlooked. ⬙

The first major consequence of the SAS axiom is the following, the ASA property.

> ***ASA Congruence Criterion*** If under some correspondence between their vertices, two angles and the included side of one triangle are congruent, respectively, to the corresponding two angles and included side of another, the triangles are congruent under that correspondence.

Proof Suppose that in $\triangle ABC$ and $\triangle XYZ$, we have $\angle A \cong \angle X$, $\angle B \cong \angle Y$, and $\overline{AB} \cong \overline{XY}$, as shown in Figure 3.12. If $AC = XZ$, then the triangles would already be congruent by SAS. Hence, suppose $AC \neq XZ$. For sake of argument assume that $AC > XZ$. By the segment construction theorem, there exists an interior point D on $\overline{AC}$ such that $AD = XZ$. By the ray/point duality theorem, $(\overrightarrow{BA}\,\overrightarrow{BD}\,\overrightarrow{BC})$ and therefore $m\angle ABD < m\angle ABC$. But in $\triangle ABD$ and $\triangle XYZ$, we have $AB = XY$, $AD = XZ$, and $\angle A \cong \angle X$. By SAS, $\triangle ABD \cong \triangle XYZ$ and by CPCPC $m\angle ABD = m\angle XYZ = m\angle ABC$, contradicting $m\angle ABD < m\angle ABC$. We must conclude that $AC = XZ$ and therefore, $\triangle ABC \cong \triangle XYZ$. ▚

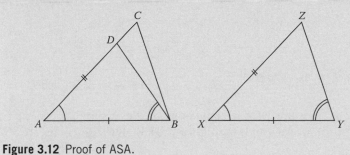

Figure 3.12 Proof of ASA.

One of Euclid's first propositions was the isosceles triangle theorem: *If two sides of a triangle are congruent, then the angles opposite are congruent.* His argument used a special configuration that resembled a highway bridge, and thus it became known as the *pons asinorum*—"bridge of asses"—the implication being that any student of geometry who could not cross this bridge was considered unable to go on in the study of geometry. (See Problem 13 for this proof.) Example 1 above shows how one might prove the theorem using a much simpler construction (but necessitating the use of the crossbar theorem). However, an extraordinarily simple proof was discovered in the 1950s by computer analysts working in the field of artificial intelligence. It turns out that this same proof was used by Hilbert much earlier.

An isosceles triangle in unified geometry is defined as follows:

> **Definition** A triangle having two congruent sides is called an **isosceles** triangle, with the two congruent sides called its **legs**, the remaining side called the **base**, the angles opposite the legs, the **base angles**, and the angle opposite the base, the **vertex angle**. (See Figure 3.13.)

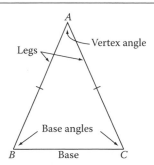

Figure 3.13 An isosceles triangle.

Isosceles Triangle Theorem   If two sides of a triangle are congruent, the angles opposite are congruent, and conversely. (See Figure 3.14.)

Proof   By SAS, $\triangle ABC \cong \triangle ACB$. Hence $\angle B \cong \angle C$. For the converse, use ASA. ⬛

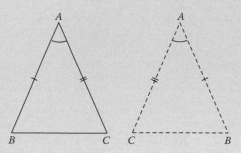

Figure 3.14 Hilbert's proof.

Finally, to prove what is commonly known as the SSS congruence theorem, we need a simple preliminary result, illustrated in Figure 3.15.

(1)   If $P \in$ Int $\triangle ABC$, then either $PA \neq CA$ or $PB \neq CB$ (or both).

To prove this, suppose that $PA = CA$ and $PB = CB$, with $P \in$ Int $\triangle ABC$ (Figure 3.15), defined in Problem 18, Section **2.6**. Then, P is an interior point of all three angles of $\triangle ABC$. By the crossbar theorem, ray CP meets side AB at some interior point W, forming the angles marked 1 through 6 in the figure. By the isosceles triangle and linear pair theorems,

$$m\angle ACB = \angle 1 + \angle 2 = \angle 3 + \angle 4 = (180 - \angle 5) + (180 - \angle 6) =$$

$$= 360 - (\angle 5 + \angle 6) = 360 - m\angle APB > 180$$

(since A, P, and B are noncollinear and $m\angle APB < 180$.) That is, $m\angle ACB > 180$, a contradiction. ⬛

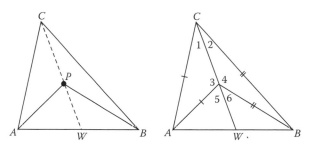

Figure 3.15 A preliminary step for SSS.

SSS Congruence Criterion   If under some correspondence between their vertices, the three sides of one triangle are congruent to the three corresponding sides of another, the triangles are congruent under that correspondence.

Proof Suppose that in triangles *ABC* and *XYZ*, *AB* = *XY*, *BC* = *YZ*, and *AC* = *XZ*. If the triangles are not congruent, then no pair of corresponding angles can be congruent (otherwise SAS implies congruence). We can therefore find two angles of one triangle each less in measure than those of the corresponding angles of the other, say $m\angle X$ < $m\angle A$ and $m\angle Y$ < $m\angle B$ (Figure 3.16). Within $\angle CAB$ construct $\angle UAB \cong \angle X$ and within $\angle CBA$ construct $\angle VBA \cong \angle Y$, where ray *AU* lies between rays *AC* and *AB* and ray *BV* lies between rays *BC* and *BA* (angle construction theorem, Section **2.5**). An application of the crossbar theorem then shows that rays *AU* and *BV* meet at an interior point *P* of $\angle ACB$ and also of $\angle A$ and of $\angle B$. Hence, $P \in$ Int $\triangle ABC$. But by ASA, $\triangle ABP \cong \triangle XYZ$ and both *PA* = *ZX* = *CA* and *PB* = *ZY* = *CB* follow by CPCPC, contradicting **(1)**. ◣

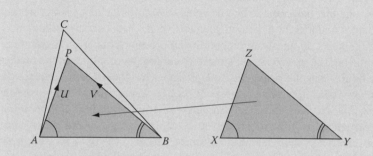

Figure 3.16 Proof of SSS congruence.

3.4 Euclid's Superposition Proof: An Alternative to Axiom 12

Euclid demonstrated the SAS congruence theorem (which we assumed as **Axiom 12**) from his basic assumptions. His argument has come to be known as the *method of superposition*. It is instructive to review the passage in the Elements where this takes place. Here is Euclid's argument for Proposition I.4 (the SAS Postulate) as quoted from *The Thirteen Books of Euclid's Elements, Book 1*, pp. 247–248 (Heath 1956); the accompanying diagram shows Euclid's special drawing for the argument. [Given: $\triangle ABC$ and $\triangle DEF$ with *AB* equal to *DE*, *AC* equal to *DF*, and $\angle A$ equal to $\angle D$.]

I say that the base *BC* is also equal to the base *EF*, the triangle *ABC* will be equal to the triangle *DEF*, and the remaining angles will be equal to the remaining angles, respectively, namely those which the equal sides subtend, that is, the angle *ABC* to the angle *DEF*, and the angle *ACB* to the angle *DFE*.

For, if the triangle *ABC be applied to* the triangle *DEF* and if the point *A be placed on* the point *D* [author's italics] and the straight line *AB* on *DE*, then the point *B* will also coincide with *E*, because *AB* is equal to *DE*.

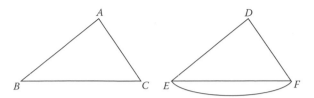

Again, *AB* coinciding with *DE*, the straight line *AC* will also coincide with *DF*, because the angle *BAC* is equal to the angle *EDF*; hence the point *C* will also coincide with point *F* because *AC* is again equal to *DF*.

But *B* also coincided with *E*; hence the base *BC* will coincide with base *EF*... and will equal it. Thus the whole triangle *ABC* will coincide with the whole triangle *DEF*....

This bit of poetry provides the basis for a geometric style transformation, one that has come to be known as a *motion*. Euclid's passage actually provides a model for a modern geometric proof, just as many arguments in the *Elements* do. In his argument, Euclid presumes that a geometric figure can be *moved*, as you would a piece of cardboard, and that, in the process, the figure does not change its size or shape. In modern terms, we are presumably talking about a correspondence between points in the plane that preserves *distance*, that is, an **isometry** (metric-preserving). Since we want to formalize Euclid's proof however, we cannot merely assume the existence of isometries (where one would ultimately have to use the SSS congruence criterion). We therefore take the following as the definition of a motion.

Definition   A **motion** is a correspondence $P \to P'$ between the points of $\mathcal{P}$ that preserves distance and angle measure. That is, for any three points A, B, and C and their correspondents A', B', and C', $AB = A'B'$, and $m\angle ABC = m\angle A'B'C'$.

It is clear that a motion is betweenness preserving, hence maps segments into segments, rays into rays, and lines into lines. Suppose that a motion fixes two distinct points A and B on some line l (i.e., $A \to A$ and $B \to B$). Then, for any point $X \in l$, X and its image X' must both belong to either ray AB or its opposite ray AC since $\angle BAX \cong \angle B'A'X' \cong \angle B'AX'$. In either case, we have $AX = A'X' = AX'$, which implies by (**7**), Section **1.8**, $X = X'$. That is, every point X on line l is fixed. Assuming the motion is nontrivial, no other point except those on line l can be fixed: for otherwise, assuming that $Y = Y' \notin l$, every point on the extended sides of $\triangle ABY$ is fixed, and hence every point in the plane would be fixed (do you see why?), and the motion would be the trivial "identity" leaving every point unchanged. A motion having a line of fixed points (and no other fixed points) is called a **reflection** in that line, with that line as **axis**.

A reflection is thus a special kind of motion, but one that has a line of fixed points. The *composition* or *product* of two reflections $f \circ g$ (one followed by the other) need not be a reflection, but since both f and g are motions and preserve distance and angle measure, their product also preserves distance and angle measure. Thus, the product of two or more reflections is a motion. The substitute axiom for the SAS postulate is the following.

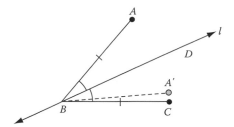

Figure 3.17 An angle bisector as line of reflection.

Axiom M   *There exists a nontrivial reflection in any given line.*

You should have no trouble proving that if $P \to P'$ is a reflection in line l as axis (hence a motion) and P is any point not on l, then the angles that P and P' make with the axis are congruent. Since $P \neq P'$, by the angle-construction theorem, P' lies on the opposite side of l as P. Moreover, l passes through a point M midway between P and P' [i.e., (PMP') and $PM = MP'$] and $\overleftrightarrow{PM} \perp l$. (Thus, if $PP' < \alpha$, line l is the *perpendicular bisector* of segment PP'.) Furthermore, suppose l contains the bisector $\overrightarrow{BD}$ of $\angle ABC$ and that $AB = BC$ (Figure 3.17). Then, since $B \to B$ and $D \to D$, $m\angle ABD = m\angle A'BD$, and by the angle-construction theorem, ray BA' coincides with ray BC and A' lies on ray BC. But also, $BA' = BA = BC$ by hypothesis and by **(7)**, Section **1.8**, $A' = C$. Thus $A \to C$. This fact will be used in the SAS theorem below.

We are now ready for the main theorem (Euclid's proposition I.4), which is implied by **Axiom M**, and a rigorous argument can be given for it.

SAS Congruence Criterion from Axiom M   If two sides and the included angle of one triangle are congruent to the corresponding two sides and included angle of another triangle, then the triangles are congruent.

Proof   Suppose that in triangles ABC and DEF we have $AB = DE$, $AC = DF$, and $m\angle BAC = m\angle EDF$ (Figure 3.18). Our goal is to find a motion that maps $\triangle ABC$ to $\triangle DEF$. Let M be the midpoint of segment AD (or if $AD = \alpha$ take M as any point midway between A and D), and let l_1 be the line perpendicular to line AM at M. Then, under the reflection $P \to P'$ in line l_1 (by **Axiom M**), $A \to A' \equiv D$, $B \to B'$, and $C \to C'$. (The shaded triangles in the figure show the images of the given triangle under the reflections to be considered.) Now take the line l_2 containing the bisector of $\angle B'DE$ (unless $B' = E$) and consider the reflection in l_2. By previous observations, $B' \to B'' \equiv E$ (and $D' \to D'' \equiv D$). Now, so far, we have the product $P \to P' \to P''$ of the two reflections in l_1 and l_2 mapping A to D, B to E, and C to C''. Next, consider line $l_3 = \overleftrightarrow{DE}$ (if $C'' \neq F$). This line contains the bisector of $\angle FEC''$ since by hypothesis $m\angle EDF = m\angle BAC = m\angle B''A''C'' = m\angle EDC''$. (Or, if rays EF and EC are opposite rays, then $l_3 \perp \overleftrightarrow{FD}$ and D is midway between F and C''.) Thus, if $P'' \to P'''$ denotes the reflection in line l_3, then C'' maps to F, and D and E are left fixed. The product of the three reflections then maps $\triangle ABC$ to $\triangle DEF$. Since motions preserve distance and angle measure, it follows that $\triangle ABC \cong \triangle DEF$. ⧊

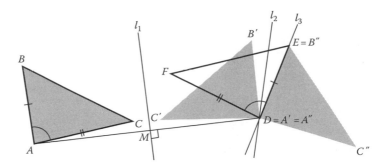

Figure 3.18 Superposing △*ABC* to △*DEF*.

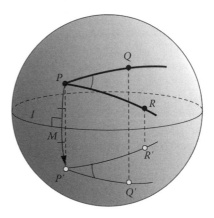

Figure 3.19 Spherical reflection.

This approach to the SAS postulate has certain advantages. One is that we have succeeded in making one of Euclid's highly visual proofs valid axiomatically. Another is that in models of non-Euclidean geometry, it is often easier to establish a reflection in a line than to directly prove the SAS postulate.

In spherical geometry, for example, plausibility arguments have already been provided for all the previous axioms up through **Axiom 11**, leaving the SAS Postulate to verify. Consider any line *l* (we can use the equator for simplicity), as in Figure 3.19. For any two points *P* and *Q*, we project orthogonally on the sphere through point *M* through *l* and on to point *P'* such that $PM = MP'$, and similarly to obtain *Q'*. (*P'* and *Q'* become the reflections of *P* and *Q* in the equatorial plane, lying directly below.) It is evident that the spherical distances *PQ* and *P'Q'* are equal. If *R* is a third point, then it can also be observed that $m∠QPR = m∠Q'P'R'$. Thus, the general mapping $P → P'$ is a reflection in line *l*. One concludes that **Axiom M** holds and that the SAS postulate is valid for spherical geometry.

3.5 Locus, Perpendicular Bisectors, and Symmetry

A concept that portrays a highly visual aspect of geometry is that of **locus**. The conventional rendering of this concept is a *path* or *curve* along which a point is constrained to move under a given condition or set of conditions, analogous to the action of a particle in physics under certain forces acting on that particle. For example, given two (fixed) points A and B, we could require that a point P always satisfies the condition $PA + PB = 1$. Those familiar with this relation would recognize it as the condition for an ellipse, so in this case P would trace out an ellipse. Or, consider the simpler requirement: $PA = 3$. In this case, the "locus of P" would be a circle centered at A having radius 3. For more involved examples, see *Sketchpad Experiment 7*, Appendix A.

The traditional proof that some curve or path C is the "locus" of a point requires that one must establish two things: (1) Every point P on the curve C satisfies the condition(s) of the locus, and (2) any point P satisfying the condition(s) of the locus lies on curve C. As with many geometric ideas, this concept can be given a strictly mathematical version that, while seemingly less spectacular, removes some of the mystery that perhaps surrounds it. We can simply agree that a **locus** is a *set of points* satisfying certain properties. Consider, for example, the locus

$$S = \{P : P \text{ satisfies conditions A, B, and C}\}$$

and suppose we want to prove that S is some curve C. This makes the mandates (1) and (2) clear: The standard way to prove that two sets C and S are the same is to prove that each is a subset of the other: (1) $C \subseteq S$ and (2) $S \subseteq C$. That is, one must prove

1. If $P \in C$ then $P \in S$, that is, P must satisfy properties (A), (B), and (C).
2. If $P \in S$ and therefore satisfies properties (A), (B), and (C), then $P \in C$.

The first useful example of locus in our development involves a familiar definition.

Definition 1   A line that both bisects and is perpendicular to a segment is called the **perpendicular bisector** of that segment.

The standard theorem in Euclidean geometry is that the *locus of points equidistant from two distinct points A and B is the perpendicular bisector of segment AB*. This is also true for unified geometry.

Theorem 1: Perpendicular-Bisector Theorem

Given any two points A and B, $0 < AB < \alpha$, the locus of points equidistant from A and B is the perpendicular bisector of segment AB.

Proof   In Figure 3.20, let l be the perpendicular bisector of segment AB, intersecting that segment at its midpoint M.

1. If P is any point on l and $PM < \alpha$, then $AM = MB$, $m\angle AMP = m\angle PMB = 90$, and $PM = PM$. By SAS, $\triangle AMP \cong \triangle BMP$ and $PA = PB$ by CPCPC. Hence P is equidistant from A and B.

2. Conversely, suppose that $PA = PB$. We must prove that P lies on line l. By the isosceles triangle theorem, $\angle PAM = \angle PBM$ and $AM = MB$. Therefore, provided $PA < \alpha$ $\triangle PAM \cong \triangle PBM$ by SAS and $m\angle AMP = m\angle PMB$ (SSS can also be used since $PM = PM$). By the criterion for perpendicularity (Theorem **4** in Section **2.4**), line PM is perpendicular to line AB. Since the perpendicular to a line at a point on that line is unique, line $PM \equiv l$ and $P \in l$, as desired. (The cases $PM = \alpha$ and $PA = \alpha$ will be left for the reader.) ◣

(1) Prove that if $P \in l$ then P is equidistant from A and B

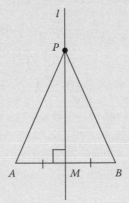

(2) Prove that if P is equidistant from A and B then $P \in l$

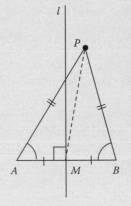

Figure 3.20 Proof of perpendicular-bisector theorem.

Symmetry in geometry, as in nature, is a dominant feature. A figure is *symmetric*, provided there is a *line of symmetry*: In Figure 3.21, line l is a line of symmetry for each of the two figures—a leaf and a polygonal region. In geometry, a *line of symmetry* for two points P and Q is the

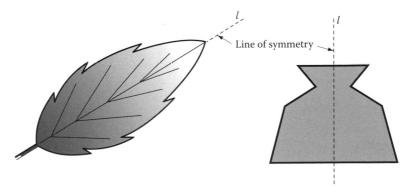

Figure 3.21 Symmetry in line l.

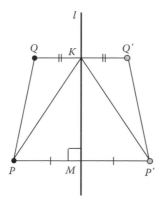

Figure 3.22 Isometric property of reflections.

perpendicular bisector of the segment joining those two points, and P and Q are said to be *reflections* of each other in that line. Figure 3.22 shows how to construct for each given point P in the plane the image point P' that defines, in general, the **reflection** in line l: with $\overleftrightarrow{PM}$ the perpendicular to line l at M, (existence proven later), one takes $P'M = MP$. For Q any other point and Q' constructed in the same manner, where $\overleftrightarrow{QK} \perp l$ at K and $QK = KQ'$, it is an elementary exercise to prove that $\triangle KMP \cong \triangle KMP'$ and $\triangle PKQ \cong \triangle P'KQ'$ and to show that $PK = P'K$ and thus $PQ = P'Q'$. Thus, reflections are *distance-preserving*, proving that nontrivial distance-preserving mappings in the plane (*isometries*) exist for unified geometry. By SSS, such mappings are also angle-measure preserving.

The properties of reflections in Euclidean geometry are interesting, a few of which will be mentioned here without proof. (These ideas appear in more detail in Chapter 8 on transformation theory.) Given two congruent triangles, there exist three or fewer reflections whose product (one mapping followed by another) maps the first triangle onto the second (as in the preceding section). Related closely to this, if any distance-preserving mapping (isometry) is given, there exist three or fewer reflections whose product is the given isometry. This endows reflections with a prominent place in geometry. In particular, translations and rotations for Euclidean geometry can be defined in terms of reflections: A **translation** is the product of two reflections in lines that are parallel, while a **rotation** is the product of two reflections in lines that intersect.

We close this section with a key fundamental result, requiring extra care for a bounded metric.

Theorem 2   There exists a line perpendicular to a given line and passing through a given point not on the given line. That perpendicular is unique unless $\alpha < \infty$ and the distance from the given point to every point on the given line is $\alpha/2$.

Proof   Consider $P \notin l$ and choose any point $Q \in l$, as shown in Figure 3.23. By hypothesis, $PQ < \alpha$ since otherwise line l passes through P.

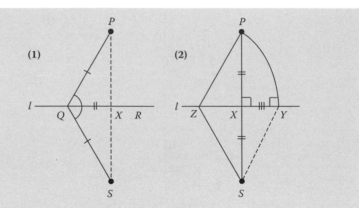

Figure 3.23 Perpendiculars of *l* passing through *P*.

1. *Existence.* If $\overleftrightarrow{PQ} \perp l$, we are finished. Otherwise, choose any point R on l distinct from Q such that $QR < \alpha$. Construct $\angle SQR \cong \angle PQR$ on the opposite side of l as P, with $SQ = QP$. Since P and S lie on opposite sides of line l, there exists a point X on l such that (PXS). Now X either lies on ray QR (the case shown in the figure) or on the opposite ray QR'. But in either case $\triangle PQX \cong \triangle SQX$* by SAS and $\angle PXQ \cong \angle SXQ$ by CPCPC. Therefore, $\overleftrightarrow{PX} \perp l$.
2. *Uniqueness.* If there were a second line $PY \perp l$ for $Y \in l$, draw segment SY (Figure 3.23(2)). We proceed to show that $\alpha < \infty$ and that $PZ = \alpha/2$ for $Z \in l$. We have $PY < \alpha$ and $SY < \alpha$ since P, $S \notin l$; also $XY < \alpha$ (or else line PX passes through Y and line PY would coincide with line PX). Hence by SAS $\triangle PXY \cong \triangle SXY$ and $m\angle XYS = m\angle XYP = 90$, which implies line $YS \perp l$ at Y. Since the perpendicular to l at Y is unique, lines PY and YS coincide. Hence lines PX and PY have the points P and S in common, which by **Axiom 4** implies $PS = \alpha < \infty$. Thus P and S are extreme opposites with (PXS), (PYS), and, for $Z \in l$, (PZS). Therefore, $PX = XS = \alpha/2$, $PY = YS = \alpha/2$, and by congruent triangles (either $\triangle ZPX \cong \triangle ZSX$ or $\triangle ZPY \cong \triangle ZSY$ by SAS), $PZ = ZS = \alpha/2$, as desired. ⬧

Corollary A   If $\alpha = \infty$, there exists a unique perpendicular to a given line and passing through a given external point.

Three results that apply to a bounded metric will be stated, which will be much used later. Because the proofs basically follow that for the previous theorem, they are regarded as corollaries, and their proofs will be left as problems (see Problems 17–19). The North Pole and Equator on the earth's surface serve as a model for these results.

---

* For a bounded metric, one must deal with the possibility that $QX = \alpha$, in which case P, Q, and X are collinear, and $\triangle PQX$ does not exist (as well as $\triangle SQX$). In this case, choose Q', such that $(QQ'X)$; it follows that $\triangle QQ'P \cong \triangle QQ'S$ and $PQ' = SQ'$, so that Q' takes the place of Q in the above argument.

Corollary B   If $PQ = \alpha/2$ for some Q on l and line PQ is perpendicular to l at Q, then for every point X on l, $PX = \alpha/2$ and line PX is perpendicular to l.

Corollary C   If the legs of an isosceles triangle each have length $\alpha/2$, then the base angles are both right angles, and conversely.

Corollary D   If a right triangle has legs (two sides including the given right angle) of length $\alpha/2$, then it is a **tri-rectangular** triangle (having all three angles right angles and all three sides of length $\alpha/2$).

A much used concept in geometry is the distance *from a point to a line*, which directly involves the use of Theorem 2. First, a definition.

Definition 2   If l is any line and A an external point, the **foot** of A on l is a point F on l such that line AF is perpendicular to l at F, and $AF \le \alpha/2$. (See Figure 3.24.)

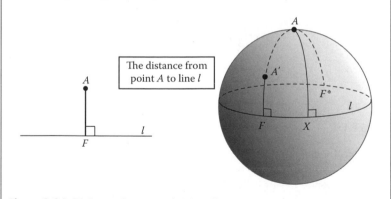

Figure 3.24 Distance from a point to a line.

In the case $\alpha = \infty$, there exists exactly one perpendicular to l through A, so the foot of A is unique. However, if $\alpha < \infty$, then there are two cases: (1) every line through A is perpendicular to l and F can be chosen as any point on l, in which case $AF = \alpha/2$, or (2) the perpendicular is unique, but intersects l at two points F and F^*, in which case the restriction $AF \le \alpha/2$ makes F unique. Thus, in unified geometry, the foot of A on l always exists and is unique, provided $AP \ne \alpha/2$ for at least one point P on l. In any case, the **distance** from point A to line l, sometimes called **perpendicular distance**, is defined to be $d = AF$, where F is a foot of A on l. (If A lies on l, we define its distance to l to be zero.)

Figure 3.24 shows an example on the sphere, where a point (the North Pole) can have infinitely many feet on a line (the equator). Thus, AX, as shown, equals $\alpha/2$ for all X on l and the distance from A to l is $\alpha/2$. However, the point A' that lies half way between A and the equator has the unique foot F as shown, and here the distance from A' to l is $\alpha/4$.

Theorem 3   The distance from any point *A* to any line *l* is the minimum value of *AX* for all *X* on *l*.

The proof of this theorem awaits the section on inequalities in triangles, and the result that the hypotenuse of a sufficiently small right triangle is greater than either leg (Section 3.7).

Problems (Section 3.5)

No betweenness proofs are required for Problems 1–6. In each case, you are to use the figure to solve for the variables and measures that are requested. In Problems 7–9, use betweenness relations evident from the figures.

Group A

1.   Find *x* (Figure P.1).

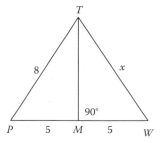

Figure P.1

2.   Find *x* and *y* (Figure P.2).

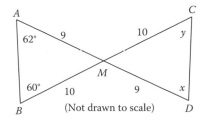

Figure P.2

3.   If $\overrightarrow{AB} \perp \overrightarrow{AD}$ find *x* and *y* (Figure P.3).

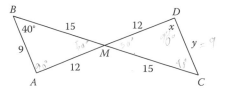

Figure P.3

4. Find x and $m\angle JHM$ (Figure P.4).

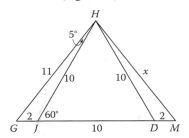

Figure P.4

5. If $\angle B \cong \angle C$ find x, y, AB and $m\angle B$ (Figure P.5).

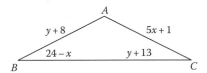

Figure P.5

6. Find x and y (Figure P.6).

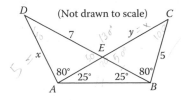

Figure P.6

7. In Problem 3, prove that $m\angle MDC = 90$ and $CD = 9$.

8. In Problem 4, prove that $HG = HM$.

9. In Problem 6, prove that $AD = 5$ and $EC = 7$.

Group B

10. In equilateral triangle, RST points K, L, and M are located on the sides, as shown in Figure P.10, such that $RK = SL = TM$. Prove that $\triangle KLM$ is equilateral.

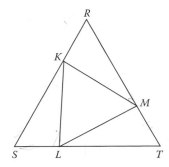

Figure P.10

11.   Points *A* and *B* lie on the sides of square *PQRS* as shown in Figure P.11. (You may assume that a square has four right angles and four congruent sides.) If $PA = PB$ and $AB = AR$, prove that $\triangle ABR$ is equilateral.

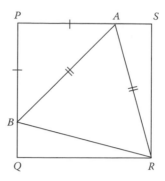

Figure P.11

12.   Diagram from a Martin Gardner Puzzle Martin Gardner, the editor of the "Mathematical Games" section of the *Scientific American* for many years, once posed the problem of adding rods of the same length to a square linkage of four rods, all hinged at the ends, in order to render the whole assemblage immobile. (See Figure P.12.) The idea was to use as few additional rods as possible. The solution was discovered by several readers: no fewer than 23 rods will suffice. The diagram of the solution is shown in Figure P.12, far right. Four segments (dashed lines) have been added to complete Figure P.12 (not part of the solution), forming 16 triangles. You are to name the triangles by number that are congruent in Figure P.12. What congruence criteria apply? (See Gardner, M., *Sixth Book of Mathematical Games from Scientific American.*)

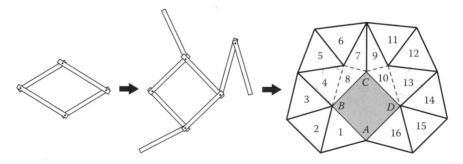

Figure P.12

13.   Figure P.13 for this problem shows the diagram for Euclid's *pons asinorum*. Construct a proof based on this diagram to show that if $\overline{AB} \cong \overline{AC}$, then $\angle ABC \cong \angle ACB$, using only the SAS postulate.

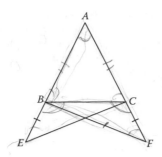

Figure P.13

14. Are the two triangles as shown in Figure P.14 congruent? Why or why not?

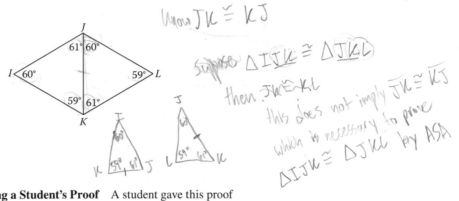

Not

know JK ≅ KJ

Suppose ΔIJK ≅ ΔJKL

then JK̄ ≅ KL

this does not imply JK̄ ≅ KJ

which is necessary to prove

ΔIJK ≅ ΔJKL by ASA

Figure P.14

15. **Critiquing a Student's Proof** A student gave this proof involving Figure P.15 for this problem:

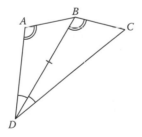

Figure P.15

Given: Ray $\overrightarrow{DB}$ bisects $\angle ADC$ and $\angle BAD \cong \angle DBC$.

Proof $\angle ADB \cong \angle CDB$ by definition of angle bisector, $\overline{BD} \cong \overline{BD}$ (reflexive law), $\angle ADB \cong \angle BDC$ (given). Therefore, $\triangle ABD \cong \triangle CBD$ by ASA and $BD = CD$.

What is wrong with this proof?

16. The universal peace symbol consists of a circle and 3 lines drawn from the center to points equally spaced on the circle, as shown in Figure P.16. (That is, the angles at the center are congruent.) Prove that $\triangle ABC$ is equilateral, thus providing an axiomatic proof that equilateral triangles exist in unified geometry.

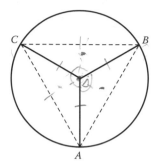

Figure P.16

17. Prove Corollary B. [**Hint:** Construct point S on ray PQ such that (PQS) and $QS = QP$. Then, every line through P passes through S.]

18. Prove Corollary C. [**Hint:** If $\triangle ABC$ is isosceles with base BC and such that $AB = AC = \alpha/2$, consider A^*, the extremal opposite of point A. Then (ABA^*) and (ACA^*); continue. For the converse, assume the base angles are right angles and use Theorem 2.]

19. Prove Corollary D.

20. Suppose that (AEC) and (DEB) in Figure P.20 and that $AE = EB$ and $\angle DAB \cong \angle CBA$. Prove that $AD = BC$ and that E lies on a common perpendicular to lines AB and CD.

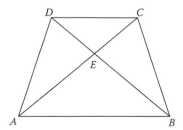

Figure P.20

The remaining problems are for the case $\alpha = \infty$ only. Accordingly, the following basic property may be used: Given three collinear points A, B, and C, either (ABC), (BAC), or (ACB). Also, the perpendicular to a line from an external point exists and is unique.

21. **Proclus' Proof**   The Greek philosopher Proclus (410–485) proved that it is impossible to draw from a point to a line more than two line segments of the same length (as suggested in Figure P.21). Help analyze his proof by answering the questions (a) through (d) below.

(a) If $AP = AQ$, what is true about $\angle 1$ and $\angle 2$?

(b) If $AQ = AR$, what is true about $\angle 3$ and $\angle 4$?

(c) If $AP = AR$, what is true about $\angle 1$ and $\angle 4$?

(d) If (PQR) (one such betweenness relation must occur), what do you conclude about $\angle 2$ and $\angle 3$? Do any of these conclusions lead to a contradiction?

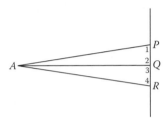

Figure P.21

22. **Kites and Darts**   Any four-sided figure whose adjacent sides are congruent is a **kite**, or **dart**, depending on its angles. As shown in Figure P.22, $ABCD$ has $AB = AD$ and $BC = CD$. The *dart* is distinguished from the *kite* by the condition that instead of all four angles at A and C being acute for the kite, at least two of them, say $\angle ACB$ and $\angle ACD$ are both obtuse for the dart. Without using the SSS congruence criterion, prove that $\angle BAC \cong \angle DAC$ and $\angle BCA \cong \angle DCA$ in all cases, for both the kite and the dart. [**Hint:** Use the Perpendicular Bisector Theorem.]

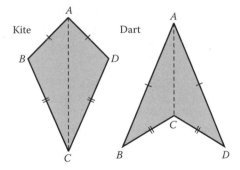

Figure P.22

23. **SSS Congruence Criterion via Kites and Darts**   Prove that in $\triangle ABC$ and $\triangle XYZ$, if $AB = XY$, $BC = YZ$, and $AC = XZ$, then $\triangle ABC \cong \triangle XYZ$. [**Hint:** Construct $\triangle ACD$ with D on the opposite side of line AC as B, congruent to $\triangle XYZ$ using the segment- and angle-construction theorems; use the result of Problem 22.]

Group C

24. Prove each of the following ($\alpha = \infty$):
   (a) No triangle can have two right angles.
   (b) No triangle can have two obtuse angles. (Construct perpendiculars at the vertices where obtuse angles have been assumed to exist, and show that the perpendiculars meet at some point using betweenness considerations.
   (c) The base angles of an isosceles triangle are acute.

25. **Triangle Inequality Using Isosceles Triangles** Complete the details in each of the following steps, which will establish the triangle inequality for $\alpha = \infty$. You will need the result of Problem 24.
   (1) For an indirect proof, suppose that in $\triangle ABC$, $AB + BC \leq AC$. If a contradiction is obtained, then $AB + BC > AC$, as desired.
   (2) If $AB + BC = AC$, construct point D on side AC such that (ADC) and $AB = AD$. This forms two isosceles triangles having acute base angles (why?)
   (3) If $AB + BC < AC$, in addition to point D constructed in Step (2), construct E such that (EBC) and $EC = DC$. What do you observe? (Figure P.25)

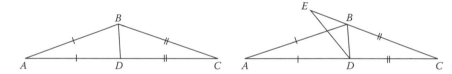

Figure P.25

3.6 The Exterior Angle Inequality

One of the most prominent and frequently used theorems in unified geometry will be established in this section. In Figure 3.25 is shown a triangle with certain angles labeled 1, 2, and 3. In Euclidean geometry, the so-called *exterior angle theorem* is the proposition

$$m\angle 3 = m\angle 1 + m\angle 2$$

The best that can be accomplished in unified geometry, however, is that for all *sufficiently small* triangles, the exterior angle $\angle ACD$ has measure *greater* than that of either $\angle A$ or $\angle B$. Euclid proved both these theorems.

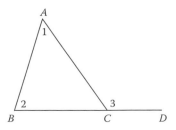

Figure 3.25 Euclid's exterior angle theorem.

The former (equality) depends on the properties of parallelism, so was proved much later in the development. The inequality however, was proven following the SAS, ASA, and SSS theorems, as we shall do here. Euclid recognized its importance in geometry.

Moment for Reflection

In Figure 3.26 is shown a tri-rectangular triangle on a sphere—a triangle having three right angles. Point A is the North Pole, and line BC the equator, with $BC = \alpha/2$. What do you observe concerning exterior angle $\angle ACD$ and the angles at A and B? Is the relation $m\angle A + m\angle B = m\angle ACD$ valid? What about the result $m\angle ACD > m\angle B$, a theorem of unified geometry for small enough spherical triangles? Related to this, do you think the hypotenuse of a right triangle in spherical geometry is the longest side?

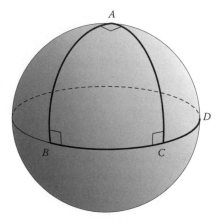

Figure 3.26 Exterior angles in spherical geometry.

Definition   In triangle ABC, if D is any point such that (BCD), as in Figure 3.25, then $\angle ACD$ is called an **exterior angle** of the triangle, and the angles at A and B are said to be the **opposite interior** angles corresponding to that exterior angle.

Euclid's proof of the weaker theorem on exterior angles can be used here for unified geometry, provided one restricts the triangles to having sides of length less than $\alpha/2$. A preliminary result is needed, which is intuitively self evident: *If the sides of a triangle are small, then so is any cevian.* More precisely, if in $\triangle ABC$ the sides are each less than $\alpha/2$ in length, then so is any cevian AD, for $D \in \overline{BC}$. To see this, observe that in Figure 3.27 the sides AB and AC of $\triangle ABC$ are extended to the extremal opposite A^* of point A; then cevian AD passes through A^* and cuts the median segment PQ at some point E, with $AE = \alpha/2$, from which it follows that $AD < AE = \alpha/2$. A proof not involving A^* and not relying as much on intuition appears in Problem 14 below.

(1)   **Special Cevian Inequality** If $AB < \alpha/2$, $AC \leq \alpha/2$, and (BDC), then $AD < \alpha/2$.

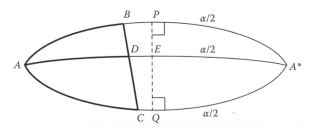

Figure 3.27 Cevian inequality.

Theorem 1: *Exterior Angle Inequality*

In any triangle having sides of length less than $\alpha/2$, the measure of an exterior angle is greater than that of either opposite interior angle.

Proof Let the triangle be $\triangle ABC$, with (BCD) and exterior angle $\angle ACD$, as in Figure 3.28. It must be proven that $m\angle ACD > m\angle A$, $m\angle B$. Let M be the midpoint of side AC. By **(1)**, $BM < \alpha/2$, so we may use the segment-doubling theorem to locate point E such that (BME) and $BM = ME$. Since rays $\overrightarrow{MA}$, $\overrightarrow{MC}$ and $\overrightarrow{MB}$, $\overrightarrow{ME}$ are pairs of opposite rays, the angles at M are vertical angles, hence $\angle AMB \cong \angle CME$. By SAS, $\triangle AMB \cong \triangle CME$, and by CPCPC,

$$m\angle ACE = m\angle MCE = m\angle A$$

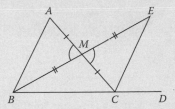

Figure 3.28 Angle ACD > angle A.

To finish the argument, we must prove that ray CE lies between rays CA and CD (as is seemingly clear from Figure 3.28). Consider

$$\text{Int}\angle ACD = H(A, \overleftrightarrow{CD}) \cap H(D, \overleftrightarrow{CA})$$

Since (BME), E lies in the same half plane of line CD as M, and (AMC) implies the same for A, thus $E \in H(A, \overleftrightarrow{CD})$. Similarly, (BMC) and (BCD) imply that E and D lie in the opposite half plane of line CA as B, or on the same side of line CA, or $E \in H(D, \overleftrightarrow{CA})$. Thus, $E \in \text{Int}\angle ACD$. Hence, by **Axiom 10**,

$$m\angle ACD = m\angle ACE + m\angle ECD > m\angle ACE = m\angle A$$

(from the first part of the argument).

It remains to prove $m\angle ACD > m\angle B$. Locate any point F on line AC such that (ACF). This makes $\angle BCF$ an exterior angle (Figure 3.29). The preceding argument showed that, in general, an exterior angle of a triangle has measure greater than that of the adjacent interior angle (the angle that shares a side with the exterior angle. Thus, $m\angle BCF > m\angle B$. But $\angle ACD$ and $\angle BCF$ are vertical angles, thus congruent, hence $m\angle ACD = m\angle BCF > m\angle B$, as desired. ◣

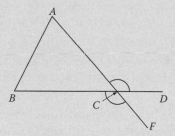

Figure 3.29 Angle ACD > angle B.

The above proof of the exterior angle inequality depended heavily on the betweenness relation $(\overrightarrow{CA}\ \overrightarrow{CE}\overrightarrow{CD})$ in Figure 3.28. But a proof without betweenness considerations is not valid (this applies to Euclidean geometry as well). One can see here that Euclid's proof, as originally given, is incomplete. But you must admit that the idea of the proof itself was ingenious.

A number of familiar results now follow—all implied by the exterior angle inequality. The proofs are mostly one-liners, so they will be left as nice exercises. The first involves the angle sum of two angles of a triangle. (We are not yet close to any estimates about the angle-sum of *all three* angles of a triangle, which, of course, is 180 in Euclidean geometry.)

Corollary A If the sides of a triangle are of length less than $\alpha/2$, then the sum of the measures of any two of its angles is less than 180.

[*Hint for proof* In order to prove $m\angle 1 + m\angle 2 < 180$ in Figure 3.30, use $m\angle 3 > m\angle 1$.]

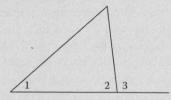

Figure 3.30 $m < 1 + m < 2 < 180$.

Corollary B Any triangle having sides of length less than $\alpha/2$ has at most one right or obtuse angle.

Corollary C   The base angles of an isosceles triangle having sides of length less than $\alpha/2$ are acute.

We define a **right triangle** in unified geometry to be any triangle having at least one right angle. By previous results, notably Corollaries **C** and **D** of Theorem **2**, Section **3.5**, triangles in unified geometry can have more than one right angle. If there is only one right angle, the side opposite is referred to as the **hypotenuse** (as usual) and the other two sides, the **legs**. A consequence of Corollary C corresponds to a familiar proposition in geometry.

Corollary D   If a right triangle has sides of length less than $\alpha/2$, there is only one right angle, and the other two angles are acute.

Note that in the case of an unbounded metric, these corollaries apply to arbitrary triangles regardless of the lengths of their sides—results that are quite familiar in Euclidean geometry. Later, we examine the possibilities regarding how long the sides of a right triangle can be relative to α in the case $\alpha < \infty$. It is an interesting exercise to establish the fact that the lengths of the sides of any right triangle in unified geometry cannot all be greater than $\alpha/2$. (Can you prove it? Try devising a plausibility argument for this on the earth's surface, using, say, the North pole and equator, then using ideas from this, write a proof for axiomatic unified geometry.)

Another result will be stated as a theorem since it is used so frequently. The proof is left as Problem 13 below.

Theorem 2   If $\angle ABC$ is acute and $AB < \alpha/2$, the foot D of A on line BC falls on ray BC.

Corollary E   In $\triangle ABC$ having sides of length less than $\alpha/2$ and has acute angles at B and C, the foot of A on line BC falls on side $\overline{BC}$.

Note: So long as they do not involve parallelism or concepts derived from it, the theorems of both Euclidean and spherical geometry are pretty much the same for all sufficiently small figures (e.g., lying in a spherical octant—a portion of the sphere bounded by a tri-rectangular triangle, as illustrated in Figure 3.26, where all distances are less than $\alpha/2$). Since the proofs are virtually the same for both geometries, with details regarding betweenness the only difference, such details will often be left to the problem sections. ⊠

3.7 Inequalities for Triangles

The exterior-angle theorem leads to several classical inequalities involving triangles. The first of these is the triangle inequality. It should be noted that this result embodies practically all the previous material developed up to this point. A chain of only the essential propositions used for this inequality would include many of the previously proven theorems as well

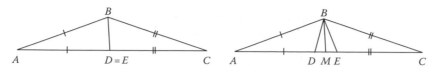

Figure 3.31 Proof of triangle inequality.

as all the axioms. It is to be especially noticed that in spite of the restrictions necessary for the exterior angle inequality used to prove the triangle inequality, *all the major inequalities established in this section are valid for all triangles.*

A preliminary triangle inequality is established first, which leads to the general case. (See Figure 3.31.) The method of proof suggested previously in Problem 25, Section **3.5** will be adapted for use here, which is valid for arbitrary α.

(1)   In $\triangle ABC$, if $AB \leq \alpha/2$ and $BC \leq \alpha/2$, then $AB + BC > AC$.

Proof   First, if both AB, $BC = \alpha/2$, then $AB + BC = \alpha > AC$. Thus, we may assume that at least one of AB, BC is less than $\alpha/2$. Suppose $AB + BC \leq AC$. Construct segment $AD \cong AB$ on $\overline{AC}$ and $CE \cong BC$ on $\overline{DC}$ with either $D = E$ or $(ADEC)$, the two cases shown in Figure 3.31. If $(ADEC)$, then it follows that $(\overrightarrow{BA}\ \overrightarrow{BD}\ \overrightarrow{BE}\ \overrightarrow{BC})$ (why?). In either case, $\triangle ABD$ and $\triangle BEC$ are isosceles triangles with congruent base angles and by **(1)**, Section **3.6**, BD, $BE < \alpha/2$. By the isosceles triangle theorem and the exterior angle inequality, we obtain

$$m\angle ABC \geq m\angle ABD + m\angle EBC = m\angle ADB +$$

$$m\angle BEC \geq m\angle ADB + m\angle BDC = 180,$$

That is, $m\angle ABC \geq 180$, a contradiction. (If $DE \geq \alpha/2$, construct the midpoint M of segment DE and apply the exterior angle inequality to triangles BME and BDM to obtain $m\angle BEC > m\angle BMC > m\angle BDC$.) ∖

In order to extend this result to arbitrary triangles for a bounded metric, several cases involving the possible lengths of the sides must be considered. (See Problem 18 for details.) Thus, the result **(1)** may be proven for all triangles, and in general, if A, B, and C are not collinear, then $AB + BC > AC$. If A, B, and C are collinear, it is found that $AB + BC \geq AC$. (One can use the general theorem on collinear points in Section 1.8 to show that if (ABC) does not hold, the strict inequality holds. This makes a good problem for you; see Problem 19.) This analysis then establishes:

Triangle Inequality   If A, B, and C are any three distinct points in unified geometry, then $AC \leq AB + BC$, with equality only when (ABC) holds.

The triangle inequality is the embodiment of the ancient geometric precept:

A line is the shortest distance between two points

where *distance* is assumed to be taken along any arc or curve in the plane. An explicit proof of this for a polygonal path is easily obtained.

Polygonal Inequality   For any positive integer $n \geq 3$, if $P_1, P_2, \ldots, P_n$ are any n points, then $P_1 P_n \leq P_1 P_2 + P_2 P_3 + P_3 P_4 + \cdots + P_{n-1} P_n$.

Proof   By repeated use of the triangle inequality, we obtain

$$P_1 P_3 \leq P_1 P_2 + P_2 P_3$$

$$P_1 P_4 \leq P_1 P_3 + P_3 P_4 \leq (P_1 P_2 + P_2 P_3) + P_3 P_4$$

$$P_1 P_5 \leq P_1 P_4 + P_4 P_5 \leq (P_1 P_2 + P_2 P_3 + P_3 P_4) + P_4 P_5 \ldots$$

and so on, the process ending (by mathematical induction) in the result we wanted to prove. ◣

One of Euclid's propositions (I, 18) is concisely put:

In any triangle the greater side subtends the greater angle.

This implied inequality will be indispensable in our work with triangles and quadrilaterals. Euclid uses Proposition (I, 18) to prove its converse by a simple elimination process, which we also use. In order to give a proof that applies to all triangles, however, the order of Euclid's two propositions will be reversed, and the triangle inequality will be used.

In stating these inequalities, it will be convenient to use standard notation, where, for example, A, B, and C denote the measures of $\angle BAC$, $\angle ABC$, and $\angle ACB$, respectively, of a given triangle ABC. It will also be convenient to introduce an inequality for angles and segments: $\angle A > \angle B$ iff $m\angle A > m\angle B$, and $\overline{BC} > \overline{AC}$ iff $BC > AC$. The first result then becomes

(2)   If $A > B$,   then $a > b$

Proof   Let $A > B$ in $\triangle ABC$. We must prove that $a > b$, or, in Figure 3.32, that $BC > AC$. Construct point D on side BC such that $\angle DAB \cong \angle B$. Thus, by the isosceles triangle theorem and the triangle inequality, we have

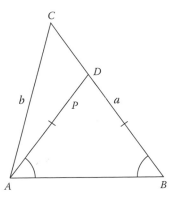

Figure 3.32 Proof of scalene inequality.

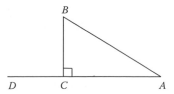

Figure 3.33 Scalene inequality applied to a right triangle.

$$BC = BD + DC = AD + DC > AC \ \text{\large$\searrow$}$$

There are two immediate corollaries, the first of which is the converse of **(2)**.

(3)   If $a > b$,   then $\angle A > \angle B$

Proof   Suppose $a > b$. By the trichotomy, either $A > B$, $A = B$, or $A < B$. But the cases $A \le B$ imply that either $a = b$ (isosceles triangle theorem) or $a < b$ (inequality **(2)**), both contradicting the hypothesis $a > b$. Hence, we conclude $A > B$, as desired. $\searrow$

It is convenient to combine these two results into the single Euclidean-like proposition which will be used repeatedly from now on.

Scalene Inequality   In any $\triangle ABC$, $BC > AC$ iff $m\angle A > m\angle B$.

Using the scalene inequality, Corollary **D**, Section **3.6** can be extended to the following (see Figure 3.33 for the proof, where $\angle DCB$ is an exterior angle of $\triangle ABC$):

> ***Theorem 1***   If the sides of a right triangle ABC with right angle at C are each less than $\alpha/2$, the angles at A and B are acute, and the hypotenuse AB is the longest side.

For the case $\alpha = \infty$, this theorem applies to all right triangles. For finite α, we proceed to consider those cases when the "legs" AC and BC of right triangle ABC are either $< \alpha/2$ or $\ge \alpha/2$. As we shall see, in most situations, the "hypotenuse" AB is at least as large as either "leg" (terms which are not uniquely defined when there is more than one right angle). Refer to Figure 3.34 for the following analysis.

1. If BC and AC are both of length $< \alpha/2$, locate D such that (CAD) and $CD = \alpha/2$; the cevian inequality **(1)** Section **3.6** then implies $AB < \alpha/2$, and Theorem 1 shows that the angles at A and B are acute and AB is the longest side.
2. If $BC < \alpha/2$ and $AC = \alpha/2$, by Corollary **B**, Section **3.5**, $AB = \alpha/2 = AC > BC$ and $\angle CBA$ is a right angle. By the scalene inequality, $m\angle CAB < m\angle C = 90$.

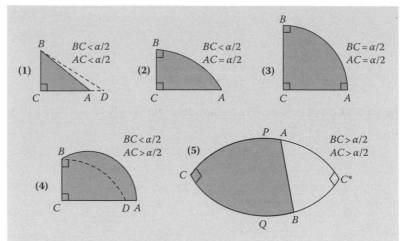

Figure 3.34 The five cases of Theorem 1.

3. If both BC and AC are of length $\alpha/2$, then again by Corollary **B**, Section **3.5**, $AB = \alpha/2 = AC = BC$, and all three angles are right angles, the case of a tri-rectangular triangle.

4. If $BC < \alpha/2$ and $AC > \alpha/2$, locate D such that (CDA) and $CD = \alpha/2$; then in $\triangle BCD$, case (2) above shows that $BD = \alpha/2$ and $m\angle CBD$ is a right angle. Since (CDA), $(\overrightarrow{BC}\ \overrightarrow{BD}\ \overrightarrow{BA})$ follows and $m\angle CBA > m\angle CBD = 90$. By the scalene inequality, $CA < AB$. Now if $AB \leq \alpha/2$, the cevian inequality implies $BD < \alpha/2$, a contradiction, proving that $AB > \alpha/2$. Thus, $BC < AB < AC$; by the scalene inequality, $m\angle BAC < m\angle C = 90$.

5. If both AB and AC are $> \alpha/2$, locate C^* the extremal opposite of point C; then $\triangle ABC^*$ has legs AC^* and $BC^* < \alpha/2$ and by Case (1) applied to $\triangle ABC^*$ the angles at B and C of are acute, so that $\angle CAB$ and $\angle CBA$ are both obtuse. Also, $AB < \alpha/2 < AC, BC$.

Thus, we have proven

> **Theorem 2** Let $\triangle ABC$ be any right triangle with right angle at C. Then the following is true:
>
> **(a)** If $AC, BC < \alpha/2$, then $AB < \alpha/2$, AB is the longest side, and the angles at A and B are both acute.
>
> **(b)** If $AC = \alpha/2$ and $BC \leq \alpha/2$, then $AB = \alpha/2$ and $\angle B$ is a right angle; $\angle A$ is acute if $BC < \alpha/2$. If $BC = \alpha/2$, then $AB = AC = BC = \alpha/2$ and $\triangle ABC$ is a tri-rectangular triangle.
>
> **(c)** If $AC > \alpha/2$ and $BC < \alpha/2$, then $AB > \alpha/2$, $\angle A$ is acute, $\angle B$ is obtuse, and $BC < AB < AC$.
>
> **(d)** If $AC > \alpha/2$ and $BC > \alpha/2$, then $AB < \alpha/2 < AC, BC$, and the angles at A and B are obtuse.

Corollary If two sides of a right triangle are of length less than $\alpha/2$, the third side is also of length less than $\alpha/2$. Alternatively, if one side of a right triangle is greater than $\alpha/2$, then exactly one other side is greater than $\alpha/2$, and the third side is less than $\alpha/2$.

The last result for this section is sometimes referred to as the "alligator" or "hinge" theorem, because it indicates that *the larger the angle, the wider the opening*. We call it the *SAS Inequality*, because, like the SAS congruence theorem, it involves two different triangles.

SAS Inequality Suppose that two triangles have a pair of corresponding sides congruent. If the included angles in the two triangles are unequal, the side opposite the greater angle is the greater side, and conversely.

Proof

1. Suppose that in $\triangle ABC$ and $\triangle XYZ$, we have $AB = XY$, $AC = XZ$, but $\angle A > \angle X$. Then, we want to prove $BC > YZ$. As shown in Figure 3.35, construct ray AD between rays AB and AC such that $\angle BAD \cong \angle X$, with $AD = XZ$. By SAS, $\triangle ABD \cong \triangle XYZ$ and $BD = YZ$. We will be finished if we can show that $BC > BD$. Suppose that the bisector AE of $\angle DAC$ has been constructed, as in Figure 3.35, with E on $\overline{BC}$. [Note that $(\overrightarrow{AB}\ \overrightarrow{AD}\ \overrightarrow{AC})$ and $(\overrightarrow{AD}\ \overrightarrow{AE}\ \overrightarrow{AC})$ are valid and imply $(\overrightarrow{AB}\ \overrightarrow{AD}\ \overrightarrow{AE}\ \overrightarrow{AC})$ and (BEC).] Since $AD = XZ = AC$, by SAS, $\triangle ADE \cong \triangle ACE$ and $DE = EC$. By the triangle inequality, we have

$$BC = BE + EC = BE + ED > BD = YZ.$$

2. [*Converse*] Suppose $BC > YZ$; we must prove that $\angle A > \angle X$. By the trichotomy, either $m\angle A < m\angle X$, $m\angle A = m\angle X$, or $m\angle A > m\angle X$. Thus, either $BC < YZ$ (by the first part), $BC = YZ$ (by congruent triangles) or $\angle A > \angle X$. Since the first two possibilities contradict the hypothesis ($BC > YZ$), we must have $\angle A > \angle X$, as desired. ◥

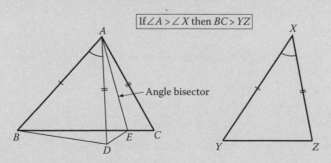

Figure 3.35 SAS inequality.

Problems (Section 3.7)

Group A

1. Using the exterior angle inequality, if (BDC) and the sides of the triangle are less than $\alpha/2$, the range of the variable x is $61 < x < 133$. (Figure P.1)

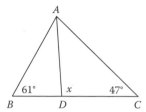

Figure P.1

2. Using the triangle inequality, find upper and lower bounds for x in Figure P.2 for this problem as $m\angle A$ varies. (Note that $x + 47 > 61$ by the triangle inequality.)

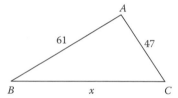

Figure P.2

3. In Figure P.3, order the quantities x, y, and z from least to greatest.

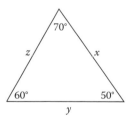

Figure P.3

4. In Figure P.4, order the quantities x, y, and z from least to greatest.

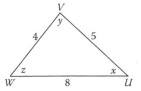

Figure P.4

5.   Name the longest and shortest segments in Figure P.5.

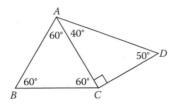

Figure P.5

6.   Given $m\angle BAC < 90$ the sides of $\triangle ABC$ as marked, where $6 < a/2$, name the smallest and largest angles (Figure P.6).

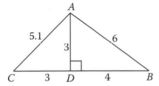

Figure P.6

7.   In Figure P.7 for this problem, which angle is smallest?

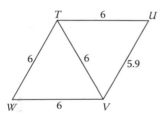

Figure P.7

8.   In Figure P.8, which of the angles at A, B, and C is the largest?

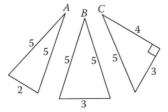

Figure P.8

9.   Using the inequality theorems of this section, find the maximum and minimum lengths of the sides of the two triangles in Figure P.9. (To show that $IJ < LK$, see figure for a hint.)

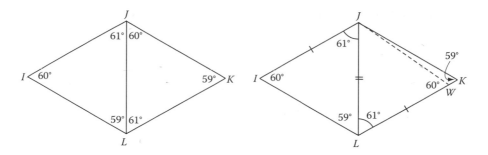

Figure P.9

10. Which of the following number triples can be the lengths of the sides of a triangle in unified geometry?
    (a) 8, 5, 12
    (b) 6, 14, 21
    (c) 300, 1, 1

11. If two sides of a triangle have lengths 1 and 250, what are the lower and upper bounds for the length of the third side?

Group B

12. **Concept Used in Proof of Exterior Angle Inequality**
    Complete the details in the following proof of the cevian inequality (**1**). Let the cevian be AD, where $D \in \overline{BC}$ and suppose $AB < \alpha/2$, $AC \leq \alpha/2$.
    (1) Extend segments AB and AC to points B' and C', respectively, such that $AB' = AC' = \alpha/2$. By hypothesis, (ABB') and either (ACC') or $C' = C$. Supply details validating this.
    (2) Ray AD meets segment $B'C'$ at some point D'. (Why?) Similarly, ray AD meets segment $B'C$ at some point E (needed in Step 3 below).
    (3) It follows that (ADD'). [Show that (ADE) from (ABB') and, if $C' \neq C$ (AED') from (ACC').]
    (4) Since $AD' = \alpha/2$, then $AD < \alpha/2$.

13. Prove Theorem **2**, Section **3.6** and Corollary **E**, making use of the exterior angle inequality (Figure P.13).

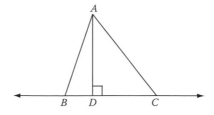

Figure P.13

14. Using the inequality theorems, give a one-line proof of the SSS congruence criterion.

15. If you are told that in $\triangle ABC$, the angles at B and C are right angles and that $BC = \alpha/4$, could you deduce the value of $m\angle A$?

16.* In Problem 15, if $BC = p\alpha$, where p is a dyadic rational, could you deduce the value of $m\angle A$?

17. A Simple Proof for the Triangle Inequality If $\alpha = \infty$, a remarkably elegant proof of the triangle inequality can be had by using the inequalities of this section. In $\triangle ABC$, let $\overrightarrow{BD}$ be the bisector of $\angle B$, as shown in Figure P.17. Then, $AB + BC > AD + DC$. (Why?) Complete the details to arrive at the inequality $AB + BC > AC$.

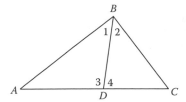

Figure P.17

18. Extension of Triangle Inequality to Arbitrary Triangles Starting with an arbitrary triangle ABC, complete the details in the following proof outline of the triangle inequality, $AB + BC > AC$:

(1) If AB and BC are both $\leq \alpha/2$, **(1)** holds. So assume $AB > \alpha/2$ or $BC > \alpha/2$. (If both, the inequality follows without further proof—why?)

(2) Case 1: $AB > \alpha/2$, $BC \leq \alpha/2$. If $AC \leq \alpha/2$, the inequality follows, so assume that $AC > \alpha/2$. With A^* the extremal opposite of A, as shown in Figure P.18, we obtain for $\triangle A^*BC$, with $A^*B < \alpha/2$, $BC \leq \alpha/2$, and $CA^* < \alpha/2$: $BC + CA^* > A^*B$, or $BC + \alpha - AC > \alpha - AB$. Simplify.

(3) Case 2: $A'B \leq \alpha/2$, $BC > \alpha/2$. Again, assume that $AC > \alpha/2$. Let C^* be the extremal opposite of C. Then, in $\triangle ABC^*$, $AB + AC^* > BC^*$ or $AB + \alpha - AC > \alpha - BC$.

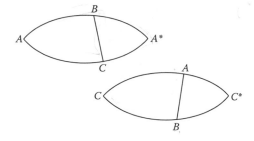

Figure P.18

19. Prove that if A, B, and C are any three distinct collinear points such that (ABC) does not hold, then $AB + BC > AC$. Use this to complete the proof of the triangle inequality.

20. Suppose $\triangle ABC$ is a right triangle having sides of length $< \alpha/2$ and right angle at C, with $m\angle B \geq 45$. If D is located on the hypotenuse such that $AD = AC$, show that $DB \leq DC$. (See Figure P.20.) [**Hint:** Prove that $\angle 1 < \angle 4$.]

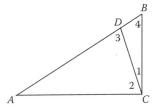

Figure P.20

21. Given: $\triangle ABC$ is equilateral, with sides of length less than $\alpha/2$. If F lies on side BC, prove that $AC > AF > FC$. (See Figure P.21.)

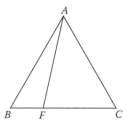

Figure P.21

Group C

22.* Suppose that in triangles ABC and XYZ each having sides less that $\alpha/2$, we have $\angle A \cong \angle X$, $\angle B \cong \angle Y$, but $AB < XY$. Prove that $AC < XZ$ and $BC < YZ$. [**Hint:** As shown in Figure P.22, construct point W on side XY such that $XW = AB$, and ray WP such that $\angle XWP \cong \angle B \cong \angle XYZ$. Why must ray WP meet segment XZ at an interior point?] **Note:** This result is useful in proving the Steiner–Lehmus theorem for unified geometry. (See Problem 25).

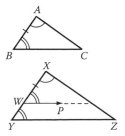

Figure P.22

23. The result of this problem dispels and erroneous belief that trisecting one side of a triangle will produce a trisection of the angle opposite—often naively proposed as a compass/straight-edge construction for trisecting an angle. With $\alpha = \infty$, prove that in Figure P.23 shown, with congruent segments as marked, $\angle 1 = \angle 3 < \angle 2$. [**Hint:** Construct segment AF on AD congruent to segment AL; why is $FL < DL$?]

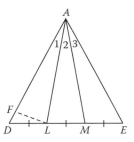

Figure P.23

24. Prove the following theorem (used later in a proof involving polygons, Section 4.4):

> ***Theorem 3*** If two sides of a triangle are each of length less than $\alpha/2$, and the included angle is acute, then the length of the third side is also less than $\alpha/2$ (Figure P.24).

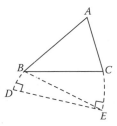

Figure P.24

(*Hint for proof* Extend the two given sides to a length equal to $\alpha/2$; this forms an isosceles triangle in which the legs are of length $\alpha/2$, the base angles are right angles, and the base is of length less than $\alpha/2$ (Why?). Now use **(1)**, Section **3.6** twice.)

25. Steiner–Lehmus Theorem in Unified Geometry *If the bisectors of two angles of a triangle having sides of length less than $\alpha/2$ are congruent, the triangle is isosceles.* (This is the converse of a routine problem in elementary geometry, which you should try for warm-up.) The following steps provide a proof of this theorem for unified geometry; you are to fill in the details, as indicated. (See Figure P.25.)

(1) Suppose bisectors $\overline{BD}$ and $\overline{CE}$ are congruent, but that $AC > AB$. Hence, $m\angle ABC > m\angle ACB$ and $m\angle ABD > m\angle ACE$. Give details and reasons.

(2) Construct $\angle FBD \cong \angle ACE$, with (AFD). (Prove that such a point F exists.)

(3) Segments BF and CE intersect at an interior point G on each segment. (Prove this.)

(4) $FB < FC$ by Scalene Inequality. (Provide missing details.)

(5) Observe triangles CFG and BDF and show that $BD < CG$, thus contradicting the hypothesis $CE = BD$. (For this part, use the result of Problem 22 above.)

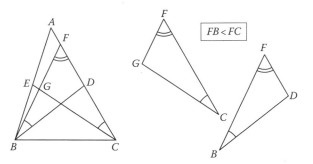

Figure P.25

3.8 Further Congruence Criteria

The congruence criterion indicated by AAS (two angles and a side opposite one of the angles) will now be established. The usual proof of this theorem is based on the Euclidean concept of the angle sum of a triangle (=180) to conclude that if two angles of one triangle are congruent to two angles of another, then the third pair of angles must be congruent, and to apply the SAS criterion. But this proof is not valid for spherical geometry since, for example, there exist triangles with three right angles, and the angle sums of right triangles can vary at least from 270 to 360 (can you see why?). One must find a proof that is valid in unified geometry.

AAS Congruence Criterion   For any two triangles having sides less than $\alpha/2$, if two angles and any side of one triangle are congruent, respectively, to the corresponding two angles and side of the second, the triangles are congruent.

Proof   Suppose in $\triangle ABC$ and $\triangle XYZ$, we have $\angle A \cong \angle X$, $\angle B \cong \angle Y$, and $BC = YZ$, as shown in Figure 3.36. The conclusion is trivially true by ASA if $\angle C \cong \angle Z$, so assume $\angle C > \angle Z$ (the proof for the case $\angle C < \angle Z$ is logically equivalent). Then, construct $\angle DCB \cong \angle Z$ such that (ADB). By ASA, $\triangle DBC \cong \triangle XYZ$ and $\angle BDC \cong \angle X \cong \angle A$. But by the exterior angle inequality, $\angle BDC > \angle A$, a contradiction. ◣

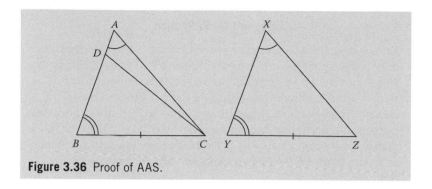

Figure 3.36 Proof of AAS.

A list of the possible combinations of three letters from the two-element set {A, S} may be made (AAA, AAS, ASA,..., SSS) each of which represent a potential congruence criterion. Eliminating triples that duplicate each other (such as SAA, which duplicates the equivalent criterion AAS), there are a total of 6 of these combinations. Four of these have already been considered and represent valid congruence criteria (as we have seen, SAS or **Axiom 12**, implies ASA, SSS, and AAS). The two remaining are those represented by AAA and SSA, illustrated in Figure 3.37.

The symbol AAA implies a congruence criterion involving just the *three pairs of congruent angles* in two triangles. This is clearly false in Euclidean geometry since we can have two *similar* triangles, one that is twice the size of another, yet having the three pairs of corresponding angles congruent. So we can eliminate AAA as a criterion for unified geometry.

The remaining criterion SSA leads to a more complicated situation. This implied criterion involves *two sides* and an *angle opposite one of them* in each of two triangles. Attempting to prove this criterion is related to the well-known *ambiguous case* in trigonometry, where one considers a hypothetical triangle in which two sides and an angle opposite one of them are given, and it is required to solve for the remaining parts of the triangle, if it exists. There can be either no solution, exactly one solution, or two solutions. This fact indicates that SSA also cannot represent a criterion for congruence in general. But under certain restrictions, it does lead to a qualified basis for congruence. A special case gives us the well-known HL congruence criterion for right triangles.

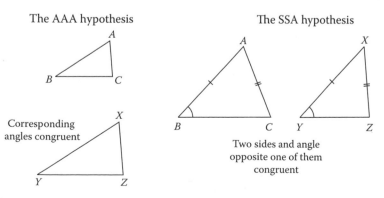

Figure 3.37 Two possible congruence criteria.

Moment for Reflection

Try constructing a counterexample to the proposed SSA congruence criterion using the example shown in Figure 3.37 as a starting point. Does the diagram suggest that an isosceles triangle is involved in some way? A common illustration of the ambiguous case in trigonometry is shown here, where an arc of a circle is drawn, missing the opposite side of the hypothetical triangle. (Could there be more than one value for c, and why?) These ideas should help you obtain an example where the SSA hypothesis fails to produce congruent triangles.

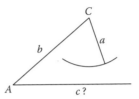

SSA Congruence Theorem   In triangles ABC and XYZ, if $AB = XY$, $AC = XZ$, and $\angle B \cong \angle Y$, then either $\angle C$ and $\angle Z$ are supplementary, or $\triangle ABC \cong \triangle XYZ$.

Proof   (See Figure 3.38.) If the triangles are congruent, there is nothing to prove, so suppose $BC \neq YZ$, and for sake of argument, assume that $BC > YZ$. Construct segment BD on segment BC congruent to segment YZ, with (BDC). Then, by SAS, $\triangle ABD \cong \triangle XYZ$, and $AD = XZ = AC$. Thus, $\triangle ADC$ is isosceles, and $m\angle C = m\angle ADC = 180 - m\angle ADB = 180 - m\angle Z$, proving that $\angle C$ and $\angle Z$ are supplementary. ◣

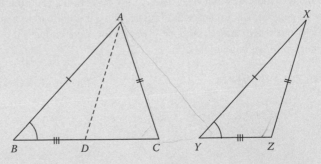

Figure 3.38 Proof of special SSA congruence.

The following is a corollary of the SSA congruence criterion involving right triangles—the one mentioned above. Note that all three of the congruence theorems for right triangles require the sides to be less than $\alpha/2$.

HL Congruence Criterion for Right Triangles   For any two right triangles having sides less than $\alpha/2$, if the hypotenuse and leg of one are congruent, respectively, to the corresponding hypotenuse and leg of the other, the triangles are congruent.

Proof   The hypothesis is shown for two right triangles (Figure 3.39). Since the sides are less than $\alpha/2$, the angles at A, C, X, and Z are acute, so there can be no supplementary pairs. Comparing Figure 3.39 to Figure 3.38, the hypothesis satisfies that of SSA, hence we conclude $\triangle ABC \cong \triangle XYZ$. ⬙

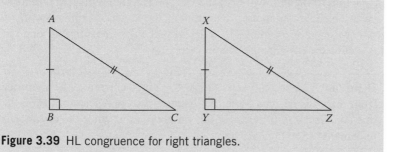

Figure 3.39 HL congruence for right triangles.

The other two congruence theorems for right triangles are simple corollaries of the AAS congruence criterion, illustrated in Figures 3.40 and 3.41, so they are stated without proof.

LA Congruence Criterion for Right Triangles   For any two right triangles having sides less than $\alpha/2$, if a leg and the angle opposite in one triangle are congruent, respectively, to the corresponding leg and angle opposite of the other triangle, the triangles are congruent.

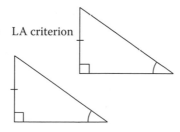

Figure 3.40 LA congruence.

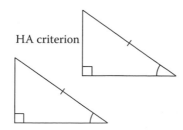

Figure 3.41 HA congruence.

HA Congruence Criterion for Right Triangles   For any two right triangles having sides less than $\alpha/2$, if the hypotenuse and adjacent angle in one triangle are congruent, respectively, to the corresponding hypotenuse and adjacent angle of the other triangle, the triangles are congruent.

Moment for Reflection

Evidently, counterexamples can be constructed for HL, LA, and HA if the restrictions on the sides be removed. Try doing so. (See Problem 6.)

An important application of the HL and HA theorems involves a classical problem in geometry: determining the locus of points *equidistant from the sides of an angle*. In order to make this meaningful, one must first define the distance from a point to a ray.

Definition   If h is any ray and A any point not on h, the **distance** from A to h is defined to be the distance d from A to a foot F on the line containing h, provided F lies on ray h; otherwise, the distance is undefined.* (See Figure 3.42.)

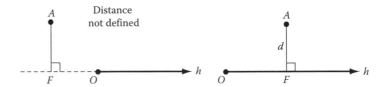

Figure 3.42 Distance from a point to a ray.

Now consider any nondegenerate, nonstraight angle, $\angle ABC$, and its bisector $\overrightarrow{BD}$ (Figure 3.43). If P is any point on the bisector, let Q be a foot of P on line AB. By the theorem stated in Problem 24, Section **3.7**, Q lies

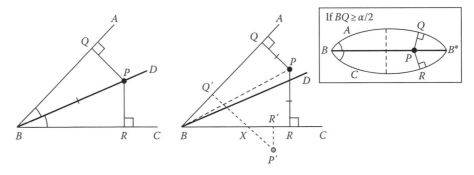

Figure 3.43 Equidistant locus.

---

* One could take $d = AO$ in this case, where O is the endpoint of h (or $d = AO^*$ if $\alpha < \infty$), in agreement with the metric space concept (the minimal distance from a point to a set). However, leaving this distance undefined makes the geometry problem simpler.

on ray *BA* (since ½∠*ABC* < 90) and the distance from *P* to ray *BA* is *PQ*. Similarly, the foot *R* of *P* on line *BC* lies on ray *BC*, and the distance from *P* to ray *BC* is *PR*. You should be able to finish the argument from this point on to show that *PQ* = *PR*, and thus *P* is equidistant from the sides of the angle. What congruence theorem was used? The first half of the following theorem has now been proven:

> **Angle-Bisector Theorem**   The locus of points equidistant from the sides of a nondegenerate, nonstraight angle is the angle bisector.

In order to finish the proof, try proving that if *P* is equidistant from the sides of ∠*ABC*, it must lie in the interior of the angle and also on its bisector. (See diagram on right in Figure 3.43; this appears as Problem 18 below.)

Problems (Section 3.8)

Group A

1. A different way to construct a counterexample to an unrestricted SSA criterion is shown in Figure P.1, where △*PQR* is isosceles and $\overline{PS}$ is drawn to base $\overline{QR}$ such that *S* is not the midpoint of $\overline{QR}$. Explain how this counterexample works. Are the supplementary angles in Figure P.1 the ones mentioned in the SSA theorem?

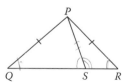

Figure P.1

2. *Given*: ∠*A* ≅ ∠*C* and ∠*ADB* ≅ ∠*CDB* (Figure P.2). *Prove*: $\overrightarrow{BD}$ bisects ∠*ABC*. Assume betweenness relations from the figure.

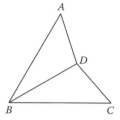

Figure P.2

3. *Given*: $\overrightarrow{FP} \perp \overrightarrow{PB}$, $\overrightarrow{FQ} \perp \overrightarrow{BQ}$ and $\overrightarrow{BF}$ bisects ∠*PBQ* (Figure P.3). *Prove*: $\overline{FP} \cong \overline{FQ}$. (Assume betweenness relations from the figure.)

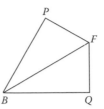

Figure P.3

4. *Given*: $\angle X$ and $\angle Y$ are right angles and $GX = GY$ (Figure P.4). *Prove*: $\overrightarrow{BG}$ bisects $\angle XBY$. Assume betweenness relations from the figure.

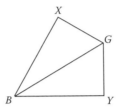

Figure P.4

5. In elementary geometry texts in which the SSA criterion is not included, a different proof of the HL congruence criterion is required. Find such a proof. (See Figure P.5 for a hint.)

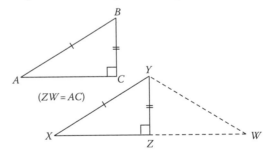

Figure P.5

6. Show that pairs of isosceles triangles whose base angles are right angles such as those shown in Figure P.6 constitute counterexamples for each of the congruence criteria established in this section.

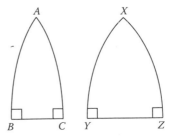

Figure P.6

7. Explain why the restricted SSA criterion cannot be used to prove that $\triangle ABD \cong \triangle ACD$ in Figure P.7.

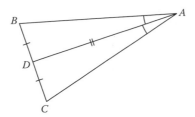

Figure P.7

8.   If two distinct lines have a common perpendicular, then they possess a pair of acute congruent alternate interior angles with respect to some transversal. To prove this, assume that line *AB* is the common perpendicular and that *M* is the midpoint of segment *AB*. Observe line *CD* in Figure P.8, passing through point *M*. Assume $\alpha = \infty$.

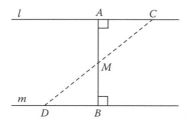

Figure P.8

9.   Prove that if two lines have a pair of acute congruent alternate interior angles with respect to some transversal, they have a common perpendicular—the converse of the theorem of Problem 8. See Figure P.9 (assume $\alpha = \infty$).

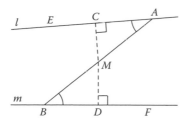

Figure P.9

10.   In Figure P.10, *M* is the midpoint of segment $\overline{AB}$, and the angles at *C* and *D* along transversal $\overleftrightarrow{CD}$ are congruent as marked. Then if (*CMD*), prove that *M* is also the midpoint of segment $\overline{CD}$.

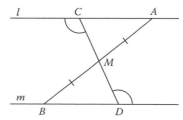

Figure P.10

11. In elementary geometry texts ($\alpha = \infty$), the scalene inequality is presented before the triangle inequality and proven in that order using the exterior-angle inequality. Reconstruct the proofs here. [**Hint:** In Figure P.11 prove that if $AC > BC$, then $\angle CBA > \angle CAB$. To prove the triangle inequality, construct point E such that (ACE) and $CE = CB$, then show that $\angle ABE > \angle E$.]

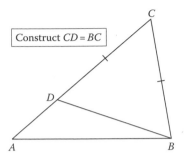

Construct $CD = BC$

Figure P.11

Group B

12. Show that the only situation in which a line can bisect both a side of a triangle and the angle opposite is when the triangle is isosceles. Assume $\alpha = \infty$.

13. Prove that the angle sum for triangles is 180 iff it is a constant for all triangles. [**Hint:** Study Figure P.13 to obtain two independent expressions involving the numbered angles as marked and the assumed constant angle sum γ for all triangles.]

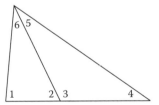

Figure P.13

14. In Figure P.14, $KF = FL$ and $\angle KML \cong \angle LNK$. Assuming the betweenness relations from the figure, prove:
   (a) $KM = NL$
   (b) $\angle KNM \cong \angle LMN$.

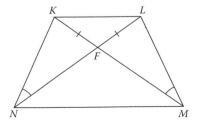

Figure P.14

15. Find an example of two triangles on the unit sphere lying in the same hemisphere whose angle sums are different. You may use an intuitive argument, but be as precise as you can.

16. Prove the following congruence criterion for triangles having two pairs of congruent sides and one pair of congruent angles, as indicated in Figure P.16. (This theorem appears in at least one high school geometry text.)

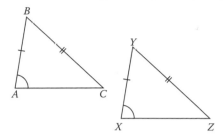

Figure P.16

SsA Congruence Criterion Suppose that in triangles ABC and XYZ, $\overline{AB} \cong \overline{XY}$, $\overline{BC} \cong \overline{YZ}$, $\angle A \cong \angle X$, and $BC \geq BA$. Then $\triangle ABC \cong \triangle XYZ$.

17. If $\triangle ABC$ is isosceles with legs $\overline{AB}$ and $\overline{AC}$ suppose that $\overline{BE} \perp \overline{AC}$ and $\overline{CF} \perp \overline{AB}$ Prove that ray AG bisects $\angle BAC$. (See Figure P.17.)

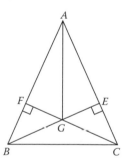

Figure P.17

Group C

18. Provide the remainder of the proof of the angle-bisector theorem by showing that if $P'R' = P'Q'$ in Figure 3.43 then $P' \in \text{Int} \angle ABC$, and that $\angle PBQ \cong \angle PBR$. Which congruence theorem was used this time?

3.9 Special Segments Associated with Triangles

It is the purpose of this section to formally introduce a few of the segments commonly associated with triangles in elementary geometry and to extend these ideas to unified geometry.

Definition 1   In triangle ABC, let D, E, and F be the feet of the respective vertices A, B, and C on the opposite (extended sides); L, M, and N the respective *midpoints* of the sides opposite A, B, and C; and X, Y, and Z the respective points where the *angle bisectors* of the angles of $\triangle ABC$ meet the sides. With these constructions having been made, the segments

$$\overline{AD}, \overline{BE}, \overline{CF}$$

are called the **altitudes** of $\triangle ABC$,

$$\overline{AL}, \overline{BM}, \overline{CN}$$

are called the **medians** of $\triangle ABC$, and

$$\overline{AX}, \overline{BY}, \overline{CZ}$$

are called the **angle bisectors** of $\triangle ABC$. (See Figure 3.44.) In addition, consider the **perpendicular bisectors** l, m, and n of the sides of $\angle ABC$, defined in Section 3.5.

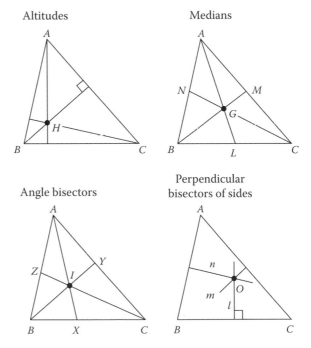

Figure 3.44 Special segments for $\triangle ABC$.

In Euclidean geometry each of the above four triples of lines are *concurrent*, and the four resulting points of concurrency each have interesting, significant properties of their own. For example, the points of concurrency of the altitudes (H), the medians (G), and the perpendicular bisectors of the sides (O) can be shown to lie on a line, called the **Euler line** of the triangle (this will be proved later for Euclidean geometry). In non-Euclidean geometry, the concurrency property for these sets of lines is true in general only for the angle bisectors and the medians—the reason for this departure from Euclidean geometry will be made clear in due time. It is premature to attempt a proof that the medians are concurrent in non-Euclidean geometry; that will also occur later, in Chapter 9.

The theorem for the angle bisectors of a triangle requires the concept of the interior of a triangle introduced previously. (Recall that this is the set of points which lie in the interiors of all three angles of the triangle, $\angle A$, $\angle B$, and $\angle C$.)

Theorem 1   The angle bisectors of a triangle with sides $< \alpha/2$ are concurrent in a point I that lies in the interior of the triangle and is equidistant from the sides.

Proof   Consider $\triangle ABC$ and its angle bisectors $\overline{AX}$ $\overline{BY}$ and $\overline{CZ}$ (Figure 3.45). It follows by the crossbar principle that since these bisectors are interior to the angles they bisect, any two bisectors meet at an interior point of each bisector and at an interior point of $\triangle ABC$. Suppose AX and BY meet at point I. Let R, S, and T be the feet of I on the sides of the triangle, as shown in Figure 3.45. By the angle-bisector theorem, I is equidistant from sides AB and AC; thus, $IS = IT$; I is also equidistant from sides AB and BC, and thus $IT = IR$. Hence $IR = IS$ and I is equidistant from sides BC and AC and thus lies on the third bisector, CZ. ◣

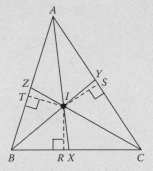

Figure 3.45 Existence of incenter.

Definition 2   The point of concurrency *I* of the three angle bisectors of a triangle is called the **incenter** of the triangle. Being equidistant from the three sides of the triangle, *I* is the center of a circle that is tangent to the sides at the points *R*, *S*, and *T* (Figure 3.46) the latter called the **incircle** of the triangle. (A formal treatment of circles and tangents will appear in the next chapter.)

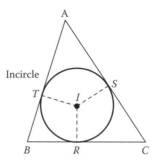

Figure 3.46 Incenter and incircle.

Theorem 2   The perpendicular bisectors of the sides of any triangle are concurrent at a point *O* that is equidistant from the vertices, provided any two of them intersect.*

(*Hint for proof*   Use a pattern analogous to that of Theorem 1, to be left as Problem 13.)

Definition 3   The point of concurrency of the perpendicular bisectors of the sides of a triangle is called the **circumcenter** of the triangle. Being equidistant from the vertices of the triangle, it is the center of a circle passing through the vertices, called its **circumcircle**. (See Figure 3.47.)

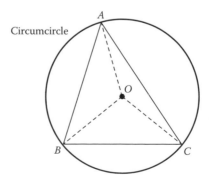

Figure 3.47 Circumcenter and circumcircle.

---

* In unified geometry, the perpendicular bisectors need not intersect. (See Chapter 5.)

Problems (Section 3.9)

Group A

1. Show that the four points of concurrency H, G, I, and O coincide for an equilateral triangle.

2. Prove that if any two of the points H, G, I, and O coincide, the triangle is equilateral.

3. Prove that the medians are concurrent in any isosceles triangle, and similarly for the altitudes.

4. In Figure 3.46, prove that $AT = AS$, $BT = BR$, and $CR = CS$.

5. Figure P.5 shows $\triangle ABC$ and its incircle, where Q, R, and S are the points of contact with the sides of the triangle. Use the result of Problem 4 to prove in standard notation that $AT = AS = s - a$, $BT = BR = s - b$, and $CR = CS = s - c$. [Recall that s is the semiperimeter $= \frac{1}{2}(a + b + c)$.]

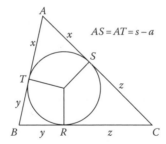

Figure P.5

6. In Chapter 4, we find that the tangent line to a circle at point P is the *perpendicular to the radius* joining P and the center. Using this, prove the standard result (Figure P.6) that if lines AP and AQ are tangent to circle C at P and Q, then $AP = AQ$. (See figure for a hint.)

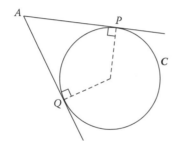

Figure P.6

7. Suppose that the pentagon *ABCDE* is circumscribed about circle *C*, as shown in Figure P.7, and that *AB* = 7, *BC* = 10, *CD* = 12, and *DE* = 8. If *AP* = 3, find *AE*, using the betweenness relations evident from the figure. [**Hint:** Use the result of Problem 6 to conduct a "walk" about the polygon along its sides, finding segment lengths as you go.]

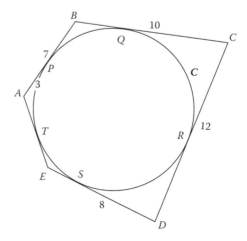

Figure P.7

Group B

8. Prove that if the sides of a polygon *ABCDE* (formal proof in Chapter 4) are each tangent to a circle at points that are equally spaced on the circle, then *ABCDE* is regular (sides congruent and angles congruent).

9. The medians to the legs of an isosceles triangle are congruent. Prove.

10. Prove that if a triangle has sides less than $\alpha/2$, two altitudes are congruent iff the triangle is isosceles.

11. *Sketchpad Experiment 6* in Appendix **A** explores the collinearity of the concurrency points *H*, *G*, and *O* of the altitudes, medians, and perpendicular bisectors of the sides in Euclidean geometry.

Group C

12. Using properties of parallelograms, prove in Euclidean geometry that if two medians of a triangle are congruent, the triangle is isosceles.

13. Prove that the perpendicular bisectors of the sides of a triangle are concurrent if any two of them intersect.

14. Prove the following proposition, the converse of the Corollary C of the exterior inequality theorem: *If the base angles of an isosceles triangle are acute, the legs are each less than α/2.*

15. Is the segment joining the midpoints of two sides of a triangle "parallel" to the third side? Is its length equal to one-half the length of the third side? Illustrate your answers by sketches, using spherical geometry.

16. **Equidistant Loci for Two Nonconcurrent Rays** Define the distance from a point to a ray as suggested in the footnote to the definition in Section 3.8 and determine the locus of points in the Euclidean plane equidistant from two rays h and k having different endpoints. (See the example illustrated in Figure P.16; part of the locus is shown.)

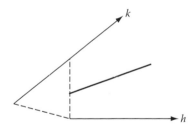

Figure P.16

17. **Undergraduate Research Project** Formulate a general definition for the distance from a point to a segment, then determine the locus of points in the Euclidean plane equidistant from two segments AB and CD. (See example in Figure P.17; part of the locus is shown.) Investigate whether *Sketchpad* can help solve this problem.

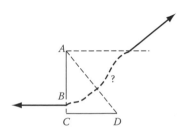

Figure P.17

18. **Undergraduate Research Project**   For what disjoint sets
S and T in the Euclidean plane does the problem of determining the locus of points equidistant from S and T have a
feasible solution? For example, if $S = P$ (a point) and $T =$ line
l (a line), the locus is the familiar parabola having focus P
and directrix l. If S and T are disjoint segments, one obtains
a curve that consists of a combination of rays, segments,
and arcs of parabolas. If S and T are disjoint nonconcentric
circles, one can get either a straight line (if the circles have
equal radii) or a hyperbola otherwise. Verify these results
and extend the investigation further.

4

Quadrilaterals, Polygons, and Circles

The fundamental properties of polygons and circles commonly studied in elementary geometry may be derived axiomatically with relative ease. The purpose of the present chapter is to explore that derivation briefly in the context of unified geometry. The polygon has an important special case in the quadrilateral—the object of discussion for the opening sections.

4.1 Quadrilaterals

Of great importance in geometry, topology, and analysis is the concept of a *polygonal path*—a set consisting of those segments (*sides*) joining a sequence of distinct points (*vertices*). Our focus at first is on *closed* polygonal paths of *order four* (four-sided polygonal paths where the first and last vertices coincide). We might note that, if we are given four points, there are precisely three distinct closed polygonal paths of order four having those points as vertices, and the location of those points relative to each other falls into two classes: (1) when the paths determined by the points are not self-intersecting (called *simple*), as shown in Figure 4.1, and (2) when some paths self-intersect, as shown in Figure 4.2. Since a closed polygonal path can enclose a well-defined interior only if it is not self-intersecting, we are only interested in paths that do not self-intersect. In the two figures there are only four of these, which are called *quadrilaterals*: all three in Figure 4.1 and one in Figure 4.2.

Figure 4.1 Simple closed paths of order four.

Figure 4.2 Closed paths of order four.

Definition 1 Let P_0, P_1, P_2, P_3, P_4 be 5 distinct points in the plane such that for each integer i and j in the range $[0, 4]$, $P_iP_j < \alpha$. A **polygonal path** of **order four** joining P_0 and P_4 is the set

$$[P_0P_1P_2P_3P_4] = \overline{P_0P_1} \cup \overline{P_1P_2} \cup \overline{P_2P_3} \cup \overline{P_3P_4}$$

with the given points called its **vertices**, and the segments joining them, its **sides** (Figure 4.3). Sides that share a vertex are called **adjacent** (or **consecutive**). [**Consecutive** or **adjacent vertices** are pairs (P_i, P_{i+1}) for $0 \le i \le 3$.] The path is **closed** if $P_0 = P_4$ and **simple** if it is not self-intersecting (i.e., pairs of nonadjacent sides do not meet).

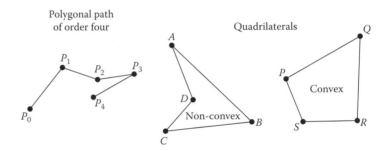

Figure 4.3 Polygonal paths and quadrilaterals.

Definition 2 A **quadrilateral** $\lozenge ABCD$ is a simple, closed polygonal path $[ABCDA]$ of order four such that no three vertices are collinear (Figure 4.3). The **angles** of $\lozenge ABCD$ are defined to be $\angle BAD \equiv \angle A$, $\angle ABC \equiv \angle B$, $\angle BCD \equiv \angle C$, and $\angle CDA \equiv \angle D$. **Opposite** sides, vertices, and angles are the pairs (AB, CD), (BC, DA), (A, C), (B, D), and $(\angle A, \angle C)$, $(\angle B, \angle D)$, respectively. The **diagonals** are segments joining opposite vertices.

A very important distinction is evident in the diagrams of Figures 4.1 and 4.2, also illustrated in Figure 4.3. In working with quadrilaterals in geometry, it is important to have each diagonal essentially *lying between* the sides adjacent to it. This happens in the case of the quadrilateral shown in Figure 4.2 and $\lozenge PQRS$ shown in Figure 4.3. We call such a quadrilateral *convex* because its interior is a convex set, unlike that of $\lozenge ABCD$

in Figure 4.3, and all those in Figure 4.1. Because of the difficulty in formulating a simple definition for the interior of a given quadrilateral, the following definition has been adopted.

Definition 3   A **convex quadrilateral*** is a quadrilateral having the property that the closed half plane determined by any of its sides completely contains the quadrilateral.

Figure 4.4 shows an example of a convex quadrilateral and illustrates the useful property mentioned above. This property is the first result to be established for convex quadrilaterals.

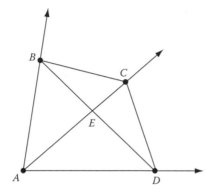

Figure 4.4 Theorem 1.

Theorem 1   If $\Diamond ABCD$ is a convex quadrilateral, then the four betweenness relations $(\overrightarrow{AB}\ \overrightarrow{AC}\ \overrightarrow{AD})$, $(\overrightarrow{BA}\ \overrightarrow{BD}\ \overrightarrow{BC})$, $(\overrightarrow{CB}\ \overrightarrow{CA}\ \overrightarrow{CD})$, and $(\overrightarrow{DA}\ \overrightarrow{DB}\ \overrightarrow{DC})$ hold.

Proof   It suffices to prove the first of these, namely $(\overrightarrow{AB}\ \overrightarrow{AC}\ \overrightarrow{AD})$, since by relabeling points, the others follow from this one case. By definition, C and D lie on the same side of line AB, the line containing ray $AB \equiv h$. If $AC = u$ and $AD = v$, as shown in Figure 4.5, then by Theorem 7, Section **2.4**, either (huv) as in Figure 4.5a, the desired result, (hvu) as in Figure 4.5b, (uhv), or $(hu'v)$. If (huv), then by the converse of **Axiom 10**, v passes through an interior point of $\angle hu \equiv \angle BAC$, and by the crossbar theorem v cuts segment BC at a point W such that (BWC). But this would put B and C on opposite sides of line AD, contradicting hypothesis. A similar contradiction is reached if either (uhv) or $(hu'v)$. Therefore, (huk) or $(\overrightarrow{AB}\ \overrightarrow{AC}\ \overrightarrow{AD})$. ◣

---

* A convex quadrilateral is a special type of quadrilateral, so the term *convex* modifies the term *quadrilateral* in the usual sense. But the meaning of the term *convex* here is different from the general definition of convexity since quadrilaterals, like triangles, are not by themselves convex sets.

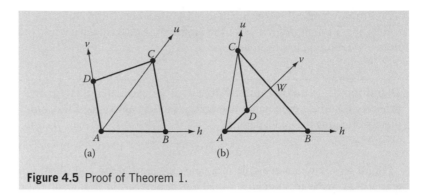

Figure 4.5 Proof of Theorem 1.

Corollary   The diagonals of a convex quadrilateral intersect at an interior point on each diagonal.

(The proof will be left as Problem 10.)

The next theorem shows that one need only prove that a quadrilateral lies to one side of the line containing a side for only *three* sides of the quadrilateral, instead of all four as in the definition. This will be used later to prove that Saccheri quadrilaterals are convex.

Theorem 2   If ◊*ABCD* has the property that *C* and *D* lie on the same side of line *AB*, *D* and *A* lie on the same side of line *BC*, and *A* and *B* lie on the same side of line *CD*, then points *B* and *C* lie on the same side of line *DA*, and ◊*ABCD* is a convex quadrilateral.

Proof   Suppose *B* and *C* lie on opposite sides of line *AD*. Then there exists *E* on $\overleftrightarrow{AD}$, such that (*BEC*). By Theorem **2**, Section **1.8**, either (*AED*), (*ADE*), (*DAE*), or (*AE*D*) (when $\alpha < \infty$ and *E** is the extremal opposite of *E*). In the first case, pictured in Figure 4.6a, *A* and *D* lie on opposite sides of line *BC*, contradicting the hypothesis. If (*ADE*), as in Figure 4.6b, then *D* ∈ Int∠*ACB* and by the crossbar theorem, ray *CD* meets segment *AB* at a point *W* such that (*AWB*). This puts *A* and *B* on opposite sides of line *CD*, contrary to hypothesis. If (*DAE*), one obtains in a similar manner *W* on segment *CD*, which puts *C* and *D* on opposite sides of line *AB* contrary to hypothesis. Finally, if (*DE*A*), then *EE** = α and line *BC* passes through both *E* and *E**, and therefore *D* and *A* lie on opposite sides of line *BC*, contrary to hypothesis. ◣

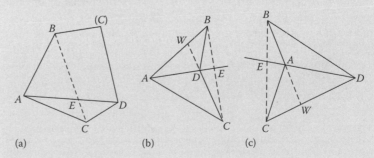

Figure 4.6 Proof of Theorem 2.

4.2 Congruence Theorems for Convex Quadrilaterals

Analogous to the correspondence of parts in two triangles, the notation $ABCD \leftrightarrow XYZW$ indicates the correspondence between the vertices, sides, and angles of quadrilaterals $ABCD$ and $XYZW$. Two quadrilaterals are said to be **congruent** iff under some correspondence between their vertices, corresponding sides and corresponding angles are congruent. The notation

$$\Diamond ABCD \cong \Diamond XYZW$$

will be used to assert that quadrilaterals $ABCD$ and $XYZW$ are congruent under the correspondence $ABCD \leftrightarrow XYZW$.

As with triangles, congruence among certain corresponding parts of pairs of convex quadrilaterals implies their congruence. There are two major such theorems that are used frequently throughout, only one of which will be proved here (the other being left as a problem).

SASAS Congruence Criterion   Two convex quadrilaterals are congruent if, under some correspondence, three consecutive sides and the two adjacent angles included by those sides of one quadrilateral are congruent, respectively, to the corresponding three sides and two included angles of the other.

Proof   In Figure 4.7, let $AB = XY$, $\angle B \cong \angle Y$, $BC = YZ$, $\angle C \cong \angle Z$, and $CD = ZW$. By SAS, $\triangle ABC \cong \triangle XYZ$. Therefore, $AC = XZ$ and $m\angle 1 = m\angle 2$ (as marked in the figure). Since $(\overrightarrow{CB}\ \overrightarrow{CA}\ \overrightarrow{CD})$ and $(\overrightarrow{ZY}\ \overrightarrow{ZX}\ \overrightarrow{ZW})$, we obtain $m\angle 3 = m\angle 4$, in addition to $CD = ZW$ by hypothesis. By SAS, $\triangle ACD \cong \triangle XZW$. Then $\angle D \cong \angle W$, $AD = XW$, and $m\angle DAB = m\angle DAC + m\angle CAB = m\angle WXZ + m\angle ZXW = m\angle WXY$ or $\angle A \cong \angle X$, as was to be proven. ◤

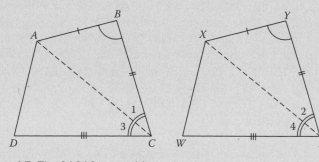

Figure 4.7 The SASAS proposition.

ASASA Congruence Criterion   Two convex quadrilaterals are congruent if, under some correspondence, three consecutive angles and the two included sides of one quadrilateral are congruent, respectively, to the corresponding three angles and included sides of the other.

(The proof of this is left as Problem 13.)

There are a number of other congruence theorems of this type, such as those symbolized by SASAA and SSSSA. Some, however, are valid only for $\alpha = \infty$ in Euclidean geometry, such as SSAAA. You are asked to find a counterexample for the proposed congruence criterion SAASA in Problem 22. On a different tack, just as there were 5-con triangles (noncongruent triangles having 5 congruent pairs of sides and angles), there are **7-con** quadrilaterals (noncongruent convex quadrilaterals having 7 congruent pairs of sides and angles). Problem 11 asks you to find such a pair in Euclidean geometry.

4.3 The Quadrilaterals of Saccheri and Lambert

Let A and B be any two points, $AB < \alpha$, and erect perpendiculars to line AB at A and B, respectively. On the perpendicular at B, locate any point C, $BC < \alpha/2$, and on the other perpendicular, locate D on the same side of $\overleftrightarrow{AB}$ as C such that $AD = BC$. The resulting quadrilateral $ABCD$ is called a **Saccheri quadrilateral**, having **base** $\overline{AB}$, **legs** $\overline{BC}$ and $\overline{AD}$, **summit** $\overline{CD}$, **base angles** $\angle A$ and $\angle B$, and **summit angles** $\angle C$ and $\angle D$ (see Figure 4.8). These quadrilaterals are named after Girolamo Saccheri (pronounced "sack-er'-ee") (1667–1733), a pioneer in non-Euclidean geometry.

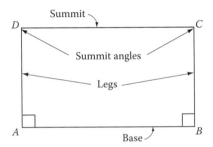

Figure 4.8 A Saccheri quadrilateral.

It easily follows that $\lozenge ABCD$ is a quadrilateral: First, no three of the points A, B, C, and D are collinear, by construction, and segments AD and BC cannot meet, as AB and CD cannot. Next, it follows by Theorem **2**, Section **4.1** that it is a convex quadrilateral: If line BC crosses side AD at some point F, then $\overline{FA} \perp \overline{FB}$, which implies that $DA > FA = FA = \alpha/2$, contradicting $DA < \alpha/2$. Similarly, B and C lie on the same side of line AD (same analysis). Finally, C and D lie on the same side of line AB by construction, which proves

---

Theorem 1: A Saccheri Quadrilateral Is a Convex Quadrilateral

---

In Euclidean geometry, a Saccheri quadrilateral is necessarily a rectangle since the summit angles are right angles, but this is not true in unified geometry: on the sphere, we can have a Saccheri quadrilateral with obtuse summit angles, such as illustrated in Figure 4.9: here, since ΔNCD is an isosceles triangle having legs less that $\alpha/2$, the base angles are acute, which implies that $\angle BCD$ and $\angle ADC$ are (congruent) obtuse angles. In fact, those summit angles at C and D can take on any value between 90 and 180 by letting point C vary on arc NB.

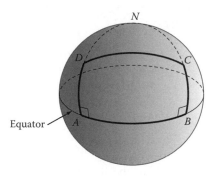

Figure 4.9 Saccheri quadrilateral on the sphere.

Moment for Reflection

This discussion brings up the question whether a rectangle can exist in unified geometry. (A *rectangle* is taken to mean *any quadrilateral with four right angles*.) It turns out that there do not, but there do exist quadrilaterals in unified geometry having *three* right angles. The construction that follows later proves this. See if you can deduce this existence without reading any further.

We can essentially split in half any Saccheri quadrilateral to form two congruent quadrilaterals having three right angles. Such a quadrilateral is called a **Lambert quadrilateral**, after Johann Lambert (1728–1777), another pioneer in non-Euclidean geometry. The next several propositions lead up to this important fact.

Theorem 2 In any Saccheri quadrilateral, the summit angles are congruent, as are the diagonals. The summit is not congruent to the base unless the quadrilateral is a rectangle.

Proof Using the notation of Figure 4.10, we have $\overline{AD} \cong \overline{BC}$, $\angle A \cong \angle B$, $\overline{AB} \cong \overline{BA}$, $\angle B \cong \angle A$, and $\overline{BC} \cong \overline{AD}$. Therefore, regarding $\Diamond ABCD$ and $\Diamond BADC$ as separate quadrilaterals (see inset), by SASAS, $\Diamond ABCD \cong \Diamond BADC$ and by CPCPC, $\angle D \cong \angle C$. Now consider the diagonals $\overline{AC}$ and $\overline{BD}$. We have $\overline{BC} \cong \overline{AD}$, $\angle B \cong \angle A$, and $\overline{AB} \cong \overline{BA}$. Therefore, $\triangle ABC \cong \triangle BAD$ by SAS, and thus $AC = DB$ as desired. Finally, suppose $CD \cong AB$; it must be proven that all four angles of $\Diamond ABCD$ are right angles. Consider $\triangle ABC$ and $\triangle DCB$. We have $AB = DC$ (assumed), $BC = CB$ (identity), and $AC = DB$ (proved above). By SSS, $\triangle ABC \cong \triangle DCB$, and, by CPCPC, $\angle ABC \cong \angle DCB$. Thus, $m\angle C = m\angle B = 90$. By the above, $m\angle D = m\angle C = 90$, and $\Diamond ABCD$ is a rectangle. ◤

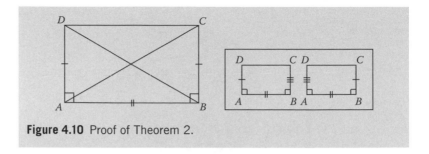

Figure 4.10 Proof of Theorem 2.

The midpoints M and N of the base and summit, respectively, of a Saccheri quadrilateral produce two further quadrilaterals, which are congruent and turn out to be a very special and important type of quadrilateral. First, one must justify that each of these congruent quadrilaterals is convex.

More generally, consider points M and N lying on opposite sides of convex quadrilateral $ABCD$ (Figure 4.11). It follows that $◊ADNM$ is a convex quadrilateral: Since C and B lie on the same side of line AD, then M and N lie on the same side. Likewise, N and D lie on the same side of line $AB \equiv$ line AM, and M and A lie on the same side of line CN. By Theorem **2**, Section **4.1**, $◊AMND$ is a convex quadrilateral. By change of notation, $◊BMNC$ is also a convex quadrilateral. This is the basis for the following result and the argument following.

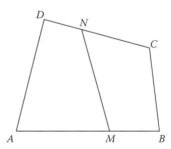

Figure 4.11 Three convex quadrilaterals.

Theorem 3   If M and N are the respective midpoints of the base and summit of a Saccheri quadrilateral $◊ABCD$, then convex quadrilaterals $◊BMND$ and $◊AMNC$ are formed. Furthermore, line MN is the perpendicular bisector of both the base and summit (see Figure 4.12).

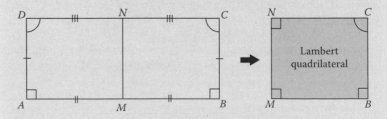

Figure 4.12 Construction for a Lambert quadrilateral.

Proof Observing the hypothesis, with $AM = MB$, $\angle A \cong \angle B$, $AD = BC$, $\angle D \cong \angle C$, and $DN = NC$, we note that $\lozenge AMND \cong \lozenge BMNC$ by SASAS and therefore $\angle AMN \cong \angle BMN$, $\angle DNM \cong \angle CNM$. Therefore, $MN \perp AB$ and $MN \perp CD$, as desired. ◣

Note: Thus has emerged a special construction in unified geometry for a Lambert quadrilateral, a convex quadrilateral having *three* right angles. The fourth angle may or may not be a right angle. We call it the **summit angle**. A **base** of a Lambert quadrilateral is defined to be either of the sides included by two of the three right angles and a **leg** to be either of the two remaining two sides. (Observe that a Lambert quadrilateral can also be constructed independently of Theorem 3, having a given base $< \alpha/2$.) ◣

Historical Note

Saccheri and Lambert were forerunners of non-Euclidean geometry. They were the first to start with a hypothesis contrary to Euclidean geometry and to derive logical results from it. Having shown that the *Hypothesis of the Obtuse Angle* is impossible, Saccheri next assumed the *Hypothesis of the Acute Angle*, hoping to reach a contradiction also for this case (these terms are defined below). This would have shown that the *Hypothesis of the Right Angle* was the only valid hypothesis, thus Euclidean geometry. His work in 1733 established important facts about absolute geometry, some of which appear in this section, but the sought-after contradiction never showed up. In desperation, thinking there had to be something wrong, Saccheri appealed to the "intuitive nature of the straight line." Had he had the courage to admit that he could not find a discrepancy, he would have achieved historical prominence as the discoverer of the geometry known today as *hyperbolic geometry*. Lambert undertook a similar investigation, revealing unusual flashes of insight. Among them was a plausibility argument showing that if the angle sums of triangles are not 180, then there exists an absolute unit of measure (a result we obtain later). This was to be proven 60 years later by both Bolyai and Lobchevski, the recognized founders of hyperbolic geometry.

Saccheri dreamed of finding a proof for Euclid's parallel postulate by essentially assuming our **Axioms 1–12** (with $\alpha = \infty$). Such a proof would mean that the parallel postulate can be logically established as a theorem from the other axioms of geometry and that Euclid's immortal Fifth Postulate is not necessary. Saccheri's method was to work with the quadrilateral named after him and investigate systematically the three possible cases (illustrated in Figure 4.13):

(a) Measure of summit angles > 90—*Hypothesis of the Obtuse Angle*
(b) Measure of summit angles = 90—*Hypothesis of the Right Angle*
(c) Measure of summit angles < 90—*Hypothesis of the Acute Angle*

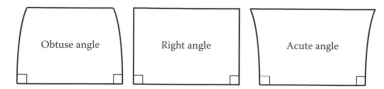

Figure 4.13 Picturing the three hypothesis.

He dismissed the *Hypothesis of the Obtuse Angle* in short order since he was assuming that lines were "unbounded"—that is, $\alpha = \infty$. But the *Hypothesis of the Acute Angle* caused considerable difficulty. He finally convinced himself that he had obtained a contradiction of this hypothesis, thus settling the issue.

Saccheri entitled his work *Euclides ab omne naevo vindicatus* ("Euclid freed of every flaw"). It is ironic that in this effort Saccheri believed he had rendered Euclid's *Elements* more perfect, but as it turns out, had Saccheri actually succeeded, his work would have dealt a terrible blow to Euclid, and it would have caused much damage to mathematics in general. It would have meant ultimately that Euclidean geometry *contains contradictory results*. This is because models for **Axioms 1–12** exist in Euclidean geometry for which the hypothesis of the acute angle is valid; if Saccheri's acute angle hypothesis were contradictory, then the model, within Euclidean geometry, would be contradictory. Misguided though he was, Saccheri very nearly discovered hyperbolic geometry—and his work anticipated that discovery by about 100 years.

It is remarkable that by simply keeping quiet on the issue, Euclid avoided the pitfall of so many able mathematicians who followed him. The *Elements* contain not a word or remark as to why the parallel postulate—stated in such an elaborate manner—might be necessary, or independent of the other postulates, and no attempt at proof was made. Euclid's first critic on this issue occurred as early as the fifth century: Proclus (410−485) believed that the fifth postulate was too complex to be an axiom and proceeded to give a rather clever argument, proving it as a theorem. His reasoning was flawed, however. (We examine this argument in Chapter 5.)

A few miscellaneous results will be mentioned without proof. They are useful, but rather elementary in nature. First, we showed above that a Saccheri quadrilateral could be divided in half to produce two congruent Lambert quadrilaterals. The reverse procedure is to start with a given Lambert quadrilateral and construct from it a Saccheri quadrilateral by essentially "doubling" it, as illustrated in Figure 4.14. Since the sides of a Lambert quadrilateral are of length less than $\alpha/2$ (Problem 20 below), this doubling procedure is valid (by the segment-doubling theorem): points E and F are located so that A and D are the midpoints of segments BE and CF and then one draws segment EF. It is routine to prove the resulting quadrilateral $\lozenge BEFC$ is a Saccheri quadrilateral. We state this as a theorem using intuitive terminology; Figure 4.14 shows a hint for proving both parts.

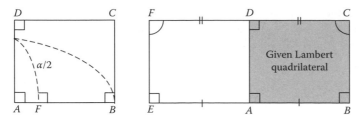

Figure 4.14 "Doubling" a Lambert quadrilateral.

Theorem 4   A given Lambert quadrilateral has sides and diagonals of length less than $\alpha/2$ and is one-half of a Saccheri quadrilateral.

A quadrilateral that lurks in the background as we examine various Lambert quadrilaterals is the rectangle, all four of whose angles are right angles. Although an object of unified geometry by definition, it presumably can exist only in Euclidean geometry. Evidently, all Lambert quadrilaterals in non-Euclidean geometry have either obtuse or acute summit angles and cannot be a rectangle (see what you can observe on the unit sphere). A distinguishing characteristic of such a quadrilateral is stated in the following theorem of unified geometry, again left as a problem (Problem 23). The Figure 4.15 gives a hint.

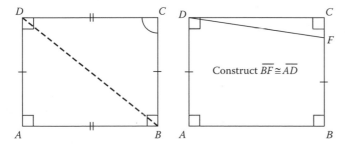

Figure 4.15 Proving Theorem 5.

Theorem 5   A Lambert quadrilateral is a rectangle iff its opposite sides are congruent.

The last result of this section exhibits the genius of Saccheri's work. An important construction due to him shows a connection between the angle-sum of triangles and the summit angles of a Saccheri quadrilateral. Saccheri's original construction applied to any triangle, but there are many betweenness cases to be dealt with, which become particularly troublesome for $\alpha < \infty$. The result to be considered here applies only to right triangles, which will be sufficient for later work.

Saccheri's Construction   Let right triangle ABC be given, with right angle at C (Figure 4.16) and sides of length less than $\alpha/2$. Locate the midpoints M of side BC and N of side AC and draw line MN. Drop perpendiculars to line *MN* from points *A*, *B*, and *C*, producing the feet *A'*, *B'*, and *C'* of *A*, *B*, and *C* on line *MN*. Thus, a quadrilateral *A'B'BA* is formed adjacent to the hypotenuse of △*ABC*.

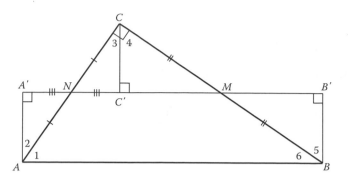

Figure 4.16 Proving Saccheri's Theorem.

Saccheri's Theorem   The quadrilateral *A'B'BA* is a Saccheri quadrilateral, with base *A'B'* and summit *AB*. Moreover, the angle-sum of △*ABC* equals twice the measure of either summit angle of ◊*A'B'BA*.

Proof   Since the legs *MC* and *NC* of right triangle *CNM* are each less than $\alpha/4$, by the triangle inequality, $MN < \alpha/2$ and the angles at *M* and *N* are acute. Hence, by Corollary **E** in Section **3.6**, (*NC'M*). Also, $NC' < NC < \alpha/4$ and $C'M < CM < \alpha/4$. To avoid tedious betweenness arguments, construct point *A'* on line *MN* such that (*A'NC'*) and $A'N = NC' < \alpha/4$, and draw segment *AA'*; by SAS △*ANA'* ≅ △*CNC'* and *A'* will be the foot of *A* on line *MN*. Similarly, (*CMB'*) and △*CC'M* ≅ △*BB'M*. Thus $A'C' + C'B' = A'N + NC' + C'M + MB' < 4(\alpha/4) = \alpha$, so by Corollary **A**, Section **1.8**, (*A'C'B'*) holds (details left for Problem 16). Hence (*A'NC'B'*) and (*A'C'MB'*). Now $AA' = CC' = BB'$, so since *A* and *B* lie on the same side of line *MN*, ◊*A'B'BA* is a Saccheri quadrilateral with summit *AB*. Finally we show that ray *AC* lies between rays *AA'* and *AB*, and that ray *BC* lies between rays *BA* and *BB'*. Since ◊*A'B'BA* is convex, ($\overrightarrow{AA'}\ \overrightarrow{AB'}\ \overrightarrow{AB}$), and we have ($\overrightarrow{AA'}\ \overrightarrow{AN}\ \overrightarrow{AB'}$) from (*A'NB'*), so that ($\overrightarrow{AA'}\ \overrightarrow{AN}\ \overrightarrow{AB'}\ \overrightarrow{AB}$) or ($\overrightarrow{AA'}\ \overrightarrow{AN}\ \overrightarrow{AB}$). Similarly, ($\overrightarrow{BA}\ \overrightarrow{BM}\ \overrightarrow{BB'}$). Thus, the angle-sum of △*ABC* is given by

$$m\angle 1 + (m\angle 3 + m\angle 4) + m\angle 6 = (m\angle 1 + m\angle 2) + (m\angle 5 + m\angle 6)$$

$$= m\angle A'AB + m\angle ABB'$$

which completes the proof. ◣

Corollary A   Under the respective hypotheses of the right angle, obtuse angle, or acute angle, the angle sum of any right triangle having sides of length less than $\alpha/2$ is either equal to 180, greater than 180, or less than 180.

The result in the above corollary can be extended to an arbitrary triangle by dropping the perpendicular to the longest side (using Corollary **E**, Section **3.6**) and summing the angle measures of the two right triangles thus formed (see Problem 17). Thus Saccheri's theorem on angle-sums of triangles has been established for all triangles whose sides are of length less than $\alpha/2$.

Now note also that in the proof of Saccheri's theorem, we obtained the betweenness relation $(A'C'B')$, and M and N are midpoints of their respective segments. Hence

$$A'B' = A'C' + C'B' = 2MC' + 2C'N = 2MN$$

By the inequalities to be provided in Problem 19, the following result holds (Figure 4.16):

> **Corollary B**   Under the respective hypotheses of the right angle, obtuse angle, or acute angle, if M and N are the midpoints of the legs AC and BC of right triangle ABC having sides of length less than $\alpha/2$, either $MN = \frac{1}{2}AB$, $MN > \frac{1}{2}AB$, or $MN < \frac{1}{2}AB$.

Problems (Section 4.3)

Group A

1. Verify the statement made above that there are only three distinct closed paths determined by four distinct points.

2. In terms of the definition for a convex quadrilateral given above, why is $\lozenge ABCD$ shown in Figure P.2 not a convex quadrilateral?

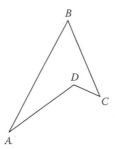

Figure P.2

3. In $\lozenge ABCD$ of Problem 2, point D lies in the interior of $\angle ABC$. Answer each of the following:
    (a) Does A lie in the interior of $\angle BCD$?
    (b) Does B lie in the interior of $\angle ADC$?
    (c) Does C lie in the interior of $\angle BAD$?

4.   In ◊*ABCD* of Problem 2, are the following true or false?
     (a) $m\angle ABC = m\angle ABD + m\angle DBC$.
     (b) $m\angle BAD = m\angle BAC + m\angle CAD$.

5.   State, without proof, the SASAA congruence criterion in terms of the specific convex quadrilaterals ◊*ABCD* and ◊*XYZW*.

6.   State, without proof, the (proposed) SSASA congruence criterion in terms of the specific convex quadrilaterals ◊*ABCD* and ◊*XYZW*.

7.   In quadrilateral ◊*ABCD*, you are given that $\overline{AD} \cong \overline{BC}$ and $\angle A \cong \angle B$. Prove $\overline{AC} \cong \overline{BD}$ (Figure P.7).

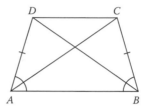

Figure P.7

8.   An equilateral convex quadrilateral ◊*KLMN* is given in Figure P.8. Show:
     (a) The diagonals are perpendicular.
     (b) Opposite angles are congruent.

Note: This quadrilateral is a parallelogram in Euclidean geometry, the familiar *rhombus*. Why is it not a parallelogram in unified geometry? ◥

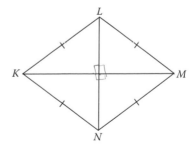

Figure P.8

Group B

9.   Prove that if the hypothesis SSASA holds for two convex quadrilaterals, either the quadrilaterals are congruent or a pair of corresponding, con-congruent angles are supplementary.

10.* Prove the corollary of Theorem 1, Section **4.1**: The diagonals of a convex quadrilateral intersect.

11. Construct a pair of 7-con convex quadrilaterals in Euclidean geometry. (The use of similar triangles is allowed.)

12. Show that if $\alpha < \infty$ the summit angles of a Saccheri quadrilateral are obtuse, implying the Hypothesis of the Obtuse Angle.

13. Prove the ASASA congruence criterion for convex quadrilaterals.

14. Prove the SASAA congruence criterion for convex quadrilaterals having sides of length less than $\alpha/2$. Give a counterexample on the sphere if the sides are allowed to exceed $\alpha/2$.

15. Draw a representative figure in the Euclidean plane to show that, unlike triangles, specifying the measures of the angles and one side does not determine a unique convex quadrilateral. (In fact, four angles and *two* sides is insufficient.) Thus, there is no SAAAA congruence criterion.

16. Use Corollary **A** of Section **1.8** to prove that $(A'C'B')$ holds. (Show that either $(A'B'C')$ or $(B'A'C')$ together with $(A'NC')$, $(NC'M)$, and $(C'MB')$, which were already established above, leads to a contradiction.)

17. Show that if all right triangles having sides less than $\alpha/2$ have angle-sum greater than 180 (or $\leq$180), then so do all triangles having sides less than $\alpha/2$ (Figure P.17). (Drop the perpendicular to the side adjacent with two acute angles and proceed, using Corollary **E**, Section **3.6**.)

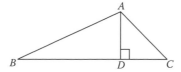

Figure P.17

18. Examine various right triangles on the unit sphere to test the plausibility of Corollary B ($MN > \frac{1}{2} AB$).

19. Prove that under the Hypothesis of the Obtuse Angle, the summit of a Saccheri quadrilateral having sides less than $\alpha/2$ is less than the base. [**Hint:** Use SAS inequality and Corollary A above. Show that $\angle 3 < \angle 1$ in Figure P.19. Note that this reasoning requires only minor changes to prove the opposite inequality under the Hypothesis of the Acute Angle.]

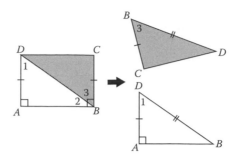

Figure P.19

20. Prove that a quadrilateral having three right angles necessarily has sides of length less than $\alpha/2$. (Use the hint provided by Figure 4.14.) Using facts concerning right triangles, also show that each diagonal has length less than $\alpha/2$.

Group C

21. Prove the SSAAA congruence criterion for convex quadrilaterals in Euclidean geometry.

22. Find a counterexample for the proposed SAASA congruence criterion for convex quadrilaterals.

23. If $\lozenge ABCD$ is a convex quadrilateral, one may define its interior as the set $\text{Int}\angle A \cap \text{Int}\angle C = \text{Int}\angle B \cap \text{Int}\angle D$. Prove this equality, and show that $\text{Int}\lozenge ABCD$ is a convex set. (See Problems 13, 14, Section **4.4** below for a method of defining the interior of a quadrilateral in general.)

24. Prove that a Lambert quadrilateral is a rectangle iff the opposite sides are congruent. (See Figure 4.15.)

4.4 Polygons

Much of the discussion and results on quadrilaterals may be generalized. Only a brief development of an introductory nature will be undertaken here, however, primarily to indicate the validity of such ideas in unified geometry.

For each integer $n > 1$, one defines a **polygonal path of order** n with, **vertices** $P_0, P_1, P_2, \ldots, P_n$, as the set of points

$$[P_0 P_1 P_2 \cdots P_n] \equiv \overline{P_0 P_1} \cup \overline{P_1 P_2} \cup \overline{P_2 P_3} \cup \cdots \cup \overline{P_{n-1} P_n}$$

with the segments $P_i P_j$ called its **sides**. As with four-sided paths, a polygonal path is **closed** iff $P_0 = P_n$, and **simple**, if it is not self-intersecting (the same definition as before). A **polygon** is then defined to be any simple, closed polygonal path of order $n \geq 3$ such that no three *consecutive* vertices are collinear. (See Figure 4.17.) (**Consecutive** vertices are defined cyclically as three vertices having respective indices $i - 1$, i, and $i + 1$ for

$0 \leq i < n$, where P_{-1} is defined to be P_{n-1}.) The terminology regarding sides and angles will be adopted as in quadrilaterals. *Consecutive sides* and *consecutive angles* follow from the concept of consecutive vertices in the obvious manner. A polygon having n sides is called an ***n*-gon**.

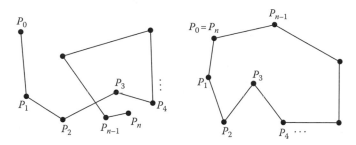

Figure 4.17 Polygonal paths and polygons.

As in the case of quadrilaterals, an important class of polygons are the **convex polygons**, defined as those having the property that they are entirely contained by the closed half plane determined by any side. It is a routine matter to show by the postulate of Pasch that the ordering of the vertices of a convex polygon determines the betweenness relations of the rays determined by those vertices. That is, if $P_1 P_2 \ldots P_n$ s a convex n-gon and the rays with endpoint P_1 passing through P_r, P_s, and P_t, where $1 < r < s < t$, then ray $P_1 P_s$ lies between rays $P_1 P_r$ and $P_1 P_t$, as shown in Figure 4.18, and similarly for the other vertices.

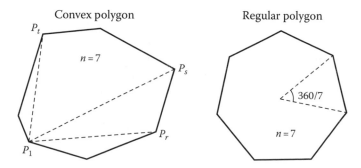

Figure 4.18 Two kinds of polygons.

In turn, the *regular* polygons form an important subclass of convex polygons, illustrated in Figure 4.18. A convex polygon is said to be **regular** provided that all the sides, and all the angles are congruent.

Moment for Reflection

One might consider the possibility of defining a regular polygon without the condition of convexity, since congruence properties might well imply this property. See if you can draw a nonconvex polygon that has all its sides congruent, and all its angles congruent.

Historical Note

Carl Frederick Gauss—known as the *prince of mathematics*—is regarded by most scholars as the greatest mathematician of all time. He made outstanding contributions to all fields of mathematics, filling many volumes, not the least of which included his desk drawer collections where he kept unpublished work. Gauss was born in Germany in 1777 into a poor and unfettered family. His father was, from time to time, a bricklayer, gardener, and construction worker. Having noticed Gauss's genius, the Duke of Brunswick helped pay for Gauss to attend Caroline College, then later, the University of Göttingen, from 1795 to 1798. In spite of his other accomplishments of greater importance, Gauss seemed proudest of the result concerning the 17-sided regular polygon, which he proved could be constructed with compass and straight-edge—the first advance in geometry since the time of Euclid—which he had discovered while a student at Göttingen. He asked that a diagram showing this polygon be carved on his tombstone, a request that was never carried out. He made outstanding contributions in the area of number theory, including a result (later generalized by Hilbert) known as the Reciprocity Theorem, and in theoretical physics. It was he who first rigorously proved that a polynomial of arbitrary degree can be factored into either linear or quadratic factors over the reals. He also was a distinguished astronomer, plotting exact orbits of asteroids and planets. In 1806, he became the director of the newly built observatory at the University of Göttingen, a position he held until his death in 1885.

The study of regular polygons in antiquity was largely focused on whether they could be physically constructed using the ancient tools, consisting of just the compass and unmarked straight-edge. This problem has become the subject of many probing investigations in number theory and algebra. Euclid's compass, straight-edge construction of the regular pentagon, which involves the golden ratio, suggests the difficulty of the problem in general. Archimedes knew that there was a problem with the construction of a regular seven-sided polygon. In fact, it is now known that for most integers, there *does not* exist such a construction for a regular n-gon.

The following theorem was discovered by Gauss when he was a student at Göttingen, and only 19 years of age, the result which he remained proudest of throughout his life. This shows that there exists a Euclidean construction for the 17-sided regular polygon, the next prime number after 5 for which a Euclidean construction exists. (The actual steps of construction were either known by Euclid, or could be easily deduced for the values $n = 3, 4, 5, 6, 8, 10, 12, 15$, and 16.) This interesting theorem will be stated without proof.

Theorem of Gauss on Regular Polygons   A regular polygon having n sides, may be constructed with the Euclidean tools iff n is either a power of 2, or the product of a power of 2 and distinct **Fermat primes**, that is, distinct primes of the form $F_m = 2^{2^m} + 1$, where m is a positive integer.

Note: Fermat primes were named after the great number theorist Pierr de Fermat (1601–1665). He conjectured around 1640 that F_m is prime for all values of m. These numbers become incredibly large very quickly with increasing m. For example, $F_5 = 4{,}294{,}967{,}297 = 641 \times 6{,}700{,}417$—a product of two primes. (Thus, Fermat's conjecture is false.) The values $m = 3, 4$ yield the Fermat numbers $F_3 = 257$ and $F_4 = 65{,}537$ (which are prime), so that, according to Gauss's theorem, regular polygons having 257 and 65,537 sides can be constructed with compass and straight-edge. Two mathematicians, Richelot and Schwendennwein, published specific instructions for the construction of a 257-sided regular polygon in 1898 (Coxeter 2001, p. 27). Obviously, a Euclidean construction for the 65,537-sided polygon would be largely unfeasible, even though it is known to exist. ◥

We confine our discussion on polygons to their existence in unified geometry. This construction also proves that a regular polygon can be inscribed in a given circle (its vertices lie on the circle). The converse property is also true: given a regular n-gon, a circle can be passed through its vertices, provided the sides are small enough.

Consider point O in the plane and any integer $n \geq 3$. Construct segment OP_1 of length less than $\alpha/2$, and ray OP_2 such that $m\angle P_1OP_2 = 360/n$ (≤ 120) and $OP_2 = OP_1$, as illustrated in Figure 4.19. Construct ray OP_3 on the opposite side of line OP_2 as P_1 such that $m\angle P_2OP_3 = 360/n$ and $OP_3 = OP_1$. Continue constructing rays OP_4, OP_5, …, OP_n such that for $3 \leq i \leq n - 2$, ray OP_{i+1} lies on the opposite side of line OP_i as P_{i-1}, $m\angle P_{i-1}OP_i = 360/n$, and $OP_{i+1} = OP_1$. It follows that $m\angle P_nOP_1 = 360 - (n - 1)\, m\angle P_1OP_2 = 360/n = m\angle P_1OP_2$ (see Problem 18, Section **2.6**). By construction, the points P_1, P_2, P_3, …, P_n lie on the circle having center O and radius P_1P_2, and all central angles have measure $360/n$ and are congruent.

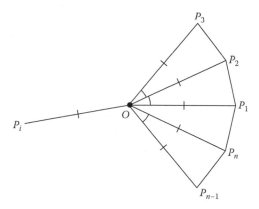

Figure 4.19 Regular polygons exist.

To show that $[P_1P_2P_3 \ldots P_nP_1]$ is a convex polygon, suppose that P_3 lies on the opposite side of line P_1P_2 as point O (Figure 4.20). Then, there exists W on line P_1P_2 such that (OWP_3) and, by construction, ray P_2W is opposite ray P_2P_1. By Theorem 3 (Problem 24, Section **3.7**), the sides of $\triangle OWP_2$ are less than $\alpha/2$. Since $\triangle OP_1P_2$ is an isosceles triangle having legs less than $\alpha/2$, $\angle OP_2W$ is obtuse, and by the scalene inequality, $OP_3 > OW > OP_2 = OP_1$, a contradiction. Hence P_3 lies on the O-side of line P_1P_2. In the same manner, P_4 lies on the O-side of line P_2P_3, thus on the O-side of line P_1P_2 and so on. Hence all the vertices and sides of the polygon lie in the closed half plane of line P_1P_2, which contains O, and the same is true with respect to any other side P_iP_{i+1}. Thus, $[P_1P_2P_3 \cdots P_nP_1]$ is a convex polygon inscribed in a circle. It remains to show that it is a regular polygon.

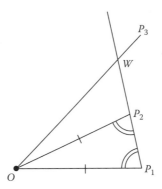

Figure 4.20 Proof that polygon is convex.

If $n = 3$, $m\angle P_1OP_2 = 120$ and $\Delta P_1P_2P_3$ is an equilateral, equiangular triangle (proof?), and if $n = 4$, $m\angle P_1OP_2 = 90$, and it follows routinely that $\Diamond P_1P_2P_3P_4$ is a regular quadrilateral. In both these cases, the vertices lie on a circle centered at O having radius less than $\alpha/2$.

Existence of Regular Polygons   The above construction produces a regular n-gon whose vertices lie on a circle having a given center O and given radius less than $\alpha/2$. Accordingly, a regular n-gon may be inscribed in a given circle for any integer $n \geq 3$.

Proof   This has already been established for $n = 3, 4$, so consider $n \geq 5$. In order to eliminate cumbersome subscripts, let A, B, C, and D be any four consecutive points of the polygon. It suffices to prove in Figure 4.21 that $AB = BC = CD$ and $\angle ABC \cong \angle BCD$, because this result then applies to any 4 consecutive vertices of the polygon, showing that any two sides are congruent, as well as any two angles. Since $OA = OB = OC = OD$ and $\angle AOB \cong \angle BOC \cong \angle COD$, by SAS $\Delta AOB \cong \Delta BOC \cong \Delta COD$ and $AB = BC = CD$. It readily follows that $[OABCO]$ is a convex quadrilateral, hence diagonals OB and AC meet at a point W such that (AWC), hence $(\overrightarrow{BA}\ \overrightarrow{BO}\ \overrightarrow{BC})$. Similarly, $(\overrightarrow{CB}\ \overrightarrow{CO}\ \overrightarrow{CD})$. Thus, we obtain

$$m\angle ABC = 2m\angle OBC = 2m\angle OCB = m\angle BCD$$

ending the proof. ◣

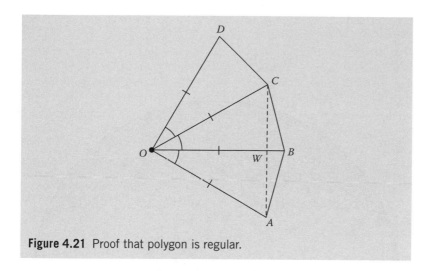

Figure 4.21 Proof that polygon is regular.

The above existence theorem asserts that a regular n-gon can be inscribed in a given circle. The converse of this is the following: *Given a regular n-gon, there exists a circle passing through its vertices.* This is true, but its proof involves proving that the perpendicular bisector of two consecutive sides meet at some point O—the hard part—and then proving that O is equidistant from the remaining vertices of the polygon (see Problem 12). A preliminary result is to first show that if n is even and $k = n/2$, the perpendicular bisector l of side P_1P_2 either meets the opposite side $P_{k+1}P_{k+2}$ (and that of P_2P_3 meets side $P_{k+2}P_{k+3}$), or, if n is odd and $k = (n + 1)/2$, l passes through P_{k+1}. Since these facts are intuitively clear and will not be used later, we leave the proof for a problem. The end result is:

> **Corollary** A given regular polygon possesses a center O and can be inscribed in a circle.

An interesting diagram—found frequently as a poster in geometry class-room exhibits—is shown in Figure 4.22. It illustrates some of the general results which can be established for unified geometry (notably, in Problems 4 and 11), as well as calculations which can be made in Euclidean trigonometry for regular polygons. The formulas needed can be deduced as follows (taken from a typical high school geometry text), where A measures the central angle, s the side, r the radius of the circumscribed circle, a the apothem, P the perimeter, and K the area:

$$A = \frac{180}{n}(n-2), \quad s = 2r\cos\tfrac{1}{2}A, \quad a = r\sin\tfrac{1}{2}A, \quad P = ns,$$

$$K = \tfrac{1}{2}Pa = \tfrac{1}{2}nr^2\sin\tfrac{1}{2}A$$

Figure 4.22 The regular 24-gon and its diagonals.

	Data		
Number of diagonals: 252		Side:	0.261 052 3844
Radius:	1 (unit)	Apothem:	0.991 444 8614
Diameter:	2	Perimeter:	6.265 257 227
Central angle:	15°	Area:	3.105 828 541
		(Pi = 3.141 592 654)	
	Approximation to pi : ½ Perimeter = 3.132 626 614		

Problems (Section 4.4)

Group A

1. The very first proposition proven in the *Elements* was the existence of an equilateral triangle. Euclid's proof was essentially a compass, straight-edge construction, and went like this (see Figure P.1): Given any segment *AB*, describe a circle with center *A* and radius *AB* passing through *B*. Now, with center *B* and radius *AB* describe a circle passing through *A*. From the point *C*, in which the two circles cut one another, draw the segments *CA* and *CB*, thereby forming a triangle *ABC*. It is seen that *AC = AB = BC* and Δ*ABC* is equilateral. What essential detail was missing?

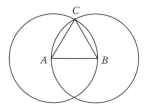

Figure P.1

2. In Euclidean geometry, the angle-sum of a triangle is 180. Using the construction given for a regular *n*-gon above, where each central angle has measure 360/*n*, deduce the measures of the angles of a regular:
   (a) Pentagon (5-gon).
   (b) Hexagon (6-gon).
   (c) Octagon (8-gon).
   (d) Decagon (10-gon).
   (e) Dodecagon (12-gon).

3. In each of the cases of *n*-gons in Problem 2, verify the familiar formula (*n* − 2)180 for the angle-sum of a convex *n*-gon in Euclidean geometry.

4. Prove in Euclidean geometry that a convex *n*-gon has angle-sum (*n* − 2)180. (Use the properties of convex *n*-gons and triangles having angle-sum 180. See Figure P.4 for a hint.) Does your proof work for unified geometry? Obtain the best result you can for this case.

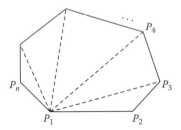

Figure P.4

5. In Euclidean geometry consider a regular 500-sided regular polygon having unit radius. Find the measure of each angle, length of each side, the apothem, perimeter, and area. Compare the area of the polygon with that of its circumcircle ($K = \pi r^2$), that is, the value of π.

6. Prove that in Euclidean geometry each angle of a regular *n*-gon has measure $(n - 2)180/n$.

Group B

7. Use the theorem of Gauss to determine which of the regular *n*-gons in Euclidean geometry can be constructed by compass and straight-edge for $3 \le n \le 100$.

8. Use the theorem of Gauss to show that there is no compass/straight-edge construction of an angle of measure 20°, and that, therefore, an angle of measure 60° cannot be trisected by compass/straight-edge.

9. State and prove the SASASAS congruence criterion for convex pentagons.

10. State and prove the SASASASAS congruence criterion for convex hexagons.

11. How many diagonals does a convex octagon have? Deduce the general formula $n(n - 3)/2$ for a convex *n*-gon.

12. Assuming the perpendicular bisectors to two consecutive sides *AB* and *BC* of a regular polygon *ABCDE* … meet at *O* (Figure P.12), prove that *ON* is the perpendicular bisector of side *CD*, where *N* is the midpoint of side *CD* and that *O* is equidistant from vertices *A*, *B*, *C*, and *D*. Thus, show that this behavior continues for the remaining vertices of the polygon, proving part of the corollary.

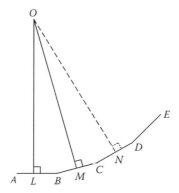

Figure P.12

Group C

13. **Parity Property of Polygonal Paths**   For any segment
$\overline{AB}$ and polygonal path $P = [P_1 P_2 P_3 \dots P_n]$, the intersection
of $\overline{AB}$ and P obviously consists of a finite number of points
or line segments. Each such isolated point or segment is
called a *component* of P on $\overline{AB}$. A component C can be a
single interior point C of a side of P lying on $\overline{AB}$, a single
vertex of P, or a segment $P_iP_j(i < j)$ where i and j are indices
between 1 and n. We say that P *crosses* segment $\overline{AB}$ *at a
component* C iff the vertices of P immediately preceding
and immediately following the points of C lie on opposite
sides of line AB. The *crossing number* of a component C is
0 if P does not cross $\overline{AB}$ at C and 1 otherwise. (See Figure
P.13 for examples.) Define the crossing number of P with a
line segment to be the sum of the crossing numbers for each
component of P on that segment, and the crossing number
of P with a triangle to be the sum of the crossing numbers of
P with the three sides of that triangle. If we take an *exterior
point* of a triangle ABC to be any point not in its interior or
on any side, prove the following theorem by induction on
$n \geq 2$, starting with a single segment $P = P_1P_2$:

> **Theorem**   Suppose that c is the crossing number of a simple
> polygonal path $P = [P_1P_2P_3 \dots P_n]$ with a triangle whose vertices
> do not lie on P. Then c is even if both endpoints P_1 and P_n are
> either interior, or exterior, points of the triangle, and odd if one of
> the endpoints is an interior point and the other an exterior point
> of the triangle (Figure P.13).

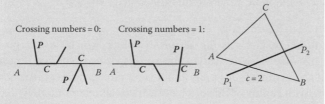

Figure P.13

14.* **Interior and Exterior of a Polygon**   Let P be any polygon, and for any two points not lying on P, define the relation $A \equiv B$ iff the sum of the crossing numbers of the components of P on $\overline{AB}$ is even.

(a) Using the theorem of Problem 13, show that this relation is an equivalence relation having two equivalence classes, E_1 and E_2.

(b) Define the **exterior** of P (Ext P) as the *unbounded* equivalence class, defined as the one containing a point outside a circle of radius $< \alpha/2$ containing P, and the **interior** of P (Int P) as the other equivalence class.

(c) Show that if A and B belong to Int P, then A and B can be joined by a polygonal path lying entirely in Int P.

(d) Show that if $\overline{AB}$ does not contain a vertex of P, then $\overline{AB}$ intersects an even number of sides of P iff A and B are both interior, or both exterior, points of P, and intersects an odd number of sides otherwise.

15. **Undergraduate Research Project**   Define the *internal angle measure* of an angle of a polygon P at vertex A as

$$m*\angle A = m\angle A, \quad \text{if some transversal of } \angle A \text{ lies in Int } P$$

$$= 360 - m\angle A, \quad \text{otherwise}$$

(A *transversal* of an angle is a segment joining points on the interior of its sides.) Consider the

> **Conjecture**   *The sum of the internal angle measures of any polygon having n sides equals* $(n-2)180$.

Try proving this conjecture for a quadrilateral and pentagon, then consider an inductive proof for an n-gon. [**Suggestion:** Use induction on the number $n - i$, where i = number of *interior angles* of P (angles for which some transversal lies in Int P). If $n - i = 0$, then P is a convex polygon, and the conjecture is true.]

4.5 Circles in Unified Geometry

Many familiar properties of circles in Euclidean geometry carry over to spherical geometry. Some of those properties will now be considered, which are valid more generally in unified geometry.

Definition   A **circle** is the set of all points, which lie at a positive, fixed distance $r < \alpha/2$, called the **radius**, from some fixed point O, called the **center**. If P is any point on a circle, the segment OP is a (**geometric**) **radius** of this circle, of which there are infinitely many.

The numerous other terms associated with circles, such as **diameter**, **chord**, and **tangent**, will be provided by a pictorial glossary (Figure 4.23); any formal definition that might be needed is clear from the figure and can be easily formulated. (Point O is the center of the circle in the two diagrams.)

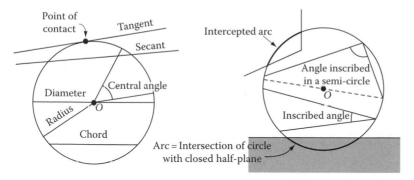

Figure 4.23 Pictorial glossary of terms associated with circle.

Note: If O is the center of circle C having radius r, and if $\alpha < \infty$, it is clear that C is also the locus of points at a fixed distance $\alpha - r$ from O^*, the extreme opposite of point O, as illustrated in Figure 4.24 in the model, where O is in the position of the North Pole. Without restricting the size of the radius as in the definition given above, such a set of points would have two centers and two radii, an undesirable situation. Furthermore, if $r = \alpha/2$, then a circle would also be a (straight) line. (Recall Corollary **B**, Section **3.5**.) ◥

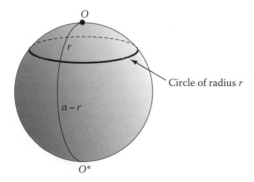

Figure 4.24 A circle on a sphere.

A few elementary properties of circles, which will be found useful, are listed below, the proofs being left to the reader. (See problem section.)

- The perpendicular bisector of any chord of a circle passes through the center.
- A line that passes through the center of a circle and bisects a chord is perpendicular to that chord.
- The foot of the perpendicular of the center of a circle on any extended chord is its midpoint.

- Congruent chords of a circle are equidistant from its center, and conversely, if two chords are equidistant from the center, they are congruent.
- Congruent central angles subtend congruent chords, and conversely, congruent chords are subtended by congruent central angles.

These theorems, all valid in unified geometry, show that there is considerable agreement between Euclidean and spherical geometry involving circles. But there are some Euclidean properties of circles that are not valid on the sphere (thus not valid in unified geometry).

Example 1

Consider $\angle BAC$ inscribed in a semicircle, as shown in Figure 4.25. In Euclidean geometry, $\angle BAC$ is a right angle. Show that in axiomatic spherical geometry $m\angle BAC > 90$.

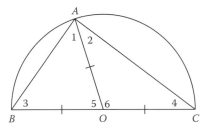

Figure 4.25 Example 1.

Solution

If follows that the angle-sums of triangles ABO and AOC are each greater than 180 (apply Corollary **A** and the results in Problems 12 and 17 of Section **4.3**, and if AB or $AC \geq \alpha/2$, use the medians from O). Since triangles ABO and AOC are isosceles, $\angle 1 = \angle 3$ and $\angle 2 = \angle 4$. Thus,

$$2 \cdot m\angle BAC = 2(\angle 1 + \angle 2) = \angle 1 + \angle 3 + \angle 2 + \angle 4 =$$

$$(\angle 1 + \angle 3 + \angle 5) + (\angle 2 + \angle 4 + \angle 6) - (\angle 5 + \angle 6) >$$

$$180 + 180 - 180 = 180$$

That is, $2 \cdot m\angle BAC > 180$ or $m\angle BAC > 90$.

Example 2

Prove the property in the above list that if $\overline{AB}$ is a chord of circle O and $\overleftrightarrow{OP} \perp \overleftrightarrow{AB}$ at P, then P is the midpoint of $\overline{AB}$. (See Figure 4.26.)

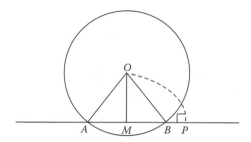

Figure 4.26 Example 2.

Solution

In order to avoid a tedious betweenness proof that $P \in \overline{AB}$, let M be the midpoint of segment AB. Since $OA = OB$, by properties of isosceles triangles, $\overline{OM} \perp \overline{AB}$ and since $OA < a/2$, M is the unique foot of O on line AB. Hence $P = M$ and we are finished.

In order to delve deeper into the behavior of circles, several results will be established concerning what may be thought of as describing the *shape* of a circle in unified geometry, as compared with curves in general. The first is clearly characteristic of circles. It was actually introduced in an earlier problem (Problem 21, Section **3.5**) and proven there, although circles were not mentioned.

(1)   A given line cannot contain three distinct points of a given circle.

A corollary of this is that given a line and a circle, the line either does not meet the given circle, passes through exactly one point of the circle (when the line is tangent to the circle), or passes through exactly two points of the circle (when the line is a secant).

A second result establishes the convexity of the **interior points** of a circle C, defined to be those points X in the plane such that $OX < r$, where O is the center and r is the radius of C. Denote this set by Int C. The **closed interior** of a circle (called a **disk**) is the set of points $C \cup$ Int $C \equiv$ INT C. This set is characterized by the condition $OX \leq r$ for any of its members X. If $OX > r$, then X is said to lie **outside** of, or **exterior** to, circle C.

(2)   If C is any circle, both Int C and INT C are convex sets.

Proof   The argument is virtually the same for both Int *C* and INT *C*. Let *AB* be any segment such that *A*, *B* ∈ Int *C*, with *X* such that (*AXB*), as shown in Figure 4.27. Then, one of the angles at *X* is obtuse or right, so either *OX* < *OA* or *OX* < *OB*. In either case, *OX* < *r* and *X* ∈ Int *C*. [If *AB* ≥ α/2, this argument shows that the median *OM* < *r*, and that if (*AXM*), *OX* < *OA* or *OM*, or if (*MXB*), *OX* < *OM* or *OB*.] ◣

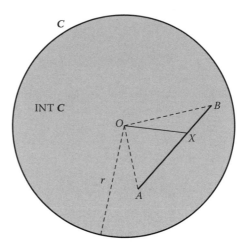

Figure 4.27 Convexity of Int *C* and INT *C*.

The next proposition is a preliminary result, which apparently has no relevance to circles, but will be used to provide a more complete description of the topology of circles in unified geometry. Note how it fits into one of the first axioms assumed for unified geometry (**Axiom 5**). The scalene inequality applies to various triangles involved, one of whose angles is always acute, the other obtuse.

Theorem 1   Suppose that the foot *C* of point *O* on line *AB* ≡ *l* lies between *A* and *B* (Figure 4.28). Further, let *OA* and *OB* be each less than α/2. Then for any point *X* on line *l*, either (1) *OX* > *OA* > *OC* if (*XAC*), (2) *OC* < *OX* < *OA* if (*AXC*), (3) *OC* < *OX* < *OB* if (*CXB*), and (4) *OX* > *OB* > *OC* if (*CBX*).

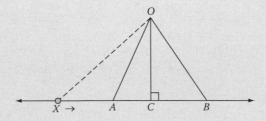

Figure 4.28 Proof of Theorem 1.

Thus, we see that if O is the center of circle C, with r its radius, and if line l is a secant intersecting the circle at A and B (as shown in Figure 4.29), then for any point X on l not on chord AB, $OX > OA = OB = r$, while if (AXB), then $OX \leq r$.

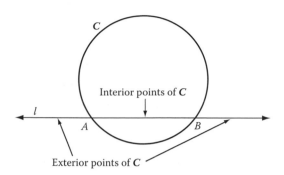

Figure 4.29 A secant line for C.

> ***Corollary A***   If $\overline{AB}$ is a chord of a circle and $X \in \overleftrightarrow{AB}$, then $OX \leq r$ if $X \in \overline{AB}$, and $OX > r$ otherwise. Thus, for any point X on line AB distinct from A and B, $X \in \text{Int } C$ if $X \in \overline{AB}$, and $X \in \text{Ext } C$ otherwise.

The proof of Theorem 1 can be adapted to cover the special case $A = B$, the case when l is tangent to C at A (*tangent* being defined as a line that meets the circle at precisely one point).

> ***Corollary B***   The points of a line that is tangent to a circle at some point A, except for A, lie exterior to the circle.

The last theorem dealing with the shapes of circles is a direct result of Theorem 1. If A, B, and C are any three distinct points of a circle (C), consider $\triangle ABC$. By results already proven,

(3) $\text{Int } \triangle ABC \subseteq \text{Int } C$

Hence no point of C can lie in the interior of $\triangle ABC$. Furthermore, if P lies in the interior of any of the three angles vertical to those of $\triangle ABC$, again $P \notin C$. (In Figure 4.30a, $OP > OQ \geq r$ by Corollary A.) It follows that no point of C can lie in any of the four shaded regions shown in Figure 4.30a. (These four regions are among the seven regions into which any triangle divides the plane.) Therefore, the points of C lie in the complement of these four or in the three (closed) regions shown in Figure 4.30b. Note that, after we prove the corollary below, C is thus divided into precisely three arcs A_1, A_2, and A_3 as shown in Figure 4.30. This proves the following theorem.

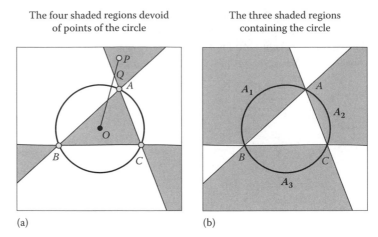

The four shaded regions devoid
of points of the circle

The three shaded regions
containing the circle

(a)   (b)

Figure 4.30 Inscribed triangle and circular arcs.

Theorem 2   Given any circle and three distinct points A, B, and C on it, no point of C can lie in the four shaded regions of Figure 4.30a and therefore C (not including A, B, or C) lies in the complementary set, the shaded regions of Figure 4.30b.

Corollary C   Any circular arc may be characterized as that part of a circle lying in the closed interior of an inscribed angle.

Proof   Recall the definition of an arc: the intersection of a circle and a closed half plane. It suffices to consider only the *interior* of the arc since we have only to add the endpoints to obtain the desired result. Let $\angle ABC$ be an inscribed angle (Figure 4.31) and Int $\overparen{AC} \equiv C \cap H$, where H is the open half plane determined by line AC not containing point B. Consider, then, the two sets (a) $C \cap H$, and (b) $C \cap \text{Int}\angle ABC$. First, let $X \in C \cap H$. If various half planes defined by lines BA and BC are as indicated in the figure, and since $H_1 \cup H_2$ equals the entire plane (less line AB), we have $X \in C \cap H \subseteq C \cap (H_1 \cup H_2)$ and either $X \in C \cap H_1$ or $X \in C \cap H_2$. But $X \notin H_1$, since otherwise, X would lie in the interior of the angle vertical to $\angle BAC$, contrary to the result in Theorem 2. Hence, $X \in H_2$. In a similar manner, $X \in H_3$ and therefore, $X \in H_2 \cap H_3 = \text{Int}\angle ABC$, or $X \in C \cap \text{Int}\angle ABC$. Hence,

$$C \cap H \subseteq C \cap \text{Int}\ \angle ABC$$

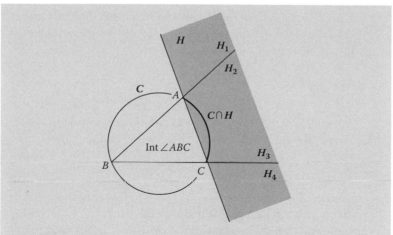

Figure 4.31 Arc intercepted by inscribed angle.

To reverse the inclusion, let $X \in C \cap$ Int $\angle ABC$ to prove that $X \in C \cap H$. It suffices to prove $X \in H$, since $X \in C$ by assumption. But if X lies in the opposite half plane of H, then it follows that $X \in$ Int $\triangle ABC$, again contradicting Theorem 2, since $X \in C$. Therefore, $X \in C \cap H$, as desired. ⟍

Another proposition that is both geometric and topological and addresses the fact that a circle is a continuum, like lines. Accordingly, since there are no "gaps" on a circle, if a line passes through an interior point of a circle, and since that line contains both points interior to and exterior to the circle, there must exist a point of crossing between it and the circle. The proof of this "obvious" fact is somewhat technical, and although important for foundations, it may be skipped without affecting your understanding of later topics.

Theorem 3: Secant Theorem

If A is an interior point of a circle C and l any line passing through A, line l must intersect C at two distinct points and is thus a secant of C. (See Figure 4.32.)

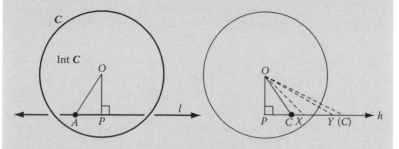

Figure 4.32 Proving line l intersects circle C.

Proof Let P be the foot of the perpendicular from O on line l; since OA and OP are both $< \alpha/2$, right triangle OAP has $OP < OA < r$, and $P \in$ Int C.

1. Let h be either one of the two rays on l having endpoint P, and consider the set $h \cap \text{Int } C$ (nonempty since P is a member) and the corresponding nonempty set of reals

$$S = \{PX : X \in h \cap \text{Int } C\}$$

For any $X \in h \cap \text{Int } C$, the triangle inequality implies $PX < XO + OP < r + r = 2r$ and thus S is a bounded, nonempty set. Hence the least upper bound exists, and we define

$$c = \text{LUB}\{PX : X \in h \cap \text{Int } C\}$$

Locate C on ray h such that $PC = c$. It remains to show that $OC = r$ and thus prove that C lies on the circle.

2. If $OC < r$ then for some positive ε, $OC = r - \varepsilon$. Choose $X \in h$ such that (PCX) and $CX < \varepsilon$, as indicated in the figure. Since $PX > PC = c$ and c is an upper bound for the numbers in S, it must be that $PX \notin S$ and $X \notin h \cap \text{Int } C$. Since $X \in h$, $X \notin \text{Int } C$ and $OX \geq r$. Using the triangle inequality, we obtain the contradiction

$$r = OC + \varepsilon > OC + CX > OX \geq r$$

3. If $OC > r$ then for some positive ε, $OC = r + \varepsilon$. By definition of the *least* upper bound, there must exist $Y \in h \cap \text{Int } C$ such that $PY > c - \varepsilon$. Thus, $PY \leq \text{LUB } \{PX\} = c = PC$ and $Y \in h = \overrightarrow{PC}$; thus $Y \in \overrightarrow{PC}$. By the triangle inequality, and since $OY < r$, we obtain the contradiction

$$r + \varepsilon = OC \leq OY + YC < r + PC - PY < r + c - (c - \varepsilon) = r + \varepsilon$$

Therefore, $OC = r$, as desired. To obtain a second point in $l \cap C$, use the opposite ray h'. ◨

A straightforward consequence is

Corollary D If A is an interior point of a circle and B an exterior point, with $AB < \alpha$, the segment AB intersects the circle in exactly one point.

The corollary is a special case of a more general theorem that applies to a much larger class of closed curves than circles, one that includes triangles, polygons, and ellipses, for example. They are called *simple closed curves*. Such a curve can be described as *any continuous path beginning and ending at the same point that never crosses itself*. The theorem states that (1) any closed curve C in the plane and lying in a half plane defines two disjoint regions in the plane called the *interior* and *exterior* and (2) any curve or arc joining a point in the interior with one in the exterior, intersects C in at least one point. This is a remarkable result when one realizes that there is no requirement that C encloses a convex region. So it could be quite complex like that illustrated in Figure 4.33. This diagram consists of a map of the hedge that forms the famous maze in the gardens at Hampton Court Palace, England. In this example, it is even difficult to determine which points of the plane lie interior to **C** and which lie in its exterior. (An interesting problem, if you enjoy solving mazes, is to see if points *A* and *B* can be joined by a curve that does not cross the boundary of the maze, and to also determine whether point *B* is actually an interior point of **C**.) The tools of geometry seem not to be of much use in proving a theorem of this type. It took the clever work of C. Jordan (1828–1922) to find a proof. His theorem is a landmark for topologists, and it is appropriately called the **Jordan closed curve theorem**.

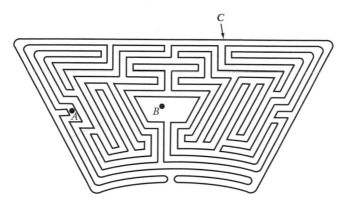

Figure 4.33 Hedge at Hampton Court Palace.

Problems (Section 4.5)

Group A

1. In a circle having center *O*, suppose ∠*AOB* and ∠*COD* are two central angles, which subtend chords $\overline{AB}$ and *CD*. Prove that if ∠*AOB* ≅ ∠*COD*, then $\overline{AB} \cong \overline{CD}$, and conversely (Figure P.1).

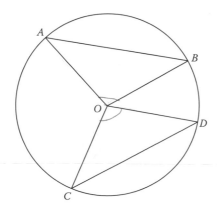

Figure P.1

2. Prove that if two chords are congruent, they are equidistant from the center.

3. Prove that if two chords are equidistant from the center, they are congruent.

4. Prove that if chord AB is greater than chord CD of circle C, the distance from the center of C to $\overline{AB}$ is less than that to $\overline{CD}$.

5. If chord AB is twice the distance from the center as chord CD, then $AB = \frac{1}{2}CD$. True or false?

6. Secants of a circle are drawn, passing through P, and intersecting the circle to form segments having lengths as indicated in Figure P.6. Find x and prove your answer is correct.

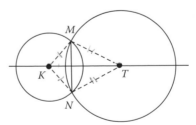

Figure P.6

7. Two circles intersect, forming a common chord MN. Prove that the line of centers KT is the perpendicular bisector of segment MN (Figure P.7).

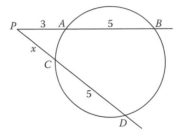

Figure P.7

Group B

8. *Sketchpad Experiment* 7 in Appendix **A** shows how to create an animation for circles tangent to the sides of an angle, leading to a geometric property of circles true in unified geometry.

9. *Sketchpad Experiment* 8 in Appendix **A** leads to a property of tangents provable in unified geometry.

10. Prove the following corollary to the secant theorem: If $A \in$ Int C and $B \in$ Ext C, any segment AB joining A and B must intersect C in exactly one point.

11. Line l intersects a circle C at point P such that the radius OP is perpendicular to l. Prove that l is tangent to C at P (Figure P.11).

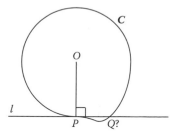

Figure P.11

12. Line l is tangent to circle C at point P. Prove that the radius OP is perpendicular to l (Figure P.12). This and the previous problem together establish the important.

> **Theorem 4** A line is tangent to a circle iff it intersects the circle and is perpendicular to the radius drawn to the point of contact.

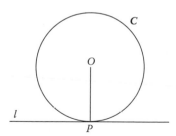

Figure P.12

13. Using the theorem of the previous problem prove:

> **Theorem 5** If two tangents to a circle intersect at P and meet the circle at points of contact A and B, then $PA = PB$. (See Figure P.13.)

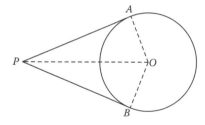

Figure P.13

14.   If two tangents to a circle intersect at *P* and meet the circle at points of contact *A* and *B*, show that the ray from *P* through the center of the circle bisects ∠*APB* (see Figure P.14).

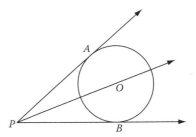

Figure P.14

15.   Suppose that circles *C* and *D* have two tangent lines in common: line *l* tangent internally at *A*, and line *m* tangent externally at *B* and *C*, as shown. Show that if *m* intersects segment *BC* at *M*, then *M* is its midpoint. (Figure P.15).

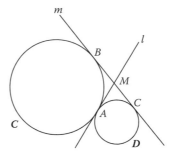

Figure P.15

16.   If the angle bisectors of a quadrilateral are concurrent, show that it may be circumscribed about a circle (i.e., there exists a circle in its interior that is tangent to all four sides) (Figure P.16).

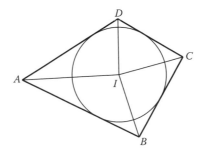

Figure P.16

17. If a quadrilateral having sides of length a, b, c, and d (consecutively) is circumscribed about a circle (as in Problem 16), then $a + c = s = b + d$, where s is the semiperimeter of the quadrilateral. Prove. (See Figure P.17 for a hint.)

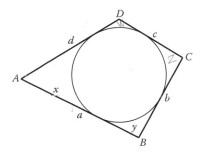

Figure P.17

18. Quadrilateral $ABCD$ is inscribed in a circle, with AB a diameter, and $AD = BC$. Prove that the diagonals are congruent and that the base angles $\angle DAB$ and $\angle ABC$ are congruent (Figure P.18). (In Euclidean geometry, $\lozenge ABCD$ would be an isosceles trapezoid, a figure that is considered in Chapter 5.)

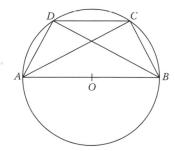

Figure P.18

19. Hexagon $UVWRST$ has two sides as chords of a circle and the other four sides as tangents as shown in Figure P.19. If line UR passes through the center, prove that chords VW and TS are congruent.

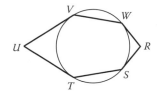

Figure P.19

20.   Show that the maze shown in Figure 4.33 and the diagram in Figure P.20 for this problem are topologically equivalent (by shrinking and pulling). Where are points A and B located relative to this diagram?

Figure P.20

21.   Prove your answer to Problem 5 if you have not already done so.

Group C

22.   Prove that the property in Problem 16 characterizes a circumscribable quadrilateral. That is, prove that if a quadrilateral can be circumscribed about a circle, then its angle bisectors are concurrent.

23.   Experiment with the result of Problem 14(d), Section 4.4, using the diagram of Figure 4.33, regarded as a polygon. This shows how to quickly determine whether A and B both lie in the interior or exterior of the maze (P) or any puzzle of this type.

24.   Find an easy way to solve this puzzle: In the spiral curve shown in Figure P.24, determine whether points A and B lie on the same side of the curve. Does the curve have an interior and exterior? If so, where do A and B lie, inside or outside the curve? (See Problem 23.)

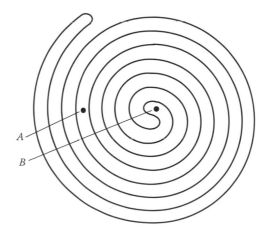

Figure P.24

25.   **The Two-Circle Theorem**   Prove the following theo-
       rem, used by Euclid in his very first proposition (without
       proof) showing the construction of an equilateral trian-
       gle, obtained by intersecting two circles having the same
       radius.

> **Theorem 6**   Suppose that two circles C and D centered at O
> and Q have radii r and s such that $s \leq r$ and $r - s < OQ < r + s$.
> Then the circles intersect at two distinct points P and R, having
> a common secant PR.

       [**Hint:** Let AB be the diameter of C lying on line OQ; with
       Q on ray OB, define c as the least upper bound of $\{m \angle XOB:$
       $X \in$ Int D, $X \in C\}$, construct $\angle COB$ having measure c,
       $C \in C$, and show that $C \in D$. Use the tangent to C at C
       to show that points interior and exterior to D lie on that
       tangent on both sides of C.]

26.   **Converse of Triangle Inequality**   Using Theorem 6
       (Problem 25), prove that, if a, b, and c are any three posi-
       tive reals less than $\alpha/2$ satisfying the triangle inequality
       (i.e., $a < b + c$, $b < a + c$, and $c < a + b$), then there is a
       triangle having sides of length a, b, and c.

5

Three Geometries

Having formulated the axioms that reflect our intuitive knowledge of Euclidean and spherical geometry, we undertook in the preceding chapters a modest development of what has been called *unified geometry* from those axioms. In that development, it was tacitly assumed that if α (the least upper bound of distances) is finite, spherical geometry is obtained, and if α is infinite, Euclidean geometry results. It may come as a surprise to the reader to learn that along with Euclidean and spherical geometry, you were also studying a third geometry just as logical and just as important. This "hidden" geometry was the one which early nineteenth century mathematicians were trying to disprove and believed was inconsistent. Some were great giants of the past, notably, Gauss and Legendre. If you have trouble believing the validity of that third geometry, then you are in good company.

As we shall see in the first section, properties of parallelism and a bounded metric are not independent concepts. In particular, if the metric is bounded, then parallel lines do not exist, and an unbounded metric implies that they do. The boundedness or unboundedness of the metric also influences what can be said regarding the angle-sum of triangles. Let's proceed.

5.1 Parallelism in Unified Geometry and the Influence of α

Euclid defined parallel lines as "two lines which do not meet however far extended," which is half definition and half exposition, as if we were starting with a segment or ray, extending it, and observing whether it would ever intersect another extended segment or ray. In our treatment, we do not have to worry about segments, rays, or their extensions. Simply put (but appropriate only for the geometry of a single plane):

Definition 1   Two lines are said to be **parallel** iff they have no point in common.

Given a line and a point not on it, it seems significant whether we can construct a line through that point which is parallel to the given line, and whether that line is unique or not. Thus, three logical cases emerge, where l is the line and P a point not on it (illustrated in Figure 5.1). These are variations of Playfair's parallel postulate for Euclidean geometry, as proposed by John Playfair (1746–1819).

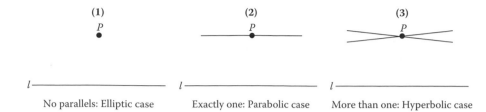

Figure 5.1 Lines through P parallel to l.

(a) There exist no lines through P parallel to l (**elliptic** parallel postulate).

(b) There exists exactly one line through P parallel to l (**parabolic** parallel postulate).

(c) There exist two or more lines through P parallel to l (**hyperbolic** parallel postulate)

These three starkly different parallel postulates give rise to three totally different geometries. It will be subsequently revealed that these three postulates correspond precisely to the three Saccheri hypotheses, the hypothesis of the obtuse, right, and acute angles, respectively. The parallel postulate that we are most familiar with, no doubt, is the parabolic postulate of parallels; it leads to **parabolic geometry**—better known as Euclidean geometry. The elliptic postulate leads to spherical or **elliptic geometry**, also sometimes called *Riemannian geometry*, while the hyperbolic postulate leads to **hyperbolic geometry**, sometimes known as *Lobachevskian* geometry, after one of the discoverers of hyperbolic geometry, N. Lobachevski (1793–1856). It is the geometry which the average person probably has trouble accepting. However, it is as consistent as Euclidean geometry itself, as later developments will show (the construction of a model in Euclidean geometry appears in Chapter 9).

Note: The use of the names of the conic sections to designate the three geometries has become traditional, having been first introduced by F. Klein (1849–1925) who wanted to put the three geometries on common ground. The terms might be suggested by analogy, using the coordinate equation for a conic section (ellipse, parabola, or hyperbola) given in xy-coordinates by

$$ax^2 + bxy + cy^2 + dx + ey + f = 0$$

The *discriminant* of this equation is defined as $\Delta = b^2 - 4ac$. The equation represents an *ellipse*, *parabola*, or *hyperbola* in the xy-plane depending on whether Δ is *negative*, *zero*, or *positive*, which, in turn, corresponds to whether the companion equation

$$ax^2 + bx + c = 0$$

has *no real solution*, *exactly one real solution*, or *more than one real solution* in x. ◼

The following results are propositions of axiomatic unified geometry. The first two are logical consequences concerning parallelism that are derived from either the boundedness or unboundedness of the metric, while the remainder deal with the angle-sums of triangles.

Theorem 1   If $\alpha < \infty$, each pair of lines intersects.

Proof   Let l and m be any two lines. Choose A, B, and C any three points on m such that $AB = \alpha$, and let D be any point on l. (See Figure 5.2, left.) Then it follows that (ACB) and (ADB). Thus, $AC < AB = \alpha$ and $AD < AB = \alpha$ so that $\overleftrightarrow{AC} = \overleftrightarrow{BC} = m$. If l passes through A, B, or C, we are finished. Otherwise, the postulate of Pasch implies, as shown in the second diagram in Figure 5.2, that l passes through either a point E such that (AEC), or a point F such that (BFC). In either case, however, l intersects m. ◣

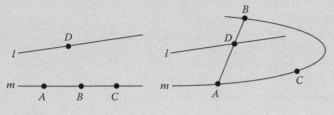

Figure 5.2 Proof of Theorem 1.

The next theorem requires special terminology, no doubt familiar to most readers.

Definition 2   If a line intersects two other lines, it is called a **transversal**. If a transversal t intersects lines l and m at A and B, and if (CAD) on l, (EBF) on m, and $(HABG)$ on t, as shown in Figure 5.3, the pairs $(\angle DAB, \angle ABE)$ and $(\angle CAB, \angle ABF)$ are called **alternate interior angles** with respect to t, while $(\angle BAD, \angle GBF)$, $(\angle HAD, \angle ABF)$, $(\angle HAC, \angle ABE)$, and $(\angle CAB, \angle EBG)$ are called **corresponding angles**. Finally, the pairs $(\angle BAD, \angle ABF)$ and $(\angle CAB, \angle ABE)$ are referred to as **interior angles on the same side of the transversal**.

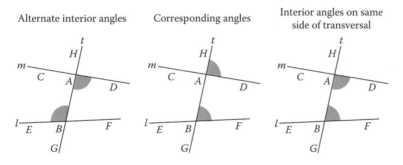

Figure 5.3 Angle pairs and transversals.

Theorem 2   If $\alpha = \infty$, parallel lines exist. In particular, if two lines are cut by a transversal such that a pair of alternate interior angles is congruent, the lines are parallel.

Proof   Let l be a given line and A any point not on l, and choose any line t passing through A and a point B on l. With F any point on l, $F \neq B$, construct ray AC on the opposite side of t as F such that $\angle CAB \cong \angle ABF$, as in Figure 5.4. Let m be the line through A and C. Then by the linear pair theorem, the other pair of alternate interior angles are congruent, as indicated. Now suppose lines l and m meet at W (as shown in the two diagrams at right). One immediately observes a contradiction to the exterior angle inequality (Section 3.6). Therefore, $l \parallel m$. ◥

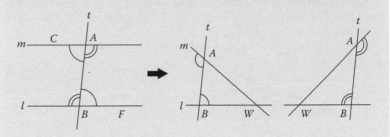

Figure 5.4 Showing that lines m and l are parallel.

Two immediate corollaries follow, which will be quite useful in later work.

Corollary A   If $\alpha = \infty$ and two lines are cut by a transversal such that a pair of corresponding angles are congruent, the lines are parallel.

Corollary B   Lines perpendicular to the same line are parallel.

Note: Under the hypothesis $\alpha = \infty$, the converse of Theorem 2 is essentially: *If parallel lines are cut by a transversal,* then a pair of alternate interior angles is congruent. This is equivalent to Euclid's fifth postulate concerning parallelism, and it is equivalent to the parabolic parallel postulate, that only one line through P is parallel to l in Figure 5.1. Thus, the parabolic postulate, which produces all of Euclidean plane geometry, is the *converse of a theorem provable in unified geometry*! This is a very significant fact, one that perhaps partly explains what made it so difficult for nineteenth century geometers to accept an alternative geometry, and to hold to the notion that the geometry we now call hyperbolic must be contradictory. Philosophers of the day, such as Immanuel Kant, who insisted that the geometry of the physical world could only be Euclidean geometry, also played a role. All other hypotheses, and the geometries derived from them, were believed to be meaningless since they could not be realized in the "real world." ◣

Now we turn our attention to the angle-sum of triangles.

Theorem 3 If $\alpha < \infty$, the angle-sum of any right triangle is greater than 180.

Proof This has already been proven for right triangles having legs of length less than $\alpha/2$ (Saccheri's theorem, Corollary **A**, Section **4.3**), so let $\triangle ABC$ be a right triangle with right angle at C and at least one leg, say BC, of length $\geq \alpha/2$, as shown in Figure 5.5. Since $m\angle C = 90$, it suffices to prove that $m\angle A + m\angle B > 90$. As indicated in the figure, construct segment CD on leg BC of length $\alpha/2$ (where point D may coincide with B if $BC = \alpha/2$). Recalling Corollary **B** of Section **3.5**, $DA = \alpha/2$ and $\overleftrightarrow{DA} \perp \overleftrightarrow{AC}$. Since either $B = D$ or (BDC), it follows that $m\angle BAC \geq m\angle DAC = 90$ and we have the angle-sum

$$m\angle A + m\angle B > m\angle A \geq 90$$

as desired. ◣

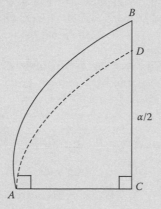

Figure 5.5 Angle-sum for arbitrary right triangles.

Saccheri was able to eliminate with ease the hypothesis of the obtuse angle, as mentioned before. His proof made the tacit assumption that the line was "unbounded," meaning essentially that distance was unbounded ($\alpha = \infty$). Thus, Saccheri's argument showed that an unbounded metric led to a denial of the hypothesis of the obtuse angle. This result is also a theorem of unified geometry, and for its proof an ingenious argument due to A. Legendre (1752–1833) will be employed. Legendre was a pioneer in absolute geometry who obtained through many editions of his geometry book a multitude of interesting results, two of which appear here. The argument for the first one actually duplicates the construction that was used for proving the exterior angle inequality. It should be pointed out that Saccheri also discovered this theorem.

Legendre's First Theorem   If $\alpha = \infty$, the angle-sum of any triangle is less than or equal to 180.

Proof   Let $\triangle ABC$ be any triangle with angle-sum $S = m\angle A + m\angle B + m\angle C$. Assume, contrary to the desired conclusion, that $S > 180$, and set $c = S - 180 > 0$. We proceed to construct a sequence of triangles all having the same segment BC as base, whose angle-sum $= S$. Let M be the midpoint of $\underline{AC}$ and extend BM its own length to point D, with M the midpoint of $\overline{BD}$ (Figure 5.6). By SAS, $\triangle ABM \cong \triangle CDM$ and $\angle BAM \cong \angle DCM$, $\angle ABM \cong \angle CDM$. Then if S' is the angle-sum of $\triangle BCD$ and S that of $\triangle ABC$, we have (making use of the labels shown in Figure 5.6),

$$S = (\angle 1 + \angle 2) + \angle 3 + \angle 4$$

$$= \angle 6 + \angle 2 + \angle 5 + \angle 4$$

$$= m\angle BDC + m\angle DBC + m\angle BCD = S'$$

Thus, the angle-sums of $\triangle ABC$ and $\triangle DBC$ are the same even though the triangles themselves are not congruent. Suppose we repeat the construction for $\triangle DBC$, with N the midpoint of $\overline{CD}$ and $\overline{BE}$. We thus construct $\triangle BCA_2$, with the angle-sum of $\triangle BCA_2 =$ that of $\triangle BCA_1$ (which equals S), where $A_1 = D$ and $A_2 = E$. Continue with the construction of triangles BCA_3, BCA_4, …, BCA_n such that the angle-sums of all these triangles are each equal to S.

Figure 5.6 suggests that $m\angle A_n$ becomes arbitrarily small as n increases. This is indeed true, but it requires proof. For convenience, let $\theta_1 = m\angle A_1 = m\angle A_1 BA$ and for each $n > 1$, $\theta_n = m\angle A_n = m\angle A_n BA_{n-1}$, as indicated in Figure 5.6. By betweenness considerations, it follows that

$$m\angle ABC > \theta_1 + \theta_2 + \theta_3 + \cdots + \theta_n$$

Now suppose that for all n, $\theta_n \geq a$ for some positive constant a. Then, with $b = m\angle ABC$, we would have

$$b > \theta_1 + \theta_2 + \theta_3 + \cdots + \theta_n \geq a + a + \cdots + a = na \quad \text{or} \quad b > na$$

for all integers n, which would be a violation of Archimedes' principle (Appendix B). Thus, since a was arbitrary, there exists n such that

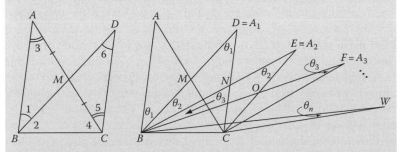

Figure 5.6 Triangles having identical angle-sums.

$\theta_n < c \equiv S - 180$ (defined above). With $W = A_n$, we then observe from Figure 5.6 for $\triangle BCW$, whose angle-sum is S,

$$S = m\angle WBC + m\angle BCW + \theta_n < m\angle WBC + m\angle BCW + (S - 180)$$

which reduces to

$$0 < m\angle WBC + m\angle BCW - 180 \quad \text{or} \quad m\angle WBC + m\angle BCW > 180$$

a contradiction (the sum of the measures of any two angles of a triangle is *less than* 180). The contradiction then proves that $S \le 180$. ⬟

Corollary   If $\alpha = \infty$, the Hypothesis of the Obtuse Angle is false.

Proof   In Figure 4.8 diagonal AC is drawn, producing two triangles (ABC and ACD) with angle-sum ≤ 180. It follows that $m < D \le 90$. ⬟

Moment for Reflection

In Figure 5.7 is shown a construction similar to that used in the proof of Legendre's theorem: M is the midpoint of both segments BD and AC, and N is the midpoint of segments CD and BE. Also, certain angle measures are as marked in the figure: $m\angle A = 64$, $m\angle ABC = 66$, $m\angle MBC = 36$, and $m\angle NBC = 20$. Making use of congruent triangles, calculate the following:

  (a) Angle-sum of $\triangle ABC$.
  (b) Measures of angles at C and D.
  (c) Angle-sum of $\triangle BCD$.
  (d) Angle-sum of $\triangle BCE$.

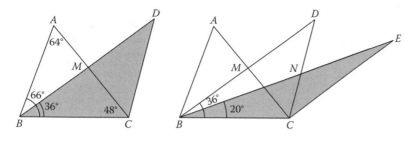

Figure 5.7 Testing a theory.

Did you notice anything? Would Legendre's proof be valid if $\alpha < \infty$? Why, or why not?

Another theorem of unified geometry which was proved by Legendre will be used later. Its interesting proof will be left for you to work out (see Problems 6–9, Section **5.4**).

> ***Legendre's Second Theorem*** If a single triangle has angle-sum equal to 180, a rectangle having arbitrarily large sides exists, all triangles have angle-sum equal to 180, and all Saccheri quadrilaterals are rectangles.

5.2 Elliptic Geometry: Angle-Sum Theorem

The basis for elliptic geometry is the elliptic parallel postulate, defined earlier. This axiom will now be assumed in the context of unified geometry (Axioms 1–12), which defines **elliptic geometry** (or the **elliptic plane**). Thus, there can be no parallel lines, and by Theorem 2 of Section **5.1**, $\alpha < \infty$. There is evidently an elliptic plane for each value of α, corresponding to the geometries of spheres having different radii. Indeed, as mentioned earlier, the geometry of a sphere of radius r is a model for unified geometry with $\alpha = \pi r$. In this section and the following (which deal with elliptic geometry), we take the canonical value $\alpha = \pi$, which corresponds to the unit sphere.

The first two theorems are essentially restatements of previously proven results.

> ***Theorem 1*** Each two lines in the elliptic plane intersect at a pair of extreme points, lying at a distance π from each other.

> ***Theorem 2: Angle-Sum Theorem for Elliptic Geometry***
>
> The angle-sum of any triangle in the elliptic plane is greater than a straight angle.
>
> *Proof* We already know that every right triangle has angle-sum > 180 (Theorem **3** in Section **5.1**). If a triangle has two obtuse angles, the result follows, so assume there is at most one obtuse angle, and the other two are acute angles. Drop perpendicular *CD* to the longest side, *AB* say, of triangle *ABC*. Then add the angle-sums of triangles *ACD* and *DCB*. You are encouraged to take the proof from this point on. ◄

A concept that is peculiar to non-Euclidean geometry—in particular, elliptic geometry—is defined as follows. It is the embodiment of the angle-sum theorem.

> **Definition** The **excess** of a **triangle** Δ, denoted $\varepsilon(\Delta)$ is the difference between its angle-sum and 180. That is, $\varepsilon(\Delta) = S - 180$, where S is the angle-sum of triangle Δ. More generally, the **excess** of an *n*-sided **convex polygon** *P* is the number defined by $\varepsilon(P) = S - (n - 2)180$, where S is the angle-sum of the polygon.

Example 1

Each angle of a regular decagon has a measure of 150. To find its excess, first calculate its angle-sum: $S = 10 \cdot 150 = 1500$. Hence by definition, $\varepsilon(P) = 1500 - (10 - 2)180 = 60$.

Example 2

The excess of a pentagon is 26. Three of its angles are congruent, having measure 110, and a fourth angle has measure 118. Show that the fourth and fifth angles are also congruent.

Solution

If S is the angle-sum, by definition of excess, $26 = S - 3 \cdot 180$, so $S = 26 + 540 = 566$. The sum of the four angles given is $3 \cdot 110 + 118$ or 448, leaving the difference $566 - 448 = 118$ for the fifth angle, showing that it is congruent to the fourth angle.

Example 3

In Figure 5.8 is shown a regular hexagon that is adjacent to a regular quadrilateral, which are, in turn, adjacent to two congruent isosceles triangles. Angle measures are as marked. Find the excess of the hexagon, quadrilateral, and the two triangles.

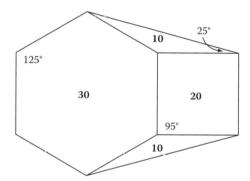

Figure 5.8 Example 3.

Solution

$$\varepsilon(\text{Hexagon}) = 6 \cdot 125 - (4)\,180 = 750 - 720 = 30;$$

$$\varepsilon(\text{Quadrilateral}) = 4 \cdot 95 - (2)180 = 380 - 360 = 20;$$

$$\varepsilon(\text{Triangle}) = 2 \cdot 25 + (360 - 125 - 95) - 180 = 10.$$

Moment for Reflection

Find the angles of the outer hexagon of Example 3, then calculate its excess. Do you notice any relationship between your answer and the excesses of the three inside polygons? (Try adding excesses.)

Example 3 and the Moment for Reflection that follows it introduce an interesting phenomenon in elliptic geometry, that of *additivity of excess*; it is partly justified in the next result.

Theorem 3   If in $\triangle ABC$ a point D is taken on the interior of side BC, with $\varepsilon_1 = \varepsilon(\triangle ABD)$ and $\varepsilon_2 = \varepsilon(\triangle ADC)$, then $\varepsilon_1 + \varepsilon_2 = \varepsilon(\triangle ABC)$.

Proof   Using the notation shown in Figure 5.9, we have

$$\varepsilon_1 = \angle 1 + \angle 2 + \angle 3 - 180 \qquad \varepsilon_2 = \angle 4 + \angle 5 + \angle 6 - 180$$

Hence

$$\varepsilon_1 + \varepsilon_2 = (\angle 1 + \angle 2 + \angle 3 - 180) + (\angle 4 + \angle 5 + \angle 6 - 180)$$

$$= m\angle B + \angle 2 + \angle 5 + m\angle A + m\angle C - 360.$$

But $\angle 2 + \angle 5 = 180$, so the above reduces to $\varepsilon_1 + \varepsilon_2\ m\angle A + m\angle B + m\angle C - 180 = \varepsilon(\triangle ABC)$. ◣

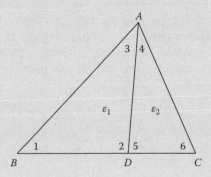

Figure 5.9 Excess is additive.

The more general result, may be stated as follows which we include without proof.

Theorem 4: Additivity of Excess

If $P_1, P_2, P_3, \ldots, P_n$ are convex polygons whose interiors and sides make up the interior and sides of another convex polygon P (in terms of set union) and do not overlap, then with $\varepsilon_i = \varepsilon(P_i)$ for $1 \le i \le n$,

$$\varepsilon_1 + \varepsilon_2 + \varepsilon_3 + \cdots + \varepsilon_n = \varepsilon(P)$$

It is thus apparent that the theory of excess leads to a rather unusual concept for area that involves angle measure instead of lengths of sides. Since area is characterized by its additive property, one can define the **area** of a convex polygon to be a *constant times excess*; that is, for any convex polygon P,

(1)   Area $P = k\varepsilon(P)$

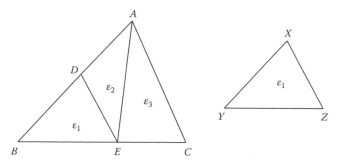

Figure 5.10 Proving the AAA theorem for elliptic geometry.

where k is some constant. It turns out that if $k = \pi/90$ this formula for area coincides with the traditional one for a spherical triangle **P** on the unit sphere, assuming what is normally taken as the concept for surface area from calculus (where, for example, $S = 4\pi r^2$ is the area of a sphere of radius r). For details, see Problem 15 below.

A result of the additive property of excess leads to another perhaps peculiar phenomenon of elliptic geometry, and a surprising criterion of congruence. The ideas involving similar triangles so useful in Euclidean geometry are not valid for elliptic geometry, as we now show.

Suppose two non-congruent triangles ΔABC and ΔXYZ have corresponding angles congruent. (Thus they would be similar triangles in Euclidean geometry.) Since no two corresponding sides of the two triangles can be congruent, it follows logically that two of the sides of one of the triangles are each greater than the corresponding sides of the other. We may thus assume without loss of generality that $\overline{AB} > \overline{XY}$ and $\overline{BC} > \overline{YZ}$ (Figure 5.10). One can then construct segments $\overline{BD} \cong \overline{XY}$ on $\overline{AB}$ and $\overline{BE} \cong \overline{YZ}$ on $\overline{BC}$. With ε_1, ε_2, and ε_3 denoting the excesses of the triangles as indicated in the figure, we have, by Theorem 3, $\varepsilon_1 + \varepsilon_2 + \varepsilon_3 = \varepsilon(\Delta ABE) + \varepsilon_3 = \varepsilon(\Delta ABC) > \varepsilon_1$. Thus, by hypothesis, $\varepsilon(\Delta ABC) = \varepsilon(\Delta XYZ) > \varepsilon_1$. But by SAS, $\Delta DBE \cong \Delta XYZ$ and it follows that $\varepsilon(\Delta XYZ) = \varepsilon_1$, a contradiction. This proves that similar triangles (triangles having congruent corresponding angles) do not exist *unless they are congruent*. This proves

Theorem 5: AAA Congruence Criterion

If two triangles have the three angles of one congruent, respectively, to the three angles of the other, they are congruent.

One final result will be made use of later. Intuitively, it states that if the sides of a triangle or polygon decrease in size, the angle-sum becomes, in the limit, the value given by Euclidean geometry (i.e., *elliptic geometry is, in the small, Euclidean-like*).

Theorem 6 If the sides of a triangle become arbitrarily small, the angle-sum of the triangle converges to 180. (More precisely: Given $p > 0$ there exists $q > 0$ such that if the sides of ΔABC are of length less than q, the angle-sum S of ΔABC differs from 180 by less than p.)

Proof Consider any right triangle PQR having sides less than $\pi/2$, with right angle at Q. For any integer $n > 0$, construct n points Q_1, Q_2, $Q_3, \ldots, Q_n = R$ on segment QR such that the angles determined at P are congruent: $\angle QPQ_1 \cong \angle Q_1PQ_2 \cong \angle Q_2PQ_3 \cong \ldots$ (see Figure 5.11). It follows that the excesses ε_1, ε_2, $\varepsilon_3, \ldots$ of the corresponding triangles ΔQPQ_1, ΔQ_1PQ_2, $\Delta Q_2PQ_3, \ldots$ become progressively larger with increasing n: $\varepsilon_1 < \varepsilon_2 < \varepsilon_3 < \cdots$ [To justify this, observe ΔPVT, where $\angle PVT$ is constructed congruent to $\angle PVU$ (since $\angle PVU < 90 < \angle PVW$); then $\Delta PVT \cong \Delta PVU$ by ASA and $\varepsilon_k = \varepsilon(\Delta PVU) < \varepsilon(\Delta PVW) \equiv \varepsilon_{k+1}$.] It follows by additivity of excess that

$$n\varepsilon_1 = \varepsilon_1 + \varepsilon_1 + \varepsilon_1 + \cdots + \varepsilon_1 < \varepsilon_1 + \varepsilon_2 + \varepsilon_3 + \cdots + \varepsilon_n = \varepsilon(\Delta PQR) \equiv \varepsilon$$

or $\varepsilon_1 < \varepsilon/n$

Hence, if n is chosen so that $np > 2\varepsilon$, then

(2) $$\varepsilon_1 < \frac{\varepsilon}{n} < p/2$$

Take q to be the minimum length of the sides of ΔPQQ_1 and let ΔABC have sides of length less than q. Since at most one angle of ΔABC is obtuse we may assume that if D is the foot of A on line BC, then (BDC). It follows that the excesses of each of the right triangles ABD and ADC are less than ε_1, and by **(2)**

$$0 < S - 180 = \varepsilon(ABC) = \varepsilon(\Delta ABD) + \varepsilon(\Delta ADC) < 2\varepsilon_1 < p \text{ ⟍}$$

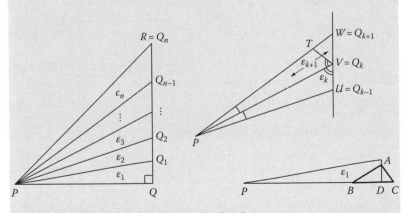

Figure 5.11 Angle-sum for very small triangles.

5.3 Pole-Polar Theory for Elliptic Geometry

One of the properties of elliptic geometry proved earlier under the hypothesis $\alpha < \infty$ involved extremal points, where it was shown that *every line passing through one of two extremal points also passes*

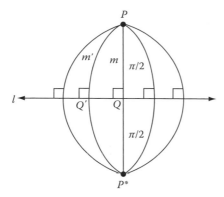

Figure 5.12 The poles of line *l*.

through the other (Theorem **1**, Section **2.6**). The fact that each pair of lines in the elliptic plane meets at a pair of extremal points gives rise to a "pole-polar" theory. Begin with a line *l* and any two points $Q, Q' \in l$ (Figure 5.12). Construct the perpendiculars *m* and *m'* to *l* at *Q* and *Q'*. These two perpendiculars meet at distinct points *P* and *P** such that $PP^* = \pi$. Triangle PQQ' has right angles at *Q* and *Q'*, hence is isosceles. By the converse part of Corollary **C** in Section **3.5**, $PQ = PQ' = \pi/2$; similarly, $P^*Q = P^*Q' = \pi/2$. Furthermore, every line through *P* is perpendicular to *l*, and passes through *P**; these lines evidently account for every perpendicular to *l* (by the uniqueness of the perpendicular to *l* at a given point on *l*). This analysis proves:

> ***Theorem 1*** The family of lines perpendicular to any given line *l* pass through the same two extremal points *P* and *P**, and every line passing through *P* and *P** is perpendicular to *l*.

> ***Corollary*** Each line in elliptic geometry is a "circle" of radius $\pi/2$ whose center is one of the poles of the line.

The points *P* and *P** mentioned in the previous theorem are called the **poles** of line *l*, and line *l* is called the **polar** of the two extremal points *P* and *P**. We have thus shown that every line has a unique pair of poles. Conversely, suppose we are given two extremal points *P* and *P** and want to determine the polar line of (*P, P**). Simply determine the midway point *Q* on any line *m* through *P* and *P** (thus, (*PQP**) and $PQ = QP^* = \pi/2$) and consider the perpendicular *l* to *m* at *Q* (Figure 5.13). Then since $PQ = \pi/2$, every line passing through *P* is perpendicular to *l*, and also passes through *P**. Thus the poles of *l* are *P* and *P**, and *l* is unique with this property, *l* is the (unique) polar of *P* and *P**.

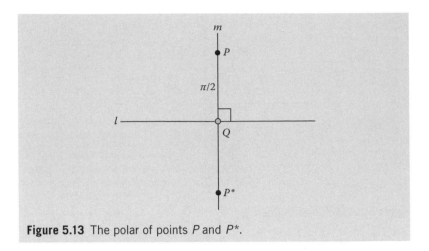

Figure 5.13 The polar of points *P* and *P**.

Theorem 2 Each line in the elliptic plane possesses a unique pair of extremal points as poles, and conversely, each pair of extremal points *P* and *P** possess a unique polar line, having *P* and *P** as its poles.

Note: The fact that each line has two poles gives rise to the term *double* elliptic geometry, as discussed earlier. By identifying extremal points one obtains *single* elliptic geometry, where two points always determine a unique line and every line has exactly one pole. Studying such a geometry would thus require a complete overhaul of the above axioms. It would be interesting to formulate an axiomatic system for synthetic geometry that includes single elliptic geometry as a special case, as well as the other three geometries. ◥

A fundamental property of poles and polars is the following.

Theorem 3 If *P* is a pole of line *l*, then the polar of any point on *l* passes through *P*, and dually, the poles of any line passing through *P* lie on *l*.

Proof

1. Let Q be any point on l (Figure 5.14) and determine $Q^* \in l$ such that $QQ^* = \pi$. The perpendicular m to l at the point R midway between Q and Q^* is thus the polar of Q and Q^* (by construction given earlier). Since $m \perp l$, by Theorem 1 m passes through P.

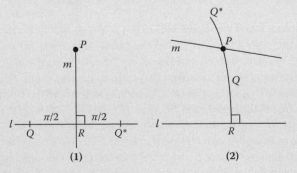

Figure 5.14 Proof of Theorem 3.

2. Conversely, let m be any line through P, and let Q and Q^* be the poles of m. Then $PQ = \pi/2$ and line PQ meets l at some point R. Since P is a pole of l, $PR = \pi/2$. Since Q, Q^* are poles of m, $PQ = PQ^* = \pi/2 = PR$ and R coincides with either Q or Q^*; thus l passes through Q and Q^*. ◥

The pole-polar theory

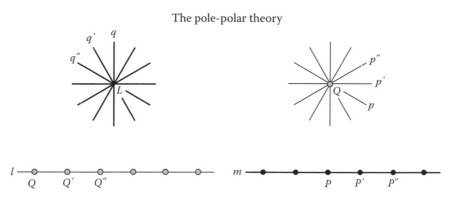

Figure 5.15 Concurrent lines and collinear points.

A dynamic view of the pole-polar theory—the result of Theorem 3—is illustrated in Figure 5.15. If point Q varies on some line l, the polars of these points (and their extremal opposites) are concurrent, and the point of concurrency on a given side of l is a pole of l. On the other hand, if lines p, p', p''... are concurrent in some point Q, the polars P, P', P''... (and P^*, P'^*, P''^*...) are collinear, and the line of collinearity is the polar of Q and Q^*. Thus, *the locus of the poles of concurrent lines is a line* (the polar of the points of concurrency).

5.4 Angle Measure and Distance Related: Archimedes' Method

In elliptic geometry (as in hyperbolic geometry), angle measure is tied inextricably to distance, unlike Euclidean geometry where all distances can be proportionately increased or decreased without affecting angle measure (as in a scale of miles on a map). That is, for example, in Euclidean geometry, if the sides of a 30–60 right triangle are doubled, one still has a 30–60 right triangle. But not in elliptic geometry—the angles change, and the resulting triangle is no longer a right triangle.

The following result will bear this out. While routine in character, the proof will be patterned after a type of argument that was used by the ancients (Archimedes in particular). This method will be repeated often in many of the developments which follow. While the use of limits might seem appropriate for such material, we prefer to keep to the synthetic method as much as possible. It is sometimes referred to as the "method of exhaustion," and shows in some sense the *power of geometry over analysis*.

However, one must use at various points the so-called *Archimedean property*, which was originally a geometric concept, but taken over by analysts and algebraists. This property can be stated strictly in terms of algebra and made meaningful in an ordered field, as follows: Given two positive numbers a and b, there exists an integer n such that $na > b$. This property is the basis for the theory of limits, since it shows, for example, that the limit of $1/n$ equals 0 as n becomes infinite. The geometric version as employed frequently by Archimedes is this: If congruent segments of length a be laid end to end along a straight line, at some point that segment will exceed any given line segment of length b on that line. That Archimedes found this property significant and recognized its axiomatic quality in a half-mechanical, half-geometric environment typifies his brilliance. (In our work we do not have to assume this as an additional axiom because the coordinatization of lines already takes on the properties of the real number system.) It should be pointed out that Legendre's argument which appeared earlier concerning the angle-sum of triangles used the Archimedean property.

Historical Note

Archimedes (287–212 BC) is known as the Father of Physics. He was the greatest scientific and mathematical intellect in the ancient world, and, according to some, in the modern world as well. He was born of low estate in the ancient Greek city of Syracuse, but he was soon recognized by King Heron as a valuable asset because of his ingenious inventions, particularly those of military application. Tradition has it that the king once asked Archimedes to test his gold crown for its authenticity. As the story goes, a flash of insight struck Archimedes as he was bathing, and the problem was solved: simply place the crown in a bucket of water, weigh the crown, weigh the water displaced by the crown, and compare the two. Thus was born the concept of specific gravity for each of the chemical elements. It was said that this excited him so much that he immediately ran into the street shouting "Eureka! I have found it!" The concept of physics he discovered was named after him and is called *Archimedes' principle*, studied today by every beginning student of physics. He is also famous for the law of levers, and it is said that he once boasted "give me a place to stand and a rod long enough, and I can lift the whole world!" But he was proudest of his work in geometry, including the development (and proof) of the formula for the volume and surface area of a sphere. In a siege on Syracuse he was killed by a soldier who ignored the orders of the commander to spare his life—the noble spirit of a genius thus cut short.

Now we proceed to make a few observations, based on the SSS criterion for triangles. Let A be one of the poles of line l, and take any two points B and C on l such that $BC < \pi$. (See Figure 5.16.) Begin by bisecting segment BC with midpoint M, then bisecting the sub-segments BM and MC with midpoints K and L. Since $AB = AK = AM = AL = AC = \pi/2$, and

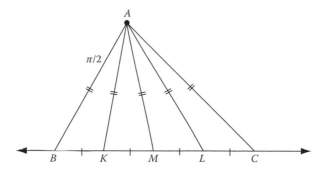

Figure 5.16 Midpoints and congruent angles.

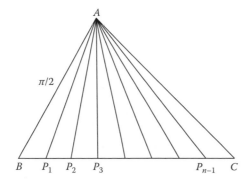

Figure 5.17 Arbitrary number of midpoints.

$BK = KM = ML = LC$, we have by SSS four congruent triangles, which imply the conclusion

$$\angle BAK \cong \angle KAM \cong \angle MAL \cong \angle LAC$$

Thus, each time a segment is bisected on the base line l, a corresponding bisection occurs in the angle opposite, A. This idea is inherent in the following result: If segment BC be divided into n congruent segments BP_1, P_1P_2, $P_2P_3, \ldots, P_{n-1}C$, and A is a pole of line BC, as shown in Figure 5.17 then

(1) $$\angle BAP_1 \cong \angle P_1AP_2 \cong \angle P_2AP_3 \cong \cdots \cong \angle P_{n-1}AC$$

From this can be proved the main idea in this section.

Theorem If B, C, and D are any three points on line l such that (BCD), and if A is a pole of l, then

(2) $$\frac{m\angle BAC}{m\angle BAD} = \frac{BC}{BD}$$

Proof Let $t = m\angle BAC/m\angle BAD$ and suppose $BC/BD \neq t$. Assume first that $BC/BD > t$, or $BC > t \cdot BD$. Construct segment BE on segment BD so that $BE/BD = t < BC/BD$ (thus $BE < BC$ and (BEC) as shown in Figure 5.18). Now bisect segment BD, then bisect the two resulting sub-segments on BD, bisect the four resulting sub-segments, and continue this bisection process indefinitely. At some stage, one of the bisecting midpoints, say P, will fall on segment EC.* Suppose there are m congruent sub-segments of length a from B to P and n from B to D.

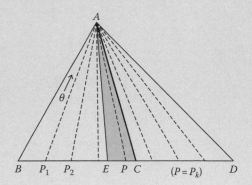

Figure 5.18 Proof of theorem.

Then

$$BP = ma, \quad BD = na$$

If θ is the measure of the angle at A created by the first sub-segment, by **(1)**,

$$m\angle BAP = m\theta, \quad m\angle BAD = n\theta$$

By algebra, $BP/BD = ma/na = m/n$ and $m\angle BAP/m\angle BAD = m\theta/n\theta = m/n$. Therefore

$$\frac{BP}{BD} = \frac{m\angle BAP}{m\angle BAD}$$

But now we obtain the contradiction

$$t = \frac{BE}{BD} < \frac{BP}{BD} = \frac{m\angle BAP}{m\angle BAD} < \frac{m\angle BAC}{m\angle BAD} = t$$

A similar contradiction is obtained if we assume that $BD/BC < t$. Therefore, $BC/BD = t$, as desired. ◣

Now in **(2)** let $BD = \pi/2$. It follows that $m\angle BAD = 90$ (why?). Hence we obtain

$$\frac{m\angle BAC}{90} = \frac{BC}{\pi/2}$$

and the result is

(3)
$$BC = \frac{\pi}{180}\, m\angle BAC.$$

---

* Here is where the Archimedean property comes in. Do you see why?

Thus, as mentioned earlier, this shows that angle measure and distance are directly connected in elliptic geometry. Indeed, if radian measure be used, the above result simplifies to

(3′) $BC = m\angle BAC$

Example

Let $\triangle ABC$ be determined on the unit sphere such that A is the North Pole and line BC is the equator, as illustrated in Figure 5.19. If $m\angle A = 75°$, then it is clear (by definition of angle measure) that $\angle BOC = 75°$, and from the traditional arc-length formula $s = r\theta$ where θ is measured in radians, $s = BC = 1 \cdot \theta = 75°$ [in radians] = $5\pi/12$. On the other hand, the above formula (3) gives us

$$BC = \frac{\pi}{180}\, m\angle A = \frac{\pi}{180} \cdot 75 = \frac{5\pi}{12}$$

in agreement.

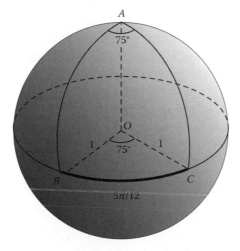

Figure 5.19 Testing equation (3′).

Problems (Section 5.4)

Group A

1. Laser beams are directed into outer space in the same plane from two points A and B 100 miles apart, as shown in Figure P.1, one making an angle of measure 89.5 and the other, 89.9 with line AB.
    (a) How can you be sure that the beams will cross paths at some point? Use anything you know from Euclidean geometry, but state the theorem you are using.
    (b) Use trigonometry to determine how far away from point A the two beams will meet (in the Euclidean plane).

Figure P.1

2. Prove Corollaries A and B in Section 5.1 ($\alpha = \infty$).

3. In Figure P.3 for this problem, the two non-triangular polygons are regular polygons and the excesses are as indicated. Find the measures of the angles of each of the three polygons.

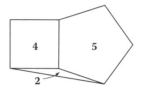

Figure P.3

4. Two regular pentagons have equal excesses. Show they are congruent. [**Hint:** Use central angles.]

5. In Figure P.5 shown, carry out the following
   (a) Find $m\angle MTP$ and calculate the excesses of the triangle and the two quadrilaterals.
   (b) Calculate the excess of pentagon $MNPQL$.
   (c) Test the additivity property of excess for this example.

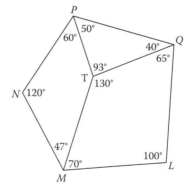

Figure P.5

Group B

The next sequence of problems (6–9) is devoted to a proof of Legendre's Second Theorem. Since the proof is to be valid in unified geometry ($\alpha = \infty$) no theorem of Euclidean geometry can be used.

6.* Give the following arguments for the first stages of Legendre's Second Theorem (Figure P.6):

(a) Show that if $\alpha = \infty$ and $\triangle ABC$ has angle-sum 180, then a right triangle having angle-sum 180 exists. (Assuming the angles at B and C are acute, show that $\alpha = \infty$ and $\triangle ABD$ has angle-sum 180, where D is the foot of A on line BC.)

(b) Construct a quadrilateral as shown in Figure P.6, and prove it is a rectangle.

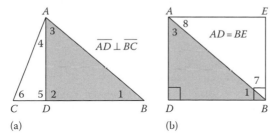

(a)   (b)

Figure P.6

7. Using the single rectangle constructed in Problem 6, show that a rectangle having arbitrarily large sides exists. (Place two rectangles side-by-side, resulting in a rectangle, and continue with this process.)

8.* Let $\triangle PQR$ be any right triangle, with right angle at R. Using the result of Problem 7, show that the angle-sum of $\triangle PQR$ is 180.

9. Using the result of Problem 8, show that all triangles have angle-sum 180, and all Saccheri quadrilaterals are rectangles, completing the proof of Legendre's Second Theorem.

10. If the three angles of a triangle on the unit sphere have measures 90°, 90°, and 10°, respectively, find the three sides to 4-decimal accuracy.

11. If the three angles of a triangle on the unit sphere have measures 90°, 90°, and 4°, respectively, find the three sides to four-decimal accuracy. If this were a problem in Euclidean geometry (where we must adjust the angles slightly, say to 88°, 88°, and 4°), could you solve this problem? Discuss.

Note: In Chapter 9, we show how to find the sides of a triangle in elliptic geometry if the angle measures are any given values A, B, and C with sum > 180. [See (7), Section **9.1**.]

Group C

12. Show that the entire elliptic plane can be expressed as the sum of eight tri-rectangular triangles and their interiors. (How many tri-rectangular triangles does it take to make a half-plane?) Thus, using additivity of excess as extended to these eight triangles, show that if the parameter for area is taken to be $k = \pi/180$, the area of the elliptic plane is 4π (the area of the unit sphere).

13. Adapt the above argument for the general elliptic plane (α arbitrary), thus generalizing **(3)**.

14. Establish the following exterior angle inequality for elliptic geometry: If $\triangle ABC$ has sides of arbitrary length and $\angle ACD$ is an exterior angle, then

(4) $$m\angle ACD < m\angle A + m\angle B$$

Group C

15. On any sphere, a **lune** is the intersection of two hemispheres, as shown in Figure P.15, left diagram.

    (a) Show that on the unit sphere the area of a lune of angle $A \equiv m\angle A$ is $\pi A/90$ [**Hint:** Use the standard area formula $S = 4\pi r^2$. What proportion of a sphere is a lune of angle A?]

    (b) Use **(a)** and a careful analysis of lunes and congruent triangles in the figure to find a formula for the (ordinary) area of $\triangle ABC$. [**Hint:** The union of the regions marked K, I, II, and III make up a hemisphere; the lunes involved are shown to the right.]

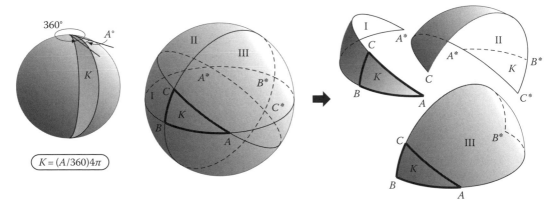

Figure P.15

5.5 Hyperbolic Geometry: Angle-Sum Theorem

The parallel postulate for hyperbolic geometry assumes that given a point A and line l not passing through it, there exist two or more lines through A which are parallel to l. Thus, **hyperbolic geometry** (or the **hyperbolic plane**) is defined as the geometry resulting from Axioms 1–12, with this hyperbolic parallel postulate. The immediate consequence, due to Theorem **1**, Section **1.5**, is that the metric is unbounded ($\alpha = \infty$). Other results presented in Section 5.1 then become valid, such as Legendre's two theorems.

A characteristic property of hyperbolic geometry, as we shall see, is that the angle-sum of all triangles is *less than* 180. Although other aspects of the hyperbolic plane may seem odd, this property by itself

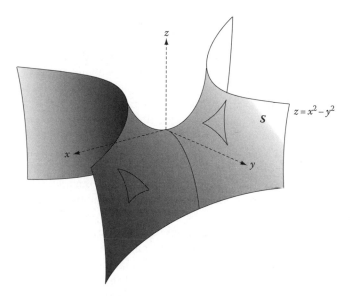

Figure 5.20 Geometry on a hyperbolic paraboloid.

is not so unusual. It is easy to visualize surfaces (different from the sphere) on which the angle-sum of triangles is less than 180. For example, consider the *saddle surface S* shown in Figure 5.20. (If you are familiar with three-dimensional coordinates, its equation is $z = x^2 - y^2$, and it is one of the hyperbolic paraboloids.) A geometry can be created on this surface, much as we have done for spherical geometry, by defining the metric in terms of arc-length for curves on S, taking as segments arcs having the shortest lengths, and geodesics (lines) as extensions of segments. Angles between rays are measured in the usual manner in terms of angles between the arcs representing those rays. The other aspects of geometry follow. It is easy to see that triangles on S have angle-sums less than 180. So, geometries for which the angle-sum of triangles is less than 180, which we derive in this section for hyperbolic geometry, might not seem so impossible after all. The real question we may ask, however, is whether the axioms for unified geometry are satisfied on such surfaces.

The basic tenets of hyperbolic geometry available for use are as follows (since $\alpha = \infty$):

- Two distinct points determine a unique line.
- Three distinct collinear points always satisfy one of three betweenness relations.
- All properties derived from the congruence theorems, and all inequalities, which now apply to all triangles.

To begin the analysis, the angle-sum for triangles will be considered. By Legendre's First Theorem, we have the inequality for any triangle ABC,

(1) $m\angle A + m\angle B + m\angle C \leq 180$

Historical Note

Although Bolyai and Lobachevski are now given credit for the "strange world" of hyperbolic geometry, their splendid accomplishments at the time were met with total indifference by the mathematical community. János Bolyai was born in 1802, and, taught by his father, he had already mastered the calculus by the time he was 13. When Gauss could not sponsor his education, Bolyai was forced to attend military school. A flamboyant Hungarian, the story is told that he once accepted the challenge from 13 officers to a duel if he could play his violin between each duel; he did so and defeated them all. His magnificent work in hyperbolic geometry included, among other things, the construction of a surface he named a *horosphere* in hyperbolic 3-space on which the geometry turns out to be the *ordinary Euclidean plane*! His work consisted of only 26 pages, printed in 1832 as an appendix to a geometry text written by his father, who sent this work to Gauss for his reaction. Gauss responded that he could not praise this work without praising his own similar work (which he had never published). After this rebuff, Bolyai published nothing else, although he left behind some 45,000 pages of manuscript (a fact that has been only recently revealed). In his last years, he learned of Lobachevski's work, and died in seclusion in 1860.

In contrast to Bolyai's free spirit, Nicholai Lobachevski (1793–1856) was a reserved scholar. He was appointed professor and rector of the University of Kasan (Russia) in 1827. His work in geometry, like Bolyai's, courageously challenged the Kantian doctrine of the universe. It included a complete theory of parallels, formulas for hyperbolic trigonometry, and the construction of a surface in hyperbolic space he called *pangeometry*, which was essentially Bolyai's horosphere. His development was published in 1829, just three years before Bolyai's, both men having worked independently. During his final years he suffered blindness and had to dictate his remaining projects. However, he held on to the belief that someday his work would win approval, as it eventually did.

The strict inequality is to be established in **(1)**, which is a characteristic property of hyperbolic geometry. If equality holds, even for a single triangle, then by Legendre's Second Theorem all triangles have angle-sum 180, and all Saccheri quadrilaterals are rectangles. It is easy to see that, in this case, any Lambert quadrilateral is a rectangle as well, since it may be "doubled" to form a Saccheri quadrilateral. Thus we have the implication as illustrated in Figure 5.21.

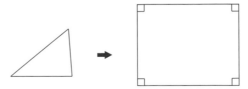

Figure 5.21 Legendre's Second Theorem.

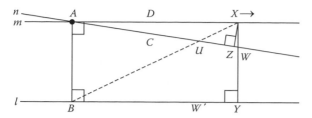

Figure 5.22 Proclus' argument.

At this point we explore an interesting, but ancient argument proposed by Euclid's first critic, Proclus (AD 410–485), which is incorrect for the purpose for which it was designed, but can be repaired for use here, enabling us to avoid a rather tedious argument. Proclus was the first notable scholar who claimed that Euclid's Fifth Postulate was unnecessary. He believed his argument accomplished that claim, but he overlooked two small details; see if you can spot them.

Proclus' Argument (Paraphrased)   Let point A be any point, and l any line not passing through A. With B the foot of A on l, consider line m, the perpendicular to line AB at A (Figure 5.22). Then $m \parallel l$. Suppose that line n is any other line passing through point A, distinct from m. Choose ray AC on n lying between ray AD on m and ray AB. With point X on ray AD, Y the foot of X on l, and Z the foot of X on n, by the crossbar theorem ray AC meets segment BX at U and by the Postulate of Pasch applied to $\triangle BXY$, either meets segment XY at some point W, or segment BY at some other point W'. Therefore, as AX becomes large without bound, XZ also becomes large without bound (a property of angles). Then as XZ becomes large, so does XW ($>XZ$). Thus the betweenness relation (XWY) cannot hold indefinitely, and at some point we must conclude that for some position of X far enough out, ray AC, and line n, meets segment BY at W'.

Conclusion   Every line besides m passing through A meets line l; thus m is the only line through A parallel to l. The Fifth Postulate is "proved."

Moment for Reflection

Does Proclus' argument actually show that only one line through A can be parallel to line l? Two major assumptions can be seen in the argument, not explicitly proven. What are they? One of them is actually correct in general, the other only if the geometry is Euclidean.

Proclus' argument yields the contradiction in hyperbolic geometry that there cannot be more than one line parallel to l and passing through A. This argument is valid if these two things are true: (a) If $XZ \to \infty$ as $AX \to \infty$, and (b) if XY is bounded as $AX \to \infty$. Showing that (a) holds is your Problem 15, Section **5.10**. For (b), note that if $\triangle ABC$ has angle-sum 180 in **(1)**, then $\Diamond ABYX$ is a rectangle, hence $XY = AB$ for all X on ray AD. Thus Proclus' argument is valid, producing a contradiction of the

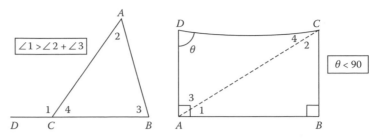

Figure 5.23 Proofs of Corollaries A and B.

hyperbolic parallel postulate. We conclude, therefore, that equality does not hold in **(1)** for any triangle in hyperbolic geometry. Thus,

---

Angle-Sum Theorem for Hyperbolic Geometry   The angle-sum of any triangle in hyperbolic geometry is less than a straight angle.

Two elementary corollaries based on the diagrams in Figure 5.23 follow.

Corollary A: Exterior Angle Inequality

The measure of an exterior angle of a triangle in hyperbolic geometry is greater than the sum of the measures of the opposite interior angles.

Proof $\angle 1 + \angle 4 = 180 > \angle 2 + \angle 3 + \angle 4$,   or $\angle 1 > \angle 2 + \angle 3$. ◣

Corollary B   Every Saccheri quadrilateral has acute summit angles. Thus the hypothesis of the acute angle holds in hyperbolic geometry.

Proof $m\angle 1 + m\angle 2 < 90$ and $\theta + m\angle 3 + m\angle 4 < 180$

Since $\angle 2 + \angle 4 = \theta$, summing gives

$$m\angle 1 + m\angle 3 + \theta + m\angle 2 + m\angle 4 < 270$$

$$90 + 2\theta < 270$$

$$2\theta < 180$$

and $\theta < 90$ as desired. ◣

5.6 A Concept for Area: AAA Congruence

Analogous to elliptic geometry and based on the angle-sum theorem for hyperbolic geometry, we define the *defect* of a triangle to be the amount its angle-sum S misses 180 by. That is, its defect is the positive number $180 - S$. More generally,

> **Definition**   The **defect** of a convex polygon P having n sides is the number
>
> **(1)** $\delta(P) = (n-2)180 - S$
>
> where S is the angle-sum of P. For triangles, $n = 3$ and the formula for $\delta(P)$ becomes $180 - S$, or
>
> **(2)** $\delta(\triangle ABC) = 180 - m\angle A - m\angle B - m\angle C.$

Just as in elliptic geometry, defect is additive, as evidenced by

> **Theorem 1**   If (BDC) in $\triangle ABC$, with $\delta_1 = \delta(\triangle ABD)$ and $\delta_2 = \delta(\triangle ADC)$, then
>
> $$\delta_1 + \delta_2 = \delta(\triangle ABC).$$

The proof is virtually the same as that for the additivity of excess, so it will be omitted (you should try showing this). Thus, as in elliptic geometry, one may define for any convex polygon P in the hyperbolic plane

(3)   Area $P = k\delta(P)$

where k is any appropriate constant for the entire hyperbolic plane. The value for k can be any positive real; in the unit disk model for hyperbolic geometry, $k = 1$.

Example 1

Suppose that a regular *decagon* has angles of measure 140. To find its area, first calculate its angle-sum: $S = 10 \cdot 40 = 1400$. Hence by definition, Area $P = k[(10 - 2)180 - 1400] = 40k$.

Example 2

The defect of a pentagon is 20. Three of its angles are congruent, having measure 116 each, and a fourth angle has measure 86. Show that the fourth and fifth angles are also congruent.

Solution

If S is the angle-sum, by definition of excess, $20 = 3 \cdot 180 - S$, so $S = 540 - 20 = 520$. The sum of the four given angles is thus $3 \times 116 + 86$ or 434, leaving the difference $520 - 434 = 86$ for the fifth angle, showing that it is congruent to the fourth angle.

Example 3

In Figure 5.24 is shown a regular hexagon that is adjacent to a regular quadrilateral, which are, in turn, adjacent to two congruent isosceles triangles. Angle measures are as marked. Find the defect of the hexagon, quadrilateral, and the two triangles.

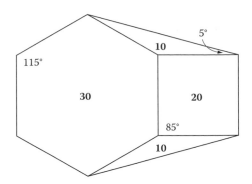

Figure 5.24 Example 3.

Solution

δ(Hexagon) = (4)180 − 6·115 = 720 − 690 = 30;
δ(Quadrilateral) = (2)180 − 4·85 = 360 − 340 = 20; the vertex angle of the triangle being 160, δ(Triangle) = 180 − 2·5 − 160 = 10.

Moment for Reflection

Find the angles of the outer hexagon of Example 3, then calculate its defect. Do your calculations lead to any significant conclusions?

A result of the additive property of defect leads to the peculiar phenomenon of hyperbolic geometry (analogous to elliptic geometry) that *there exist no non-congruent similar triangles*. Suppose two non-congruent triangles $\triangle ABC$ and $\triangle XYZ$ have congruent angles. We may assume that $AB > XY$ and $BC > YZ$ and construct segments $\overline{BP} \cong \overline{XY}$ on $\overline{AB}$ and $\overline{BQ} \cong \overline{YZ}$ on $\overline{BC}$. It follows that $\triangle PBQ \cong \triangle XYZ$ with the defects of the triangles as indicated in Figure 5.25, we have, by Theorem 1, $\delta_1 = \delta(\triangle XYZ) = \delta(\triangle ABC) = \delta_1 + \delta_2 + \delta_3 > \delta_1$, a contradiction. Thus

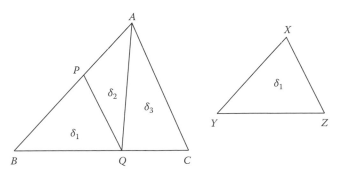

Figure 5.25 Proof of AAA congruence.

AAA Congruence Criterion for Hyperbolic Geometry If two tri-angles have the three angles of one congruent, respectively, to the three angles of the other, they are congruent.

Another result analogous to elliptic geometry will be needed later. It shows that hyperbolic geometry, like elliptic geometry, is Euclidean *in the small*. That is, if the sides of a triangle or polygon decrease in size, the angle-sum becomes, in the limit, the value given by Euclidean geom-etry. Since the proof is substantially the same as in elliptic geometry, it is unnecessary to repeat the details here; we refer the reader to the theorem in Section **5.2** for that proof.

> ***Theorem 2*** If the sides of a triangle become arbitrarily small, the angle-sum of the triangle converges to 180.

5.7 Parallelism in Hyperbolic Geometry

We now consider how parallel lines behave in hyperbolic geometry. The basis for this analysis is that used for the proof of the angle-sum theorem, which employed Proclus' argument. Let l be a line, A any point not on it and $AB \perp l$ at B; construct a Lambert quadrilateral $\lozenge BCDA$, with right angles at B, C, and D, where $\angle BAD$ is known to be an acute angle (Figure 5.26).

Suppose an arbitrary point X is taken on ray $h \equiv \overrightarrow{BC}$, and let $BX \to \infty$. Ray AX then seems to approach ray $AD \equiv k$ as its limiting position. In order to carefully investigate this conjecture, define the set of numbers $\{\theta = m\angle BAX\}$. In view of the inequality $m\angle BAX < 90$ for all $X \in h$, this set is bounded, so one may define the real number

(1) $\gamma = \mathrm{LUB}\{\theta = m\angle BAX : X \in h\} \leq 90$

Now construct the ray $k^* \equiv \overrightarrow{AK}$ such that $m\angle BAK = \gamma$ and lying on the same side of line AB as h. (Note that $\gamma < 90$ since either k^* lies between rays AB and k, or $k^* = k$, as will be shown momentarily.)

A few interesting facts about ray k^* emerge. First, in view of its defini-tion as the limiting position of rays which intersect h, it might seem sur-prising that k^* itself does *not* intersect h. However,

(2)   Ray k^* is parallel to ray h

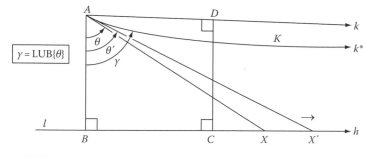

Figure 5.26 The angle of parallelism.

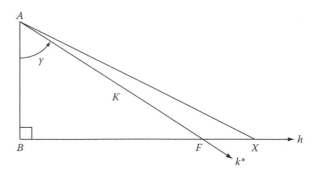

Figure 5.27 Proof of **(2)**.

Proof Suppose $k*$ meets h at some point F, as in Figure 5.27. Choose X on h such that (BFX). We then obtain the following contradiction

$$\gamma = m\angle BAF < m\angle BAX \leq \gamma \text{ \textbackslash}$$

Another fact is

(3)   Any ray u between ray AB and $k*$ meets h

Proof As shown in Figure 5.28, suppose $u = \overrightarrow{AL}$ lies between rays AB and $k* = \overrightarrow{AK}$. If $\omega = m\angle KAL$, then $m\angle BAL = \gamma - \omega < \gamma$. Since γ is the *least* upper bound for the set $\{m\angle BAX\}$ for $X \in h$, then $\gamma - \omega$ cannot be an upper bound. Accordingly, there exists $X \in h$ such that $m\angle BAX > \gamma - \omega = m\angle BAL$. Thus ray u must lie between rays AB and AX. By the crossbar principle, u must meet segment BX, hence ray h, as desired. $\text{\textbackslash}$

Note: An intuitive rendering of **(2)** and **(3)** is that $k*$ is the "first" parallel ray from A as ray u rotates in the counterclockwise direction, starting with $u = \overrightarrow{AB}$. This proves that k cannot lie between rays AB and $k*$, justifying a statement about $k*$ made earlier. (One can also show that $k* \neq k$.) $\text{\textbackslash}$

Definition 1 The ray $k*$ constructed as above is said to be **asymptotically parallel** (sometimes, **limit parallel**) to ray h, (Figure 5.26). The angle $\angle BAK$, as well as its measure γ, is called the **angle of parallelism** with respect to segment AB. The line containing ray $k*$ is also said to be **asymptotically parallel** to the line containing ray h (line l).

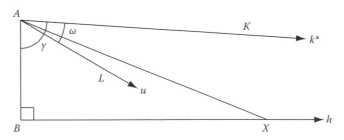

Figure 5.28 Proof of **(3)**.

The results **(2)** and **(3)** actually characterize asymptotic parallels, as the following result shows. This result provides a strictly synthetic approach for dealing with asymptotic parallels without using the value γ, which turns out to be quite useful.

Theorem 1   Let $AB \perp h$ at B (Figure 5.29). The ray $AW \equiv w$ is asymptotically parallel to ray h iff the following is true:

(a) w lies on the same side of line AB as h,
(b) w is parallel to h, and
(c) every ray u between w and $v \equiv \overrightarrow{AB}$ meets h.

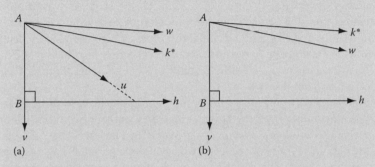

(a) (b)

Figure 5.29 Proof of Theorem 1.

Proof   Certainly the asymptotic parallel k^* has the three properties mentioned. Conversely, suppose ray w has these properties. If $w \neq k^*$, then either (wk^*v) as in Figure 5.29a, or (k^*wv) as in Figure 5.29b. But in either case, the properties which w and k^* have produce the contradiction that either k^*, or w, meet s ray h. Therefore, $w = k^*$ and w is asymptotically parallel to h. ∖

Recall that the asymptotic parallel k^* is regarded as the first parallel ray that is parallel to h as a ray u concurrent with h rotates counterclockwise. If, however, $u \equiv \overrightarrow{AY}$ rotates clockwise, with Y on the ray h' opposite h, starting with $\overrightarrow{AB}$, then one obtains the first ray parallel to h' lies on the other side of line AB (Figure 5.30). To be more precise, let Y be a point on h' and let $BY \to \infty$, as previously with BX. Construct the angle of parallelism $\angle BAK'$ on the h' side of line AB having measure

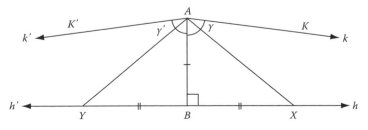

Figure 5.30 Angles of parallelism on either side of AB.

(4) $\gamma' = \text{LUB}\{m\angle BAY : Y \in h'\} \leq 90$

Then $\angle BAK'$ is the angle of parallelism for h', and $k'^* = \overrightarrow{AK'}$ is limit-parallel to h'. By symmetry it seems clear enough that $\gamma' = \gamma$. To prove this, note that one can construct congruent triangles ABY and ABX by letting $BY = BX$, so that $m\angle BAY = m\angle BAX$. Then the two sets of numbers $\{m\angle BAY\}$ and $\{m\angle BAX)\}$ are the same, so that, indeed, $\gamma' = \gamma$.

Thus we have the two limit parallels k and k' to h and h' on the two sides of line AB, with congruent angles of parallelism. Suppose we now consider *all* the lines passing through point A.

These lines fall into precisely three classes, as shown in Figure 5.31: The **intersectors** (those lines which intersect line l), the two **asymptotic parallels** (the lines containing the rays from A that are asymptotically parallel to h and h'), and the remaining lines, which are also evidently parallel to l (said to be **hyperparallel**, or **divergently parallel**, to l). Thus it follows that there are an infinite number of lines passing through A parallel to line l, not just two.

Another interesting fact concerning parallelism in hyperbolic geometry is a theorem proven by Hilbert. It asserts that any two lines that are hyperparallel *possess a common perpendicular*. Thus in Figure 5.31, each of the parallels lying between the two asymptotic parallels has a perpendicular line in common with line $l = h \cup h'$ (the converse is clear since the perpendicular to line AB at A evidently lies between the limit parallels and is thus hyperparallel to l). Moreover, two hyperparallel lines are *divergent* in the sense that as we move out further on the two lines, the distance PQ (Figure 5.32) from one line to the other not only is not constant, but

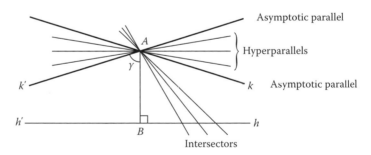

Figure 5.31 The lines through point A.

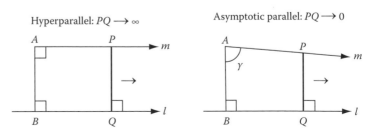

Figure 5.32 Two types of parallelism.

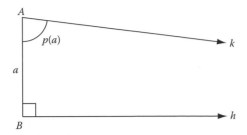

Figure 5.33 Angle of parallelism as a function of distance.

it grows *large without bound* as $AP \to \infty$. In stark contrast, the distance from an asymptotic limit parallel to *l converges to zero* as $AP \to \infty$. Thus emerges the reason for the terms being used. The proofs of these two phenomena will appear in the following problem section (Problems 16, 17).

Note: Virtually all this work is valid in unified geometry, with $\alpha = \infty$. The angle of parallelism can still be defined (as above) with measure $\gamma \leq 90$, which changes very little in the ensuing results, even if $\gamma = 90$. The only need to have $\gamma < 90$ occurred in the discussion surrounding Figures 5.31 and 5.32, where hyperparallels were involved. An interesting exercise in logic is to see what implications the above analysis brings if $\gamma = 90$. Is it all trivial, or are there any new results for unified geometry? ◤

We end this section by considering a function introduced by both Bolyai and Lobachevski in their work on hyperbolic geometry. Consider the angle of parallelism γ with respect to segment $AB \perp h$ (where h can be taken on either side of line AB). If $\overline{CD}$ is any segment congruent to $\overline{AB}$ and $\overline{CD} \perp \overline{CE}$, let γ' be the angle with respect to segment CD and ray $CE \equiv h'$. By construction of the appropriate congruent angles (as in Figure 5.30) one can observe that $\gamma' = \gamma$. Thus, the angle of parallelism depends only on the distance $AB \equiv a$. One can then meaningfully define the function $p(a) \equiv \gamma$ as the *angle of parallelism with respect to the distance a*, illustrated in Figure 5.33. An important property which $p(a)$ has involves what happens when the distance a increases.

> **Theorem 2**   The function p defined above is nonincreasing. That is, given $a < b$, $p(a) \geq p(b)$.
>
> *Proof*   In Figure 5.34 let X and Y be points on the perpendicular to h at B such that (BXY), with $BX = a < b = BY$. Also, let rays $k = XK$ and $k' = YK'$ be the asymptotic parallels to h, with $\gamma = p(a)$ and $\gamma' = p(b)$. We are to prove that $\gamma \geq \gamma'$. Suppose $\gamma < \gamma'$. Then construct ray YF on the K'-side of line YB such that $\angle FYB \cong \angle KXB$ and $(\overrightarrow{YK'}\ \overrightarrow{YF}\ \overrightarrow{YB})$ by the angle construction theorem. By **(3)** ray YF must meet h at some point W, and by the postulate of Pasch applied to triangle YBW, line XK must intersect one of the sides YW or BW. But by the corollary of Theorem 2, Section 5.1, lines XK and YF are parallel, so $\overleftrightarrow{XK}$ meets $\overline{BW}$, hence k meets h, contradicting **(2)**. Therefore, $\gamma \geq \gamma'$. ◤

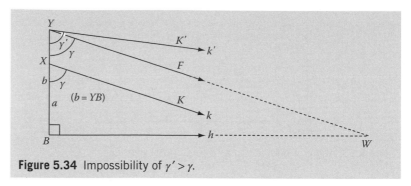

Figure 5.34 Impossibility of $\gamma' > \gamma$.

In the next section the strict inequality $p(a) > p(b)$ will be established. Thus, p is strictly decreasing and has a unique inverse $a = p^{-1}(\gamma)$. The implication is that one can find a distance (or segment AB) such that an angle of any given positive measure is the angle of parallelism with respect to point A and perpendicular ray $h \equiv BX$. For example, there is a segment AB whose corresponding angle of parallelism is only $1°$—an incredible result. The actual length a for this case can be determined by setting $\gamma = 1/180$ (rad) and solving for a using Bolyai's formula:

(5) $$\gamma = p(a) = 2\tan^{-1} e^{-a}$$

This relation was proved independently by both Bolyai and Lobachevski in their work on hyperbolic geometry, using hyperbolic trigonometry. A proof will appear later (see Section 9.9). This formula shows that the angle of parallelism corresponding to the distance $1/10$ ($a = 0.1$) is approximately $84°$.

Further results involving parallelism in hyperbolic geometry will be taken up in the next (optional) section. These ideas are quite interesting, and certainly corroborate Bolyai's declaration in a letter to his father in 1831 "I have discovered such magnificent things that I am myself astonished at them… Out of nothing I have created a strange new world." The reader should at least skim the pages of the next section to understand Bolyai's fascination.

5.8 Asymptotic Triangles in Hyperbolic Geometry

A concept which Bolyai invented provides another aspect of parallelism, one which makes it easier to prove symmetry and transitivity for limit parallels (there is no transitivity law for divergent parallelism). In fact, the very definition of *lines* which are limit- or divergent-parallel depends on this approach. The analysis used to define the limit parallel ray k^* to h (Figure 5.26) in the previous section will be used here in a more general situation.

Starting with any angle $\angle ABC$, which need not be a right angle, one repeats verbatim the construction in the definition in the previous section. (Figure 5.26), as illustrated in Figure 5.35, by defining the real number

$$\gamma = \text{LUB}\left\{ m\angle BAX : X \in h \equiv \overrightarrow{BC} \right\}$$

and, as before, constructing the ray $k = \overrightarrow{AK}$ on the C-side of line AB such that $\angle KAB = \gamma$. In this case, depending on circumstances, $\angle KAB$ can be an obtuse or right angle. (That is, $\gamma > 90$ is possible here.)

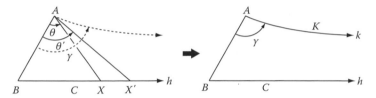

Figure 5.35 General definition of asymptotic parallelism.

The properties cited in the previous section for Theorem 1, are also true in this more general situation, so we adopt a definition that eliminates the use of the least upper bound γ.

Definition 1   Ray AC is said to be **asymptotically parallel** to ray BD, denoted by $\overrightarrow{AC} \|_a \overrightarrow{BD}$, iff the following is true:

  (a) $\overrightarrow{AC}$ and $\overrightarrow{BD}$ lie on the same side of line AB,

  (b) $\overrightarrow{AC}$ does not meet $\overrightarrow{BD}$, and

  (c)   every ray AR between rays AC and AB meets ray BD.

If rays AC and BD satisfy (a) and (b) but not (c), then ray AC is said to be **divergently parallel** to BD, denoted by $\overrightarrow{AC} \|_d \overrightarrow{BD}$.

Note that the configuration in Figure 5.35 (right diagram) consists of a segment AB and two rays, one asymptotically parallel to the other, which intuitively converge to a "point at infinity." It could be thought of as a triangle having two sides of infinite length and a vertex at infinity. Traditionally (as in Bolyai's work) a Greek character Ω (omega) is used to represent the missing vertex—the point at infinity. This is merely for convenience, and it must be remembered that Ω does not represent an ordinary point. Some authors refer to it as an "ideal point." To think of it as a point at infinity may help your intuition, but care must be taken, obviously (see Figure 5.36).

Definition 2   An **asymptotic triangle** is the set of points consisting of an ordinary segment AB and two rays AC and BD such that $\overrightarrow{AC} \|_a \overrightarrow{BD}$, called its **sides**, with A, B, and Ω its **vertices** (see Figure 5.36). Such a set is denoted by the symbol $\triangle AB\Omega$. Segment AB is called the **base**, and the two rays AC and AD, its **legs**. The **angles** (also, **base angles**) of $\triangle AB\Omega$ are the angles $\angle BAC$ and $\angle ABD$, and the **angle-sum** is $m\angle A + m\angle B$. A line is said to **contain** or **pass through** the vertex Ω iff it contains a ray PQ such that $\overrightarrow{PQ} \|_a \overrightarrow{BD}$, for some P on line AB.

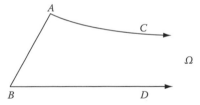

Figure 5.36 Asymptotic triangle.

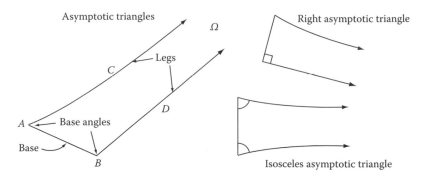

Figure 5.37 Types of asymptotic triangles.

Note that since symmetry has not yet been established (but almost certain to be true), the ordering in the notation $\Delta AB\Omega$ is intended to mean, as above, that $\overrightarrow{AC}\|_a\overrightarrow{BD}$, and not necessarily $\overrightarrow{BD}\|_a\overrightarrow{AB}$ (in which case one would assert that $\Delta BA\Omega$ is an asymptotic triangle).

An asymptotic triangle has three "sides," but only two real angles—the angles $\angle A$ and $\angle B$ in Figure 5.37. The only side whose length is defined is the base AB. Thus, the measurable parts of an asymptotic triangle $\Delta AB\Omega$ are, with standard notation, as indicated in the following:

$$\overline{AB} \;\;\rightarrow\;\; AB \equiv c$$

$$\angle A \;\;\rightarrow\;\; m\angle A \equiv A$$

$$\angle B \;\;\rightarrow\;\; m\angle B \equiv B$$

Asymptotic triangles can therefore come in all sizes and shapes (Figure 5.37). Special types of such triangles are: A **right asymptotic triangle**—an asymptotic triangle one of whose base angles is a right angle, and an **isosceles asymptotic triangle**—one whose base angles are congruent. We will say that two asymptotic triangles are **congruent** whenever, under some correspondence, the measurable corresponding parts of the two triangles are congruent.

Asymptotic triangles behave much the way ordinary triangles do in unified geometry. In fact, every property of a unified triangle ΔABC that does not involve vertex C (an ordinary point) will reflect some valid property of asymptotic triangle $\Delta AB\Omega$ in the hyperbolic plane. One notable example is the fact that, in standard notation (where A represents $m\angle CAB$, for instance),

(1) $A + B < 180$

In the special case of an isosceles asymptotic triangle, the proof of this result is quite straightforward, so we start there.

Theorem 1   If $\triangle AB\Omega$ is an isosceles asymptotic triangle, the base angles are acute.

Proof   If $A = B > 90$, then construct perpendiculars AE and BF to line AB, as in Figure 5.38. Ray AE then lies between rays AC and AB, and AF lies between AB and AD. Since $\overrightarrow{AC} \parallel_a \overrightarrow{BD}$, ray AE meets ray BD at some point P. But the crossbar principle implies that ray BF meets segment AP at an interior point. That is, lines AE and BF intersect, contradicting the fact that they have a perpendicular line in common and are parallel. Next suppose that $A = B = 90$. But this implies that the angle of parallelism at A is a right angle, again, a contradiction. Hence $A = B < 90$. ◣

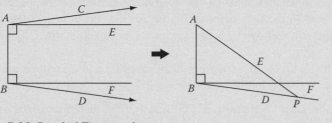

Figure 5.38 Proof of Theorem 1.

In order to prove **(1)** in general, suppose that $\triangle AB\Omega$ is any asymptotic triangle and that $A + B \geq 180$. The two cases are shown in Figure 5.39. If $A + B > 180$, then in the figure, $\angle 1 + m\angle ABD = 180 < m\angle ABD + m\angle BAC$, which reduces to $\angle 1 < m\angle BAC$. Therefore, we can construct ray AE between rays AC and AB such that $m\angle EAB = \angle 1$. But then ray AE must meet ray BD, contradicting $\overleftrightarrow{AE} \parallel \overleftrightarrow{BC}$ ($\angle EAB$ and $\angle 1$ are congruent alternate interior angles as in Theorem 2 of Section **5.1**). Therefore, we cannot have $A + B > 180$.

A similar proof for the case $A + B = 180$, shown in the diagram at right in Figure 5.39, is also valid, but it poses a technical problem. The obvious construction is to take M as the midpoint of segment AB, then drop perpendiculars MP and MQ to lines AC and BD, respectively. As in Problem 9, Section **3.8**, this will produce a common perpendicular PQ between lines AC and BD. This no doubt contradicts asymptotic parallelism, as suggested by Figure 5.31 (asymptotic parallels evidently cannot have perpendiculars in common). Specifically what needs to be concluded in Figure 5.39 is that $\triangle PQ\Omega$ is an asymptotic triangle, which would contradict $A + B < 180$ for an isosceles asymptotic triangle, already established.

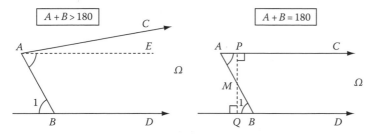

Figure 5.39 Impossibility of $A + B \geq 180$.

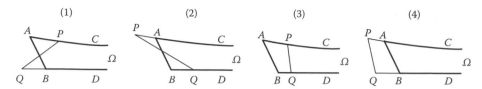

Figure 5.40 The four cases of Theorem 2.

Even though it will take some extra work to complete the proof, the results will be rewarded by other applications that give us results that are more easily obtained, one of them being *symmetry* (if $\triangle AB\Omega$ is an asymptotic triangle, then so is $\triangle BA\Omega$). One needs to establish the phenomenon exhibited in the four cases of Figure 5.40, where P and Q lie on lines AC and BD, that $\triangle PQ\Omega$ is an asymptotic triangle (i.e., if $\overrightarrow{AC} \parallel_a \overrightarrow{BD}$, then $\overrightarrow{PC} \parallel_a \overrightarrow{QD}$).

The major details in the proofs of cases (1) and (2) will be presented; cases (3) and (4) follow by simple logic.

1. In case (1) suppose (APC) and (QBD), and that $\triangle AB\Omega$ is an asymptotic triangle (Figure 5.41). Since AC (therefore PC) is parallel to BD and lies on the same side of AB as BD (and QD), it suffices to prove that any ray PW between rays PC and PQ will meet ray QD. Take U on ray PW, and consider ray AU. It follows that since $U \in \text{Int}\angle BAC$ (proof?), ray AU meets ray BD at some point V, and (AUV) holds. Applying the postulate of Pasch to $\triangle ABV$, ray PW meets either side AB or side BV, so it must meet either segment TB (hence segment QB), or segment BV. In either case, ray PW meets ray QD, as desired. ⬄

2. If (PAC) and (BQD) as in Figure 5.42, again we must prove that any ray PW between PC and PQ meets ray QD. Ray PW must meet segment AT at some point K by the crossbar theorem, and (AKB) holds. Construct ray AU on the B-side of line PC such that $\angle CAU \cong \angle APW$; since $m\angle CAT > m\angle APT > m\angle APW = m\angle CAU$, ray AU lies between rays AC and AB. Then ray AU meets ray BD at some point V, and by the postulate of Pasch applied to $\triangle ABV$, ray PW meets either side AV or side BV. But lines PW and AU make congruent corresponding angles with respect to transversal PC and by the corollary of Theorem **2**, Section **5.1**, the two lines are parallel. Thus PW cannot meet AV, so it must meet BV, hence ray QD. ⬄

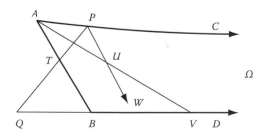

Figure 5.41 Proving case (1).

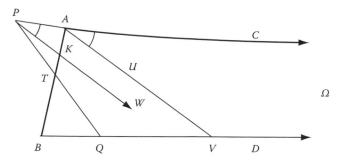

Figure 5.42 Proving case (2).

Note: At this point we can complete the proof of **(1)** since the first two cases just proven cover the needed step in the argument begun earlier for disproving the case $A + B = 180$. ◢

The proofs for cases (3) and (4) follow easily by locating points R and S strategically on lines AC and BD, and applying cases (1) and (2). The general result is:

> **Theorem 2**   If $\triangle AB\Omega$ is an asymptotic triangle, and P and Q are any two points lying on lines AC and BD, respectively, then $\triangle PQ\Omega$ is also an asymptotic triangle.

Since **(1)** is now valid, we can use it to obtain a prominent feature of asymptotic triangles analogous to ordinary triangles. In Figure 5.43, consider the exterior angle $\angle EBA$ of asymptotic triangle $AB\Omega$. From **(1)** it follows that $m\angle 1 = 180 - B > A$. Thus,

(2)   An exterior angle of an asymptotic triangle has measure greater than that of the opposite base angle.

(One obtains an even stronger analogy by regarding the "angle" at Ω as having measure zero, so that $m\angle 1 > m\angle \Omega$—the "opposite interior angle.")

A corollary of Theorem 2 is an important property of asymptotic parallelism, which can now be proved without difficulty.

(3)   If $\triangle AB\Omega$ is an asymptotic triangle, the lines l and m which contain the legs AC and BD have no perpendicular line in common.

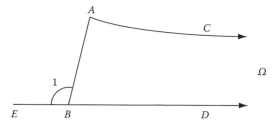

Figure 5.43 Exterior angle inequality: $\angle 1 > \angle A$.

Proof   Suppose line PQ is a perpendicular in common with lines AC and BD, with P and Q on AC and BD. Then by Theorem 2, $\Delta PQ\Omega$ is an asymptotic triangle having supplementary base angles, contradicting (**1**). ◥

For the symmetry property mentioned earlier, this is proved first for isosceles asymptotic triangles, then extended to the general case.

Theorem 3   If $\overrightarrow{AC} \parallel_a \overrightarrow{BD}$ and $\angle CAB \cong \angle ABD$, then $\overrightarrow{BD} \parallel_a \overrightarrow{AC}$.

Proof   Let $u \equiv \overrightarrow{BW}$ be any ray between rays BD and BA, to prove that u meets ray AC (Figure 5.44). On the C-side of line AB construct $\angle BAF \cong \angle ABW$. Since $m\angle BAF = m\angle ABW < m\angle ABD$, then $m\angle BAF < m\angle BAC$ (using the isosceles triangle hypothesis) and ray AF lies between rays AB and AC. Since $\overrightarrow{AC} \parallel_a \overrightarrow{BD}$, ray AF meets ray BD at some point P. Locate Q on ray AC so that $AQ = BP$. By SAS, $\Delta ABQ \cong \Delta BAP$, and by CPCPC, $m\angle ABQ = m\angle BAF = m\angle ABW$, so that ray BW coincides with ray BQ, which meets ray AC, as desired. ◥

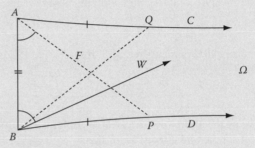

Figure 5.44 Symmetry for isosceles asymptotic triangles.

In order to extend this result to arbitrary asymptotic triangles, we need a new idea. Let $\Delta AB\Omega$ be any asymptotic triangle, with $\overrightarrow{AC} \parallel_a \overrightarrow{BD}$. Construct the bisectors of the base angles (Figure 5.45), which meet at a point W equidistant from the base AB and the legs, and, accordingly, triangle WPQ is isosceles, and it follows that $\angle QPC \cong \angle PQD$ so that asymptotic triangle $PQ\Omega$ is isosceles. Then since symmetry has been established for isosceles asymptotic triangles, $\Delta QP\Omega$ is asymptotic, and by Theorem 2, $\Delta BA\Omega$ is also asymptotic, with $\overrightarrow{BD} \parallel_a \overrightarrow{AC}$. This proves the

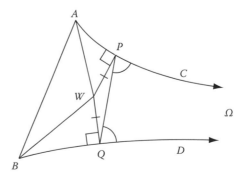

Figure 5.45 Symmetry for arbitrary asymptotic triangles.

Symmetric Law for Asymptotic Parallelism If $\overrightarrow{AC} \parallel_a \overrightarrow{BD}$, then $\overrightarrow{BD} \parallel_a \overrightarrow{AC}$.

The symmetry law for asymptotic parallelism directly implies that for divergent parallelism, if $h \parallel_d k$ where $h = \overrightarrow{AC}$ and $k = \overrightarrow{BD}$ then (a) and (b) of Definition 1 are satisfied, but not (c). If k were not divergently parallel to h, then since (a) and (b) of Definition 1 are satisfied for k and h, (c) must also be satisfied for k and h, in that order, hence $k \parallel_a h$. Thus, by symmetry, $h \parallel_a k$, and (c) is, on the contrary, satisfied. This contradicts our assumption and proves

Symmetric Law for Divergent Parallelism If $\overrightarrow{AC} \parallel_d \overrightarrow{BD}$, then $\overrightarrow{BD} \parallel_d \overrightarrow{AC}$.

Definition 3 Lines l and m are said to be **asymptotically parallel** ($l \parallel_a m$) iff l and m contain asymptotically parallel rays h and k. Lines l and m are **divergently parallel** ($l \parallel_d m$) iff l and m contain divergently parallel rays h and k.

Note that built into this definition is the symmetry of both asymptotic and divergent parallelism for lines. It may surprise the reader to learn that there is *no transitive law for divergent parallelism*. But a simple counter example is provided by Figure 5.31 (can you find it?). However, asymptotic parallelism is transitive. This rather tedious proof will be omitted here; it is not actually used for the propositions of this section.

Transitive Law for Asymptotic Parallelism If $\overrightarrow{AC} \parallel_a \overrightarrow{BD}$ and $\overrightarrow{BD} \parallel_a \overrightarrow{EF}$, then $\overrightarrow{AC} \parallel_a \overrightarrow{EF}$. Thus, for lines, if $l \parallel_a m$ and $m \parallel_a n$, then $l \parallel_a n$.

Two congruence theorems for asymptotic triangles may be established. If so inclined, you will be allowed to "discover" one of these for yourself in the following Moment for Reflection.

Moment for Reflection

Consider the two asymptotic triangles shown in Figure 5.46. Suppose that $AB = XY$ and $\angle B \cong \angle Y$, as indicated. Does it seem that $\angle A$ ought to be congruent to $\angle X$? Let's analyze this carefully. Assume, for sake of argument, that $m\angle A > m\angle X$, and construct ray AT between rays AP and AB such that $\angle BAT \cong \angle X$. What must happen? The figure indicates where to proceed from here. See if you can obtain a contradiction.

If $m\angle BAC > m\angle YXU$

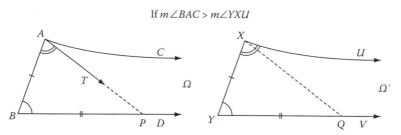

Figure 5.46 Base-angle congruence?

If you were successful in the above analysis, then you proved the following congruence theorem.

> ***Theorem (BA Congruence Criterion)***   If two asymptotic triangles have congruent bases and a pair of congruent base angles, they are congruent, with the remaining pair of angles congruent.

The other congruence criterion is as follows:

> ***Theorem (AA Congruence Criterion)***   If two asymptotic triangles have the base angles of one congruent to the base angles of the other, they are congruent.

Proof   Suppose $\triangle AB\Omega$ and $\triangle XY\Omega'$ have $\angle A \cong \angle X$ and $\angle B \cong \angle Y$, as in Figure 5.47; then $\overline{AB} \cong \overline{XY}$ must be shown. Suppose that $AB > XY$. Locate E on $\overline{AB}$ such that $AE = XY$. Let ray $\overrightarrow{EF}$ be constructed asymptotically parallel to ray BD. We will prove that $\overrightarrow{AC} \parallel_a \overrightarrow{EF}$. By the BA congruence criterion, $\triangle EB\Omega \cong \triangle XY\Omega$ and $\angle BEF \cong \angle YXU \cong \angle BAC$ (by hypothesis).

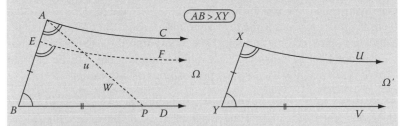

Figure 5.47 AA congruence.

Hence lines AC and EF are cut by a transversal such that a pair of corresponding angles are congruent, thus are parallel by the corollary of Theorem 2, Section 5.1. Then properties (a) and (b) of the definition of asymptotic parallelism hold. For (c), let ray $u \equiv \overrightarrow{AW}$ be any ray between rays $\overrightarrow{AE}$ and $\overrightarrow{AC}$, to show that u meets ray $\overrightarrow{EF}$. Since $\overrightarrow{AC} \parallel_a \overrightarrow{BD}$, u meets ray BD at some point P, and by the postulate of Pasch applied to $\triangle ABP$, ray EF meets either side AP or BP. But ray EF is parallel to ray BP by construction, hence must meet segment AP. That is, u = ray AP meets ray EF proving that $\triangle AE\Omega$ is an asymptotic triangle. But $\angle EAC$ and $\angle AEF$ are supplementary, contradicting **(1)**. Hence $AB = XY$, as desired. ◥

A corollary of AA congruence has implications for a result mentioned earlier. Recall the angle of parallelism function $p(a)$ defined in Section 5.2. Theorem **2**, Section **5.2** established the inequality $p(a) \geq p(b)$ if $a < b$. The strict inequality can now be proven.

> ***Corollary***   If $p(a) = p(b)$, then $a = b$. Thus if $a < b$, then $p(a) > p(b)$.

> *Proof*   If $p(a) = p(b)$, then the right angled asymptotic triangles which they define (Figure 5.48) are congruent by the above theorem, hence $a = b$. ◥

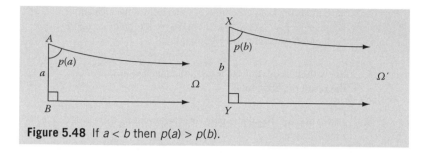

Figure 5.48 If $a < b$ then $p(a) > p(b)$.

Problems (Section 5.8)

Group A

1. If an equilateral triangle has angles of measure 55 each, find its defect.

2. Two regular polygons having defects 6 and 18 are shown in Figure P.2, with a triangle whose defect is 4. Find the angle measures of the triangle.

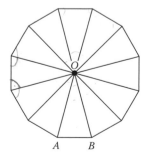

Figure P.2

3. The defect of the regular dodecagon shown is 12. If O is its center, find the measure of
   (a) each angle of the polygon.
   (b) each angle of the congruent sub-triangles (such as $\triangle OAB$) (Figure P.3).

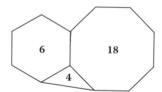

Figure P.3

4. In elliptic geometry there exist equilateral triangles whose angles are right angles. In hyperbolic geometry, it can be shown that there exist equilateral triangles whose angles have measure 45. By the imagery of assembling such equilateral triangles, show that there exists a regular octagon each of whose angles are right angles.

5. Show that the area of the hyperbolic plane is infinite. (Use the result of Problem 4.)

6. Prove that two regular n-gons are congruent iff their areas are the same.

7. One of the characteristics of hyperbolic geometry is that every angle has infinitely many interior points each of which do not lie on a **transversal** of the angle (a segment extending from one side of the angle to the other), as illustrated in Figure P.7. Find an example of an angle and such a point in Figure 5.31.

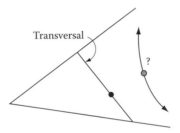

Figure P.7

8. The phenomenon discussed in Problem 7 takes on dramatic proportions when one considers the following construction. Locate such a point P within right angle A not lying on a transversal of $\angle A$, as shown in Figure P.8, and let points X and Y be chosen on the sides such that $AX = AY$, creating an isosceles right triangle. As $AY = AX \to \infty$, what becomes of line XY? Discuss.

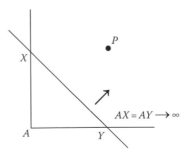

Figure P.8

9. Using Bolyai's formula **(5)**, show that unit distance corresponds to an angle of parallelism of 40.395°, and then find the distance a that will produce an angle of parallelism of 1° (Figure P.9).

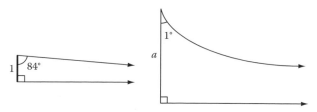

Figure P.9

Group B

10.* Show that given any three non-intersecting lines *l*, *m*, and *n* such that no line intersects all three of them, then each of the lines lies between the other two (i.e., *l* lies in both the half-plane **H** determined by *m* containing *n*, and in the half-plane **H'** determined by *n* containing *m*, and similarly for *m* and *n*). This seemingly incredible result can occur in the hyperbolic plane, and can be seen without difficulty in the model considered later (Chapter 9).

11. **Equilateral Triangles in Hyperbolic Geometry** The angles of an equilateral triangle in the hyperbolic plane measure less than 60, and the angles decrease to zero as the sides increase without bound. For example, each angle of a unit equilateral triangle measures approximately 53°; if one doubles the size of the triangle, the angles decrease to approximately 38° (Figure P.11). The connecting relation between side *a* and angle *A* of an equilateral triangle can be shown in hyperbolic trigonometry to be sec *A* = 1 + sech *a*, or equivalently,

$$(4) \qquad\qquad \sec A = 1 + \frac{2e^a}{e^{2a}+1}$$

Use this to find the measure of each angle of an equilateral triangle whose sides are 3 units in length. Experiment with your pocket calculator to find how large *a* must be in order for the angles to measure approximately 1°. (See Problem 9, Section **6.12**.)

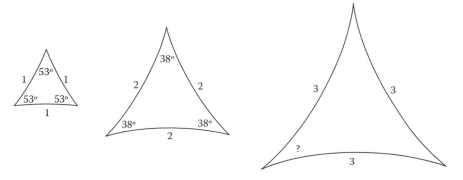

Figure P.11

12. Using **(4)** of Problem 11, find what happens to *A* as *a* approaches zero.

13. Taking $k = \pi/180$, what is the maximal area of an equilateral triangle in hyperbolic geometry? Is it unbounded like it is in Euclidean geometry?

14. Use analysis based on the Figure P.14 to show that $\theta < \omega/2^n$ in the nth step (where $\omega = \angle 1 > 2(\angle 2)$, $\angle 2 > 2(\angle 3)$,...). From this, prove that a right triangle can be constructed having arbitrary large sides whose area is less than π. (Observe $\triangle ABX$ in Figure 5.26 as $BX \to \infty$.) In the process, show that the area of any right triangle cannot exceed the value π, a phenomenon that Gauss noticed, which he observed could be a reason to disqualify hyperbolic geometry.

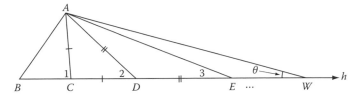

Figure P.14

15. The following result is needed for the next two problems and for Proclus' argument which was used earlier in proving the angle-sum theorem. Complete the details in the steps given in the following proof, valid in unified geometry with $\alpha = \infty$.

> ***Theorem*** Let $\angle BAC$ be a given angle, with X any point on side AB. If $AX \to \infty$ and Z is the foot of X on side AC, then $XZ \to \infty$.

1. Let ABC be any right triangle with M the foot on leg AC of the midpoint N of hypotenuse AB (see Figure P.15). Construct P on ray MN such that N is the midpoint of segment PM. (Give reasons for these steps.)
2. $\triangle NMA \cong \triangle NPB$, so that $\angle NPB$ is a right angle.
3. $\lozenge PMCB$ is a Lambert quadrilateral, with $BC > PM = 2 \cdot NM$.
4. Given $\angle BAC$ and X on side AB, and its foot Z on ray AC. Let n be a positive integer, and consider the points $B = B_1, B_2, B_3, \ldots, B_n$ on side AC such that B is the midpoint of $\overline{AB_2}$, B_2 is the midpoint of $\overline{AB_3}, \ldots$, and let C_i be the foot of B_i on $\overline{AC}$ for each i. Then as $AX \to \infty$, $XZ > B_n C_n \geq 2^n\, MN \to \infty$.

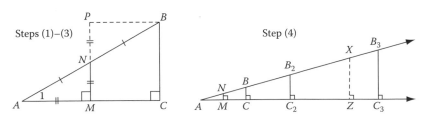

Figure P.15

16. **Divergent Parallelism** If $\overrightarrow{AC} \parallel_d \overrightarrow{BD}$ and X is any point on ray AC, with Y its foot on ray AD. If $AX \rightarrow \infty$, then $XY \rightarrow \infty$. Prove this, by constructing ray AE asymptotically parallel to ray BD, showing that ray AE lies between rays AX and AB, and using the result of Problem 16 applied to $\angle CAE$ (Figure P.16).

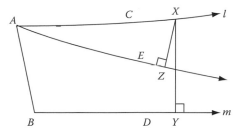

Figure P.16

Group C

17. **Asymptotic Parallelism** Let $\overrightarrow{AC} \parallel_a \overrightarrow{BD}$, where $\overrightarrow{AB} \perp \overrightarrow{BD}$, and let ε be a given positive real. There exists a point X on ray AC such that the perpendicular distance from X to ray BD is less than ε (i.e., if Y is the foot of X on line BD, as $AX \rightarrow \infty$, $XY \rightarrow 0$). The proof is outlined in the following; you are to fill in the details (Figure P.17).

   1. Let $UV = \varepsilon$; construct rays UW and VK perpendicular to segment UV. Since $\overrightarrow{UW} \parallel_d \overrightarrow{VK}$, there exists a perpendicular segment XY from ray UW to VK such that $XY > AB$. (See Problem 16.)
   2. Construct P on ray BA such that $BP = XY$ and ray PQ on the C-side of line AB such that $\angle BPR \cong \angle UXY$. Locate R on ray PQ such that $PR = XU$, and S on ray BD such that $BS = VY$. Then $RS = UV$ and line RS is the common perpendicular to lines PQ and RS.
   3. Since (PAB), ray AC intersects segment RS at an interior point. (Apply the Postulate of Paseh to $\triangle PBR$ and use Theorem **2**, Section **5.8**.) (Figure P.17)

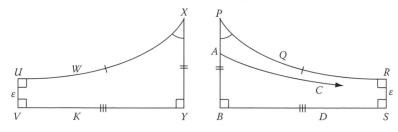

Figure P.17

18. **"Concurrent" Perpendicular Bisectors** Prove that if the perpendicular bisectors of two of the sides of $\triangle ABC$ have a common perpendicular m, then the perpendicular bisector of the third side is also perpendicular to line m, and all three perpendicular bisectors are pairwise divergently parallel. [**Hint:** In Figure P.18, let m be the common

perpendicular bisectors of sides AC and BC; drop perpendiculars from the three vertices A, B, and C to line m.
Prove that $\Diamond A'C'CA$ is a Saccheri quadrilateral.]

Note: Points A, B, and C lie on what is known as an **equidistant locus E**
with respect to line m, whose points are at a constant distance from m.
(See Figure P.18, right diagram). This locus itself cannot be a line
(why not?). ◣

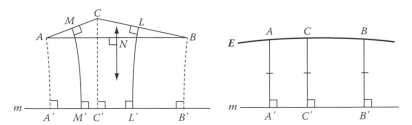

Figure P.18

19. **Postulate of Pasch for Asymptotic Triangles** Prove
     that if a line passes through an interior point of one of the
     sides of an asymptotic triangle and does not pass through
     any of its vertices (including Ω), it passes through an inte-
     rior point of one of the other two sides. [**Hint:** If line l
     passes through $P \in \overline{AB}$, choose $Q \in l$ on the C-side of
     line AB. Construct $\overrightarrow{PR}\|_a \overrightarrow{BD}$, and consider the cases
     $(\overrightarrow{PA}\ \overrightarrow{PQ}\ \overrightarrow{PR})$ or $(\overrightarrow{PR}\ \overrightarrow{PQ}\ \overrightarrow{PB})$. If l passes through $P \in \overline{AC}$,
     choose $Q \in l$ on the B-side of line AC and consider the
     cases $(\overrightarrow{PA}\ \overrightarrow{PQ}\ \overrightarrow{PB})$ or $(\overrightarrow{PB}\ \overrightarrow{PQ}\ \overrightarrow{PC})$.]

20. Prove that the perpendicular bisector of the base AB of
     an isosceles asymptotic triangle $AB\Omega$ is asymptotically
     parallel to either leg AC or AD (hence passes through Ω).
     What theorem of unified geometry does this resemble?
     [**Hint:** Use the result of Problem 19.]

21. **Hilbert's Lemma** Suppose that ray AP is divergently
     parallel to ray BQ, with $\overline{AB} \perp \overline{BQ}$, and points C, D, and E
     such that (ACP), (BDQ), (AEB), $\overline{CD} \perp \overline{DQ}$, and $EB = CD$
     (Figure P.21). Then if ray ER is constructed on the P-side
     of line AB such that $\angle BER \cong \angle DCP$, ray ER will intersect
     ray AP. Complete the details in the following proof.
     (1) Construct rays $\overrightarrow{DS}\|_a \overrightarrow{CP}$ and $\overrightarrow{BT}\|_a \overrightarrow{ER}$. By the
          BA congruence criterion, $\triangle CD\Omega \cong \triangle EB\Omega'$, and by
          CPCPC, $\angle EBT \cong \angle CDS$.
     (2) $\angle TBD \cong \angle SDQ$ and $\overrightarrow{BT}\| \overrightarrow{DS}$. Hence ray BT is diver-
          gently parallel to ray DS.
     (3) Ray ER intersects segment CD at an interior point W.
          (Apply postulate of Pasch to triangles ACB and BCD.)
          Similarly, ray BT intersects segment CD at an interior
          point U.
     (4) Ray BT does not pass through C, D, or Ω, hence meets
          ray CP at some point V (postulate of Pasch applied to
          $\triangle CD\Omega$, established in Problem 19).

(5) By the ordinary postulate of Pasch applied to $\triangle CUV$, since line ER is parallel to line BT, ray ER meets segment CV, hence ray AC, at some point K, as desired.

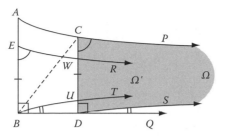

Figure P.21

22. **Hilbert's Theorem** If lines l and m are divergently parallel, they have a perpendicular line in common. The method of proof is to find congruent, perpendicular segments KL and MN from l to m which will produce a Saccheri quadrilateral, and thus a common perpendicular between l and m (Theorem **3**, Section **4.3**). Show that the following constructions performed at certain points in the argument will produce such segments. Choose any two points A and C on line l and let B and D be their feet on line m. If $AB = CD$, we are finished. So assume $AB > CD$ and construct $\overline{EB}$ on $\overline{AB}$ congruent to $\overline{CD}$, and also ray ER on the same side of line AB as C such that $\angle BER \cong \angle DCP$, where (ACP). By Hilbert's lemma rays AP and ER intersect at some point K, so locate M on ray AC such that $CM = EK$, and let N be the foot of M on line m. Use SASAA to show that $\lozenge KEBL \cong \lozenge MCDN$ and $KL = MN$ (Figure P.22).

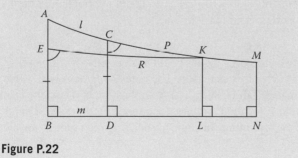

Figure P.22

5.9 Euclidean Geometry: Angle-Sum Theorem

Within the context of unified geometry, the basis for **Euclidean geometry** is Playfair's parallel postulate: If A is a point and l a line not passing through it, there exists a unique line m through A parallel ($\parallel$) to l. The first several results deal with the familiar theorems for parallel lines involving transversals (see previous discussion on this in Section **5.1**).

Theorem 1 If parallel lines are cut by a transversal, each pair of alternate interior angles thus formed are congruent.

Proof   In Figure 5.49 we have $l \parallel m$ and transversal t intersecting l and m at A and B, respectively, with alternate interior angles $\angle CAB$ and $\angle ABF$. Construct ray AC' on the C-side of line AB such that $\angle C'AB \cong \angle ABF$. By Theorem **2**, Section **5.1**, $\overleftrightarrow{AC'} \parallel m$. Since there can be only one parallel to m through A, line AC' must coincide with line l and $C' \in l$. Since C' and C lie on the same side of A, rays AC and AC' coincide. Therefore, $\angle CAB = \angle C'AB \cong \angle ABF$. ◣

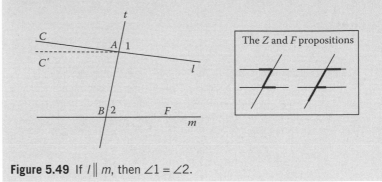

Figure 5.49 If $l \parallel m$, then $\angle 1 = \angle 2$.

Since each pair of alternate interior angles always makes a pair of vertical angles with one of a pair of corresponding angles (such as $\angle CAB$ and $\angle 1$ in Figure 5.49, with $\angle 1$ and $\angle 2$ a pair of corresponding angles), there is an obvious corollary (using also the linear pair theorem of Section **2.4**).

Corollary A   If two parallel lines are cut by a transversal, each pair of corresponding angles are congruent, and interior angles on the same side of the transversal are supplementary.

Note: The two preceding results could be referred to as "Z" and "F" propositions for a fairly obvious reason. The letter Z (perhaps backwards, upside down, or rotated) can be seen in any application of Theorem 1, and similarly the letter F for the corollary. (See inset, Figure 5.49.) This is often a helpful aid for beginning students, particularly high school students. ◣

Another straightforward result is:

Corollary B   Two lines perpendicular to the same line are parallel, and a line that is perpendicular to one of two parallel lines is perpendicular to the other.

The angle-sum theorem for Euclidean geometry now follows. We give the traditional proof involving parallelism.

Angle-Sum Theorem for Euclidean Geometry   The angle-sum of any triangle equals the measure of a straight angle.

Proof   Let $\triangle ABC$ be any given triangle (Figure 5.50). Construct the line through A parallel to line BC. Then, in terms of the angles as indicated in the figure, $\angle 1 + \angle 2 + \angle 3 = 180$ (application of the linear pair theorem). But $\angle 1$ and $\angle A$ are alternate interior angles along a transversal, so $\angle 1 = m\angle A$ (Theorem 1). Similarly, $\angle 3 = m\angle C$. By substitution,

$$m\angle A + m\angle B + m\angle C = 180$$

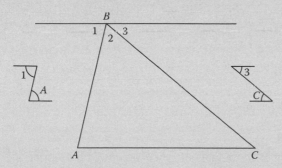

Figure 5.50 Proof of angle-sum theorem.

Exterior Angle Theorem   The measure of an exterior angle of a triangle equals the sum of the measures of the two opposite interior angles.

The applications of these results are extensive, showing up in many different situations. We give a few examples of typical problems in this area.

Example 1

If ray *PT* is the bisector of $\angle QPR$ in $\triangle PQR$ (Figure 5.51), find the measures of the two unknown angles indicated by *x* and *y* if $m\angle Q = 50$ and $m\angle R = 70$.

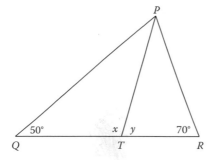

Figure 5.51 Example 1.

Solution

By the angle-sum theorem, $m\angle P = 180 - (50 + 70) = 60$ and $m\angle TPR = \frac{1}{2}(60) = 30 = m\angle TPQ$. By the exterior angle theorem, $x = 30 + 70 = 100$ and $y = 30 + 50 = 80$.

Example 2

A routine problem of a type often found in high school geometry texts is the following: If the angles of $\triangle KLP$ are as indicated in Figure 5.52, find the numerical measure of each angle, and show that $\triangle KLP$ is a right triangle.

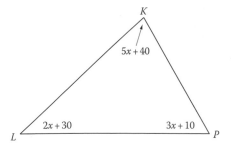

Figure 5.52 Example 2.

Solution

The sum of the angles of $\triangle KLP$ results in the equation

$$(5x + 40) + (2x + 30) + (3x + 10) = 180$$
$$10x + 80 = 180$$

Solving for x, we find $x = 10$. Hence $m\angle K = 5(10) + 40 = 90$ (right angle), $m\angle L = 50$ and $m\angle P = 40$.

Example 3

An equilateral triangle ABC has isosceles triangles ACD, ABE, and BCF constructed on its sides, as shown in Figure 5.53, with D, E and F joined to form triangle DEF. If the base angles of each of the isosceles triangles have measure 70, making them congruent, use angle analysis exclusively to show that $\triangle DEF$ is equilateral.

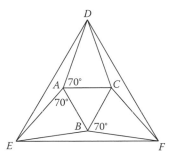

Figure 5.53 Example 3.

Solution

$m\angle ADC = 180 - 2(70) = 40$; $m\angle EAD = 360 - (70 + 60 + 70) = 160$, so that $m\angle EDA = \frac{1}{2}(180 - 160) = 10$. In the same manner $m\angle DCF = 360 - 200 = 160$ and $m\angle CDF = \frac{1}{2}(20) = 10$. Thus $m\angle EDF = 10 + 40 + 10 = 60$. The other two angles of $\triangle DEF$ are obtained in a similar manner, and found to be of measure 60, proving that $\triangle DEF$ is equiangular, thus equilateral.

Moment for Reflection

Suppose in Figure 5.54 the angle trisectors of $\triangle ABC$ are drawn, intersecting pairwise at P, Q, and R, with $m\angle PQC = 72$. If $\triangle ABC$ is isosceles with base BC and having vertex angle of measure 36 (which will determine the base angles), find the angles of $\triangle PQR$. Does your result suggest anything? Repeat when $m\angle PQC = 68$ and the vertex angle is of measure 24. (See Problem 28 below.)

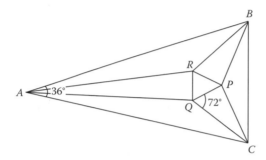

Figure 5.54 Experimenting with angle measures.

One more direct application of the angle-sum theorem will be made, often quite useful (in particular, it is needed in the solution of Problem 28 below). Consider $\triangle ABC$ and its angle bisectors, meeting at the incenter I. Then

$$m\angle BIC = 90 + \frac{1}{2}A$$

(1) $$m\angle AIC = 90 + \frac{1}{2}B$$

$$m\angle AIB = 90 + \frac{1}{2}C$$

where A, B, and C in the above denote the measures of the angles at A, B, and C, as in standard notation. (See Figure 5.55.)

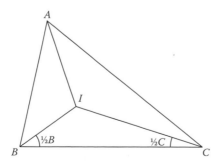

Figure 5.55 Angle measures at incenter.

Proof   By arbitrariness of the labels in the figure, it suffices to prove just the first of these relations. Thus,

$$m\angle BIC = 180 - \left(\tfrac{1}{2}B + \tfrac{1}{2}C\right)$$

$$= 180 - \tfrac{1}{2}(B + C)$$

$$= 180 - \tfrac{1}{2}(180 - A)$$

$$= 90 + \tfrac{1}{2}A \ \text{\char92}$$

5.10 Median of a Trapezoid in Euclidean Geometry

Certain properties of trapezoids will eventually lead to the important concept of similar triangles in Euclidean geometry. (The topic of parallelograms is relegated to the problem section below.) We define a **trapezoid** to be a quadrilateral having two opposite sides parallel (lying on parallel lines). The two parallel sides are called **bases**, and the other two sides, the **legs**, as shown in Figure 5.56. A quadrilateral with both pairs of opposite sides parallel is a **parallelogram,** and a parallelogram whose angles are right angles is a **rectangle**, agreeing with previous terminology. (Thus, in our treatment, a trapezoid can be a parallelogram.) An **isosceles trapezoid** is a trapezoid that has exactly two sides congruent (called the **legs**). It follows that the congruent sides must be the non-parallel sides. Finally, the **median** of a trapezoid is the segment joining the midpoints of its legs.

An interesting result is that the length of the median of any trapezoid is the *arithmetic mean of the lengths of the two bases*. That is, $m = \tfrac{1}{2}(b_1 + b_2)$, where m, b_1, and b_2 are the respective lengths of the median and two bases. We shall eventually establish this result, but a few preliminary results are needed.

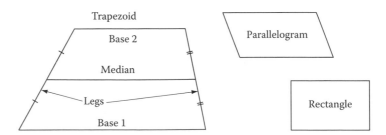

Figure 5.56 Trapezoids and parallelograms.

Theorem 1   The midpoint of the hypotenuse of a right triangle is equidistant from the three vertices. Conversely, if the midpoint of a side of a triangle is equidistant from the three vertices, the triangle is a right triangle.

Proof   Suppose $\triangle ABC$ is a right triangle with right angle at C, and with M the midpoint of hypotenuse AB. Since we already have $AM = BM$ it remains to prove $CM = AM$. First, suppose that $CM > AM$, as shown in Figure 5.57. Then by the scalene inequality applied to $\triangle ACM$, $\angle A > \angle 1$ and $\angle B > \angle 2$. We then obtain the contradiction

$$\angle A + m\angle B > m\angle 1 + m\angle 2 = 90$$

If $CM < AM$ it follows that $m\angle A + m\angle B < 90$ in the same manner. Therefore, we conclude that $CM = AM$, as desired. For the converse, consider Figure 5.57 where $AM = MC = MB$. By the isosceles triangle theorem, $m\angle ACB = m\angle 1 + m\angle 2 = A + B = \frac{1}{2} \cdot (2A + 2B) = \frac{1}{2}(A + B + m\angle 1 + m\angle 2) = \frac{1}{2}(A + B + C) = \frac{1}{2}(180) = 90$. ◥

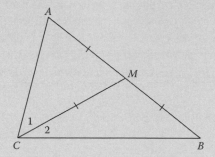

Figure 5.57 Proof of Theorem 1.

Corollary   In any right triangle, the segment joining the midpoints of a leg and hypotenuse is parallel to, and has one-half the length of the second leg.

Proof   Let L be the midpoint of leg AC and M that of hypotenuse AB. By the previous theorem, $\triangle AMC$ is isosceles with base AC (Figure 5.58). Then $\overleftrightarrow{LM} \perp \overleftrightarrow{AC}$ by standard (already proven) properties of isosceles triangles, and by Corollary B of Section **5.9**, $\overleftrightarrow{LM} \parallel \overleftrightarrow{BC}$. Likewise, if N is the midpoint of leg BC, then $\overleftrightarrow{NM} \perp \overleftrightarrow{BC}$, and again by Corollary B, $\overleftrightarrow{NM} \parallel \overleftrightarrow{AC}$. Thus $\lozenge LMNC$ is a rectangle, and $LM = CN = \frac{1}{2}CB$, as desired. ◥

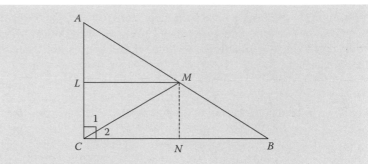

Figure 5.58 Proof of the corollary.

Theorem 2: Midpoint-Connector Theorem for Triangles

In any triangle, the segment joining the midpoints of two sides of a triangle is parallel to the third side and has length one-half that of the third side.

Proof   Let L be the midpoint of AD, where D is the foot of A on line BC, with either (BDC), (DBC), or (BCD). The first two cases are shown in Figure 5.59, the third logically equivalent to the second. All conclusions follow immediately from the corollary of Theorem 1: lines ML and NL are parallel to line CB; since there can be only one parallel through L, these two lines coincide, and we have $\overleftrightarrow{MN} \parallel \overleftrightarrow{BC}$. In the case (BDC), one obtains $(\overrightarrow{AB}\,\overrightarrow{AD}\,\overrightarrow{AC})$, hence (MLN) follows. Thus, $MN = ML + LN = \frac{1}{2}BD + \frac{1}{2}DC = \frac{1}{2}BC$. The case (DCB) follows in much the same manner. ⟍

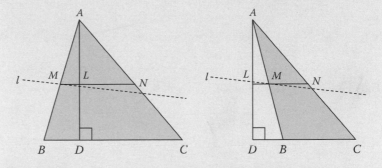

Figure 5.59 Line joining midpoints of two sides of a triangle.

Theorem 3: Midpoint-Connector Theorem for Trapezoids

In any trapezoid, the segment joining the midpoints of the two legs is parallel to either base, and has length one-half the sum of the lengths of the two bases.

Proof   Let M and N be the midpoints of legs AD and BC of trapezoid $ABCD$, with lines AB and DC parallel (Figure 5.60). If L is the midpoint of diagonal D, apply the previous theorem to $\triangle DAB$ and $\triangle DCB$; thus $\overleftrightarrow{ML} \parallel \overleftrightarrow{AB}$ and $\overleftrightarrow{LN} \parallel \overleftrightarrow{DC} \parallel \overleftrightarrow{ML}$ so that lines ML and LN coincide, with (MLN). Therefore, $\overleftrightarrow{MN} \parallel \overleftrightarrow{AB}$; also, $MN = ML + LN = \frac{1}{2}AB + \frac{1}{2}DC$. ⟍

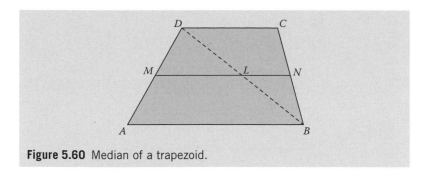

Figure 5.60 Median of a trapezoid.

In geometry the situation frequently occurs where three or more parallel lines passing through points on one line l, and intersecting another line m (Figure 5.61). This construction sets up a correspondence $f: P \rightarrow P'$ between the points of l and those of m, where $PP' \parallel k$ for $P \in l$ and $P' \in m$. This correspondence is called a **parallel projection** from l to m. Three fundamental principles relevant to this situation are characteristic of Euclidean geometry. At this point we adopt the convention of allowing a line to be *parallel to itself.*

(1)   The relation of parallelism is symmetric and transitive.

Proof   For the symmetry property let $l \parallel m$. Then either $l = m$, or l and m have no points in common. In either case, $m \parallel l$. For the transitive property, let $l \parallel m$ and $m \parallel n$; we are to prove that $l \parallel n$. If $l = n$, then $l \parallel n$ and we are finished. Otherwise, $l \neq n$; suppose l meets n at a unique point W (Figure 5.62). By hypothesis, both l and n are parallel to m, and pass through W, contradicting the parallel postulate for Euclidean geometry. Therefore, $l \parallel n$. ◧

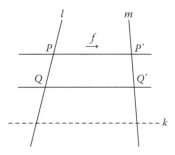

Figure 5.61 Parallel projection.

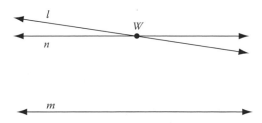

Figure 5.62 Proof of transitive law for parallel lines.

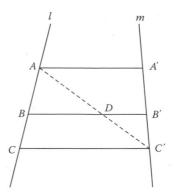

Figure 5.63 Order-preserving property of parallel projection.

(2)   The mapping f is one-to-one and order-preserving.

[The proof will be left to the reader to show that if $A \neq B$ then $A' \neq B'$ (f is one-to-one) and if (ABC) then $(A'B'C')$; apply the Postulate of Pasch, as illustrated in Figure 5.63.]

The third property is an application of similar triangles (which will be taken up in the next section) and states that the mapping f is *ratio-preserving*. That is, in Figure 5.63

(3)   The mapping f is ratio-preserving: $\dfrac{AB}{BC} = \dfrac{A'B'}{B'C'}$

5.11 Similar Triangles in Euclidean Geometry

The basis for similar triangles in the Euclidean plane emerges from the results in this section. The proof of the key theorem follows the pattern used previously in elliptic geometry (identity **(2)** in Section **5.4**). Its name reflects that of many secondary school textbooks.

> ***The Side-Splitting Theorem***   In $\triangle ABC$ let D be a point on side AB. Then the parallel to side BC passing through D cuts side AC at a point E such that
>
> **(1)** $\dfrac{AD}{DB} = \dfrac{AE}{EC}$
>
> *Proof* We first make the observation that the above ratio **(1)** is equivalent to the ratio $AD/AB = AE/AC$ by simple algebra (add 1 to both sides of the reciprocal equation $DB/AD = EC/AE$ to obtain $AB/AD = AC/AE$, or $AD/AB = AE/AC$, and reverse the steps to obtain the opposite implication). We then proceed to prove the ratio $AD/AB = AE/AC$ instead.

1. Suppose that $AD/AB > AE/AC$. Locate F on segment AB such that $AF = (AE/AC)AB$, (that is, $AF/AB = AE/AC$); since therefore $AD/AB > AF/AB$, $AD > AF$, hence (AFD), as shown in Figure 5.64. Locate the midpoints of sides AB and AC, then bisect the segments that these midpoints determine, and continue the bisection process repeatedly. At some stage, one of the midpoints say P will fall on segment DF (due to the Archimedean principle*). The lines joining corresponding midpoints are parallel to line BC by the midpoint-connector theorem.

Thus, the midpoint Q on side AC corresponding to P will fall on segment GE by the order-preserving property (2), of parallel projection. Suppose there are m parallel lines between A and line PQ, counting PQ, and n between A and line BC. If we let b and c be the lengths of each of the congruent sub-segments on AB and AC, as indicated, then

$$AP = mb, \quad AQ = mc$$

$$AB = nb, \quad AC = nc,$$

Therefore, $AP/AB = m/n = AQ/AC$. This leads to the contradiction

$$\frac{AF}{AB} < \frac{AP}{AB} = \frac{AQ}{AC} < \frac{AE}{AC} = \frac{AF}{AB}$$

2. A similar argument leads to a contradiction if $AD/AB < AE/AC$. Hence it must be concluded that $AD/AB = AE/AC$, and therefore, $AD/DB = AE/EC$. ◣

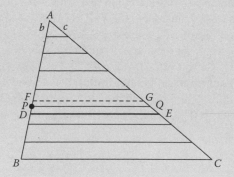

Figure 5.64 Proof of side-splitting theorem.

Corollary   With the notation of the theorem, where $\overleftrightarrow{DE} \parallel \overleftrightarrow{AB}$,

(2)
$$\frac{AD}{AB} = \frac{DE}{BC}$$

---

* The Archimedean principle shows the existence of a positive integer k such that $k\,DF > AB$, which yields the inequality $AB/2^k < AB/k < DF$. This shows that two consecutive midpoints are too close to straddle the points D and F.

Proof Locate point F on AB such that $\overline{FB} \cong \overline{AD}$, and let the parallel to line AC through F meet segment BC at G, as shown in Figure 5.65. Then by ASA, $\triangle ADE \cong \triangle FBG$ (using corresponding angles along transversal AB), and $DE = BG$. By the relation established in the theorem, we have

$$\frac{DE}{BC} = \frac{BG}{BC} = \frac{BF}{BA} = \frac{AD}{AB}$$

as desired. ◣

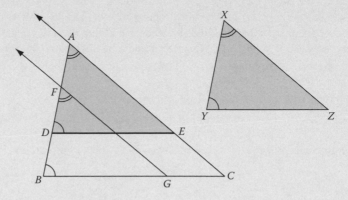

Figure 5.65 $\triangle ABC \sim \triangle XYZ$.

The triangles ADE and ABC in Figure 5.65 have some important properties. Corresponding angles are congruent, and corresponding sides are proportional, as we have just seen. If $\triangle XYZ \cong \triangle ADE$ as indicated in Figure 5.65, then the corresponding angles of $\triangle ABC$ and $\triangle XYZ$ are congruent ($\angle A \cong \angle X$, $\angle B \cong \angle Y$, and $\angle C \cong \angle Z$), and corresponding sides are proportional (i.e., by (**1**) and (**2**), $AB/XY = BC/YZ = AC/XZ$). Two triangles related in this manner are said to be **similar**, denoted by $\triangle ABC \sim \triangle XYZ$.

It is often helpful to express the proportionality of the sides of similar triangles differently: If we set the above ratios equal to k, then we have, by algebra,

$$AB = kXY, \quad BC = kYZ, \quad AC = kXZ$$

The value k is called the **constant of proportionality**, and we may then say that $\triangle ABC$ is k *times the size of* $\triangle XYZ$. (If $k = 1$ the triangles are congruent by the SSS criterion.)

There are several criteria for the similarity of triangles, requiring less than the six conditions stated in the definition, analogous to the situation for congruence. Unlike non-Euclidean geometry where congruent angles imply congruent triangles, one of the criteria is the following.

AA Similarity Criterion   If two triangles have two angles of one congruent, respectively, to the two corresponding angles of another, the triangles are similar under that correspondence.

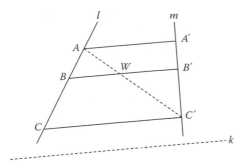

Figure 5.66 Ratios preserved under parallel projection.

Proof   Suppose in $\triangle ABC$ and $\triangle XYZ$, $\angle A \cong \angle X$ and $\angle B \cong \angle Y$ (where $ABC \leftrightarrow XYZ$). Unless the triangles are congruent, we can assume either $AB > XY$ or $AB < XY$, say the former, as in Figure 5.65. Construct segment AD on AB congruent to XY, and consider parallel lines DE and BC, E on side AC. By corresponding angles, $\angle ADE \cong \angle B \cong \angle Y$. By the ASA congruence criterion, $\triangle ADE \cong \triangle XYZ$. Then by the above corollary, the corresponding sides of $\triangle ABC$ and $\triangle ADE$ (thus of $\triangle XYZ$) are proportional. Therefore, $\triangle ABC \sim \triangle XYZ$. ◢

The other criteria for similar triangles are taken up in the problem section below. They will be stated here, with the proofs left to the reader.

SAS Similarity Criterion   If two triangles have the lengths of two sides of one proportional to the corresponding two sides of the other, and the included angles congruent, the triangles are similar under that correspondence.

SSS Similarity Criterion   If two triangles have the lengths of all three sides of one in proportion to the corresponding three sides of the other, the triangles are similar under that correspondence.

The result **(3)** Section **5.11** can now be established as a result of **(1)**, the side-splitter theorem. The details will be left as Problem 10. (For a rough proof, join A and C', with line BB' cutting $\overline{AC'}$ at W, as shown in Figure 5.66, then apply the side-splitting theorem to triangles ACC' and $AC'A'$: $AB/BC = AW/WC' = A'B'/B'C'$.)

Example 1

Prove that the medians of a triangle are concurrent.* (The point of concurrency is called the **centroid** of the triangle, and may be characterized as the point on each median 2/3 the distance from the vertex to the midpoint of the opposite side.)

Solution

By the crossbar theorem we know that any two medians meet at an interior point on each median. In Figure 5.67, let medians *AL* and

---

* This particular property is true in non-Euclidean geometry using an adaptation of Ceva's theorem. See Chapter 9, Section **9.6**.

BM meet at *G*. Draw *LW* joining *L* and *W*, where *W* is the mid-point of *MC*. Then (*AMWC*) and *AM* = *MC* = 2*MW*, or *AW* = 3*MW*. Thus by substitution, *AM/AW* = 2*MW*/3*MW* = ⅔. Since $\overleftrightarrow{BM} \parallel \overleftrightarrow{LW}$, the side-splitting theorem yields *AG/AL* = *AM/AW* = ⅔. That is, *AG* = ⅔*AL*. This argument applies to the third median *CN*, so if *CN* meets *AL* at *G′* we obtain (*AG′L*) and *AG′* = ⅔*AL* and it follows that *G′* = *G*. Hence all three medians pass through *G*.

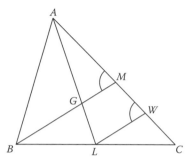

Figure 5.67 Proof that medians are concurrent.

The four classical points of concurrency associated with a triangle (as introduced in Section 3.9) can all be justified in Euclidean geometry, as indicated in the following table, and illustrated in Figure 5.68.

> *I* = **incenter** (angle bisectors—Theorem **1**, Section **3.9**)
> *O* = **circumcenter** (perpendicular bisectors of sides—Theorem **2**, Section **3.9**)
> *G* = **centroid** (medians—above example)
> *H* = **orthocenter** (altitudes—Problem **26** below)

Note that these four points coincide trivially if Δ*ABC* is equilateral, and if Δ*ABC* is isosceles with base *BC*, they all lie on median *AL*.

As one observes in Figure 5.68, it appears that *H*, *G*, and *O* are collinear. Indeed, this turns out to be the case, the line of collinearity having been first discovered (and proved) by L. Euler (1707–1783); it is appropriately named the **Euler line.**

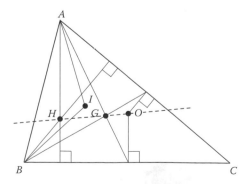

Figure 5.68 The four classical points of concurrency.

Example 2

Prove that the orthocenter H, the centroid G, and the circumcenter O are collinear, and that further, (HGO), with

(3) $$HG = 2GO$$

Solution

Assuming $\triangle ABC$ is not equilateral, then G and O are distinct points (proof?). First, consider the case when line GO meets altitude AD at a unique point H_1, as shown in Figure 5.69a. If L is the midpoint of side BC, then $\overleftrightarrow{AD} \parallel \overleftrightarrow{OL}$ and $\angle H_1AG \cong \angle GLO$ (alternate interior angles along a transversal). Since (AGL) implies (H_1GO), $\angle AGH_1 \cong \angle LGO$ as vertical angles. By AA, $\triangle AH_1G \sim \triangle LOG$. But since G is the centroid, $AG = \tfrac{2}{3}AL$ (as shown above), or $AG = 2GL$. Hence $H_1G = 2GO$. Now if H_2 is the intersection of GO with any other altitude, it follows in the same manner that (H_2GO) and $H_2G = 2GO$. Hence, $H_1 = H_2 = H$, the orthocenter of $\triangle ABC$. (This proves that the altitudes are concurrent.) If line GO is itself an altitude (and thus either H_1 or H_2 do not exist), then it follows that $\triangle ABC$ is isosceles. (For example, if $\overleftrightarrow{GO} = \overleftrightarrow{AD}$ then $D = L$, as shown in the inset.) In this case as shown in Figure 5.69b, the similar triangles have the same property as those in Figure 5.69a, where $BG = 2GM$, and the same conclusion follows: (HGO) with $HG = 2GO$.

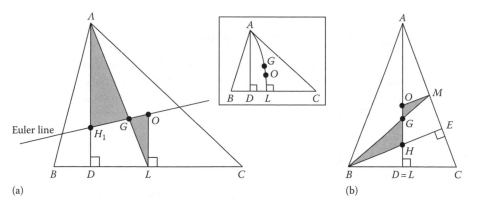

(a)   (b)

Figure 5.69 Existence of Euler line.

A final application of similar triangles will reveal a quite interesting result involving the altitudes of a triangle and the orthocenter H. The feet D, E, and F of the altitudes on the interior sides of $\triangle ABC$ having acute angles determine an inscribed $\triangle DEF$, called the **orthic triangle**.

The Orthic Triangle Theorem   In an acute-angled triangle, the two angles which each side makes with the sides of the orthic triangle are congruent to the angle opposite (see Figure 5.70). Consequently, the altitudes bisect the angles of the orthic triangle.

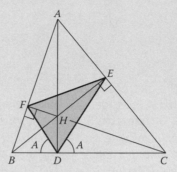

Figure 5.70 The orthic triangle.

Proof   By the AA similarity criterion, $\Delta ADC \sim \Delta BEC$. Hence $AC/BC = DC/EC$ or $AC/DC = BC/EC$. Observe the triangles EDC and BAC, having $\angle C$ in common and the including sides proportional. By the SAS similarity criterion, $\Delta EDC \sim \Delta BAC$. Hence, $m\angle EDC = m\angle BAC \equiv A$, in standard notation. Analogously, $\Delta FDB \sim \Delta CAB$ and $m\angle FDB = m\angle CAB = A$. By change of notation, the rest follows due to the arbitrary choice of altitude AD. ◣

Corollary   The orthocenter of an acute-angled triangle is the incenter of its orthic triangle.

5.12 Pythagorean Theorem

We end this chapter with probably the most famous characteristic of Euclidean geometry of all times, the Pythagorean theorem. Consider any right triangle, ΔABC, with right angle at C. We use standard notation $a = BC$, $b = AC$, and $c = AB$ (the length of the hypotenuse). Drop the perpendicular CD from C to side AB, as shown in Figure 5.71 and let $c_1 = BD$ and $c_2 = DA$. By the AA criterion for similar triangles, this creates two right triangles (ΔADC and ΔCDB) each similar to the given right triangle; the figure depicts those triangles, repositioned so as to show the similarity relations more clearly. Thus, for certain constants p and q of proportionality, we have

(1) $$a = pc_1 = qh$$

(2) $$b = ph = qc_2$$

(3) $$c = pa = qb$$

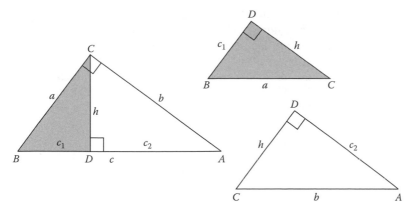

Figure 5.71 Proving the Pythagorean theorem.

Now multiply both sides of the first equation **(1)** by a and use **(3)**:

(4)
$$a^2 = (ap)c_1 = cc_1$$

Multiply both sides of the equation $b = qc_2$ in **(2)** by b and again use **(3)**:

(5)
$$b^2 = (bq)c_2 = cc_2$$

Now sum the equations in **(4)** and **(5)** to obtain the desired result:

(6)
$$a^2 + b^2 = cc_1 + cc_2 = c(c_1 + c_2) = c^2$$

Because the first general proof of **(6)** was derived by Pythagoras, or the school of scholars known as the *Pythagoreans* (and much different from the above proof), it has been named after Pythagoras, which is stated in general as follows.

Pythagorean Theorem   In any right triangle, the square of the length of the hypotenuse equals the sum of the squares of the lengths of the two legs.

   The method used here to prove **(6)** can be used to establish all the possible algebraic relations among the sides and altitude to the base of a right triangle. The reader may want to try proving some of these in the problem section below (Problems 13–16). One well-known example involves the length h of the altitude to the hypotenuse:

(7)
$$h^2 = c_1 c_2$$

   We conclude with the usual trigonometric relations that the reader is no doubt already familiar with (see Problem 33 for a formal development), gathered together in a convenient table that includes **(6)**. An analogous table will appear later for the two non-Euclidean geometries. Interestingly enough, establishing the corresponding formulas for non-Euclidean geometry can be carried out synthetically also, as will be seen in the next chapter.

The development of the usual *xy*-coordinate system, now referred to as *coordinate geometry*, and the formal development of the trigonometric functions, are also characteristically Euclidean, and cannot be carried out in spherical or hyperbolic geometry in the same way. (We shall leave this development for Problems 32 and 33 below.) This provides a foundation for the important development of vectors and vector spaces and should not be overlooked.

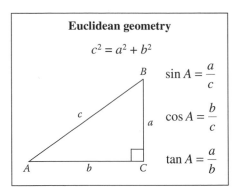

Euclidean geometry

$$c^2 = a^2 + b^2$$

$$\sin A = \frac{a}{c}$$

$$\cos A = \frac{b}{c}$$

$$\tan A = \frac{a}{b}$$

Problems (Section 5.12)

Group A

1. Figure P.1 shows a square, ◊*PQRS*, with inscribed triangle *PUV*. Find *x*, *y*, and *z*. (Proofs involving betweenness are not required for this problem.)

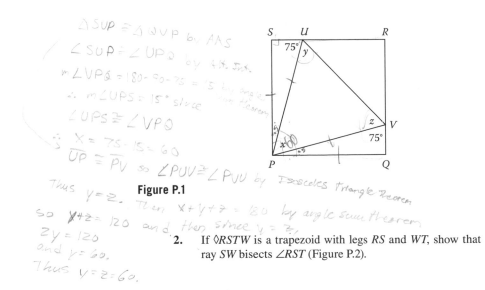

Figure P.1

2. If ◊*RSTW* is a trapezoid with legs *RS* and *WT*, show that ray *SW* bisects ∠*RST* (Figure P.2).

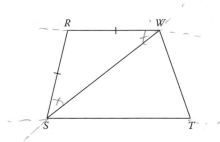

Figure P.2

3. In Figure P.3 for this problem, triangles BCD and ACD are both isosceles triangles ($AD = BD$).
    (a) Prove that the isosceles triangles are similar.
    (b) Prove that $BD^2 = AC \cdot BC$.
    (c) Find $m\angle C = \theta$ if, in addition, $AB = BD$.

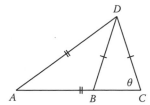

Figure P.3

Note: This construction was introduced earlier in Problems 16, 17 in Section 1.7, which showed how to obtain these two triangles. As mentioned before, the result produces a compass straight-edge construction of a regular pentagon. ◣

4. Using the formula $(n - 2)180$ for the angle-sum of a convex n-gon, find x and y, and the measures of the base angles of the isosceles triangle shown in Figure P.4.

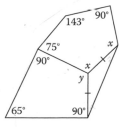

Figure P.4

5. As in Problem 4, find the measures of the base angles of the two isosceles triangles in Figure P.5 if the hexagon and octagon are regular polygons.

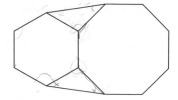

Figure P.5

6. Two congruent isosceles triangles whose base angles each have measure 75 are placed so that a leg from each triangle lie on parallel lines, with the vertices opposite coinciding as shown in Figure P.6. Prove:
    (a) △*FLM* is an equilateral triangle.
    (b) ◊*KLMN* is a square.

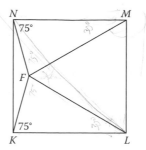

Figure P.6

7. In Figure P.7 for this problem, $\overleftrightarrow{DE} \parallel \overleftrightarrow{BC}$. Find *x* and *y*.

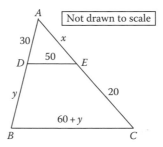

Figure P.7

8. Prove that if a line intersects one of two parallel lines it intersects the other also.

9. Prove that a circle can be passed through any three non-collinear points. (Is this equivalent to proving that the perpendicular bisectors of the sides of any triangle are concurrent?)

Group B

10. Prove that parallel projection is ratio-preserving.

11. *Sketchpad Experiment 9.* In Figure P.11 for this problem, $\overleftrightarrow{EG} \parallel \overleftrightarrow{BC}$. Certainly it is true by similar triangles that $AE/AB = AF/AD = AG/AC$, but the proportion $EF/FG = BD/DC$ is in question. Investigate, using *Sketchpad* (Appendix **A**).

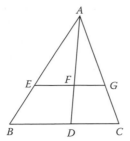

Figure P.11

12.* Prove the SAS (two sides proportional) and SSS (three sides proportional) criteria for similar triangles.

13. In Figure P.13 prove that $h^2 = c_1 c_2$ (identity **(7)** above).

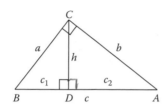

$h^2 + (c_2)^2 = b^2$

$h^2 + (c_1)^2 = a^2$

$a^2 + b^2 = c^2 = (c_1 + c_2)^2$

Figure P.13

14. *Sketchpad Experiment 10* (Appendix **A**) tests the relation in **(7)**.

15. Prove the identity $ab = ch$ from the relations **(1)**, **(2)**, and **(3)**. Note that this agrees with the two formulas for area in right triangle ABC: $K = \frac{1}{2}ab$ (base a, altitude b) and $K = \frac{1}{2}ch$ (base c, altitude h).

16. Prove the identity $abc = h(a^2 + b^2)$ from the relations **(1)**, **(2)**, and **(3)**.

17. Prove that every interior point P of an angle $\angle ABC$ lies on a transversal of that angle. (See Problems 6 and 7 in Section **5.11**.)

18. Explain why all four vertices of $\Diamond XYZW$ shown in Figure P.18 for this problem are concyclic (lie on a circle).

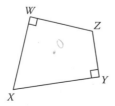

Figure P.18

19.   Proof without Words   Explain how Figure P.19 proves this proposition: *The geometric mean g of two positive numbers a and b is less than or equal to the arithmetic mean m of a and b, with equality only if a = b.*

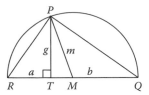

Figure P.19

20.   Proof without Words   Explain how Figure P.20 proves this proposition: *The sum of the vertex angles of a five-pointed star equals the measure of a straight angle.**

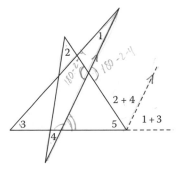

Figure P.20

21.   *Sketchpad Experiment 11.* Use this experiment to investigate the result in Problem 20 for six-pointed stars.

22.   Prove the following identity for the segments *BD* and *DC* as determined by the angle bisector *AD* of $\triangle ABC$ by observing a triangle similar to $\triangle ABD$, as shown in Figure P.22. (See *Sketchpad Experiment 12.*)

$$(8) \qquad \frac{CD}{DB} = \frac{b}{c}$$

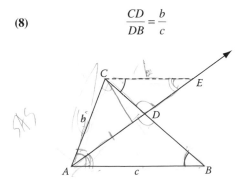

Figure P.22

---

* Proof by Fouad Nakhli, found in *Proof without Words: Exercises in Visual Thinking,* by Roger B. Nelsen (MAA Publication, 1993).

23. **Properties of Parallelograms**   Prove each of the following.
   (a) The opposite sides of a parallelogram are congruent.
   (b) The opposite angles of a parallelogram are congruent, and consecutive angles are supplementary.
   (c) The diagonals of a parallelogram bisect each other.
   (d) If the opposite sides of a convex quadrilateral are congruent, the quadrilateral is a parallelogram.
   (e) If two opposite sides of a convex quadrilateral are both congruent and parallel, the quadrilateral is a parallelogram.

24. Pappus' Theorem on Area   A generalization of the Pythagorean theorem discovered by the Greek geometer Pappus applies to arbitrary triangles. Given $\triangle ABC$, construct parallelograms I and II on sides AC and BC, and extend the opposite sides of these two parallelograms to the point of intersection, P (Figure P.24). If segment AQ is constructed parallel and congruent to segment CP and a parallelogram III constructed with AB and AQ as sides, then

$$\text{Area I} + \text{Area II} = \text{Area III}$$

Use *Sketchpad Experiment 13* to test this relation.

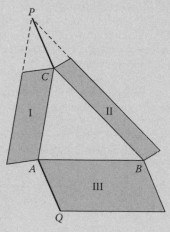

Figure P.24

25. Converse of the Pythagorean Theorem   Using the construction suggested in Figure P.25, write a proof for the following theorem and its corollary using the SAS inequality theorem.

Theorem   In $\triangle ABC$, using standard notation,

   (a) If $\angle C$ is acute, then $c^2 < a^2 + b^2$.
   (b) If $\angle C$ is a right angle, then $c^2 = a^2 + b^2$.
   (c) If $\angle C$ is obtuse, then $c^2 > a^2 + b^2$.

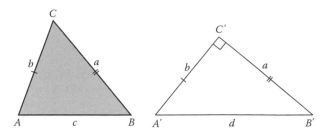

Corollary If $a^2 + b^2 = c^2$ in $\triangle ABC$ (standard notation), then $\angle C$ is a right angle.

Figure P.25

26. **Nested Triangles** Given $\triangle ABC$, a sequence of triangles $\Delta_1, \Delta_2, \Delta_3, \ldots, \Delta_n, \ldots$ can be constructed in its interior such that for each positive integer n, the vertices of Δ_{n+1} are the midpoints of the sides of Δ_n, where we define $\Delta_0 = \triangle ABC$ (see Figure P.26). Call $\Delta_1, \Delta_2, \Delta_3, \ldots$ the *(midpoint) descendants* of Δ_0. Let $\Delta_0, \Delta_1, \Delta_2, \ldots, \Delta_n$ be called the *ancestors* of Δ_{n+1} for any $n \geq 0$. Prove:

   (a) All the descendents of $\triangle ABC$ have the same centroid as $\triangle ABC$. (Use similar triangles.)

   (b) The circumcenter of $\triangle ABC$ is the orthocenter of $\triangle A_1B_1C_1$, and, in general, the circumcenter of Δ_n is the orthocenter of Δ_{n+1} for all $n \geq 0$ (assuming the orthocenter of a triangle exists).

   (c) $\triangle ABC$ is the midpoint descendant of a unique triangle, and thus possesses infinitely many midpoint ancestors, which can be denoted, respectively, by $\Delta_{-1}, \Delta_{-2}, \Delta_{-3}, \ldots$.

   (d) Using Δ_{-1} and $\Delta_0 = \triangle ABC$, prove that the altitudes AD, BE, and CF of $\triangle ABC$ are concurrent. Thus,

Theorem The altitudes of any triangle are concurrent

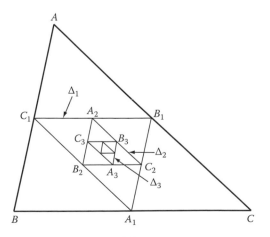

Figure P.26 The descendants of $\triangle ABC$.

Group C

27. The sequence of triangles defined in Problem 26 appear to "converge" to G, the centroid of $\triangle ABC$. In the case of an acute-angled triangle, as depicted in Figure P.26, the circumcenters O_1, O_2, O_3, ... of Δ_1, Δ_2, Δ_3, ..., also converge to G, along some curve, presumably. Show that this curve is actually a straight line—*the Euler line of $\triangle ABC$*.

28. **Morley's Theorem**   Frank Morley (1860–1937), the father of novelist Christopher Morley, discovered that the trisectors of the angles of any triangle intersect in pairs to form a perfect equilateral triangle, as shown in Figure P.28. Its proof did not appear until 1914. There are several indirect proofs, but an elementary direct proof may be constructed as follows. First, consider $\triangle PQR$ determined by the trisectors of a given triangle ABC, as shown in the diagram at left, and let X be the point of intersection of rays BR and CQ. Construct $\triangle Q'PR\,'$ such that $m\angle Q'PX = m\angle XPR' = 30$. Show that $\triangle PQ'R'$ is equilateral (note that P is the incenter of $\triangle BXC$). Let Y and Z be the points of intersection of the perpendiculars from R' and Q' to PQ' and PR' with rays BP and CP. Next, establish the following facts:
    (a) $m\angle ZR'P = m\angle ZPR' = 180 - (m\angle XPC + 30) = 60 - B/3$ [recall (1), Section **5.9**].
    (b) $\theta = 60 + B/3$.
    (c) $\varphi = 60 + C/3$.

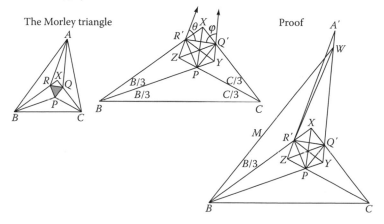

The Morley triangle

Proof

Figure P.28

(d) $\theta + \varphi = 120 + \frac{1}{3}(B + C) = 180 - A/3 < 180$. Therefore, rays ZR' and YQ' meet at some point A', and $m\angle R'A'Q' = A/3$.
(e) If $m\angle MBC = m\angle ABC \equiv B$, then ray BM meets ray YQ' at some point W. Obtain $m\angle BR'Y = 180 - B/3 - m\angle BYR' = 180 - B/3 - (90 - m\angle PQ'Y = 90 + A/3$. If $W \neq A'$, then since R' is the incenter of $\triangle WBY$, it follows from (1), Section **5.9** that $m\angle R'WQ' = \frac{1}{2}m\angle BWY = A/3 = m\angle R'A'Q'$, contradicting the exterior angle inequality. Hence $W = A'$ and $m\angle A'BC = B$.
(f) Similarly, $m\angle A'CB = C$ and points A' and A coincide. It follows that rays AR' and AQ' are the angle trisectors of $\angle BAC$, hence $Q' = Q$ and $R' = R$.

29. In Figure P.29 for this problem, $\triangle KLM$ is isosceles with base $\overline{LM}$ and base angles of measure 80. If $m\angle LMW = 70$, prove that $\overline{KW} \cong \overline{LM}$.

Figure P.29

30. In trapezoid $\Diamond ABCD$ line EF is parallel to base $\overline{AB}$, with $p = AE/AD$ and $q = ED/AD$. If $a = AB$, $b = CD$, and $m = EF$, using Figure P.30 for a hint prove from similar triangles the relation

(9) $$m = pa + qb$$

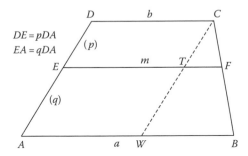

Figure P.30

31. Prove the SASA congruence criterion for trapezoids.

32. Rectangular Coordinates for Euclidean Geometry Using the ruler postulate, take two coordinated lines l and m that are perpendicular at O, the origin on each line. These lines are then taken as the *coordinate axes*, (the *x-axis* as l and the *y-axis* as m). If P is any point in the plane, drop perpendiculars PQ and PR to the x- and y-axes, respectively, $Q \in l$ and $R \in m$. If those points are $Q = Q[x]$ and $R = R[y]$ in terms of their coordinates on l and m, we then assign to P the ordered pair of reals (x, y) as its xy-coordinates and write $P = P[x, y]$. (See Figure P.32a.)

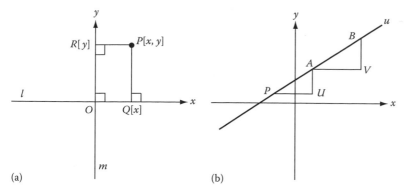

(a)   (b)

Figure P.32 Existence of Euler line.

(a) Show that given a pair of reals (a, b), there exists a unique point $P[a, b]$ having a and b as xy-coordinates.

(b) Show that any line parallel to the y-axis (called a *vertical line*) has equation $x = c$ (constant), and any line parallel to the x-axis (called a *horizontal line*) has equation $y = d$ (constant).

(c) Let line u be a non-vertical line (Figure P.32b) with $A[a_1, a_2]$ and $B[b_1, b_2]$ any two distinct points on u. If $P[x, y]$ is any other point on u, show that in either of the three cases (PAB), (APB), or (ABP),

$$\frac{x - a_1}{y - a_2} = \frac{b_2 - b_1}{a_2 - a_1}$$

It follows that a non-vertical line has equation of the form $y = mx + b$. (One can then define and develop the concept of slope from a strictly algebraic standpoint.) [**Hint:** Show that the coordinates of U and V shown in Figure P.32b are given by $[a_1, y]$ and $[b, a]$, that $\Delta PAU \sim \Delta ABV$.]

(d) Establish the *distance formula* for $P(x_1, y_1)$ and $Q(x_2, y_2)$:

$$PQ = \sqrt{(x_1 - y_1)^2 + (x_2 - y_2)^2}$$

33. **Formal Definition for the Circular Functions** (In this development we use radian measure for angles, which can be effected by simply taking the value of β–the least upper bound for angle measure–to be π.) In terms of the origin O, let θ be a given real number, positive, negative, or zero. As a real number, θ must lie on exactly one of the real intervals ... $[-3\pi, -\pi]$, $[-\pi, \pi]$, $[3\pi, 5\pi]$, $[5\pi, 7\pi]$, ... (using odd multiples of π). Hence there exists a unique integer n such that θ belongs to the interval $[(2n - 1)\pi, (2n + 1)\pi]$, that is, $-\pi < \theta - 2n\pi \leq \pi$. Thus, $\varphi \equiv \theta - 2n\pi$ is the coordinate of a unique ray h from O (Figure P.33), and h will intersect the unit circle at a unique point $P[x, y]$. Define the *sine*, *cosine*, and *tangent* functions, respectively, by

$\sin \theta = y$, $\cos \theta = x$,   and $\tan \theta = y/x$ (if $x \neq 0$)

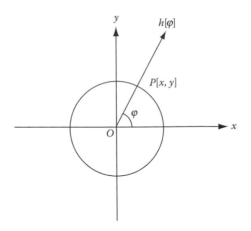

Figure P.33

(a) In Figure P.33 show that $\sin(-\varphi) = -y = -\sin\varphi$ and $\cos(-\varphi) = x = \cos\varphi$ for $-\pi < \varphi \leq \pi$ and thus prove the identities $\sin(-\theta) = -\sin\theta$, $\cos(-\theta) = \cos\theta$, and $\tan(-\theta) = -\tan\theta$.

(b) Show that if $\varphi > 0$ then $\sin(\varphi - \pi) = -y = -\sin\varphi$ and $\cos(\varphi - \pi) = -x = -\cos\varphi$, and if $\varphi < 0$ then $\sin(\varphi + \pi) = -\sin\varphi$ and $\cos(\varphi + \pi) = -\cos\varphi$.

(c) Using (b), prove the identities for any odd integer n and any integer k: $\sin(\theta + 2n\pi) = \sin\theta$, $\cos(\theta + 2n\pi) = \cos\theta$, $\sin(\theta + n\pi) = -\sin\theta$, $\cos(\theta + n\pi) = -\cos\theta$, and $\tan(\theta + k\pi) = \tan\theta$.

(d) Defining $\operatorname{cosec}\theta$, $\sec\theta$, and $\cot\theta$ in the usual manner, prove the Pythagorean identities [e.g., $\sin^2\theta + \cos^2\theta = 1$]. [**Hint:** Derive from Problem 30 the xy equation for the unit circle.]

(e) Establish the trigonometric relations exhibited in the table of this section.

34. **Polar Coordinates**   Let (r, θ) be a given pair of reals. If φ is determined on $[-\pi, \pi]$ as in Problem 31, consider the ray from the origin O of an xy-coordinate system having coordinate φ. Let P be determined on that ray such that $OP = r$ if $r \geq 0$, or on the opposite ray such that $OP = -r$ if $r < 0$. Then associate with P the pair (r, θ), a set of **polar coordinates** of P. (Conversely, if P is any point in the plane on a line l through O, and either ray h from O on line l has coordinate φ, $-\pi < \varphi \leq \pi$, assign P the polar coordinates (r, φ) where $r = OP$ if $P \in h$, or $r = -OP$ otherwise.) Establish the conversion formulas in all cases between the (rectangular) coordinates (x, y) and a pair of polar coordinates (r, θ):

$$x = r\cos\theta, \quad y = r\sin\theta$$

35. **Law of Sines**   Let triangle ABC, in standard notation, be positioned in the coordinate plane such that A is the origin, side AB lies along the positive x-axis, and C lies above the x-axis (Figure P.35). Thus, the

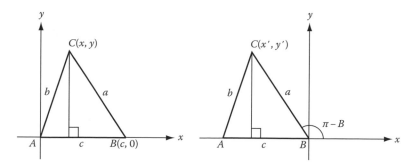

Figure P.35

coordinates of B and C are, respectively, $(c, 0)$ and (x, y), where $y > 0$. By determining the polar coordinates of C, show that the y-coordinate of C is given by $b \sin A$. Next, if B is chosen as the origin instead of A, with side AB still on the x-axis, find the polar coordinates of C for this case. Using the fact that $y = y'$ prove:

$$\frac{a}{\sin A} = \frac{b}{\sin B} = \frac{c}{\sin C}$$

36.*   Law of Cosines   Let triangle ABC be positioned along the x-axis of an xy-coordinate system as in Problem 33, with A at the origin. Determine the polar coordinates of C in terms of standard notation, and using the conversion formulas of Problem 32, along with the distance formula for $a = BC$, expand using algebra, and prove:

$$a^2 = b^2 + c^2 - 2bc \cos A$$
$$b^2 = a^2 + c^2 - 2ac \cos B$$
$$c^2 = a^2 + b^2 - 2ab \cos C$$

Inequalities for Quadrilaterals: Unified Trigonometry

In order to advance our knowledge of the two non-Euclidean geometries, it is necessary to develop a non-Euclidean "trigonometry." This necessitates the development of a few inequalities for midlines in right triangles and Lambert quadrilaterals. That will be the goal of the first part of this chapter. The development of unified trigonometry will then occupy the second half—material that includes a special "Pythagorean theorem" for the two non-Euclidean geometries.

6.1 An Inequality Concept for Unified Geometry

One of the major, defining concepts for unified geometry involves the angle-sum for triangles. In elliptic geometry, the angle-sum of a triangle is greater than 180, while in hyperbolic geometry, the angle-sum is less than 180 (and in Euclidean geometry, of course, it is 180). It is convenient at this point to introduce a special type of inequality that will apply to all three geometries simultaneously. It might be referred to as a *unified inequality*. While it does not necessarily simplify proofs (more often than not, proofs must be constructed separately for the three geometries), it does save in writing and in stating the inequalities concisely. The idea is this: We let $a * b$ denote the relation in unified geometry that represents $a < b$, $a = b$, or $a > b$, respectively, in hyperbolic, Euclidean, or elliptic geometry. Instead of the asterisk, a symbol more suggestive of inequality will be used: a "curved" inequality symbol, to distinguish this from ordinary inequality. The formal definition is then

---

Definition   If a and b are real numbers, write $a \prec b$ iff $a < b$ in hyperbolic geometry, $a = b$ in Euclidean geometry, or $a > b$ in elliptic geometry. Define $a \succ b$ iff $b \prec a$.

It is immediately clear that this new relation (inequality) has all the attributes of ordinary inequality for real numbers, since it coincides with the ordinary inequality for reals (either less than or greater than) in either hyperbolic or elliptic geometry. Of course, in Euclidean geometry, $a \prec b$ iff $a = b$ and $\prec$ represents ordinary equality. Thus, we obtain the *transitive law* for $\prec$ and a familiar addition and product property:

(1)   If $a \prec b$ and $b \prec c$, then $a \prec c$ for arbitrary reals a, b, and c.
(2)   If $a \prec b$ and $c \prec d$, then $a + c \prec b + d$
(3)   If $a \prec b$ and c is a positiveereal, then $ca \prec cb$

Proof of (2)   In hyperbolic geometry we have $a < b$ and $c < d$; therefore by ordinary inequalities, $a + c < b + d$. In Euclidean geometry, $a = b$ and $c = d$ so that $a + c = b + d$. Finally, in elliptic geometry, $a > b$ and $c > d$, which implies $a + c > b + d$. Combining the three cases, $a + c \prec b + d$. ◥

As a result of this new inequality concept, the angle-sum theorem takes on the following comprehensive form:

Angle-Sum Theorem for Unified Geometry   In any triangle ABC,

$$m\angle A + m\angle B + m\angle C \prec 180$$

Corollary A   In any right triangle ABC with right angle at C, $m\angle A + m\angle B \prec 90$.

To see how a typical proof works using $\prec$, observe the following corollary to the Angle-Sum Theorem.

Corollary B   In any Saccheri quadrilateral $ABCD$ with base $\overline{AB}$, $m\angle C = m\angle D \prec 90$ (see Figure 6.1).

Proof   One first proves that $S \prec 360$ for the angle-sum S of any convex quadrilateral. Let $\Diamond ABCD$ be a convex quadrilateral, with diagonal AC (Figure 6.2). Thus $(\overrightarrow{AD} \overrightarrow{AC} \overrightarrow{AB})$ and $(\overrightarrow{CD} \overrightarrow{CA} \overrightarrow{CB})$ both hold, and, using the notation in the figure,

$$S \equiv m\angle A + m\angle B + m\angle C + m\angle D =$$

$$(\angle 1 + \angle 2) + m\angle B + (\angle 3 + \angle 4) + m\angle D =$$

$$(\angle 1 + \angle 4 + m\angle D) + (\angle 2 + \angle 3 + m\angle B) \prec 180 + 180 = 360$$

In Saccheri quadrilateral $ABCD$ we have (since $m\angle A = m\angle B = 90$) $90 + 90 + m\angle C + m\angle D \prec 360$, or $180 + 2m\angle C \prec 360$, which produces the desired result. ◥

Figure 6.1 Saccheri quadrilateral in unified geometry.

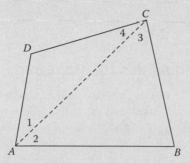

Figure 6.2 Angel-sum of $\lozenge ABCD < 360$.

Corollary C   If $\lozenge ABCD$ is a Saccheri quadrilateral with base $\overline{AB}$ and summit $\overline{CD}$, then $AB < CD$.

Proof   In Figure 6.3, $m\angle 2 + m\angle 1 < 90 = m\angle 3 + m\angle 1$ which reduces to $m\angle 2 < m\angle 3$. In $\triangle ACB$ and $\triangle CAD$ we have $CB = AD$ and $CA = AC$. By the SAS inequality, $AB < CD$. ◳

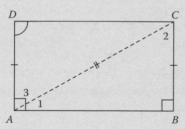

Figure 6.3 Proof of Corollary C.

Corollary D   If $\lozenge ABCD$ is a Lambert quadrilateral with right angles at A, B, and C (Figure 6.4), and having sides less than $\alpha/2$, $\angle D < 90$, $AB < CD$, and $BC < AD$.

Proof   Double the sides AB and DC to points E and F as shown in Figure 6.4. Then $AB = BE$, $BC = BC$, and $DC = CF$ so by SASAS $\lozenge ABCD \cong \lozenge EBCF$, $\angle A \cong \angle E$, $\angle D \cong \angle F$, and $AD = EF$. Thus $\lozenge AEFD$ is a Saccheri quadrilateral, and Corollaries B and C apply to give us $m\angle D < 90$ and $2AB = AE < DF = 2CD$ or $AB < CD$. By change of notation, $BC < AD$ also follows. ◳

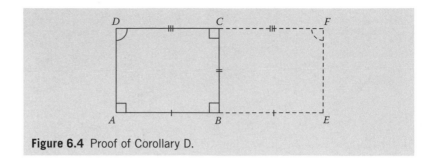

Figure 6.4 Proof of Corollary D.

6.2 Ratio Inequalities for Trapezoids

Corollaries C and D in the previous section enable us to prove a few inequalities concerning what may be called *midlines* of right triangles and certain quadrilaterals. Although stronger results may be obtained, our main interest involves only what is needed for the trigonometry introduced later. As will be seen, the limits needed involve arbitrarily small figures. So, to make the proofs in the current chapter simpler, we assume from now on that all triangles and quadrilaterals are sufficiently small to guarantee the validity of the betweenness properties used in any given argument, particularly in elliptic geometry. This means that we can assume the basic betweenness concepts that are valid in Euclidean geometry (in small regions of the plane). These comments apply to the definition of midlines.

If $\triangle ABC$ is a right triangle with right angle at C, define the **midline** at D on side AC to be the perpendicular $\overline{DE}$, where E is the foot of D on line AC (Figure 6.5). The existence of a unique midline DE for each point E on AC is clear by the postulate of Pasch. Similarly, consider the quadrilateral $ABCD$ having right angles at B and C. The perpendicular $\overline{EF}$ is the **midline** of the quadrilateral at E to side BC at F, where F is the foot of E on BC (Figure 6.6). The existence of a unique midline EF for each given point F on BC is also clear by two applications of the postulate of Pasch, as indicated in Figure 6.6.

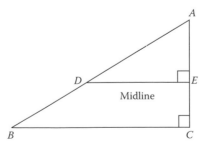

Figure 6.5 A midline of right triangle *ABC*.

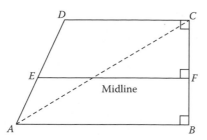

Figure 6.6 Midline of quadrilateral *ABCD*.

In Euclidean geometry the following ratios were proved in Section 5.13:

$$\frac{AD}{DB} = \frac{AE}{EC} \quad \text{and} \quad \frac{DE}{EA} = \frac{CF}{FB}$$

However, in unified geometry these ratios must be replaced by the following inequalities:

(1) $\dfrac{AD}{DB} < \dfrac{AE}{EC}$ (in Figure 6.5)

(2) $\dfrac{DE}{EA} < \dfrac{CF}{FB}$ (in Figure 6.6)

Since the results which follow are only valid for a special type of quadrilateral, a definition is in order.

Definition   If a quadrilateral ◊*ABCD* has right angles at *B* and *C*, and if $m\angle A \leq 90$ and $m\angle D \geq 90$, then ◊*ABCD* is called a **normal trapezoid**, with **base** *AB* and **altitude** *BC*, as indicated in Figure 6.7.

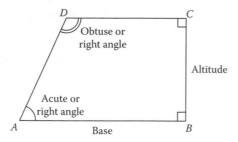

Figure 6.7 A normal trapezoid.

Note: The definition allows all four angles of a normal trapezoid to be right angles, so in Euclidean geometry, a rectangle is a normal trapezoid. In non-Euclidean geometry, a Lambert quadrilateral is a normal trapezoid, but a Saccheri quadrilateral is not: the non-right angles, being congruent, are either both acute, or both obtuse—not allowed by the definition. In the case of a Lambert quadrilateral in hyperbolic geometry with right angles at *A*, *B*, and *C*, then $m\angle A = 90$, $m\angle D < 90$, and the base is *CD*, instead of side *AB*. ◊

Midline Bisector Theorem for Unified Geometry Suppose that
◊*ABCD* is a normal trapezoid with base *AB* and altitude *BC* (Figure
6.8). If *M* is the midpoint of side *BC*, and *LM* is the corresponding
midline, then

(3) $$CD \leq AB$$

(4) $$DL < LA$$

(5) $$LM < \tfrac{1}{2}(AB + CD)$$

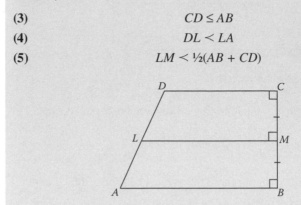

Figure 6.8 Midline bisector theorem.

Proof Each of the above is clear in Euclidean geometry (which in **(4)**
and **(5)** are equalities), so we assume the geometry is either hyperbolic
or elliptic. Observe that the hypothesis implies ∠*ALM* is obtuse in either
case: In hyperbolic geometry, ∠*DLM* is acute by the angle-sum theorem, and in elliptic geometry, ∠*ALM* is obtuse, also by the angle-sum
theorem. As illustrated in Figure 6.9, in both geometries construct the
perpendicular *l* to line *LM* at *L*, which meets segment *AB* at an interior
point *P* (since ∠*ALM* is obtuse). Extend segment *PL* its own length to
point *Q* such that *L* is the midpoint of segment *PQ*; it will be shown that
$Q \in \overleftrightarrow{CD}$. By SASAS ◊*QLMC* ≅ ◊*PLMB* so that ∠*QCM* ≅ ∠*PBM*. Thus
$QC \perp BC$ and $Q \in \overleftrightarrow{CD}$ follows. Since *Q* is on the opposite side of line *AD* as
P, and therefore as *C*, (*QDC*) holds. Observe that ◊*CBPQ* is a Saccheri
quadrilateral with *QC* = *PB* and ∠*Q* ≅ ∠*LPB*. Thus, *CD* < *QC* = *PB*
< *AB* and, since this argument applies to both geometries, **(3)** is estab-
lished. At this point we give separate arguments for the two geometries.

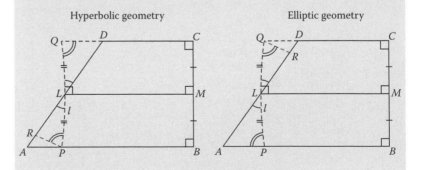

Figure 6.9 Proof of midline bisector theorem.

1. In hyperbolic geometry $\angle LPB \cong \angle Q$ are acute angles, so $m\angle LPA > 90$. Thus construct ray PR between rays PA and PL such that $\angle LPR \cong \angle Q$, with (LRA). By AAS, $\Delta LPR \cong \Delta LQD$ and $DL = LR < LA$ [thus (4)]. Also, in ΔPRA, $m\angle ARP = 180 - m\angle LRP = 180 - m\angle LDQ = m\angle LDC \geq 90$ and therefore $QD = PR < AP$. Then since $QC = PB$ and $LM < PB$, we have

$$2LM < PB + QC = AB - AP + QD + CD <$$

$$AB - AP + AP + CD = AB + CD$$

or $LM < \frac{1}{2}(AB + CD)$, and (5) is established for hyperbolic geometry.

2. In elliptic geometry, $m\angle LQD$ is obtuse and $m\angle LPA < 90$, so we can construct ray QR between rays QD and QL such that $\angle LQR \cong \angle LPA$, with (LRD). By AAS, $\Delta LQR \cong \Delta LPA$ and $DL > RL = LA$, which is (4) in elliptic geometry. In ΔQRD we have $m\angle DRQ = 180 - m\angle QRL = 180 - m\angle PAL \geq 90$, since $\angle PAL$ is acute or right, by hypothesis. Hence $QD > QR = AP$. In elliptic geometry, $LM > PB = QC$ and

$$2LM > PB + QC = AB - AP + QD + DC >$$

$$AB - AP + AP + DC = AB + DC$$

which establishes (5) for elliptic geometry. ⬥

Moment for Reflection

Suppose $\lozenge ABCD$ has right angles at B, C, and D. Then according to (3) we are supposed to have $CD < AB$. (Recall that $\angle D$ is allowed to be a right angle.) But in elliptic geometry, Lambert quadrilateral $ABCD$ satisfies the reverse inequality, $CD > AB$ (Corollary **D**, Section **6.1**). Can you explain the apparent inconsistency?

The preceding result leads to the inequality (1) mentioned earlier. A special case of this follows almost immediately. Suppose we divide segment BC into n congruent sub-segments each having length b, and determine the corresponding midlines $P_k Q_k$ at the endpoints of those sub-segments (Figure 6.10). The endpoints $P_k (1 \leq k \leq n - 1)$ of those midlines determine n corresponding sub-segments on AD having respective lengths a_1, a_2, $a_3, \ldots, a_n$. Note that $\angle DP_k Q_k$ is an acute angle and $\angle AP_k Q_k$ is obtuse, hence the midlines are the bases of normal trapezoids $\lozenge P_{k+1} Q_{k+1} Q_k P_k$ for $0 \leq k \leq n - 1$ (with $P_0 = D$, $Q_0 = C$, $P_n = A$, and $Q_n = B$). Hence, by (4),

$$a_1 < a_2 < a_3 < \cdots < a_n$$

For convenience, let $\overline{PQ}$ represent the midline $P_m Q_m$ where $1 \le m \le n - 1$. Then if $a_{m+1} = a$ we have

$$DP = a_1 + a_2 + a_3 + \cdots + a_m < a + a + a + \cdots + a = ma$$

and

$$PA = a_{m+1} + a_{m+2} + \cdots + a_n \ge a + a + \cdots + a = (n - m)a$$

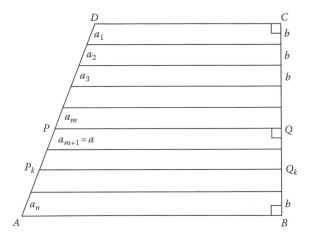

Figure 6.10 Proof of **(6)**.

Therefore,

$$\frac{DP}{PA} < \frac{ma}{(n - m)a} = \frac{m}{n - m}$$

Also, $CQ = mb$ and $QB = (n - m)b$ so that $CQ/QB = m/(n - m)$. This proves:

Theorem If $\overline{BC}$ is divided into n congruent sub-segments and Q is one of the endpoints of those sub-segments, with $\overline{PQ}$ the midline at Q, then

(6)

$$\frac{DP}{PA} < \frac{CQ}{QB}$$

Note: The algebraic form of **(6)** can be changed to one that is often more convenient. This device will be used from now. (This idea was already introduced for Euclidean geometry in Section 5.11 in the proof of the side-splitting theorem.) Starting with **(6)**, take reciprocals to obtain $PA/DP > QB/CQ$. Add unity to both sides to obtain $(DP + PA)/DP > (CQ + QB)/CQ$ or $DA/DP > CB/CQ$. Take reciprocals again to obtain

(7)

$$\frac{DP}{DA} < \frac{CQ}{CB}$$

Since these steps are reversible, it follows that the two forms **(6)** and **(7)** are equivalent and may be used interchangeably. ◣

Side-Splitter Theorem for Normal Trapezoids   Suppose $\Diamond ABCD$ is a normal trapezoid with base AB, with EF a midline drawn to altitude BC, as in Figure 6.11. Then

(8)
$$\frac{DE}{EA} < \frac{CF}{FB} \quad \text{or} \quad \frac{DE}{DA} < \frac{CF}{CB}$$

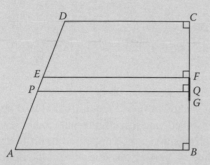

Figure 6.11 Proof of side-splitter theorem (first part).

Proof   Suppose, to the contrary, that in hyperbolic or Euclidean geometry (Case 1) $DE/DA > CF/CB$, and that in Euclidean or elliptic geometry (Case 2), $DE/DA < CF/CB$. In either case, choose G on $\overline{BC}$ such that $CG/CB = DE/DA$. Then $CG > CF$ and (CFG) in Case 1, and, $CG < CF$ and (CGF) in Case 2 (see Figure 6.11 for Case 1 only). Now determine n congruent sub-segments of $\overline{BC}$ with n sufficiently large that one of the endpoints Q of those sub-segments is an interior point of segment FG (specifically, $n > CB/FG$ is sufficient). Let $\overline{PQ}$ be the corresponding midline. Since (CFQ), then (DEP) follows. In Case 1, we obtain from (7) applied to the midline PQ, the following contradiction

$$\frac{DE}{DA} < \frac{DP}{DA} < \frac{CQ}{CB} < \frac{CG}{CB} = \frac{DE}{DA}$$

Hence, $DE/DA \geq CF/CB$. In Case 2 the reverse inequalities hold, proving that $DE/DA \leq CF/CB$. At this point, note that in Euclidean geometry this proves both $DE/DA \leq CF/CB$ and $DE/DA \geq CF/CB$, or equality. In non-Euclidean geometry we have proven $DE/DA \leq CF/CB$ for a midline of any normal trapezoid $\Diamond ABCD$. For the final step, let PQ be the midline corresponding to $Q = Q_k$, an endpoint of a congruent sub-segment of BC falling on FB (Figure 6.12). Then in $\Diamond PQCD$ we have $DE/DP \leq CF/CQ$. Thus, the use of (7) produces

$$\frac{DE}{DA} = \frac{DE}{DP} \cdot \frac{DP}{DA} \leq \frac{CF}{CQ} \cdot \frac{DP}{DA} < \frac{CF}{CQ} \cdot \frac{CQ}{CB} = \frac{CF}{CB}$$

as desired. The alternative form in (8) then follows (see Note above). ◣

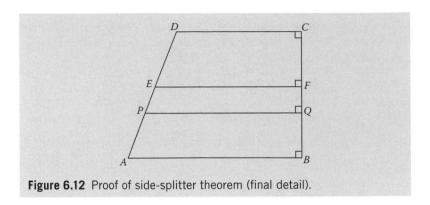

Figure 6.12 Proof of side-splitter theorem (final detail).

The next result is another ratio inequality that will be needed later. It is somewhat more involved, and is a counterpart in non-Euclidean geometry to the Euclidean proposition which appeared in Problem 30, Section **5.12**. In Euclidean geometry, given trapezoid $ABCD$ having base DC and midline EF parallel to DC, if $p = CF/CB$ and $q = 1 - p$, then $EF = pAB + qCD$. (This is an interesting problem, which you might want to attempt if you have not already done so.) The analogue of this result in unified geometry is the inequality for normal trapezoids (illustrated in Figure 6.13),

(9) $EF < pAB + qCD$ where $p = CF/CB$, $q = FB/CB$

The proof makes use of **(4)** and **(5)**.

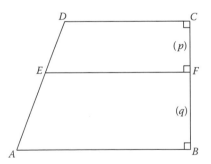

Figure 6.13 Inequality **(9)** for unified geometry.

This inequality will be proven first when p is a dyadic rational (a number of the form $m/2^n$ for positive integers m and n). For $p = \frac{1}{2} = q$, **(9)** reduces to **(5)**, which has already been proven. The goal is to increase the value of n (the *order* of p) to higher values, where **(9)** is assumed to have been proven for all m and some particular value of n, and one establishes it for order $n + 1$. Mathematical induction then guarantees the result for all n, thus for all dyadic rationals. Suppose the result has been proven for all dyadic rationals of a particular order n and that $p = m/2^{n+1}$ (of order $n + 1$). Then m must be odd, so that $m = 2k + 1$ for some integer $k \geq 0$, with $p = CQ/BC$ for corresponding midline PQ. Let UR, VS be midlines corresponding to $r = k/2^n$ and $s = (k+1)/2^n$, where $CR = rCB$ and $CS = sCB$, as in Figure 6.14. Then by hypothesis, it has been proven that

$$UR < rAB + (1-r)CD \quad \text{and} \quad VS < sAB + (1-s)CD$$

Note that by algebra $2CQ = CR + CS$, making Q the midpoint of RS. Thus **(5)** applies to the midline PQ of trapezoid $\lozenge VSRU$ and $PQ < \frac{1}{2}(UR + VS)$. Then by "substitution" (transitive law for $<$),

$$PQ < \tfrac{1}{2}(UR + VS) < \tfrac{1}{2}\big[rAB + (1-r)CD + sAB + (1-s)CD\big]$$

$$= \tfrac{1}{2}\big[(r+s)AB + (2-r-s)CD\big]$$

$$= \big[(2k+1)/2^{n+1}\big]AB + \big[1 - (2k+1)/2^{n+1}\big]CD$$

$$= pAB + qCD$$

So this proves the desired inequality $PQ < pAB + qCD$ for dyadic rationals of order $n + 1$.

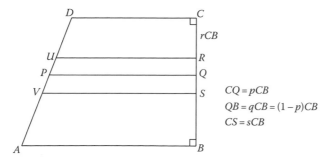

Figure 6.14

Note: If you have trouble visualizing this proof, observe Figure 6.15, where midlines UR, VS, and WT of $\lozenge ABCD$ are shown, corresponding to $p = \frac{1}{4}$, $\frac{1}{2}$, and $\frac{3}{4}$ (dyadic rationals of order 2). Start with $UR < \frac{1}{2}(CD + VS)$ and $VS < \frac{1}{2}(AB + CD)$ [true by **(5)**]. Thus $UR < \frac{1}{2}[CD + \frac{1}{2}(AB + CD)]$. Show that this reduces to $UR < \frac{1}{4}AB + \frac{3}{4}CD$. Obtain a similar result for WT. Try inserting midline PQ between, say, VS and WT such that $SQ = QT$, and obtaining the inequality $PQ < pAB + qCD$ where $p = 5/8$. After experimenting with such inequalities a few times, you should be able to appreciate the general proof above. ◣

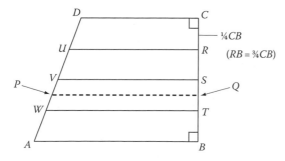

Figure 6.15 Using dyadic rationals.

Thus **(9)** is valid for all dyadic rationals p, $0 < p < 1$, and the general result now follows in much the same manner as **(8)** was established from **(7)**; the details are left as Problem 19 below.

Example 1

In Figure 6.16, $\lozenge ABCD$ has angles as indicated, and midline EF drawn as shown, with certain measurements as indicated in the figure. Show that $10 < EF < 12$, $x < 3$, and $y > 6$.

Solution

From the angle measures indicated, the geometry must be hyperbolic, and $\lozenge ABCD$ is a normal trapezoid having base AB. From **(3)**, $DC < EF < AB$ or $10 < EF < 12$. From **(8)**, $x/DA < CF/CB$, or $x/9 < 2/6 = 1/3$. Thus, $x < 3$. Since $y = AD - x$ then $y > 9 - 3 = 6$. (Note that if $\lozenge ABCD$ were a trapezoid in Euclidean geometry, then $x = 3$ and $y = 6$.)

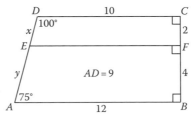

Figure 6.16 Example 1.

Example 2

In Figure 6.17, $\lozenge ABCD$ has angles as indicated, and midlines EF and GH drawn as shown. If the lengths of certain segments are as indicated in the figure, show that $x > 10$ and $y > 11$.

Solution

From the angle measures indicated the geometry must be elliptic, and $\lozenge ABCD$ is a normal trapezoid having base AB. Therefore, with $p = CF/CB = 1/3$ and $q = 2/3$, **(9)** yields $x > pAB + qCD = (1/3)12 + (2/3)9 = 4 + 6 = 10$. Similarly, with $p = CH/CB = 2/3$ and $q = 1/3$, $y > (2/3)12 + (1/3)9 = 8 + 3 = 11$.

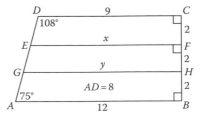

Figure 6.17 Example 2.

Example 3

In Figure 6.18, $\lozenge PQRS$ is a normal trapezoid in hyperbolic geometry, with certain measurements as indicated. From **(3)**, 20 is an upper bound for x. Show that the use of **(9)** yields a sharper upper bound for x.

Solution

Using **(9)** we find that $p = QW/QR = 12/15 = 4/5$ and $q = 1 - p = 1/5$. Therefore,

$$x < 20p + 5q = 20(4/5) + 5(1/5) = 16 + 1 = 17$$

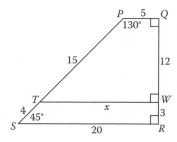

Figure 6.18 Example 3.

Example 4

Let $\triangle ABC$ be a right triangle in hyperbolic geometry, with $\angle C$ the right angle and midline DE drawn (Figure 6.19). Midline LM is drawn at the midpoint M of AE. Show that if $AL \leq LD$, then

$$\frac{AD}{DB} < \frac{AE}{EC}$$

Solution

Since $\angle B$ and $\angle ALM$ are acute in hyperbolic geometry, then $\angle MLB$ is obtuse and $\lozenge LMCB$ is a normal trapezoid, with base BC and altitude MC. Then **(8)** applies in the analysis which follows (using labels appearing in the figure).

$$\frac{q}{c} < \frac{a}{b}$$

Therefore, since $AL \leq LD$ $(p \leq q)$,

$$\frac{AD}{DB} = \frac{p+q}{c} \leq \frac{2q}{c} < \frac{2a}{b} = \frac{AE}{EC}$$

as desired.

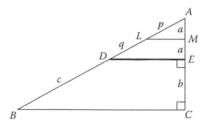

Figure 6.19 Example 4.

Since the inequalities established here for general normal trapezoids are needed primarily for Lambert quadrilaterals later on, it is useful to restate them for Lambert quadrilaterals exclusively.

Side-Splitter Theorem for Lambert Quadrilaterals   Let $\Diamond ABCD$ be a Lambert quadrilateral, with right angles at A, B, and C. If EF is a midline to side BC, as shown in Figure 6.20, the following inequalities prevail (note the *ordinary inequality* in **(10)**):

(10) $\dfrac{EA}{DA} \le \dfrac{FB}{CB}$ (equality only in Euclidean geometry)

(11) $EF < pAB + qCD$ where $p = CF/CB$ and $q = FB/CB$

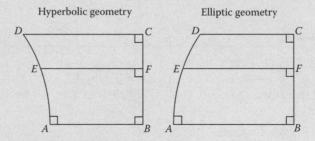

Figure 6.20 Side-splitter theorem.

Proof   In hyperbolic geometry, $\angle D$ is acute so $\Diamond ABCD$ is a normal trapezoid with base CD. Hence, by **(8)**

$$\frac{AE}{AD} < \frac{BF}{BC}$$

In elliptic geometry, $\Diamond ABCD$ is a normal trapezoid with base AB. Again by **(8)**

$$\frac{DE}{DA} > \frac{CF}{CB} \rightarrow \frac{DA-EA}{DA} > \frac{CB-FB}{CB} \rightarrow 1-\frac{EA}{DA} > 1-\frac{FB}{CB}$$

and **(10)** follows. Finally, to establish **(11)**, let $p = CF/CB$ and $q = FB/CB$. In hyperbolic geometry, by **(9)**, $EF < (BF/BC)CD + (FC/BC)AB = pAB + qCD$. In elliptic geometry, $EF > (CF/CB)AB + (FB/CB)CD = pAB + qCD$. Therefore, the comprehensive inequality **(11)** follows. ◣

6.3 Ratio Inequalities for Right Triangles

The midline inequalities for normal trapezoids may be used for quite simple proofs of those for right triangles, namely, **(1)** and **(2)** in Section **6.2** mentioned earlier. Example 4 above is suggestive of the kind of proof possible. If we first prove that $AL \leq LB$ (Figure 6.21) for midline LM, where M is the midpoint of side AC, the rest will follow exactly as in the example. Let $AX = x$ and $AY = y$, where $X \in \overline{AL}$ and Y is the foot of X on $\overline{AE}$ (as in Figure 6.21). Then by **(8)** applied to midline LM of trapezoid $\Diamond XYCB$,

$$\frac{XL}{LB} < \frac{YM}{MC} \quad \text{or} \quad \frac{p-x}{q} < \frac{a-y}{a}$$

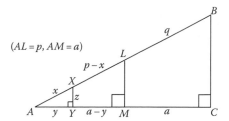

Figure 6.21 Proving a special case of **(1)**, Section **6.2**.

Taking the limit as $x \to 0$ (therefore $y \to 0$), the result is $p/q \leq 1$ or $p \leq q$, and $AL \leq LB$. Thus if LM is the midline of any right triangle ABC (right angle at C) and M is the midpoint of side AC, then

(1) $$AL \leq LB$$

Side-Splitter Theorem for Right Triangles   Consider right triangle ABC having midline $DE \perp AC$ at E on side AC (Figure 6.22). Then

(2) $$\frac{AD}{DB} < \frac{AE}{EC} \quad \text{or} \quad \frac{AD}{AB} < \frac{AE}{AC}$$

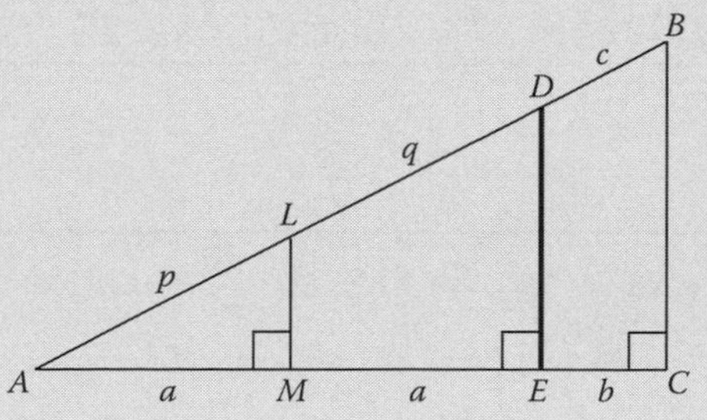

Figure 6.22 General proof of **(1)**, Section **6.2**.

Proof Construct the midline *LM* at the midpoint *M* of segment *AE*, and label distances as shown in Figure 6.22. As in Example 4, ◊*LBCM* is a normal trapezoid (base *BC*, altitude *MC*). By **(1)** $p \leq q$ and therefore $p + q \leq 2q$. By **(8)**, Section **6.2**,

$$\frac{q}{c} < \frac{a}{b}$$

Hence

$$\frac{AD}{DB} = \frac{p+q}{c} \leq \frac{2q}{c} < \frac{2a}{b} = \frac{AE}{EC} \quad \backslash\!\backslash$$

Another inequality can be proven using **(9)**, which will be left as a problem (Problem 15 below):

(3) $$\frac{DE}{BC} < \frac{AE}{AC}$$

Example 1

Using **(2)**, estimate the value for *x* in Figure 6.23 by giving upper and lower bounds, where *AB* = 30, and other measurements are as indicated.

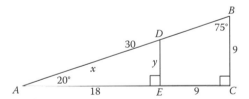

Figure 6.23 Examples 1,2.

Solution

In elliptic geometry, $x < y + 18 < 9 + 18 < 27$ (triangle inequality) and by **(2)**, $x/30 > 18/27 = 2/3$ or $20 < x < 27$.

Example 2

Estimate the value for *y* in Figure 6.23, with the same conditions as in Example 1.

Solution

By **(3)**, $y/9 > 18/27$ or $y > 6$. Thus, $6 < y < 9$.

Example 3

In Figure 6.24, two right triangles $\triangle ABC$ and $\triangle XYZ$, have $\angle A \cong \angle X$ and $XZ = 4/5AC$. In Euclidean geometry the triangles would be similar and $\triangle XYZ$ would be 4/5th the size of $\triangle ABC$, with sides in that same proportion. Show, however, that in non-Euclidean geometry $XY < 4/5AB$ and $YZ < 4/5BC$, verifying again that similar triangles as normally defined do not exist in non-Euclidean geometry.

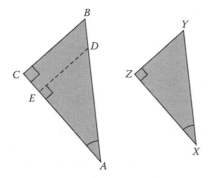

Figure 6.24 Example 3.

Solution

Locate E on AC such that $AE = XZ$, and erect the perpendicular ED at D. Thus by the LA congruence criterion, $\triangle XYZ \cong \triangle ADE$ and $AE/AC = 4/5$. By **(2)** and **(3)** $XY = AD < AB(AE/AC) = 4/5AB$, and $YZ = DE < BC(AE/AC) = 4/5BC$.

6.4 Orthogonal Projection and "Similar" Triangles

The inequalities of this chapter have applications which connect with two permanent staples of Euclidean geometry: orthogonal (parallel) projection and (again) similar triangles. In Euclidean geometry, a parallel projection from one line to another preserves ratios of lengths of segments. Furthermore, similar triangles appear when one multiplies the lengths of the sides of a given triangle by any positive constant. These phenomena are radically different in non-Euclidean geometry. Consider two lines l and m and PP', QQ', and RR' perpendiculars to m (Figure 6.25), and suppose that $\angle QPP'$ and $\angle RQQ'$ are obtuse. Then **(8)** implies that

$$\frac{PQ}{QR} < \frac{P'Q'}{Q'R'}$$

Parallel projection

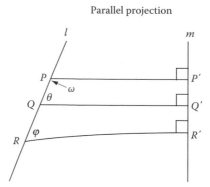

Figure 6.25 Parallel projection.

If the points on m are evenly distributed ($P'Q' = Q'R'$), then $PQ/QR < 1$ or $PQ < QR$. In hyperbolic geometry, $PQ < QR$, and in elliptic geometry, $PQ > QR$. Thus, in attempting a representative drawing in the Euclidean plane, we are led to a diagram like that shown in Figure 6.25 for hyperbolic geometry, where we must curve one of the lines in order to account for this result. Since the area (defect or excess) of $\lozenge PRR'P'$ is greater than that of $\lozenge PQQ'R'$ it follows that in the two geometries, $\varphi < \theta$ and $\varphi > \theta$ respectively. (Let the reader verify this.) If more perpendiculars are added (equally spaced on line m), Figures 6.26 and 6.27 are the results (hyperbolic and elliptic geometry, respectively). As more lines are added, the distortion becomes more pronounced. Without models, this can only be left to the imagination. In elliptic geometry, however, the fictitious drawing in the Euclidean plane of Figure 6.27 is realized in actual fact on the unit sphere, as illustrated.

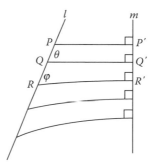

Figure 6.26 Parallel projection in hyperbolic geometry.

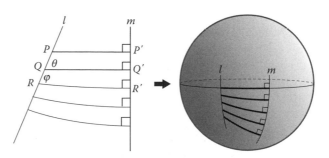

Figure 6.27 Parallel projection in elliptic geometry.

In the case of similar triangles, consider any right triangle ABC in unified geometry (Figure 6.28). Suppose we triple side AC to create a second right triangle DEF, with $DF = 3b$ (one can just as easily take $DF = kb$ in general for this analysis), and with $\angle A \cong \angle D$. In Euclidean geometry, the two triangles would be similar, with the sides of $\triangle DEF$ of length 3 times those of $\triangle ABC$. But in unified geometry, we obtain, instead, $EF > 3a$ and $DE > 3c$. The analysis in Example 3 applies here, where we construct Q on side DF such that $DQ = AC$ and erect the perpendicular QP to side DE, creating $\triangle DPQ \cong \triangle ABC$. By (2) and (3),

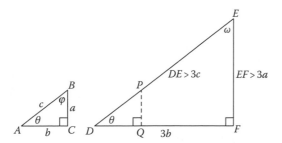

Figure 6.28 Similar triangles.

$$\frac{DP}{DE} < \frac{DQ}{DF} \quad \text{and} \quad \frac{PQ}{EF} < \frac{DQ}{DF}$$

which lead to $DE > 3c$ and $EF > 3a$. In hyperbolic geometry, we obtain $DE > 3c$ and $EF > 3a$. Moreover, since the area (defect) of $\triangle DEF$ is greater than that for $\triangle DPQ$ (and $\triangle ABC$), it follows that $\omega < \varphi$. Thus, a representative drawing in the Euclidean plane might take on the appearance of the diagram in Figure 6.29.

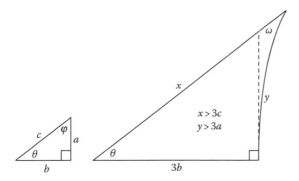

Figure 6.29 Similar triangles in hyperbolic geometry.

Similarly, in elliptic geometry $DE < 3c$ and $EF < 3a$, so that a representative drawing in the Euclidean plane appears as in Figure 6.30; here, an analysis on excess produces $\omega > \varphi$, in accord with the drawing. Again, such phenomena pictured in the Euclidean plane can be portrayed in reality on the sphere.

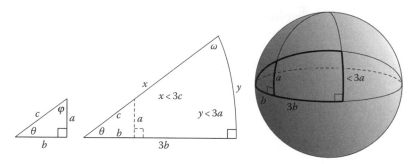

Figure 6.30 Similar triangles in elliptic geometry.

This behavior of right triangles has another very important application. Consider a fixed angle hk having vertex A and measure θ (Figure 6.31). For any positive real x, measure off segment AB of length x on ray k, and drop the perpendicular BC to ray h, forming right triangle ABC. Let AC define the function $f(x)$. In Euclidean geometry, the ratio $f(x)/x \equiv AC/AB$ defines the cosine function, and stays constant as x varies (the result of similar triangles). But in non-Euclidean geometry this ratio varies. In fact, suppose we define $F(x) = f(x)/x$, and let $x_1 < x_2$, with E the foot of D on AC, $x_1 = AD, f(x_1) = AE, x_2 = AB,$ and $f(x_2) = AC$. Then **(2)** implies:

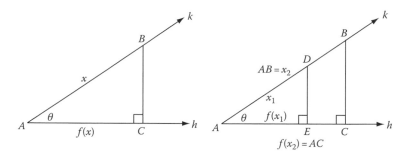

Figure 6.31 Monotone property of $F(x)$.

$$\frac{AD}{AB} < \frac{AE}{AC} \quad \text{or} \quad \frac{AE}{AD} > \frac{AC}{AB}$$

That is,

$$\frac{f(x_1)}{x_1} > \frac{f(x_2)}{x_2} \quad \text{or} \quad F(x_1) > F(x_2)$$

and F is a *monotone* function (either strictly decreasing, strictly increasing, or constant). Thus, we find that F is constant for Euclidean geometry, as expected, decreasing in hyperbolic geometry, and increasing in elliptic geometry. Figure 6.32 illustrates a representative graph of F for the three cases. Thus it is apparent that as $x \to 0$, the limit of $F(x)$ exists as either the greatest lower bound of the set $\{F(x)\}$ in hyperbolic geometry, or the least upper bound of $\{F(x)\}$ in elliptic geometry. This limit, denoted $c(\theta)$, exists for each value of θ, $0 \le \theta \le 90$, and may, or may not, bear resemblance to the cosine function in Euclidean geometry. This function is the object of a detailed study starting in the next section, where it is ultimately concluded that, indeed, $c(\theta) = \cos \theta$. By itself, this is not too surprising since very small triangles in either elliptic or hyperbolic geometry behave like Euclidean triangles (recall from Chapter 5 that angle-sums of small triangles become 180 in the limit). But proving it is another matter.

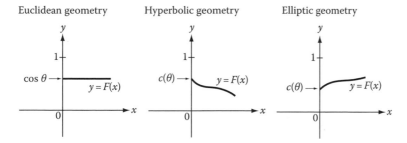

Figure 6.32 Graphs of $F(x)$.

Problems (Section 6.4)

Group A

1. In Figure P.1, $\lozenge PQRS$ is a normal trapezoid in elliptic geometry having base RS, with lengths of segments as indicated. Show that if $y = 14$, $x > 7$.

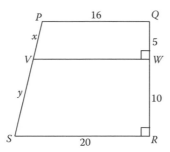

Figure P.1

2. $\lozenge PQRS$ is a normal trapezoid in elliptic geometry, as in Problem 1 with $x = 7.5$. If the polygonal triangle inequality were used it would follow that $y < 20 + 10 + 16 = 46$ (since $VW < 16$). However, from the inequalities of this section, derive the sharper inequality $y < 15$.

3. If $\lozenge ABCD$ is a normal trapezoid in hyperbolic geometry, with midlines drawn at the points of trisection of altitude BC, derive the inequalities $17 < x < 18$ and $17 < y < 19$ (Figure P.3).

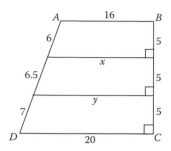

Figure P.3

The following problems (4–8) are to be solved using the figures as indicated. The value for x is to be estimated (lower and upper bounds in each case). The triangle inequality is to be used when appropriate.

4.

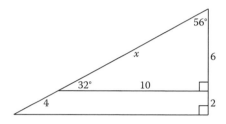

5.

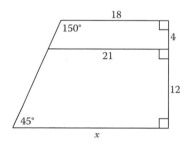

6.

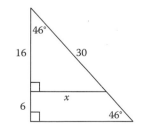

7.

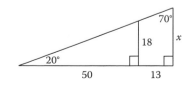

8.

9.   In Figure P.9, $\lozenge TUVW$ is a Lambert quadrilateral in elliptic geometry and RS is a midline, with distances as indicated. Show that $11 < RS < 12$. [**Hint:** Since $TW = TU$, it
follows that $WV = UV$. In relation to this, see Problem 10.]

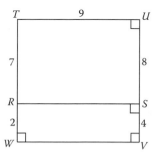

Figure P.9

Group B

10.   If $\lozenge ABCD$ is a Lambert quadrilateral with right angles at
A, B, and C, prove that if $AD = DC$ then $AB = BC$, and
conversely (Figure P.10).

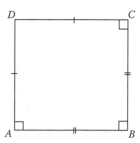

Figure P.10

11.   In Figure P.11, determine the range of values for x and y in
each diagram in order for it to be valid in:
  (a) Euclidean geometry.
  (b) Hyperbolic geometry (assume that the quadrilateral
    is a normal trapezoid in standard position).
  (c) Elliptic geometry (assume the quadrilateral is a nor-
    mal trapezoid in standard position).

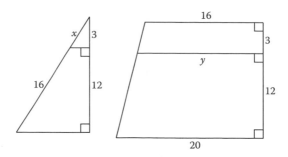

Figure P.11

12. Estimate the values for x and y giving upper and lower bounds in each geometry if the quadrilateral shown is a normal trapezoid whose base is the side of length 20 in Figure P.12.

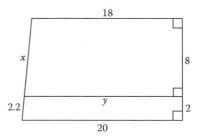

Figure P.12

13. Estimate the values for x and y giving upper and lower bounds in each geometry if the quadrilateral shown is a normal trapezoid whose base is the side of length 20 in Figure P.13.

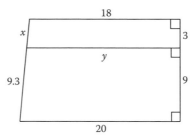

Figure P.13

Group C

14. **Important Inequality for Lambert Quadrilaterals** If $\lozenge ABCD$ is a Lambert quadrilateral with right angles at A, B, and C, and with sides less than $\alpha/4$, prove that $AB < 2CD$. [**Hint:** For elliptic geometry, let P be a pole of line AB, with $PA = PB = \alpha/2$, then locate F on segment CP such that $CF = CB$, with (BCF). Apply **(5)**, Section **6.2**.]

15.   The relatively easy method used in proving the mid-line inequality **(2)**, Section 6.3, can also be used for **(3)**. (See Figure P.15.)

    (a)   Prove that $LM \leq \tfrac{1}{2} DE$ by applying **(9)**, Section **6.2**, to midline LM of trapezoid $YXED$ and taking the limit as $x \to 0$.

    (b)   Prove $DE/BC < AE/AC$ making use of **(9)**, Section **6.2**, as applied to trapezoid $LMCB$.

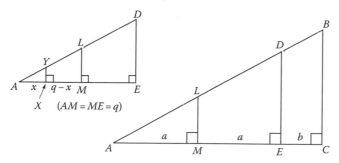

Figure P.15

16.   Right triangles ABC and XYZ with right angles at C and Z have $\angle A \cong \angle X$ and $AC < XZ$. Prove that

$$\frac{AB}{XY} < \frac{AC}{XZ} \quad \text{and} \quad \frac{BC}{YZ} < \frac{AC}{XZ}$$

17.   Ratios of segments are preserved under parallel projection in Euclidean geometry. Prove this version in unified geometry that is valid under orthogonal projection, as follows: If the foot of point P on line l is P' on line m, the mapping P to P' defines an **orthogonal projection** from l to m. Several points from l are mapped to m as shown in Figure P.17. Assuming $\angle QPP'$ is obtuse for these points, prove that

$$\frac{x}{y} < \frac{a}{b}, \quad \frac{y}{z} < \frac{b}{c}, \quad \frac{x}{z} < \frac{a}{c}$$

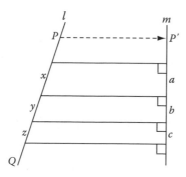

Figure P.17

18.   If $\lozenge ABCD$ is a normal trapezoid having base AB, and EF is a midline to altitude BC, prove that

$$\frac{CF}{BC} > \frac{EF - CD}{AB - CD}$$

19. To prove (**9**) for an arbitrary midline *EF*, follow this outline (you are to fill in the details, and then adapt the proof given here for hyperbolic geometry in order to extend it to elliptic geometry) (Figure P.19):

1. Suppose, to the contrary, that $EF > p(AB - CD) + CD$ where $p = CF/CB$. Locate *G* on *BC* so that $EF = p'(AB - CD) + CD$ where $p' = CG/CB$. Hence, $CG/CB > CF/CB$ and (*CFG*), as indicated in the figure. (How do you know that (*CFGB*) holds—equivalent to $CG < CB$?)

2. By successive midpoint construction on segment *BC*, and for *k* large enough, one of those midpoints, say *Q*, lies on segment *FG* such that *CQ/CB* is a dyadic rational of order *k*. Consider the midline *PQ*. By (**9**) for dyadic rationals, $PQ < (CQ/CB)(AB - CD) + CD$.

3. We then obtain the contradiction $PQ < (CG/CB)(AB - CD) + CD = EF$. (What does this contradict?) Therefore, $EF \leq p(AB - CD) + CD$.

4. To prove the strict inequality, let *Q* be a successive midpoint on segment *FB* such that *CQ/CB* is a dyadic rational. Then by (**9**) for dyadic rationals and the ≤ version of (**9**) already established,

$$EF \leq \frac{CF}{CQ}(PQ - CD) + CD, \quad PQ < \frac{CQ}{CB}(AB - CD) + CD$$

[thus $PQ - CD < (CQ/CB)(AB - CD)$], and therefore

$$EF < \frac{CF}{CQ} \cdot \frac{CQ}{CB}(AB - CD) + CD = \frac{CF}{CB}(AB - CD) + CD$$

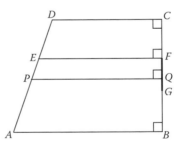

Figure P.19

6.5 Unified Trigonometry: The Functions $c(\theta)$ and $s(\theta)$

It is remarkable that the pioneers of non-Euclidean geometry, Bolyai and Lobachevski, in their pursuit of non-Euclidean geometry (which was confined to hyperbolic geometry) went as far as they did in their development. They obtained the same formulas for hyperbolic geometry we are about to develop, as well as some interesting results in three-dimensional absolute geometry. Our method, due to H.G. Forder (Forder 1953), stays two-dimensional and does not involve surface theory in any way.

At this point in the analysis, we start using radian measure for angles, in agreement with modern mathematics, and with traditional calculus courses. The chief reason for this, of course, is that the derivatives of sine and cosine are simplified by this convention. The value β introduced in the axiom for angle measure then takes on the value π instead of 180. The traditional conversion formula is, as usual,

$$\pi \text{ rad} = 180°$$

The last section justified the definition of the function $c(\theta)$ in terms of the special function $F(x)$ and its limit as $x \to 0$ for $0 < \theta < \pi/2$, where $F(x)$ is decreasing in hyperbolic geometry and increasing in elliptic geometry. Using standard notation, as exhibited in Figure 6.33, this limit may be written symbolically as

(1) $$c(\theta) = \lim_{c \to 0} \frac{b}{c} \quad \text{or} \quad c(\theta) = \lim_{B \to A} \frac{AC}{AB}$$

(the symbol $B \to A$ in the above is used to represent $AB \to 0$, widely used in metric spaces in general; the symbol $F(x) \to c(\theta)$ will also be used occasionally to represent the limit of $F(x)$). From the properties of $F(x)$ discussed in the last section (as illustrated in Figure 6.32), we have the following bounds for $c(\theta)$ when $0 < \theta < \pi/2$:

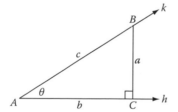

Figure 6.33 The function $c(\theta)$.

(2) $$\begin{cases} 0 < c(\theta) \le 1 & \text{in hyperbolic geometry} \\ 0 \le c(\theta) < 1 & \text{in elliptic geometry} \end{cases}$$

The companion function $s(\theta)$ is, for now, defined by the relation

(3) $$s(\theta) = c(\pi/2 - \theta), \quad 0 < \theta < \pi/2$$

As will be observed later, $s(\theta)$ turns out to be the limit of BC/AB as $B \to A$.

If $\theta = 0$ then the rays h and k coincide, hence $B = C$ in Figure 6.33 as $B \to A$. Hence by definition of $c(\theta)$, $c(0) = \lim AB/AB = 1$. If $\theta = \pi/2$ then $\angle hk$ is a right angle, and the foot of B on k is A itself, so that in this case $c(\pi/2) = \lim 0/AB = 0$. Thus one obtains the boundary conditions for $c(\theta)$ and $s(\theta)$:

(4) $$c(0) = 1, \quad c(\pi/2) = 0, \quad s(0) = 0, \quad s(\pi/2) = 1$$

It will be convenient to extend the two functions $c(\theta)$ and $s(\theta)$ to values greater than $\pi/2$. For $\pi/2 < \theta < \pi$, define $s(\theta) = s(\pi - \theta)$ and $c(\theta) = -c(\pi - \theta)$. The following identities then result.

(5) $s(\pi/2 + \theta) = c(\theta), \quad c(\pi/2 + \theta) = -s(\theta)$

Proof Since $\pi/2 + \theta > \pi/2$, $s(\pi/2 + \theta) = s(\pi - (\pi/2 + \theta)) = s(\pi/2 - \theta) = c(\theta)$. The other identity is similar. ◤

It will be very important in establishing identities for $c(\theta)$ and $s(\theta)$ to allow θ to vary as one takes the limit in obtaining $c(\theta)$ and to conclude that if $\theta \to \varphi$, then the ratio defining $c(\theta)$ has limit $c(\varphi)$. This involves an elusive double limit. Normally a tedious task in analysis often involving special theorems, it is in this case however, quite manageable on an elementary level, which, nevertheless, will be relegated to Appendix **E**. Thus in Figure 6.34, as B and C approach A as limit in such a way that $m\angle BAC$ converges to some value φ between 0 and $\pi/2$, we may assume that the following limit exists and gives the expected value:

$$\lim_{\substack{c \to 0 \\ \theta \to \varphi}} \frac{b}{c} = c(\varphi)$$

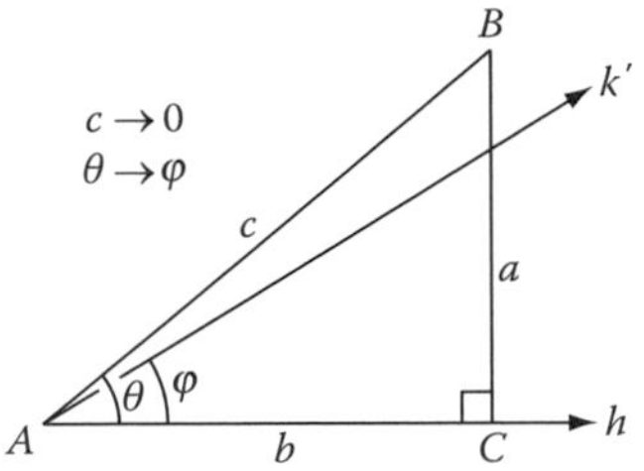

Figure 6.34 A double limit.

A corollary is that the function $c(\theta)$ is continuous (also established in Appendix **E**).

A generalization of this property is easily established, and will be quite useful. Suppose the vertices of a variable right triangle ABC approach a common point P as limit such that $\theta \equiv m\angle BAC$ converges to some value φ, as illustrated in Figure 6.35. Choose a fixed ray h with endpoint P and ray k' such that $hk' = \varphi$. Construct $\triangle PQR$ congruent to $\triangle ABC$ such that R lies on ray h and Q lies on the k'-side of h, with $\theta = m\angle QPR$. Now we have the situation of Figure 6.34, where the double limit as $Q \to P$ and $m\angle QPR \to \varphi$ equals $c(\varphi)$. Hence, as A, B, and $C \to P$ and $\theta \to \varphi$, the ratio AC/AB converges to $c(\varphi)$. This result will be referred to as the *floating triangle theorem*, or simply, *floating triangles*.

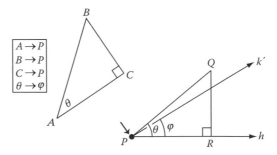

Figure 6.35 Floating triangle theorem.

The first application is to show that the function $s(\theta)$ can be expressed as a limit that resembles the familiar sine relation in Euclidean geometry (mentioned above). In $\triangle ABC$, let $B \to A$ and $C \to A$. Then by previous results, as the sides of right triangle ABC approach zero as limit, (figure below), the limit of the sum of the angle measures $\theta + \varphi + \pi/2$ is π, hence if θ is fixed in value, then φ has limit $\pi/2 - \theta$. By floating triangles,

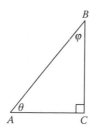

(6) $$\lim_{B \to A} \frac{BC}{AB} = c(\pi/2 - \theta) = s(\theta)$$

The floating triangle theorem is observed to apply to the function $s(\theta)$ as well. As a result, a limit version of the law of sines can be established. Let the vertices of any triangle ABC converge to some point P, and assume that the angle measures of $\angle A$ and $\angle B$ tend to the values A and B, respectively. Then it will be shown that if $s(B) \neq 0$,

(7)   **Law of Sines** $$\lim_{\substack{a \to 0 \\ b \to 0}} \frac{a}{b} = \frac{s(A)}{s(B)}$$

Proof   Let F be the foot of C on line AB and $h = CF$. The proof basically follows that of the geometric proof for the ordinary law of sines. We want to establish that $\lim h/b = s(A)$ and $\lim h/a = s(B)$. Consider the ratio h/b and the three cases: $A < \pi/2$, $A = \pi/2$, and $A > \pi/2$.

1. $A < \pi/2$: Since $m\angle CAB$ must remain less than $\pi/2$ as points $A, B, C \to P$, by Theorem **2**, Section **3.6**, F must fall on ray AB, as shown in Figure 6.36 for this case. Thus $m\angle CAF = m\angle CAB \to A$ and by floating triangles, $\lim h/b = s(A)$.

The case $A, B < \pi/2$ The case $A > \pi/2$

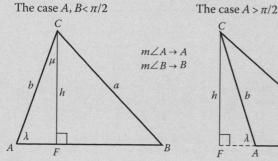

Figure 6.36 Proof of law of sines.

2. $A = \pi/2$: Consider the three angles of ΔACF, $m\angle CAF \equiv \lambda$, $m\angle ACF \equiv \mu$, and $m\angle AFC \equiv \pi/2$. By theorems in Chapter 5, the sum $\lambda + \mu + \pi/2$ has limit π. Since $\lambda =$ either $m\angle CAB$ or $\pi - m\angle CAB$ and $m\angle CAB \to \pi/2$, then $\lambda \to \pi/2$. Thus, μ must have zero limit. Hence, by floating triangles, $\lim h/b = c(0) = 1 = s(\pi/2) = s(A)$.

3. $A > \pi/2$: In this case $m\angle CAB$ must remain greater than $\pi/2$ as A, B, $C \to P$, so F must fall on the ray opposite AB and, as shown in Figure 6.36 for this case $\lambda = \pi - m\angle CAB \to \pi - A < \pi/2$. Then $\lim h/b = s(\pi - A) = s(A)$.

4. Thus it has been proven that in all cases, $\lim h/b = s(A)$. Similarly, $\lim h/a = s(B)$. Since $s(B) \neq 0$, the quotient theorem on limits applies, and we have

$$\lim \frac{a}{b} = \lim \frac{\dfrac{h}{b}}{\dfrac{h}{a}} = \frac{\lim \dfrac{h}{b}}{\lim \dfrac{h}{a}} = \frac{s(A)}{s(B)} \quad \text{\textbackslash\textbackslash}$$

Problems (Section 6.5)

Group A

1. Using Figure P.1, prove the non-increasing property of $c(\theta)$: If $\theta_1 < \theta_2$, then $c(\theta_1) \geq c(\theta_2)$.

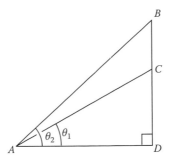

Figure P.1

2. Using the identity $s(\theta) = c(\pi/2 - \theta)$, prove the non-decreasing property of $s(\theta)$: if $\theta_1 < \theta_2$, then $s(\theta_1) \leq s(\theta_2)$.

3. **The Function $t(\theta)$**   Define $t(\theta) = s(\theta)/c(\theta)$ for each θ for which $c(\theta) > 0$ (true for all $0 \leq \theta \leq \pi/4$). Show that $t(\theta) = \lim a/b$ as $c \to 0$, with $\theta = m\angle A$ and ΔABC in Figure 6.33.

Group B

4. Using equilateral triangles and floating triangles, prove that $c(\pi/3) = \frac{1}{2} = s(\pi/6)$.

5. Using isosceles right triangles and floating triangles, prove that $t(\pi/4) = 1$.

6. Prove the second identity in **(5)**: $c(\pi/2 + \theta) = -s(\theta)$.

7. If $\pi/2 < \theta < \pi$, prove the identities:
   (a) $s(\theta - \pi/2) = -c(\theta) = c(\pi - \theta)$
   (b) $c(\theta - \pi/2) = s(\theta)$

8. Use the triangle inequality $a + b > e$ to prove
   (a) $s(\theta) + c(\theta) \geq 1$
   (b) $s(\pi/4) = c(\pi/4) \geq \frac{1}{2}$

9. In the next section, the following identity is established: $s^2(\theta) + c^2(\theta) = 1$. Use this to show that $s(\pi/4) = c(\pi/4) = \frac{\sqrt{2}}{2}$.

6.6 Trigonometric Identities

The first identity we prove will appear quite familiar, and one might expect its proof to be rather involved since the analogous identity for ordinary trigonometry depends on the Pythagorean theorem ($c^2 = a^2 + b^2$). However, observe the following analysis:

Let D be the foot of C on side AB in right triangle ABC (Figure 6.37), then let $B \to A$ (hence $C \to A$), with $\theta \equiv m\angle A$ fixed and $\varphi = m\angle B \to \pi/2 - \theta$. Hence:

$$\lim \frac{DB}{BC} = c(\lim \varphi) = s(\theta) \quad \text{and} \quad \lim \frac{BC}{AB} = s(\theta)$$

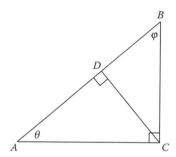

Figure 6.37 Proof of identity **(1)**.

Taking the product of these two limits,

$$\lim \frac{DB}{AB} = s^2(\theta)$$

Similarly,

$$\lim \frac{AD}{AC} = c(\theta) \quad \text{and} \quad \lim \frac{AC}{AB} = c(\theta) \quad \text{or} \quad \lim \frac{AD}{AB} = c^2(\theta)$$

Therefore, by the sum theorem on limits,

$$\lim \frac{DB}{AB} + \lim \frac{AD}{AB} = \lim \frac{DB + AD}{AB} = \lim \frac{AB}{AB} = 1$$

proving

(1) $s^2(\theta) + c^2(\theta) = 1$

which can be extended to all θ $(0 < \theta < \pi)$ using the definitions in Section 6.4.

At this point, the strict inequalities in **(2)** of Section **6.5** can be established $(0 < \theta < \pi/2)$. If $c(\theta) = 1$ for any value of θ, then by **(1)** $s(\theta) = 0$. This implies that $c(\pi/2 - \theta) = 0$, violating **(2)** in hyperbolic geometry. Similarly for elliptic geometry, $c(\theta) > 0$. Hence, $0 < c(\theta) < 1$ for both geometries. It then follows that $0 < s(\theta) < 1$ as well.

The next identity is one of the product-to-sum relations for $s(\theta)$ and $c(\theta)$; these mirror those involving sine and cosine in Euclidean trigonometry. In Figure 6.38 let concurrent rays h, k, u, and v be constructed such that $(kuvh)$ and $hu = uv = \varphi$, $uk = \theta$, and $vk = \theta - \varphi$ (thus $hk = \theta + \varphi$), where $0 < \theta + \varphi < \pi/2$. Let $B \in h$ approach A as limit. For B close enough to A, the foot M of B on u is such that ray BM will meet v and k at certain points C and D, respectively. (For this purpose, simply choose $P \in k$ such that $AP < \alpha/2$, and let Q be the foot of P on h, and with B on segment AQ; see inset in Figure 6.38.) Since M is the midpoint of $\overline{BC}$,

$$MD = \tfrac{1}{2}(BD + CD)$$

and

$$\frac{2MD}{AD} = \frac{BD}{AD} + \frac{DC}{AD}$$

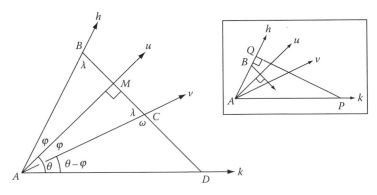

Figure 6.38 Proof of indentity **(2)**.

Now take the limit as $B \to A$, with the rays fixed in place. Note that $\lambda \to \pi/2 - \varphi \equiv \lambda'$ and $\omega \to \lim(\pi - \lambda) = \pi - \lambda' = \pi/2 + \varphi \equiv \omega'$. By definition and law of sines in $\triangle ABC$ and $\triangle ACD$,

$$2s(\theta) = \frac{s(\theta + \varphi)}{s(\lambda')} + \frac{s(\theta - \varphi)}{s(\omega')}$$

$$2s(\theta) = \frac{s(\theta + \varphi)}{s(\pi/2 - \varphi)} + \frac{s(\theta - \varphi)}{s(\pi/2 + \varphi)} = \frac{s(\theta + \varphi)}{c(\varphi)} + \frac{s(\theta - \varphi)}{c(\varphi)}$$

by **(5)**, Section **6.5**. Upon multiplying through by $c(\varphi)$, we arrive at the following identity, valid for $0 < \theta - \varphi < \theta + \varphi < \pi/2$:

(2)   Product-to-sum Identity 1 $2s(\theta)c(\varphi) = s(\theta + \varphi) + s(\theta - \varphi)$

Now consider $\theta' = \pi/2 - \theta$. Provided that $0 < \varphi < \theta'$ and $\theta' + \varphi < \pi/2$ (which can be easily verified), **(2)** applies to establish the following:

$$2s(\theta')c(\varphi) = s(\theta' + \varphi) + s(\theta' - \varphi)$$

or

$$2s(\pi/2 - \theta)c(\varphi) = s(\pi/2 - (\theta - \varphi)) + s(\pi/2 - (\theta + \varphi))$$

Since $\theta - \varphi < \pi/2$ we obtain the companion identity

(3)   Product-to-Sum Identity 2 $2c(\theta)c(\varphi) = c(\theta + \varphi) + c(\theta - \varphi)$

In order to obtain a half-angle identity, first let $\varphi \to \theta$ as limit, and use the continuity of $c(\theta)$ (established in Appendix E). The result is $2c^2(\theta) = c(2\theta) + c(0)$ or, solving the equation for $c(2\theta)$,

(4)   Double-Angle Identity $c(2\theta) = 2c^2(\theta) - 1$

Rewriting **(4)** with θ replaced by $\theta/2$, the result is the equivalent

(5)   Half-Angle Identity $c(\theta/2) = \sqrt{\dfrac{1 + c(\theta)}{2}}$

Problems (Section 6.6)

Group B

1.   Use **(4)** and $s^2(\theta) = \frac{1}{2}(2 - 2c^2(\theta))$ from **(1)** to obtain a half-angle identity for $s(\theta)$. (Replace θ by $\frac{1}{2}\theta$.)

2.   Prove the identity $s(2\theta) = 2s(\theta)c(\theta)$.

3. Prove the identity $c(2\theta) = c^2(\theta) - s^2(\theta)$.

4. With $t(\theta)$ defined as in Problem 3, Section **6.5**, prove the identity

$$t(2\theta) = \frac{2t(\theta)}{1 - t^2(\theta)}$$

5. Write the expression $s(2\theta)/(c(2\theta) + 1)$ in terms of $s(\theta)$ and $c(\theta)$, simplify, and prove the following half-angle identity for $t(\theta)$

$$t(\theta/2) = \frac{s(\theta)}{c(\theta) + 1}$$

6. Prove the identity

$$\frac{2t(\theta)}{t^2(\theta) + 1} = s(2\theta)$$

Group C

7. **Sine Addition Identity in Euclidean Geometry**   Using Figure P.7 for this problem, as labeled, prove the addition identity

$$\sin(\theta + \varphi) = \sin\theta\,\cos\varphi + \cos\theta\,\sin\varphi$$

[**Hint:** Start with the right side of the identity written in the form (verification needed) $(ED/BD)\cdot(AC/AB)$ + $(BE/BD)\cdot(BC/AB)$, and factor out the term BE/AB. Make use of similar triangles.]

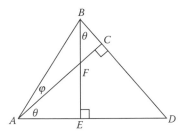

Figure P.7

8. **Cosine Addition Identity in Euclidean Geometry**   Using Figure P.7, prove the addition identity

$$\cos(\theta + \varphi) = \cos\theta\cos\varphi - \sin\theta\sin\varphi$$

[**Hint:** Start with the right side of the identity written in the form (verification needed) $(AE/AF)(AC/AB) - (FC/BF)$ (BC/AB), and factor out the term AE/AB. Make use of similar triangles.]

6.7 Classical Forms for $c(\theta)$ and $s(\theta)$

We are now in a position to prove what has been anticipated, that is, $c(\theta) = \cos\theta$. The starting point is to observe that for some arbitrarily chosen constant λ between 0 and $\pi/2$, $c(\lambda)$ is a positive number less than 1, hence there exists a unique μ, $0 < \mu < \pi/2$, such that

(1) $c(\lambda) = \cos\mu$

We now use the half-angle identity **(5)**, Section **6.6** (also valid for the cosine function) to observe that

$$c(\tfrac{1}{2}\lambda) = \sqrt{\frac{1 + c(\lambda)}{2}} = \sqrt{\frac{1 + \cos\mu}{2}} = \cos\tfrac{1}{2}\mu$$

This procedure can be repeated to obtain

$$c(\tfrac{1}{4}\lambda) = \cos\tfrac{1}{4}\mu$$

and so on. It then follows, by repeating the process, that for any integer $n > 0$, $c(\lambda/2^n) = \cos\mu/2^n$. Hence for any dyadic rational r of the form $1/2^n$,

(2) $c(r\lambda) = \cos r\mu$

To engage the reader in the next step, suppose one has proven that

(*) $c(\tfrac{1}{4}\lambda) = \cos\tfrac{1}{4}\mu, \quad c(\tfrac{1}{2}\lambda) = \cos\tfrac{1}{2}\mu, \quad \text{and} \quad c(\tfrac{3}{4}\mu) = \cos\tfrac{3}{4}\mu$

What are the results if **(3)** of Section **6.6** is applied to $c(\tfrac{5}{8}\lambda)$? You should obtain

$$2c(\tfrac{5}{8}\lambda)\,c(\tfrac{1}{8}\lambda) = c(\tfrac{5}{8}\lambda + \tfrac{1}{8}\lambda) + c(\tfrac{5}{8}\lambda - \tfrac{1}{8}\lambda) = c(\tfrac{3}{4}\lambda) + c(\tfrac{1}{2}\lambda)$$

Using the previous results (*), and since the cosine function satisfies the same product-to-sum identity, we have (writing one identity directly below the other for comparison)

$$2c(\tfrac{5}{8}\lambda)\,c(\tfrac{1}{8}\lambda) = \cos(\tfrac{3}{4}\lambda) + \cos(\tfrac{1}{2}\lambda)$$

$$2\cos\tfrac{5}{8}\lambda\cos\tfrac{1}{8}\lambda = \cos(\tfrac{3}{4}\lambda) + \cos(\tfrac{1}{2}\lambda)$$

Since $c(1/8\lambda) = \cos 1/8\mu$ follows from **(2)**, one concludes by simple algebra that

$$c(\tfrac{5}{8}\lambda) = \cos\tfrac{5}{8}\mu$$

The reader is encouraged to try this for $c(\tfrac{13}{16}\lambda)$ assuming $c(m/8\cdot\lambda) = \cos(m/8\cdot\lambda)$ is valid for $0 \le m \le 8$. This shows how one uses **(2)** for any dyadic rational r of order n to obtain it for one of order $n + 1$.

   In general, observe that if it is assumed that **(2)** has been established for all dyadic rationals of order n of the form $m/2^n$, $0 \le m \le 2^n$ (which is true for $n = 1$), and if we consider $r = m/2^{n+1}$ of order $n + 1$, where $m = 2k + 1$ (odd integer), then as in the example above, by the product-to-sum identity for both $c(\theta)$ and $\cos\theta$,

$$2c(m/2^{n+1}\lambda)c(1/2^{n+1}\lambda) = c(m/2^{n+1}\lambda + 1/2^{n+1}\lambda) + c(m/2^{n+1}\lambda - 1/2^{n+1}\lambda)$$

$$= c((k+1)/2^n\lambda) + c(k/2^n\lambda)$$

That is, by hypothesis,

$$2c(m/2^{n+1}\lambda)\cos(1/2^{n+1}\lambda) = \cos\left((k+1)/2^n\lambda\right) + \cos(k/2^n\lambda)$$

$$2\cos(m/2^{n+1}\mu)\cos(1/2^{n+1}\mu) = \cos\left((k+1)/2^n\mu\right) + \cos((k/2^n\mu)$$

and we conclude that $c(m/2^{n+1}\lambda) = \cos(m/2^{n+1}\mu)$. That is, $c(r\lambda) = \cos r\mu$ holds for a dyadic rational r of order $n + 1$. By mathematical induction, we have now established for all dyadic rationals, $0 \le r \le 1$,

(3) $c(r\lambda) = \cos r\mu$

We need to extend this to dyadic rationals $r > 1$ for which $r\lambda \le \pi/2$. This, however, is merely a matter of using the double-angle identity to prove, sequentially, $c(2r\lambda) = c^2(r\lambda) - 1 = \cos^2 r\mu - 1 = \cos 2r\mu$, $c(4r\lambda) = c^2(2r\lambda) - 1 = \cos^2 2r\mu - 1 = \cos 4r\mu,\dots$, and in general, $c(2^n r\lambda) = \cos 2^n r\mu$ for all appropriately large integers n.

Since the dyadic rationals are dense on the positive reals, we can find a sequence of dyadic rationals $\{r_n\}$ converging to any real $x > 0$, and by continuity of $c(\theta)$ and $\cos\theta$,

(4) $c(x\lambda) = \cos x\mu, \quad 0 \le x\lambda \le \pi/2$

Set $\theta = x\lambda$ and we have the equivalent result, with $\kappa = \mu/\lambda$,

(5) $c(\theta) = \cos\kappa\theta, \quad 0 \le \theta \le \pi/2$

It follows that $\kappa = 1$: set $\theta = \pi/2$ in **(5)**, with the result $0 = \cos\kappa\pi/2$; since κ is positive, $\kappa\pi/2 = (2k + 1)\pi/2$ for some integer $k \ge 0$. Hence $\kappa = 2k + 1$ and we conclude $k = 0$.

The end result is, as expected,

(6) $c(\theta) = \cos\theta \quad$ and $\quad s(\theta) = \sin\theta$

This represents the first major step in deriving the trigonometry for non-Euclidean geometry.

6.8 Lambert Quadrilaterals and the Function $C(u)$

A function which is defined in terms of Lambert quadrilaterals will eventually lead to a Pythagorean relation for unified geometry. Start out with any real number u, $0 \le u < a/2$, and construct a Lambert quadrilateral $ABCD$ with base AB of length u and right angles at A, B, and C (Figure 6.39a). Point C is thus the foot of D on the perpendicular to line AB at B. With $AD = x$ and $BC = g(x)$, define the ratio

$$G(x) = \frac{x}{g(x)} = \frac{AD}{BC}$$

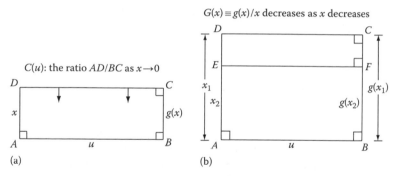

Figure 6.39 Defining the function $C(u)$.

Once the existence of the limit is proved, the function

$$C(u) = \lim_{x \to 0} G(x) = \lim_{AD \to 0} \frac{AD}{BC}$$

is then defined for each u, $0 \le u \le \alpha/2$. (It is observed that if $u = 0$, when $A = B$, then $C(0) = 1$, and when $u = \alpha/2$, when $C = B$, then $C(\alpha/2) = 0$.)

Just as the function $F(x)$ which was used to defined $c(\theta)$ was monotone, so is the function $G(x) = x/g(x)$. Let $x_1 > x_2$, with $AD = x_1$ and $AE = x_2$ as shown in Figure 6.39b. From **(10)** of Section **6.2** one obtains

$$\frac{AE}{AD} < \frac{BF}{BC} \quad \text{or} \quad \frac{AD}{BC} > \frac{AE}{BF}$$

which shows that $x_1/g(x_1) > x_2/g(x_2)$.
That is, if $x_1 < x_2$,

$$G(x_1) > G(x_2)$$

for *both hyperbolic and elliptic geometry*, its graph illustrated in Figure 6.40. Thus G is a decreasing function for decreasing x and the limit of $G(x)$ as $x \to 0$ is the GLB of $G(x)$ for $x > 0$. $C(u)$ is therefore properly defined as

(1)
$$C(u) = \lim_{AD \to 0} \frac{AD}{BC}$$

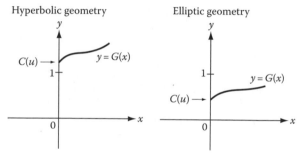

Figure 6.40 Graphs of $G(x)$.

One can observe from Figure 6.40 and the properties of $G(x)$, that for $0 \le u \le \alpha/2$,

(2) $\begin{cases} C(u) \ge 1, & \text{in hyperbolic geometry; } C(0) = 1 \\ 0 \le C(u) < 1, & \text{in elliptic geometry; } C(0) = 1, C(\alpha/2) = 0 \end{cases}$

In order to proceed, a double limit for the function $C(u)$ is needed, analogous to the situation for right triangles. The proof of this result, together with the continuity of C, is found in Appendix E. It follows that if a Lambert quadrilateral $\lozenge ABCD$ with right angles at A, B, and C converges to a segment PQ as shown in Figure 6.41, where $A, D \to P$ and $B, C \to Q$, the limit

$$\lim \frac{AD}{BC} = C(u)$$

exists, where $u = PQ$. This result will be referred to by the term *floating quadrilaterals*.

The floating quadrilateral theorem

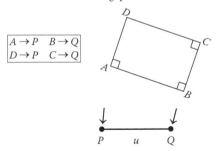

Figure 6.41 A Lambert quadrilateral converging to a segment.

The first application is to show that $C(u) > 1$ for $u > 0$ in hyperbolic geometry, which will be very important later. Observe that in Figure 6.42 $\lozenge CDEF$ is a Lambert quadrilateral with right angles at C, E, and F, and a bisection process is indicated where B_1 is the midpoint of segment CF, B_2 the midpoint of CB_1, B_3 the midpoint of CB_2, and so on. Construct further Lambert quadrilaterals, as shown, by constructing perpendiculars to line CF at B_1, B_2, $B_3, \dots$ and letting A_1, A_2, $A_3, \dots$ be the feet of D on those perpendiculars. Finally, in the inset, which shows two such Lambert quadrilaterals, the following argument shows that ray DA' meets line $A''B''$ at a point W such that $(WA''B'')$: If W is located on ray DA' such that A' is the midpoint of segment DW, and line WB'' is drawn, by SASAS $\lozenge WA'B'B'' \cong \lozenge DA'B'C$ so that $\angle WB''B'' \cong \angle DCB' = $ right angle, and W lies on the unique perpendicular to line $B''C$ at B''; $WB'' = DC > A''B''$, therefore $(WA''B'')$. Thus, because B' is the midpoint of segment CB'', in right triangle WDA'' we have

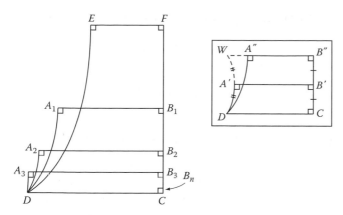

Figure 6.42 Proving $C(u) > 1$.

$$A'D = \tfrac{1}{2}WD > \tfrac{1}{2}A''D$$

Now one can obtain a series of inequalities which will lead to the desired conclusion.

$$DA_1 > \tfrac{1}{2}DE \quad \text{because } B_1 \text{ is the midpoint of } \overline{CF}$$

$$DA_2 > \tfrac{1}{2}DA_1 \quad \text{because } B_2 \text{ is the midpoint of } \overline{CB_1}$$

$$DA_3 > \tfrac{1}{2}DA_2 \quad \text{because } B_3 \text{ is the midpoint of } \overline{CB_2}$$

...

By substitution,

$$DA_2 > \tfrac{1}{2}(\tfrac{1}{2}DE) = \tfrac{1}{4}DE, \quad DA_3 > \tfrac{1}{2}(\tfrac{1}{4}DE) = \tfrac{1}{8}DE \ldots$$

and, in general

(3) $$DA_n > DE/2^n$$

Also, since B_n is defined in the nth bisection of segment CF,

(4) $$CB_n = CF/2^n$$

Dividing in **(3)** and **(4)**, one then obtains

(5) $$\frac{DA_n}{CB_n} > \frac{DE}{CF}$$

If the nth quadrilateral is denoted $\lozenge ABCD$ where $A \equiv A_n$ and $B \equiv B_n$, then as $n \to \infty$, $BC \to 0$ (or $B \to C$) and therefore $A \to D$, then by the floating quadrilaterals (with $D = P$ and $C = Q$),

$$C(u) = \lim_{\substack{A \to D \\ B \to C}} \frac{DA}{CB} \geq \frac{DE}{CF} > 1$$

This proves

(6) $\qquad C(u) > 1$ for $u > 0$ (hyperbolic geometry)

An additional related property which will also be needed is

(7) $0 < C(u_0) < 1$ for some u_0, $0 < u_0 < \alpha/2$ (elliptic geometry)

[*Proof*   By continuity of $C(u)$ at $u = 0$, since $C(0) = 1$, then as $u \to 0$ $C(u) \to 1$ so there must exist u such that $C(u) > \frac{1}{2}$ (for example).]

6.9 Identities for *C(u)*

It is quite significant, if not actually incredible, that the same product-to-sum identity that was proved earlier for $c(\theta)$ is also true for $C(u)$. The reader will no doubt conclude that this must lead to $C(u) = \cos u$, or something similar (which is half true). As a matter of fact, we shall obtain, for some constant κ, $C(u) = \cos \kappa u$ in elliptic geometry, and $C(u) = \cosh \kappa x$ in hyperbolic geometry The constant κ in both cases has geometric significance, which we will see later.

The necessary construction is shown in Figure 6.43: Let $0 < v < u < \alpha/2$ and $u + v < \alpha/2$, with points A, B, C, D such that $(ABCD)$ and $BD = u$, $AB = BC = v$, and $CD = u - v$. Erect perpendiculars to line AD at A, B, C, and D, and locate on these perpendiculars collinear points P, Q, R, and S, as in the figure, all lying on the same side of line AB. As $S \to D$, then $P \to A$, $Q \to B$, and $R \to C$, and it follows by definition of $C(u)$ that $PA/SD \to C(u + v)$, $QB/SD \to C(u)$, and $RC/SD \to C(u - v)$. As in a preceding construction, let the perpendicular to BQ at Q meet rays AP and CR at U and V, respectively, with T on line PS such that $\angle TUQ \cong \angle QVR$. By ASASA $\Diamond ABQU \cong \Diamond CBQV$ and it follows that $\Diamond ACVU$ is a Saccheri quadrilateral, with congruent summit angles at U and V, and $UQ = QV$. Then by AAS, $\triangle TUQ \cong \triangle RVQ$ and $VR = UT$. Therefore, in the two geometries we have the following:

(*) $\begin{cases} PA + RC = UP + UA + VC - VR = 2UA + (UP - UT) & \text{(hyperbolic)} \\ PA + RC = UA - UP + VC + VR = 2UA - (UP - UT) & \text{(elliptic)} \end{cases}$

This relation will be assumed in the proof of the following theorem.

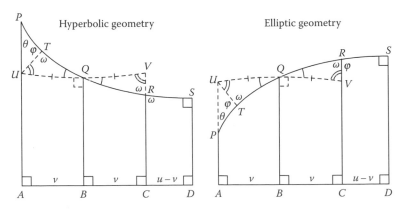

Figure 6.43 Proof of product-to-sum identity.

Theorem: *Product-to-Sum Identity for C(u)*

If u and v are any two real numbers such that $0 < u - v < u + v < \alpha/2$, then

(1) $$C(u + v) + C(u - v) = 2C(u)C(v)$$

Proof In the construction given above (Figure 6.43), we obtain from (∗), by dividing by SD,

$$\frac{PA}{SD} + \frac{RC}{SD} = 2 \cdot \frac{UA}{QB} \cdot \frac{QB}{SD} \pm \left(\frac{UP}{UT} - 1 \right) \frac{UT}{SD}$$

Note that in both hyperbolic and elliptic geometries $\angle UTP$ is obtuse and $UT < UP$ so that, in hyperbolic geometry $UT/SD < PA/SD \to C(u + v)$. In elliptic geometry $SD > QB$ and $UT/SD < UA/QB \to C(v)$, and thus UT/SD is bounded in both cases. If we show that $UP/UT \to 1$, then the expression on the far right has zero limit as $SD \to 0$. By angle analysis, as $SD \to 0$, in both geometries, $\theta \to \pi/2$, $\omega \to \pi/2$, and therefore $\varphi \to \pi/2$. Hence, by the law of sines **(7)**, Section **6.5**,

$$\lim_{PD \to 0} \frac{UP}{UT} = \frac{s(\pi/2)}{s(\pi/2)} = 1$$

Therefore, as $SD \to 0$ we obtain

$$\lim_{SD \to 0} \frac{PA}{SD} + \lim_{SD \to 0} \frac{RC}{SD} = 2 \cdot \lim_{BQ \to 0} \frac{UA}{QB} \cdot \lim_{SD \to 0} \frac{QB}{SD} \pm 0$$

and applying the definition of the function C completes the proof. ◥

The continuity of the function C allows one to essentially set $u = v$ in **(1)** (i.e., to take the limit as $v \to u$) in order to prove the additional identity

$$C(2u) + C(0) = 2C^2(u) \quad \text{or} \quad C(2u) = 2C^2(u) - 1$$

from which a "half-value" identity can be established. (To derive it, solve the above for $C(u)$, then replace u by $\frac{1}{2}u$.) These two identities will be listed together for convenience, with x and y replacing u and v to conform to notation used later, and in keeping with traditional usage.

(2) $\quad 2C(x)C(y) = C(x + y) + C(x - y) \quad (0 < x - y < x + y < a/2)$

(3) $\quad\quad\quad C(2x) = 2C^2(x) - 1 \quad (0 \le x \le \alpha/4)$

(4) $\quad\quad\quad C(\tfrac{1}{2}x) = \sqrt{\dfrac{1 + C(x)}{2}} \quad (0 \le x \le \alpha/2)$

Problems (Section 6.9)

Group B

1. Explicitly establish the boundary conditions (2), Section 6.8, by using the definition of C.

2. Establish the equivalence of the two identities

$$C(\tfrac{1}{2}x) = \sqrt{\frac{1 + C(x)}{2}} \quad \text{and} \quad C(2x) = 2C^2(x) - 1$$

3. Suppose that $C(a) = 4$. Find the values of $C(2a)$ and $C(4a)$.

4. Suppose that $C(a) = 1921$. Find the value of $C(\tfrac{1}{8}a)$.

5. Consider the function S defined by $S^2(x) = \sigma\,(1 - C^2(x))$, where $\sigma = 1$ in elliptic geometry and $\sigma = -1$ in hyperbolic geometry (necessary since $C(x) > 1$ in hyperbolic geometry). Prove the identity

$$C^2(x) + \sigma S^2(x) = 1$$

6. Prove the double-value identity for $S(x)$ where S is as defined in Problem 5:

$$S(2x) = 2S(x)C(x)$$

7. If $S(a) = \tfrac{1}{2}$ find $S(2a)$ and $S(4a)$ in elliptic geometry.

8.* If $S(a) = \tfrac{1}{2}$ find $S(2a)$ and $S(4a)$ in hyperbolic geometry.

6.10 Classical Forms for $C(u)$

The identities (2) and (4) make it possible to prove the special formulas (for $x \geq 0$ and some constant $\kappa > 0$)

(1) $C(x) = \cos \kappa x$ in elliptic geometry

(2) $C(x) = \cosh \kappa x$ in hyperbolic geometry

for $0 \leq x < \alpha/2$. The proof for (1) follows that which was given earlier for $c(x) = \cos\kappa x$, where $\alpha/2$ replaces $\pi/2$, since there is no substantial difference between $C(x)$ in elliptic geometry and $c(\theta)$. The argument for (2) is virtually the same. However, it involves the hyperbolic cosine function and its identities, so we outline its proof. Before going further, a brief introduction to the hyperbolic functions is in order (which may be review material for some readers). If you have never seen these definitions before and worked with the algebra involved, you should find this material very interesting, in fact somewhat remarkable, in that functions so different from the trigonometric functions could have such close analogies to them. This material is also needed in Chapter 9. All this material is found in any calculus text; applications in engineering mathematics often make substantial use of these functions and their identities. The interesting thing is that their definitions are so concise requiring only algebra and no geometry, in contrast with the trigonometric functions.

The *hyperbolic cosine* and *hyperbolic sine* functions are defined for all real x as follows:

(3) $$\cosh x = \frac{e^x + e^{-x}}{2}, \quad \sinh x = \frac{e^x - e^{-x}}{2}$$

where the exponential function e^x is understood (with which you have no doubt worked before). Note that by the above, the substitution of $-x$ for x proves that $\cosh(-x) = \cosh x$ and $\sinh(-x) = -\sinh x$. (Thus $\cosh x$ is an even function, while $\sinh x$ is odd.) Also, the special values $\sinh 0 = 0$ and $\cosh 0 = 1$ emerge by simple observation. The rest of the material involving identities that resemble those of sine and cosine all come directly from **(3)** and the use of elementary algebra.

The first one is:

(4) $$\cosh^2 x - \sinh^2 x = 1$$

This may be proved as follows. Start with the left side of **(4)** and plug in the forms from **(3)**:

$$\left(\frac{e^x + e^{-x}}{2}\right)^2 - \left(\frac{e^x - e^{-x}}{2}\right)^2 = \frac{e^{2x} + 2e^x \cdot e^{-x} + e^{-2x}}{4} - \frac{e^{2x} - 2e^x \cdot e^{-x} + e^{-2x}}{4}$$

$$= \frac{2e^x \cdot e^{-x} - (-2e^x \cdot e^{-x})}{4} = \frac{2+2}{4} = 1$$

It is curious that the sum of the squares of the hyperbolic functions produces an identity analogous to the double-angle formula for cosine. If a plus sign be substituted for the minus sign in the above steps, one obtains, instead of 1, the form for $\cosh 2x$. Thus emerges the identity

(5) $$\cosh^2 x + \sinh^2 x = \cosh 2x$$

Of more fundamental importance are the four addition/subtraction identities which, again, resemble those for sine and cosine.

(6) $$\sinh(x \pm y) = \sinh x \cosh y \pm \cosh x \sinh x$$

(7) $$\cosh(x \pm y) = \cosh x \cosh y \pm \sinh x \sinh$$

The proof of just one of the above will be given to illustrate. To prove the first half of **(7)**, it is easier to start with the right side and derive the left side. Thus

$$\cosh x \cosh y + \sinh x \sinh y = \frac{(e^x + e^{-x})(e^y + e^{-y})}{4} + \frac{(e^x - e^{-x})(e^y - e^{-y})}{4}$$

$$= \frac{e^{x+y} + e^{x-y} + e^{-x+y} + e^{-x-y}}{4}$$

$$+ \frac{e^{x+y} - e^{x-y} - e^{-x+y} + e^{-x-y}}{4}$$

$$= \frac{2e^{x+y} + 2e^{-x-y}}{4} = \frac{e^{x+y} + e^{-x-y}}{2} = \cosh(x+y)$$

The remaining addition identities are left to the reader (note that you can substitute $-y$ for y in the identity for $\sinh(x + y)$ to obtain that for $\sinh(x - y)$, which saves a considerable amount of work).

Even common limits and derivatives, often used in calculus, are analogous to those involving sine and cosine. For example, the derivative of $\sinh x$ is $\cosh x$. Recall the limit relation for the sine function $(\sin x)/x \to 1$ as $x \to 0$, which is used in calculus to show that the derivative of sine is cosine. It corresponds to a similar limit relation for hyperbolic sine:

(8)
$$\lim_{x \to 0} \frac{\sinh x}{x} = 1$$

This result can be obtained from a simple application of L'Hospital's Rule, which some readers may have seen. (You can also experiment with this using your pocket calculator and substituting values for x close to zero.)

The introduction to hyperbolic functions will end with a proof of the identities analogous to **(2)–(4)** of the previous section, which will actually be used in the proof of **(2)** above. The two addition identities **(7)** show that

$$\cosh(x + y) + \cosh(x - y) = \cosh x \cosh y + \sinh x \sinh y$$

$$+ \cosh x \cosh y - \sinh x \sinh y$$

$$= 2\cosh x \cosh y$$

That is,

(9) $2\cosh x \cosh y = \cosh(x + y) + \cosh(x - y)$

Setting $y = x$, then replacing x by $\frac{1}{2}x$ (analogous to an earlier argument) produces $\cosh 2x = 2\cosh^2 x - 1$ and

(10)
$$\cosh \tfrac{1}{2}x = \sqrt{\frac{1 + \cosh x}{2}}$$

Attention is finally directed to proving **(2)**, that is, $C(x) = \cosh \kappa x$. The first step is to observe that the hyperbolic cosine is a strictly increasing function with values > 1 for $x > 0$, with a well-defined inverse on real numbers greater than 1. Recall that in hyperbolic geometry $C(x) > 1$ for $x > 0$. Thus $C(\lambda) > 1$ for any particular $\lambda > 0$, and there exists a value $\mu > 0$ such that $\cosh \mu = C(\lambda)$. [Thus we are defining μ as $\cosh^{-1}(C(\lambda))$.] Beginning with $C(\lambda) = \cosh \mu$, one obtains from (4) of the previous section and (10) (just as in the earlier proof of $c(r\lambda) = \cos r\mu$),

(11) $$C(r\lambda) = \cosh r\mu$$

for all dyadic rationals of the form $r = 1/2^n$. Next, assume that (11) has been established for all dyadic rationals $r = m/2^n$, $0 \le m \le 2^n$, of order n (true for $n = 1$). Consider a dyadic rational of order $n + 1$: $r = m/2^{n+1}$ for $m = 2k + 1$ and k an integer ≥ 0. If $s = 1/2^{n+1}$, then it follows that $r + s \equiv r_1$ and $r - s \equiv r_2$ are dyadic rationals of order n, and it then follows that $C(r_1\lambda) = \cosh r_1\mu$, $C(r_2\lambda) = \cosh r_2\mu$ (which have been established previously by the induction hypothesis). Then (2) of Section 6.9 produces

$$2C(r\lambda)C(s\lambda) = C(r\lambda + s\lambda) + C(r\lambda - s\lambda) = C(r_1\lambda) + C(r_2\lambda)$$

Hence by the induction hypothesis,

(12) $$2C(r\lambda)\cosh s\mu = \cosh r_1\mu + \cosh r_2\mu$$

By (9)

$$2\cosh r\mu \cosh s\mu = \cosh(r\mu + s\mu) + \cosh(r\mu - s\mu)$$

or

$$2\cosh r\mu \cosh s\mu = \cosh r_1\mu + \cosh r_2\mu$$

Comparing this with (12), one concludes that $C(r\lambda) = \cosh r\mu$, where r is a dyadic rational of order $n + 1$. As in the proof for $c(r\lambda) = \cos r\mu$, we extend (11) to any dyadic rational $r > 1$ by using the double-value identity for both $C(x)$ (using (3) of Section 6.8) and $\cosh x$ (using (10) above).

Thus (11) is valid for dyadic rationals of all sizes and orders, a set of numbers shown to be dense on the positive real number line. Hence a sequence of such numbers can be chosen to converge to any real $u > 0$, and by continuity of both $C(u)$ and $\cosh u$, we obtain $C(\lambda u) = \cosh \mu u$ for all read u. Now set $x = \lambda u$ $(u = x/\lambda)$ in order to obtain

$$C(x) = \cosh kx \quad (x \ge 0)$$

where $\kappa = \mu/\lambda$.

6.11 The Pythagorean Relation for Unified Geometry

The ordinary Pythagorean theorem for Euclidean geometry was established in Chapter 5 from an argument involving similar triangles. In order to prove an analogous relation for the two non-Euclidean geometries, rather unusual methods are needed since similar triangles do not exist. The form we obtain is rather peculiar. It is in terms of the function $C(u)$ exclusively, and the well-known classical forms for the Pythagorean relation for hyperbolic and elliptic geometry emerge. The average reader may find all this somewhat confusing since these formulas are so different from the Euclidean formula. A number of examples will be presented later to assure the reader of their correctness and logical validity, particularly as they apply to spherical geometry where verification can be made directly.

Let ΔABC be any right triangle, with right angle at C, having sides of length a, b, and c each less than $\alpha/2$, in standard notation. For each point $P \in \overline{AC}$ construct two other "variable" right triangles PQR and XYZ each congruent to ΔABC, as follows (see Figure 6.44): Slide ΔABC (by construction) along ray AC from A to P forming ΔPQR, with segments PQ and BC meeting at point Y, then slide triangle PQR along ray QP from Q to Y forming ΔXYZ, such that (XYQ) holds. All subsequent limits will be with respect to $PA \to 0$ and $XP \to 0$, and thus triangles PQR and XYZ converge to ΔABC.

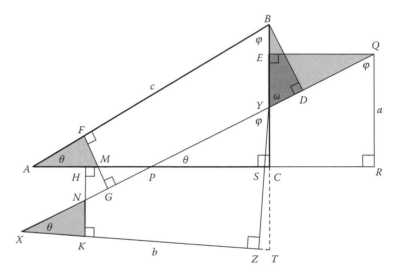

Figure 6.44 Pythagorean relation proved in unified geometry.

Let M be the midpoint of $\overline{AP}$ and N the midpoint of $\overline{XP}$, with F the foot of M on $\overline{AB}$ and H the foot of N on $\overline{AP}$ (since $\angle A$, $\angle X$, $\angle QPR$, and $\angle APX$ are acute angles). For P sufficiently close to A, ray FM meets line XY at G, and since $\Delta AFM \cong \Delta PGM$ by AAS, $\angle MGP \cong \angle MFA$ and $\overline{FG} \perp \overline{XP}$, with M the midpoint of $\overline{FG}$. Similarly, ray HN meets line XZ at K and since $\Delta HPN \cong \Delta KXN$ by AAS, it follows that $\overline{HK} \perp \overline{XZ}$ and N is the midpoint of $\overline{HK}$. Thus:

$$\frac{HK}{QY} = \frac{HK}{XP} = \frac{2NK}{2XN} = \frac{NK}{XN} \quad \text{and} \quad \frac{CR}{FG} = \frac{AP}{FG} = \frac{2AM}{2FM} = \frac{AM}{FM}$$

Thus we have, using right triangles XNK and AMF,

(1)
$$\lim_{AP \to 0} \frac{HK}{QY} \cdot \frac{CR}{FG} = s(\theta) \cdot \frac{1}{s(\theta)} = 1$$

Let D be the foot of B on line PQ, and E the foot of Q on line BC. Note that $\omega \equiv m\angle BYQ = m\angle PYC$ has limit $m\angle ABC = \varphi$. Using right triangles BYD and QYE,

(2)
$$\lim_{AP \to 0} \frac{BD}{BY} \cdot \frac{QY}{QE} = s(\varphi) \cdot \frac{1}{s(\varphi)} = 1$$

Next, we establish the inequality

(3)
$$SZ < BY < CT$$

where S and T are the intersection points shown in Figure 6.44. Note that

$$SZ = YZ - YS < YZ - YC = BC - YC = BY$$

and

$$BY = BC - YC = YZ - YC < YT - YC = CT$$

These two combined establish **(3)**. Thus

$$\frac{SZ}{HK} < \frac{BY}{HK} < \frac{CT}{HK}$$

Since $\Diamond ZKHS$ is a Lambert quadrilateral with right angles at Z, K, and H and $\Diamond CHKT$ is a Lambert quadrilateral with right angles at C, H, and K, and since HS and $KT \to AC = b$ as limit, then $\lim SZ/HK = \lim CT/HK = C(b)$ as $AP \to 0$. By the pinching theorem on limits

(4)
$$\lim_{AP \to 0} \frac{BY}{HK} = C(b)$$

The above analysis leads to the result:

Pythagorean Relation for Non-Euclidean Geometry   Let $\triangle ABC$ be any right triangle having sides less than $\alpha/2$ with right angle at C. Then

$$C(c) = C(a)C(b)$$

where, for some constant $\kappa > 0$
  $C(x) = \cos \kappa x$ in elliptic geometry
  $C(x) = \cosh \kappa x$ in hyperbolic geometry

Proof   Construct the congruent right triangles PQR and XYZ as above, with the rest of the construction carried out as before, and with **(1)**–**(4)** valid. A sequence of six telescoping ratios whose product collapses to the single ratio BD/FG will be formed so as to prove the desired result. Write

$$\frac{BD}{FG} = \frac{BD}{BY} \cdot \frac{BY}{HK} \cdot \frac{HK}{QY} \cdot \frac{QY}{QE} \cdot \frac{QE}{CR} \cdot \frac{CR}{FG} = \left(\frac{BD}{BY} \cdot \frac{QY}{QE} \right) \frac{BY}{HK} \left(\frac{HK}{QY} \cdot \frac{CR}{FG} \right) \frac{QE}{CR}$$

where we have grouped two pairs of ratios together, each of whose limits is unity by **(1)** and **(2)**. Hence

$$\lim_{AP \to 0} \frac{BD}{FG} = \lim_{AP \to 0} \frac{QE}{CR} \cdot \lim_{AP \to 0} \frac{BY}{HK}$$

Thus, by floating quadrilaterals and **(4)**,

$$C(c) = C(a) \cdot C(b) \quad \blacksquare$$

Note: It is interesting to examine the possibility of deriving the usual formula for Euclidean geometry from the above analysis since this was true in several previous situations for ratios involving midlines of normal trapezoids. However, we see the utter failure in this attempt, for in Euclidean geometry Lambert quadrilaterals are rectangles, with opposite sides congruent. It follows that $C(x) = 1$ for all x in Euclidean geometry, and the equation obtained for non-Euclidean geometry reduces to the triviality $1 = 1$, providing zero information. ◣

6.12 Classical Unified Trigonometry

Although the Pythagorean relation is in itself a remarkable analogy to the Euclidean version involving only the sides of a right triangle, and its very existence could not have been predicted, the theory can be carried much further. Additional relations exist that are also analogous to Euclidean trigonometry. The forms are strikingly parallel to those of spherical geometry, providing heuristic evidence that hyperbolic geometry is probably valid, if there be any doubt at this point.

We begin by establishing one of the three trigonometric relations for a right triangle involving an acute angle. The other two will follow from this one using algebra and the Pythagorean relation above. Again, all triangles must be restricted to having sides of length less than $\alpha/2$. This restriction will be removed later (proof outlined in Problem 19).

In order to avoid separate proofs for elliptic and hyperbolic geometry, we introduce a notation suggested previously (Problems 5 and 6 in Section **6.9**). Recall that $C(x)$ is $\cos \kappa x$ in elliptic geometry and $\cosh \kappa x$ in hyperbolic geometry. Define the companion functions as follows

$$S(x) = \begin{cases} \sin \kappa x & \text{in elliptic geometry} \\ \sinh \kappa x & \text{in hyperbolic geometry} \end{cases}$$

$$T(x) = \begin{cases} \tan \kappa x & \text{in elliptic geometry} \\ \tanh \kappa x & \text{in hyperbolic geometry} \end{cases}$$

If one also defines the special sign indicator σ to be +1 in elliptic geometry and −1 in hyperbolic geometry (as before), then there are a few (unified) identities valid simultaneously in the two geometries. Those especially needed are listed in the following; the first you should immediately recognize as the composite of the two familiar identities $\cos^2 x + \sin^2 x = 1$ and $\cosh^2 x - \sinh^2 x = 1$.

(1) $$C^2(x) + \sigma S^2(x) = 1$$

(2) $$T(x) = \frac{S(x)}{C(x)}$$

(3) $$C(x - y) = C(x)C(y) + \sigma S(x)S(y)$$

Finally, observe the limits introduced previously, $(\sin x)/x \to 1$ and $(\sinh x)/x \to 1$ as $x \to 0$; since $\cos x \to 1$ and $\cosh x \to 1$, it follows from **(2)** that

(4)
$$\lim_{x \to 0} \frac{T(x)}{x} = 1$$

Now consider any (fixed) right triangle ABC with right angle at C, and having sides of length less than $\alpha/2$. Let P be any point on segment AC approaching A as limit, and let Q be the foot of P on side AB. Set $x = AQ$, $y = QP$, and $z = AP$, as in Figure 6.45. Apply the Pythagorean relation to $\triangle ABC$ and $\triangle APQ$. Thus

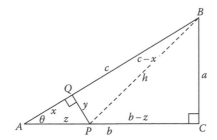

Figure 6.45 Proving $\cos A = T(b)/T(c)$.

(5)
$$C(a)C(b) = C(c)$$

(6)
$$C(x)C(y) = C(z)$$

In $\triangle PQB$ and $\triangle PBC$ we obtain

(7)
$$C(y)C(c-x) = C(h) = C(a)C(b-z)$$

Expanding the expressions on each side using the addition identity **(3)**, and using **(5)** and **(6)**, we obtain

$$C(y)\big[C(c)C(x)+\sigma S(c)S(x)\big] = C(a)\big[C(b)C(z)+\sigma S(b)S(z)\big]$$

$$C(z)C(c)+\sigma C(y)S(c)S(x) = C(c)C(z)+\sigma C(a)S(b)S(z)$$

That is,

$$C(y)S(c)S(x) = C(a)S(b)S(z)$$

Division by $C(x)C(y)C(c)$ produces

$$\frac{C(y)S(c)S(x)}{C(x)C(y)C(c)} = \frac{C(a)S(b)S(z)}{C(z)C(a)C(b)}$$

or

$$T(c)T(x) = T(b)T(z)$$

This has the equivalent form

(8)
$$\frac{T(b)}{T(c)} = \frac{T(x)}{T(z)} = \frac{\dfrac{T(x)}{x}}{\dfrac{T(z)}{z}} \cdot \frac{x}{z}$$

Take limits as $x \to 0$ and $z \to 0$; we have $T(x)/x \to 1$, $T(z)/z \to 1$, and $x/z \to c(\theta) = \cos\theta$, the final result is

(9)
$$\frac{T(b)}{T(c)} = \cos\theta$$

Thus emerges one of the well-known relations for non-Euclidean geometry (using the standard notation $A \equiv m\angle A = \theta$)

(10)
$$\begin{cases} \cos A = \dfrac{\tan\kappa b}{\tan\kappa c} & \text{in elliptic geometry} \\[3mm] \cos A = \dfrac{\tanh\kappa b}{\tanh\kappa c} & \text{in hyperbolic geometry} \end{cases}$$

The remaining relations for $\sin A$ and $\tan A$ can be routinely obtained by algebra (see Problem 11). Thus are established the classical formulas for non-Euclidean geometry, as presented in the following.

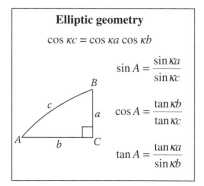

Elliptic geometry

$\cos\kappa c = \cos\kappa a \cos\kappa b$

$\sin A = \dfrac{\sin\kappa a}{\sin\kappa c}$

$\cos A = \dfrac{\tan\kappa b}{\tan\kappa c}$

$\tan A = \dfrac{\tan\kappa a}{\sin\kappa b}$

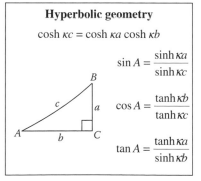

Hyperbolic geometry

$\cosh\kappa c = \cosh\kappa a \cosh\kappa b$

$\sin A = \dfrac{\sinh\kappa a}{\sinh\kappa c}$

$\cos A = \dfrac{\tanh\kappa b}{\tanh\kappa c}$

$\tan A = \dfrac{\tanh\kappa a}{\sinh\kappa b}$

In view of the total dissimilarity between the two non-Euclidean geometries, one involving a finite metric and no parallel lines, the other, an infinite metric and a complete theory of parallel lines, the analogies revealed in these two tables for the two geometries seem remarkable.

The similarity does not end there; many other significant formulas are valid. In Chapter 9 when we undertake a separate, detailed analytic treatment of elliptic and hyperbolic geometry, another striking parallel will emerge: If K is the area of $\triangle ABC$ and a, b, and c are its sides, with s the semi-perimeter, the area can be expressed in each of the two geometries as follows (with $\kappa = 1$):

$$\tan \tfrac{1}{4} K = \sqrt{\tan \tfrac{1}{2} s \tan \tfrac{1}{2}(s-a)\tan \tfrac{1}{2}(s-b)\tan \tfrac{1}{2}(s-c)}$$

(elliptic geometry)

$$\tan \tfrac{1}{4} K = \sqrt{\tanh \tfrac{1}{2} s \tanh \tfrac{1}{2}(s-a)\tanh \tfrac{1}{2}(s-b)\tanh \tfrac{1}{2}(s-c)}$$

(hyperbolic geometry)

Example 1

Establish the identity for sin A appearing in the above table.

Solution

A separate proof for each of the two geometries might seem more natural, but certainly a unified proof is more efficient. After experimenting a little with the formulas involved, it is found best to begin with the identity for sine and cosine:

$$\sin^2 A = 1 - \cos^2 A = 1 - \frac{T^2(b)}{T^2(c)} = \frac{T^2(c) - T^2(b)}{T^2(c)} = \frac{\dfrac{S^2(c)}{C^2(c)} - \dfrac{S^2(b)}{C^2(b)}}{\dfrac{S^2(c)}{C^2(c)}}$$

$$= \frac{S^2(c) - S^2(b)\cdot \dfrac{C^2(c)}{C^2(b)}}{S^2(c)} = \frac{S^2(c) - S^2(b)C^2(a)}{S^2(c)}$$

$$= \frac{\sigma S^2(c) - \sigma S^2(b)C^2(a)}{\sigma S^2(c)} = \frac{1 - C^2(c) - [1 - C^2(b)]C^2(a)}{\sigma S^2(c)}$$

$$= \frac{1 - C^2(c) - C^2(a) + C^2(b)C^2(a)}{\sigma S^2(c)} = \frac{1 - C^2(a)}{\sigma S^2(c)} = \frac{\sigma S^2(a)}{\sigma S^2(c)}$$

That is,

$$\sin^2 A = \frac{S^2(a)}{S^2(c)} \quad \text{or} \quad \sin A = \frac{S(a)}{S(c)}$$

Example 2

As shown in Figure 6.46, consider a tri-rectangular triangle *ABC* on the unit sphere (where $\alpha = \pi$, and in the forms for elliptic geometry, $\kappa = 1$). Thus, $AB = BC = AC = \pi/2 \approx 1.5708$. Consider *D* and *E* on $\overline{AB}$ and $\overline{AC}$ such that $e = AD = AE = d = 1$. Then $\triangle ADE$ is an isosceles right triangle, with acute base angles.

Furthermore, $\lozenge BCDE$ is a Saccheri quadrilateral, with $DE < BC$. Verify these facts using the classical forms for elliptic geometry to find

(a) $x = DE\ (<BC?)$
(b) $\theta = m\angle ADE$ (acute angle?)

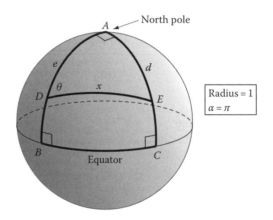

Figure 6.46 Example 2.

Solution

(a) Using the Pythagorean relation for elliptic geometry,

$$\cos x = \cos d \cos e = \cos^2 1 \approx 0.2919 \quad \text{or}$$

$$x = \cos^{-1} 0.2919 \approx 1.2746 < 1.5708 = BC$$

(b) Since $\triangle ADE$ is a right triangle with right angle at *A*, we conclude that

$$\tan\theta = \frac{\tan d}{\tan e} = \frac{\tan 1}{\sin 1} \approx 1.8508$$

so that $\theta \approx 1.075 \approx 61.6°$.

Example 3

To test the integrity of the formulas in hyperbolic geometry (with $\kappa = 1$), consider an isosceles right triangle ABC with right angle at C and legs of length 1 (Figure 6.47). The altitude CD to hypotenuse AB bisects AB as well as the vertex angle, $\angle ACB$. The classical forms for hyperbolic geometry in the above table then produce the relation

$$\frac{\sinh AD}{\sinh AC} = \sin m\angle ACD \quad (m\angle ACD = \sin \pi/4)$$

(a) Find the value of c using the Pythagorean relation for hyperbolic geometry.
(b) Show that with this value for c, $(\sinh AD)/\sinh AC$ $\sin \pi/4$.

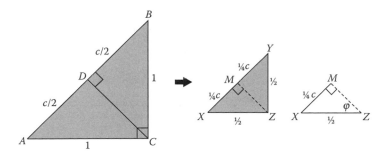

Figure 6.47 Example 3.

Solution

(a) $\cosh c = \cosh a \cosh b = \cosh^2 1 \approx 2.381098.$

   Therefore, $c = \cosh^{-1}(2.381098) \approx 1.513374.$

(b) $\dfrac{\sinh \frac{1}{2}(1.523374)}{\sinh 1} = \dfrac{\sinh 0.756687}{\sinh 1} \approx \dfrac{0.8309927}{1.1752012}$

   $\approx 0.707107 (= \sqrt{2}/2)$

Example 4

We consider here the effect of scaling up or down in geometry. If we start with the same triangle we had in Example 3, suppose the three sides are reduced by one-half in length. The result is a new triangle XYZ in Figure 6.47 whose sides are of length $x = y = \frac{1}{2}$ and $z = \frac{1}{2}c$. The question is whether $\angle Z$ is still a right angle.

Moment for Reflection

Before going further, it is instructive to attempt an answer without working out any calculations. What do you think would be pertinent? $\triangle XYZ$ is still an isosceles triangle, like $\triangle ABC$, so if the angles of $\triangle XYZ$ are not congruent to those of $\triangle ABC$, we would expect a rather uniform change—either all three angles are smaller, or all three are larger, than those of $\triangle ABC$. Can you think of anything covered previously that would have some bearing on the issue?

Example 4 (contd.)

The problem is to calculate $m\angle Z$ using the classical formulas of hyperbolic geometry (again, with $\kappa = 1$).

Solution

Since $\triangle XYZ$ may not be a right triangle, we use right triangle MXZ shown in the figure, where M is the midpoint of XY. Note that $\varphi = \frac{1}{2}m\angle Z$, so it suffices to evaluate φ.

$$\sin\varphi = \frac{\sinh XM}{\sinh XZ} = \frac{\sinh 0.3784}{\sinh 0.5} \approx 0.7436$$

Hence $\varphi = \sin^{-1}(0.7436) \approx 0.8385 \approx 48.0°$ and therefore $m\angle Z \approx 96.0°$.

Example 5

Derive the following formula for elliptic geometry relating the acute angles of a right triangle to the hypotenuse, and use it to show that the angle-sum of a right triangle is greater than π (180°). Assume that $\kappa = 1$.

(11) $$\tan A \tan B = \sec c$$

Solution

Observe from the table of formulas for elliptic geometry that

$$\tan A = \frac{\tan a}{\sin b} \quad \text{and} \quad \tan B = \frac{\tan b}{\sin a}$$

Thus,

$$\tan A \tan B = \frac{\frac{\sin a}{\cos a} \cdot \frac{\sin b}{\cos b}}{\sin a \sin b} = \frac{\sin a \sin b}{\cos a \cos b \sin a \sin b}$$

$$= \frac{1}{\cos a \cos b} = \frac{1}{\cos c} = \sec c$$

To prove that $A + B > 90°$, since $\cos c < 1$ and $\sec c > 1$, the above formula implies

$$\tan A \tan B > 1$$

Now suppose that $A + B \leq 90°$. By the increasing property of $\tan x$,

$$\tan A \tan B \leq \tan A \tan(90° - A) = \tan A \cot A = 1$$

a contradiction, thus proving that $A + B > 90°$ (and $A + B + C > 180°$).

Note: In using formulas for hyperbolic geometry for problems, if your pocket calculator does not have the hyperbolic trigonometric functions (such as the TI-83), it is helpful to enter short programs in your calculator that will evaluate the hyperbolic sine, cosine, and tangent, and their inverses, in one step. For your convenience, we display the pertinent formulas to be used in creating programs (you must follow the instructions in your calculator manual for writing programs in Basic). For the TI-83, use the command INPUT X in your program.

$$\sinh X = (e^X - e^{-X})/2 \qquad \sinh^{-1} X = \ln\left[X + \sqrt{X^2 + 1}\right],$$

$$\cosh X = (e^X + e^{-X})/2 \qquad \cosh^{-1} X = \ln\left[X + \sqrt{X^2 - 1}\right],$$

$$\tanh X = \sinh X/\cosh X \qquad \tanh^{-1} X = \ln\frac{1 + X}{1 - X},$$

(The domain for $\cosh^{-1} x$ is $1 \leq x < \infty$ and that for $\tanh^{-1} x$ is $0 \leq x < 1$; you will receive an error prompt if you try to calculate $\cosh^{-1} x$ for $x < 1$ or $\tanh^{-1} x$ for $x > 1$ on your calculator.) ⊠

Problems (Section 6.12)

In all problems in this section, take the value $\kappa = 1$ for the classical forms unless indicated otherwise. For calculations in elliptic geometry using a pocket calculator, radian mode is required.

Group A

1. Right triangle ABC has $a = 1$ and $b = 2$ (standard notation). Find c to three decimals in both elliptic and hyperbolic geometry (see Figure P.1).

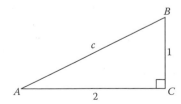

Figure P.1

2. Right triangle ABC has $a = 1$ and $b = 2$. Find the measures of the acute angles in degrees, accurate to 1/10th degree, and show that their sum S satisfies the inequality $S < 90°$ in unified geometry. (Recall the conversion formula π rad $= 180°$.)

3. An isosceles right triangle in elliptic geometry has legs of length ½ each, as shown in Figure P.3 for this problem. Find the measure of the base angles in degrees, accurate to 1/10th of a degree.

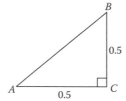

Figure P.3

4. We know that the geometry of triangles on the earth's surface whose sides lie on great circles is virtually Euclidean for "small" triangles (having sides less than a mile in length, say). Evidence of this emerged in Problem 3. Test this hypothesis for triangles in hyperbolic geometry by showing that the hypotenuse of a right triangle having legs of length 0.3 and 0.4 is approximately 0.5 (see Problem 16 for more on this).

5. With $\kappa = 1$, prove the following identity that is the analogue in hyperbolic geometry of (11) (Example 5 above):

    (12) $\tan A \tan B = \operatorname{sech} c$

6. According to the AAA congruence theorem in hyperbolic geometry, there is only one triangle having angle measures 20°, 60°, and 90°, up to congruence. Illustrate by finding the unique values of the three sides to three decimal accuracy, using (12) above.

7. Show that if an isosceles right triangle in elliptic geometry has legs of length $\pi/4$, the hypotenuse is of length $\pi/3$.

8. Give a trigonometric proof of the proposition in elliptic geometry: If $AB = \alpha/2$ and $\overleftrightarrow{BC} \perp \overleftrightarrow{AB}$, then $AX = \alpha/2$ for every point X on line BC. (With $\kappa = 1$, the value of α is π.)

9. Prove the following identity (and its inverse) that relates the angle measure A of an equilateral triangle in hyperbolic geometry to the length a of each side, which was introduced earlier in Problem 11, Section **5.8**:

    (13) $\sec A = 1 + \operatorname{sech} a$

    (14) $\cosh a = \dfrac{\cos A}{1 - \cos A}$

10.   Use **(14)** from Problem 9 to find a for an equilateral triangle having angles of measure 37.8°. (Compare your answer with the discussion in Problem 11, Section **5.8**.)

Group B

11.   Prove the remaining identity in the above tables for non-Euclidean trigonometry:

$$\tan A = \frac{\tanh \kappa a}{\sinh \kappa b} \quad \text{and} \quad \tan A = \frac{\tan \kappa a}{\sin \kappa b}$$

12.   **The Value for κ in Elliptic Geometry**   This problem outlines a proof of the result $\kappa = \pi/\alpha$ in elliptic geometry. You are to fill in the details. (See Figure P.12.)
1. Consider a tri-rectangular triangle CPQ and locate points A and B on sides PC and QC, with $x = PA$ and $y = QB$; the sides of $\triangle ABC$ have lengths less than $\alpha/2$.
2. Let $x \to 0$ and $y \to 0$, then $a \to \alpha/2$ and $b \to \alpha/2$. Since $-x - y < c - \alpha/2 < x + y$ by the polygonal inequality, $c \to \alpha/2$.
3. The result is $\cos \kappa\alpha/2 = \cos^2\kappa\alpha/2$, an equation of the type $z^2 - z = 0$. Since $\cos \kappa\alpha/2 \neq 1$ then $\cos \kappa\alpha/2 = 0$, with general solution $\kappa\alpha/2 = \pi/2, 3\pi/2, 5\pi/2, \dots$
4. If $\kappa\alpha \geq 3\pi$ choose x such that $\pi/2\kappa < x < 3\pi/2\kappa \leq \alpha/2$; $C(x)$ is then defined. But $C(x) = \cos \kappa x < 0$. The result is

(15) $$\kappa = \frac{\pi}{\alpha}$$

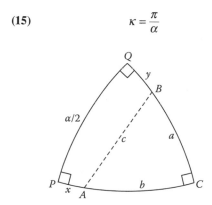

Figure P.12

13.   **Application: Navigation on Earth's Surface**   If we apply elliptic trigonometry to a sphere of radius r, since $\alpha = \pi r$ then $\kappa = 1/r$ from **(15)** (previous problem). Thus the Pythagorean theorem for a right triangle on the earth's surface becomes $\cos c/r = \cos a/r \cos b/r$, with $r = 3960$ approximately—the radius of the earth in miles. The small countries of Nairobi in Africa (A) and Ecuador in South America (C) both lie on the equator, at the respective locations 36°E and 80°W (meridians). Toronto, Canada (B) is

43° due north of Ecuador along the 80°W meridian. Thus $\triangle ABC$ is a spherical right triangle, right angle at C.

(a) Find the great-circle distance from Toronto to Nairobi. (Each degree of latitude or longitude equals approximately 70 mi on the earth's surface.)

(b) Find the angle in degrees between the great circle joining Toronto and Nairobi and the 80°W meridian joining Toronto and Ecuador.

Note: You are to use the formulas for elliptic geometry (above table) which can be shown to apply to all triangles (see Problem 19 for the general proof). See Appendix B for details concerning the geography of the earth. ⬛

14.   Using the formula $\cosh x = \frac{1}{2}(e^x + e^{-x})$, prove that if the length of the hypotenuse of an isosceles right triangle is exactly $\ln \tau$, then the length of each leg is exactly $\ln \tau$, where $\tau \equiv (1 + \sqrt{5})/2$ (the golden ratio) (Figure P.14).

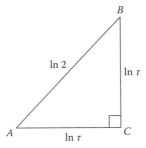

Figure P.14

15.   Obtain the general formula $\cot A = \cosh \kappa a$ [$\cot A = \cos \kappa a$] for the base angle of an isosceles right triangle in terms of the length of one leg, then use it to show that $\lim_{a \to 0} = 45°$, independent of the value of κ.

Group C

16.*   If the values for a, b, and c are "small," then powers and products of these numbers of order 4 and higher can be ignored. Show that under these conditions, the Pythagorean relation for elliptic geometry reduces to the Euclidean relation. Use power series expansions for $\cos x$.

17.   Using Figure P.17, establish the following relations giving two sides c and d of a Lambert quadrilateral $ABCD$ in hyperbolic geometry in terms of the other two sides a and b:

(16) $\sinh c = \sinh a \cosh b$

(17)
$$\tanh d = \frac{\tanh b}{\cosh a}$$

[**Hint:** Start out with $\sin \varphi = \sinh c/\sinh w = \cos \theta = \tanh a/\tanh w$.]

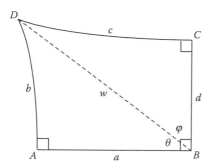

Figure P.17

18.*   Use (**16**) and (**17**) above to show that $\tanh c = \tanh a \cosh d$, and use this to show that $\lim_{a \to 0} c/a = \cosh d$, in agreement with (**1**) of Section **6.8** (in hyperbolic geometry $C(u) = \cosh u$ if $\kappa = 1$).

19.   Extend the trigonometric table for elliptic geometry to arbitrary right triangles by using the identities for sine, cosine, and tangent involving supplementary angles (such as $\tan(\pi - x) = -\tan x$) and Figure P.19, where there are only the following two cases to consider (a result of the corollary in Section **3.7**): (**1**) when two legs are greater than $\alpha/2$, and (**2**) when a leg and hypotenuse are greater than $\alpha/2$. The original formulas apply to the "supplementary" triangles created by extremal points C^* and B^* whose sides are each less than $\alpha/2$, as shown in the figure for this problem. (Use continuity to establish any formula involving sides whose lengths are precisely equal to $\alpha/2$. Thus it suffices to prove the identities for right triangles having one or more sides strictly greater than $\alpha/2$.)

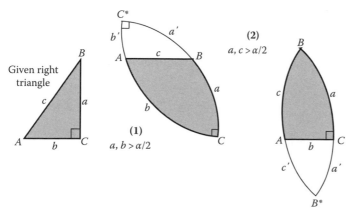

Figure P.19

Beyond Euclid: Modern Geometry

The material in this chapter and the next will extend the results in Euclidean geometry introduced in Chapter 5, to include a few classical theorems in an area that has come to be known as *modern geometry*. This theory includes a development of the Euclidean plane which has taken place over approximately the last 300 years. It goes far beyond the self-imposed bounds of the ancient Greek geometers. We have chosen to focus our attention on the two most common geometric objects: the triangle and the circle. The interplay between them is seemingly inexhaustible, with new relationships and varying viewpoints being explored and appearing in the literature year after year. We are going to consider here a few of the more prominent ones that have become famous over time, such as the nine-point circle, the collinearity theorems of Menelaus and Ceva, and systems of orthogonal circles. Properties of directed distance and circles will be considered first, to provide the background needed for the other more advanced concepts.

7.1 Directed Distance: Stewart's Theorem and the Cevian Formula

An unbounded metric ($\alpha = \infty$) lends itself to a signed distance concept on each individual line. Observe:

> **Definition** For any line l, the ruler postulate guarantees a coordinate system for the points of l. If $A[a]$ and $B[b]$ are any two points on l as indicated by their coordinates, we define the **directed** (or **signed**) **distance** from A to B to be the number
>
> $$AB = b - a$$

Note that this definition for distance allows it to be negative. This would seem to be rather a nuisance, but, as we shall see, it leads to greater preciseness with less effort. We should observe at this early stage that any two points on a line can be assigned coordinates $a = 0$ and $b > 0$—resulting in a positive directed distance *AB*. Thus, directed distance for any two points can be chosen to be positive. However, once established for those two points, this then determines the "direction"—and the sign for distance—for all other pairs of points on the given line. For convenience it will always be assumed that for any two points *A* and *B*, where no other points on line $\overleftrightarrow{AB}$ are involved, the directed distance *AB* is positive.

The following theorem summarizes the fundamental properties for directed distance; the proof of each part is a routine thought process using coordinates and the above definition, which will be left to the reader.

Theorem on Directed Distance   If *A*, *B*, *C* and *X* are any four points on a line, then

(a) $AB + BA = 0$
(b) $AB + BC + CA = 0$
(c) If $AX = AB$ then $X = B$
(d) $AB/BC > 0$ iff (ABC) holds.

Corollary   If $P_1, P_2, P_3, ..., P_n$ are *n* points on a line, *n* a positive integer > 1, then

$$P_1P_2 + P_2P_3 + \cdots + P_nP_1 = 0$$

Observe that this is an easy inductive consequence of Theorem 1(b): start with $P_1P_2 + P_2P_3 + P_3P_4 = 0$; for $n = 4$, $P_1P_2 + P_2P_3 + (P_3P_4 + P_4P_1) = P_1P_2 + P_2P_3 + (P_3P_1) = 0$, and so on. A good conceptual image for using this result is to think of "inserting" a number of points "between" two others. For example, one may "insert" an arbitrary point *B* collinear with *A* and *C* "between" *A* and *C*, and use $AB + BC$ in place of *AC* (Theorem 1(b)). Similarly, inserting *B* and *C* between *A* and *D* results in the equation

$$AB + BC + CD = AD$$

which holds by the corollary as applied to four collinear points, regardless of the order of the points. Note, however, that by our convention regarding exactly two points on a line, if *A*, *B*, and *C* are not collinear then, by the triangle inequality,

$$AB + BC > AC > 0$$

Such facts will often be used without warning in everything that follows.

Example 1

Suppose *A*, *B*, and *C* are the points on line *l* having coordinates as indicated in Figure 7.1. Find the (directed) distances needed and show that

(a) $AB + BC = AC$
(b) (ACB)

Figure 7.1 Example 1.

Solution

(a) $AB = 6 - 3 = 3$, $BC = 5 - 6 = -1$, and $AC = 5 - 3 = 2$.
Thus, $AB + BC = 3 + (-1) = 2 = AC$

(b) Since $AB = 3$ and $CB = -BC = 1$, $AB/CB = 3 > 0$ and
therefore (ACB) holds.

Two notable identities for directed distance involve collinear points A, B, and C and a fourth point P, where l denotes the line of collinearity. (The first requires P to lie on line l.)

(1)   **Euler's Identity** $AB \cdot PC + BC \cdot PA + CA \cdot PB = 0 \ (P \in l)$

(2)   **Stewart's Theorem** $AB \cdot PC^2 + BC \cdot PA^2 + CA \cdot PB^2 + AB \cdot BC \cdot CA = 0$

Proofs   For **(1)**, on the left, insert A between P and C, and also between P and B, and note the cancellation of the terms inside the brackets, due to Theorem 1(a):

$$AB \cdot PC + BC \cdot PA + CA \cdot PB$$

$$= AB(PA + AC) + BC \cdot PA + CA(PA + AB)$$

$$= AB \cdot PA + [AB \cdot AC] + BC \cdot PA + CA \cdot PA + [CA \cdot AB]$$

$$= (AB + BC + CA)PA = (0)PA = 0$$

For **(2)**, we first assume $P \in l$ where use of Euler's identity can be made. Multiply both sides of **(1)** by PC, solve for $AB \cdot PC^2$, and substitute the result into the left side Q of **(2)**. The result is

$$Q = (BC \cdot PA \cdot CP + CA \cdot PB \cdot CP) + BC \cdot PA^2 + CA \cdot PB^2 + AB \cdot BC \cdot CA$$

$$= BC \cdot PA(CP + PA) + CA \cdot PB(CP + PB) + AB \cdot BC \cdot CA$$

$$= BC \cdot PA \cdot CA + CA \cdot PB \cdot CB + AB \cdot BC \cdot CA$$

$$= BC \cdot CA(PA + BP + AB) = BC \cdot CA(0) = 0$$

For the case when $P \notin l$, let Q be the foot of P on l (Figure 7.2). Employing the Pythagorean theorem gives us

$$Q = AB(PQ^2 + QC^2) + BC(PQ^2 + QA^2) + CA(PQ^2 + QB^2) + AB \cdot BC \cdot CA$$

$$= (AB + BC + CA)PQ^2 + \{\text{expression}\,(\mathbf{2})\,\text{for}\,A, B, C, Q\}$$

$$= (0)\,PQ^2 + 0 = 0 \ \diagdown$$

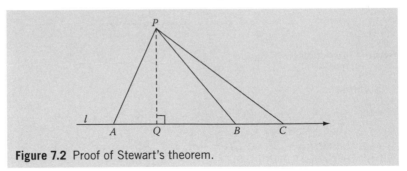

Figure 7.2 Proof of Stewart's theorem.

A test of Stewart's theorem is provided by the following example, where two Pythagorean triangles, each similar to the 3-4-5 right triangle, are placed side-by-side.

Example 2

Observe the right triangles shown in Figure 7.3. Test Stewart's theorem by calculating the signed distances *AB*, *BC*,..., and substituting into the expression **(2)**.

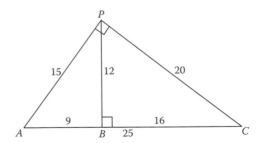

Figure 7.3 Example 2.

Solution

$$AB \cdot PC^2 = 9 \cdot 20^2 = 3600 \quad CA \cdot PB^2 = -25 \cdot 12^2 = -3600$$

$$BC \cdot PA^2 = 16 \cdot 15^2 = 3600 \quad AB \cdot BC \cdot CA = 9 \cdot 16 \cdot (-25) = -3600.$$

The sum of these four quantities is zero, in agreement with Stewart's theorem.

A segment which joins a vertex of a triangle with a point on the opposite side is called a **cevian**, a term derived from a theorem in geometry concerning such segments due to Giovanni Ceva; Ceva's theorem will be stated and proven later. The two most common cevians of a triangle are the medians and angle bisectors, introduced in Chapter 3. The next result involves arbitrary cevians.

The Cevian Formula   In $\triangle ABC$, consider the cevian $\overline{CD}$ joining C with any point D on line AB, as shown in Figure 7.4. Then if p and q are the ratios given by $p = AD/AB$, $q = DB/AB$ (i.e., $pc = AD$ and $qc = DB$ in magnitude and in sign), then

(3) $$d^2 = pa^2 + qb^2 - pqc^2$$

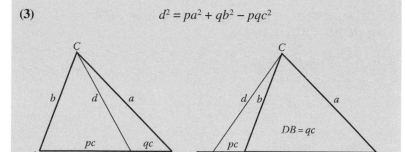

Figure 7.4 Proving the cevian formula.

Proof   Simply apply Stewart's theorem to the collinear points A, D, B, and C, a point not on line AB, playing the roles of A, B, C, and P in that order, and rearrange the terms by algebra:

$$AD \cdot CB^2 + DB \cdot CA^2 + BA \cdot CD^2 + AD \cdot DB \cdot BA = 0$$

$$pca^2 + qcb^2 + (-c)d^2 + pc \cdot qc(-c) = 0$$

$$cd^2 = pca^2 + qcb^2 - pqc^3$$

which is the desired identity upon dividing throughout by c. \

Note: An interesting aspect of (3) is that it applies even when D is not an interior point of segment AB, where we are taking advantage of signed distance (with $AB = c > 0$). \

The reader should immediately recognize the usefulness of the cevian formula for determining distances in Euclidean geometry in a variety of situations. We shall illustrate a few of these in examples and in the discussion which follows. In the first example (Figure 7.5), we know that $d = 13$ in advance by observing a familiar Pythagorean triangle, so this will provide a check on the validity of (3).

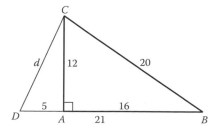

Figure 7.5 Testing the cevian formula.

Example 3

Use the cevian formula **(3)** to find the length d of cevian CD of right triangle $\triangle ABC$, as shown in Figure 7.5.

Solution

First we determine the ratios p and q: $p = AD/AB = -5/16$, $q = DB/AB = 21/16$. By **(3)**,

$$d^2 = \frac{-5}{16}a^2 + \frac{21}{16}b^2 - \frac{-5}{16}\cdot\frac{21}{16}c^2$$

$$= -\frac{5}{16}(20^2) + \frac{21}{16}(12^2) + \frac{105}{16^2}(16^2)$$

$$= -125 + 189 + 105 = 169$$

Hence, $d = \sqrt{169} = 13$.

Example 4

Use the cevian formula to find the length of the median AL of $\triangle ABC$ (Figure 7.6).

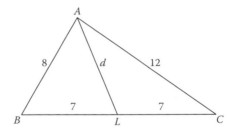

Figure 7.6 Example 4.

Solution

Since $BL = LC$ then $p = q = \frac{1}{2}$ and we have

$$d^2 = \frac{1}{2}\cdot 12^2 + \frac{1}{2}\cdot 8^2 - \frac{1}{4}\cdot 14^2 = 72 + 32 - 49$$

$$= 45 \to d = \sqrt{45} \approx 6.7$$

Example 5

Triangle ABC has sides $AB = 8$, $AC = 4$, and $BC = 9$, as shown in Figure 7.7. If an internal cevian AD from vertex A is the arithmetic mean of AB and AC, find the ratios into which point D divides segment BC and find the lengths of segments BD and DC.

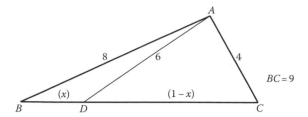

Figure 7.7 Example 5.

Solution

The arithmetic mean of 8 and 4 is 6, so set $AD \equiv d = 6$ and $p = x$, $q = 1 - p = x$, to solve for x. Thus **(3)** becomes

$$36 = x \cdot 16 + (1 - x)64 - x(1 - x)81$$

$$= 16x + 64 - 64x - 81x + 81x^2$$

which simplifies to the quadratic equation

$$81x^2 - 129x + 28 = 0 \rightarrow (3x - 4)(27x - 7) = 0$$

The solutions are $x = 4/3$ and $x = 7/27$. Note that $x = 4/3 > 1$ so this solution for D is exterior to segment BC. The solution we seek is therefore $x = 7/27$, which yields $BD = 7/3 \approx 2.33$ and $DC = 20/3 \approx 6.67$.

Moment for Reflection

What is the geometric significance of the two solutions for point D we obtained in Example 5? To help you analyze this question, imagine the actual location of point D on line AB corresponding to the other solution $x = 4/3$ (make a sketch). Is the value for AD still 6? If AD is given, will there always be two distinct solutions for point D?

7.2 Formulas for Special Cevians

Expressions for the three special segments of a triangle in terms of its sides (the medians, angle bisectors, and altitudes) will now be derived using the cevian formula. For the medians, simply put $p = q = \frac{1}{2}$ and simplify:

$$m_a = \sqrt{\frac{1}{2}b^2 + \frac{1}{2}c^2 - \frac{1}{4}a^2} \quad \text{(median to side } BC)$$

(1) $$m_b = \sqrt{\frac{1}{2}a^2 + \frac{1}{2}c^2 - \frac{1}{4}b^2} \quad \text{(median to side } AC)$$

$$m_c = \sqrt{\frac{1}{2}a^2 + \frac{1}{2}b^2 - \frac{1}{4}c^2} \quad \text{(median to side } AB)$$

Next, consider the angle bisectors, with lengths t_a, t_b, and t_c as shown in Figure 7.8. In order to determine the ratios for p and q needed here, recall the basic property (appearing in a problem in Chapter 5 (Problem 22, Section **5.12**): if the bisector of $\angle BAC$ cuts the opposite side at D (Figure 7.8) then $AB/AC = BD/DC$. That is, $c/b = pa/qa = p/q$. Since $p + q = 1$, it follows that $p = c/(b + c)$ and $q = b/(b + c)$. The cevian formula then yields

$$t_a^2 = pb^2 + qc^2 - pqa^2$$

$$= \frac{c}{b+c}b^2 + \frac{b}{b+c}c^2 - \frac{bc}{(b+c)^2}a^2$$

$$= \frac{bc}{b+c}[b+c] - \frac{bc}{(b+c)^2}a^2$$

$$= \frac{bc}{(b+c)^2}[(b+c)^2 - a^2]$$

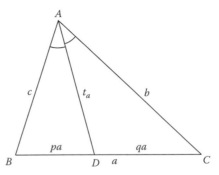

Figure 7.8 Formula for angle-bisectors.

To improve the form of this last expression, the notation involving the semi-perimeter $s = \frac{1}{2}(a + b + c)$ will be used. Note that $2s = a + b + c$ and $2s - 2a = b + c - a$. Also, we are going to use the factoring law $X^2 - Y^2 = (X + Y)(X - Y)$, with $X = b + c$ and $Y = a$. Thus from the above equation for t_a^2 we obtain

$$t_a^2 = \frac{bc}{(b+c)^2}(b+c+a)(b+c-a) = \frac{bc \cdot 2s(2s-2a)}{(b+c)^2} = \frac{4bcs(s-a)}{(b+c)^2}$$

and since $b + c = 2s - a$ (used mainly in order to gain calculational advantages) we obtain

$$t_a = 2\sqrt{bcs(s-a)} / (2s-a) \quad \text{(angle bisector to side } BC\text{)}$$

(2) $\qquad t_b = 2\sqrt{acs(s-b)} / (2s-b) \quad \text{(angle bisector to side } AC\text{)}$

$$t_c = 2\sqrt{abs(s-c)} / (2s-c) \quad \text{(angle bisector to side } AB\text{)}$$

In order to derive formulas for the lengths of the altitudes to the sides of a triangle, it is necessary to use a "trick" to obtain the ratios p and q for **(3)** (previous section). But the development we use is interesting because, in the process, an unusual non-trigonometric proof of Hero's formula for the area of a triangle emerges.

Let h_a denote the altitude AD to side BC of $\triangle ABC$ (as in Figure 7.9). Note that our method encompasses both cases when the angles at B and C are acute, or when one of them is obtuse. Since we do not know the ratios p and q, we let $p = x$ and $q = 1 - x$. Then, by the cevian formula

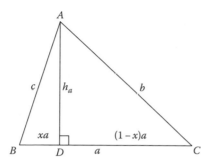

Figure 7.9 Formula for altitudes.

$$h_a^2 = xb^2 + (1-x)c^2 - x(1-x)a^2$$

which leads to the quadratic equation

$$a^2x^2 + (b^2 - c^2 - a^2)x + c^2 - h_a^2 = 0$$

Now $\overline{AD}$ is perpendicular to $\overline{BC}$ so there can be only one solution in x for the cevian, hence the discriminant of the quadratic expression in x must be zero. That is, $(B^2 - 4AC) = 0$, or

$$(b^2 - c^2 - a^2)^2 - 4a^2(c^2 - h_a^2) = 0$$

Solving this equation for h_a^2, we obtain

$$4a^2h_a^2 = 4a^2c^2 - (b^2 - c^2 - a^2)^2$$

Using difference-of-two-squares factoring and the notation for the semi-perimeter s introduced earlier, we have

$$
\begin{aligned}
4a^2h_a^2 &= \left[2ac + (b^2 - c^2 - a^2)\right]\left[2ac - (b^2 - c^2 - a^2)\right] \\
&= \left[b^2 - (a^2 - 2ac + c^2)\right]\left[(a^2 + 2ac + c^2) - b^2\right] \\
&= \left[b^2 - (a-c)^2\right]\left[(a+c)^2 - b^2\right] \\
&= (b + a - c)(b - a + c)(a + c + b)(a + c - b) \\
&= (2s - 2c)(2s - 2a)(2s)(2s - 2b)
\end{aligned}
$$

or

$$4a^2h_a^2 = 16s(s-a)(s-b)(s-c)$$

Dividing by $4a^2$ and taking the square root, the first of the following forms emerges (the other two by cyclic notation):

$$h_a = (2/a)\sqrt{s(s-a)(s-b)(s-c)} \quad \text{(the altitude to side } BC)$$

(3) $\quad h_b = (2/b)\sqrt{s(s-a)(s-b)(s-c)} \quad \text{(the altitude to side } AC)$

$$h_c = (2/c)\sqrt{s(s-a)(s-b)(s-c)} \quad \text{(the altitude to side } AB)$$

If we accept at this point the usual formula for the area of a triangle (one-half base times height), then $K = \frac{1}{2}ah_a$ and the first formula in **(3)** immediately produces the familiar Heron formula for the area of $\triangle ABC$:

(4) $$K = \sqrt{s(s-a)(s-b)(s-c)}$$

Problems (Section 7.2)

Group A

1. Consider the points, with their coordinates, lying on some line: $A[-4]$, $B[1]$, $C[6]$, and $D[7]$. Verify Euler's identity **(1)**, Section **7.1** for these four points by direct calculation.

2. Prove for any four distinct collinear points A, B, C, and P:

$$\frac{PC}{BC \cdot CA} + \frac{PA}{AB \cdot CA} + \frac{PB}{AB \cdot BC} = 0$$

3. By simplifying, prove the following identity for any five collinear points A, B, C, D, and E

$$AB \cdot CE + AC \cdot DE + AD \cdot EC + AE \cdot BD + BE \cdot CA = 0$$

4. Suppose that $O, A, B,$ and C are collinear points such that $OA + OB + OC = 0$ (Figure P.4 shows a specific example). Show that for any other point P on this line,

$$PA + PB + PC = 3PO$$

Figure P.4

5. Two Pythagorean triangles are placed side-by-side, as shown in Figure P.5. Test Stewart's theorem by calculating the signed distances involved and substituting into the expression **(2)**, Section **7.1**.

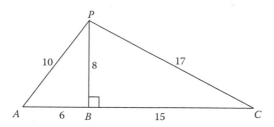

Figure P.5

6. Prove the relation, valid for all triangles,

$$m_a^2 + m_b^2 + m_c^2 = \tfrac{3}{4}(a^2 + b^2 + c^2)$$

7. Cevian AD of $\triangle ABC$ having the Pythagorean triple (5, 12, 13) as sides, has length 20 (Figure P.7). Find the appropriate ratios p and q in the cevian formula and thus determine the location of the other point D' (in terms of B and C) such that $AD' = 20$ which lies on line BC. Sketch your own diagram showing the other position possible for cevian AD.

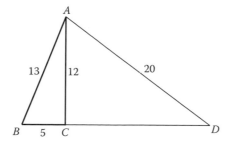

Figure P.7

8. The medians of a triangle are of length 5, $2\sqrt{13}$, and $\sqrt{73}$, respectively. Find the lengths of the sides.

9. Two medians of a triangle are congruent. Use **(1)** to show that the sides opposite are congruent. Try a synthetic proof for this proposition.

10. If a triangle has altitudes of length $\tfrac{1}{3}, \tfrac{1}{4}$, and $\tfrac{1}{5}$, respectively, show it is a right triangle.

Group B

11. The sides of a triangle are $72/\sqrt{455}$, $96/\sqrt{455}$, and $144/\sqrt{455}$, respectively. Find the lengths of the three altitudes.

12. Cevians of length 13/15 are drawn from one vertex of a unit equilateral triangle to the opposite side. Find the two positions where the cevians meet that side.

13. Using Theorem **(5)**, prove the Steiner–Lehmus theorem that if two angle bisectors are congruent, the sides

opposite are congruent. [**Hint:** Show that if $a < b$ then $t_a > t_b$; a and b can essentially be omitted from the formulas for t_a and t_b by using the product rule for inequalities. By algebra, the expression $(sa + sb - ab)(b - a)$ will follow, all other terms canceling out.]

Note: The synthetic proof of this theorem occurred in Problem 25, Section **3.7.**

14.   Establish the following identities which give the sides of a triangle in terms of its medians.

$$
\begin{aligned}
a &= \tfrac{4}{3}\sqrt{\tfrac{1}{2}m_b^2 + \tfrac{1}{2}m_c^2 - \tfrac{1}{4}m_a^2}\\
(5)\qquad b &= \tfrac{4}{3}\sqrt{\tfrac{1}{2}m_a^2 + \tfrac{1}{2}m_c^2 - \tfrac{1}{4}m_b^2}\\
c &= \tfrac{4}{3}\sqrt{\tfrac{1}{2}m_a^2 + \tfrac{1}{2}m_b^2 - \tfrac{1}{4}m_c^2}
\end{aligned}
$$

15.   Prove that $\triangle ABC$ is a right triangle iff either of the following two conditions are met:

(a) $5m_c^2 = m_a^2 + m_b^2$

(b) $(1/h_c)^2 = (1/h_a)^2 + (1/h_b)^2$

16.*   Let P be any point and $\triangle ABC$ any triangle with centroid G (Figure P.16). Using the cevian formula for cevian PG in $\triangle APL$ (L the midpoint of BC), and similarly for cevian PL in $\triangle BPC$, prove that

$$(6)\qquad PG^2 = \tfrac{1}{3}(PA^2 + PB^2 + BC^2) - \tfrac{1}{9}(a^2 + b^2 + c^2)$$

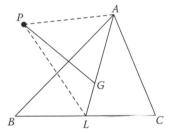

Figure P.16

17.   Using **(6)** from the previous problem, establish the following relation giving the distance from the circumcenter of a triangle to its centroid in terms of the circumradius R and the sides of the triangle:

$$(7)\qquad OG^2 = R^2 - \tfrac{1}{9}(a^2 + b^2 + c^2)$$

Group C

18.   If the incircle of $\triangle ABC$ has radius r and I is the incenter, use the fact that the sum of the three inner triangles (figure) equals the area K of $\triangle ABC$ to show that

(a) $K = sr$

(b)   Using Heron formula, establish this identity for the inradius of a triangle in terms of its sides:

(8) $\qquad r = \sqrt{\dfrac{(s-a)(s-b)(s-c)}{s}}$

(c) Using (11), derive the following identity for the inra-dius of a triangle in terms of its altitudes:

(9) $\qquad \dfrac{1}{r} = \dfrac{1}{h_a} + \dfrac{1}{h_b} + \dfrac{1}{h_c}$

19. Prove that if m_s stands for the *semi-median* $\frac{1}{2}(m_a + m_b + m_c)$, the area of $\triangle ABC$ in terms of its medians is given by

(10) $\qquad K = \frac{4}{3}\sqrt{m_s(m_s - m_a)(m_s - m_b)(m_s - m_c)}$

20. Prove that if h_s stands for the *semi-harmonic altitude* sum given by

$$\dfrac{1}{h_s} = \dfrac{1}{2}\left(\dfrac{1}{h_a} + \dfrac{1}{h_b} + \dfrac{1}{h_c}\right)$$

the area of $\triangle ABC$ in terms of its altitudes is given by

(11) $\qquad \dfrac{1}{4K} = \sqrt{\dfrac{1}{h_s}\left(\dfrac{1}{h_s} - \dfrac{1}{h_a}\right)\left(\dfrac{1}{h_s} - \dfrac{1}{h_b}\right)\left(\dfrac{1}{h_s} - \dfrac{1}{h_c}\right)}$

21. Derive the following identities that give the sides of $\triangle ABC$ in terms of its altitudes:

(12) $\qquad \dfrac{1}{a} = 2h_a\sqrt{\dfrac{1}{h_s}\left(\dfrac{1}{h_s} - \dfrac{1}{h_a}\right)\left(\dfrac{1}{h_s} - \dfrac{1}{h_b}\right)\left(\dfrac{1}{h_s} - \dfrac{1}{h_c}\right)}$

[Similar expressions for $1/b$ and $1/c$.]

7.3 Circles: Power Theorems and Inscribed Angles

The cevian formula will be used to establish the classical theorems for circles. This procedure, like that of the last section, may be referred to as the *cevian method*, a significant departure from the usual development for circles. Throughout this section and those following, it will be useful to adopt the notation $[O, r]$ to designate a *circle having center O and radius r*. A definition will be given to start the development.

Definition 1   The **power** of a point P with respect to a circle $C \equiv [O, r]$, is defined as the number

$$\text{Power } P \text{ (circle } C) = PO^2 - r^2$$

Note that the power of P is zero iff P lies on the circle, and is positive or negative depending on whether P is outside, or inside, the circle, respectively. The cevian formula provides an elegant method for finding the power of P in terms of any secant of the circle that passes through P. This will immediately lead to the classical theorems involving circles known as the **power theorems**. An example will illustrate Definition 1.

Example 1

In Figure 7.10 is shown circle **C** having center O with tangent line through P, with A as the point of contact. Show that PA^2 equals the power of P with respect to **C**.

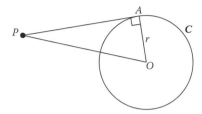

Figure 7.10 Example 1.

Solution

Draw line PO and radius OA. Since $\overline{OA} \perp \overline{PA}$, the Pythagorean theorem implies that

$$PA^2 = PO^2 - r^2 = \text{Power } P(\boldsymbol{C})$$

Moment for Reflection

The above example shows that there is a geometric construction for the power of P with respect to a circle if P lies outside the circle: Merely construct one of the tangents to the circle through P and take the square of the distance from P to the point of contact. Can you think of a way to construct the power if P lies inside the circle? Could you just draw a secant through P and work, somehow, with the distances from P to the points of intersection A and B of the secant and the circle? (Figure 7.11 illustrates.) Although this may seem useless at first, it turns out to provide an affirmative answer, and your chances of discovering the conclusion of the next theorem are quite likely. For extra help on this, it is recommended that you work through the calculations in the numerical experiment which follows.

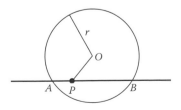

Figure 7.11 The power of an interior point.

Numerical Experiment

A circle of radius 10 is given, having center O (Figure 7.12). Chords AB and CD are perpendicular at P, where AB is a diameter and $AP = 2$ (thus $PB = 18$). A third chord EF passes through P such that $EP = 3$. Part (1) of the experiment is to find PC (=PD) and PF. Use the Pythagorean theorem in $\triangle OPD$ to find x (note that $PO = 8$). With y as indicated, determine the other segment shown as $y + 3$, then use the Pythagorean theorem to obtain the equation $8^2 - y^2 = z^2 = 10^2 - (y + 3)^2$. Expand, simplify, and solve for y; then find PF. Part (2) of the experiment is to calculate the resulting products $AP \cdot PB$, $CP \cdot PD$, and $EP \cdot PF$. Also, calculate the power of P with respect to circle O. Any conjectures? (*Sketchpad* can be used for the general situation.)

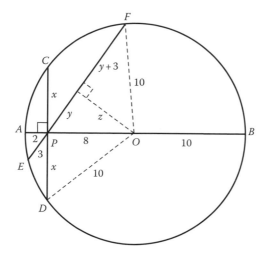

Figure 7.12 Segments on concurrent chords.

Theorem If P is any point and l any secant through P intersecting circle $\mathbf{C}$ at points A and B, then

(1) $$\text{Power } P \text{ (circle } \mathbf{C}) = PA \cdot PB$$

(*Hint for proof* In Figure 7.11, draw radii OA and OB and use the cevian formula for OP in $\triangle OAB$, taking note of signed distance. Observe that this result applies to all cases, whether P is interior or exterior to the circle; $PA \cdot PB$ is negative if P is an interior point of the circle.)

Special cases of (1) immediately lead to the three classical theorems now found in most high-school geometry texts—the so-called *power theorems* (illustrated in Figure 7.13). The proofs will be left to the reader as easy exercises. Note that the third power theorem that follows is a direct corollary of the second; this labor-saving connection between these two results often goes undetected in textbooks covering this sequence of theorems.

The power theorems

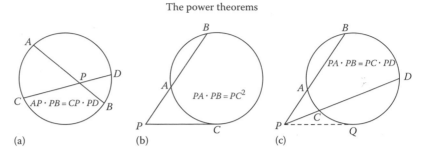

(a)   (b)   (c)

Figure 7.13 Three types of angles and intercepted arcs.

> **Two-Chord Theorem**   If two chords AB and CD meet at an interior point P of a circle, then $AP \cdot PB = CP \cdot PD$ (see Figure 7.13a).

> **Secant-Tangent Theorem**   If line PA is a secant of a circle intersecting it at A and B, and line PC is tangent at C, then $PA \cdot PB = PC^2$ (see Figure 7.13b).

> **Corollary: Two-Secant Theorem**
>
> If secants PA and PC of a circle meet that circle at points A and B, and at points C and D respectively, then $PA \cdot PB = PC \cdot PD$ (see Figure 7.13c).

Note: The usual procedure to justify these results appearing in elementary geometry textbooks is to use arcs of circles and their degree measures. One must accept the additivity of arc measure as an axiom, which is used to prove the inscribed angle theorem, from which follows the preceding power theorems, using similar triangles. Observe that our approach is independent of arc measure. It also provides a unifying concept for *all three* power theorems. ◣

The result mentioned earlier involving a circle and a right triangle will now be generalized in what is termed the *inscribed angle theorem*. First, some terminology will be defined.

> **Definition 2**   An **inscribed angle** of a circle is an angle whose vertex lies on the circle and whose sides contain chords of that circle ($\angle BAC$ in Figure 7.14). A **central angle** of a circle is an angle whose vertex is the center of the circle and whose sides contain radii of that circle ($\angle BOC$ in Figure 7.14).

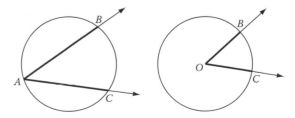

Figure 7.14 Two types of angles.

Recall that an *arc* of a circle (or *circular arc*) is the intersection of a circle and a closed half-plane determined by a secant of that circle. The arc is said to be a *semicircle* iff the secant passes through the center, and is a *major* or *minor* arc depending on whether the center does, or does not, lie in the half-plane. Another term used in the inscribed angle theorem is defined next.

Definition 3   Any angle which contains an arc of a circle in its interior and sides is said to **inscribe**, or **intercept**, that arc. (See Corollary **C** of Section **4.5**.)

Figure 7.15 shows two inscribed angles intercepting the same arc, $\angle ABC$ and $\angle ADC$. By Theorem **2**, Section **4.5**, D lies on arc BUA or arc BVC (the case shown in the figure). First, with D on arc BVC, D lies in the interior of $\angle BAC$ (Corollary **C**, Section **4.5**) and by the crossbar theorem, ray AD meets segment BC (a chord of the circle) at some point P; since D is not interior to the circle, (APD) holds. Then, as vertical angles, $\angle BPA \cong \angle DPC$. By the two-chord theorem,

$$BP \cdot PC = DP \cdot PA \quad \text{or} \quad \frac{BP}{PD} = \frac{PA}{PC}$$

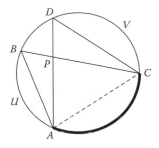

Figure 7.15 Inscribed angle theorem.

Hence by the SAS similarity criterion, $\triangle BPA \sim \triangle DPC$, and thus $\angle ABC \cong \angle ADC$. (The proof for the case when D lies on arc BUC is virtually the same.) Thus, we have established a key theorem on circles:

Inscribed Angle Theorem   Inscribed angles which intercept the same arc of a circle are congruent.

Moment for Reflection

The above theorem shows that as point B varies on the circle and stays on the same side of line AC, the measure of $\angle ABC$ is constant. But what if B moves to the other side, shown as B' in Figure 7.16? Can you see what happens in that case? (Give proofs if you can.)

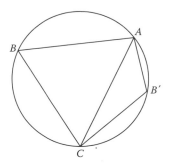

Figure 7.16 Can you find and prove relation between ∠B and ∠B'?

Another proof of a result mentioned previously thus emerges.

Corollary A   An angle inscribed in a semicircle is a right angle. Conversely, an inscribed angle that is a right angle is inscribed in a semicircle.

Proof   Consider inscribed angle ∠ABC (Figure 7.17). We prove first that if AC is a diameter of the circle, then $m\angle ABC = 90$; then we prove the converse, that if $m\angle ABC = 90$, then AC is a diameter.

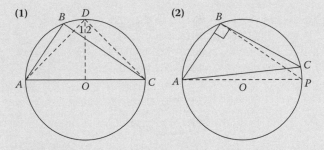

Figure 7.17 Angle inscribed in a semicircle.

1. Suppose AC is a diameter of the circle. Draw radius OD perpendicular to AC. In Figure 7.17(1), it follows that $\angle 1 = \angle 2 = 45$. By the inscribed angle theorem, $m\angle B = m\angle D = 90$.
2. If $m\angle ABC = 90$, draw diameter AP. Then by the preceding step, $m\angle ABP = 90 = m\angle ABC$, and AC coincides with diameter AP.∎

The following result will be left to the reader.

Corollary B   Let ∠ABC be an inscribed angle of a circle, with line CP tangent to the circle at C. If ∠ACP inscribes the same arc as ∠ABC, then ∠ABC ≅ ∠ACP.

(*Hint for Proof*   Draw diameter CD perpendicular to the tangent at C, making △CDA a right triangle, with complementary acute angles; use ∠B ≅ ∠D.)

The converse of the inscribe angle theorem is also useful (which generalizes the converse part of the preceding corollary) and will be established next.

Corollary C If points A and B lie on the same side of line CD, and if $\angle CAD \cong \angle CBD$, the four points A, B, C, and D lie on a circle (are **concyclic**) (see Figure 7.18).

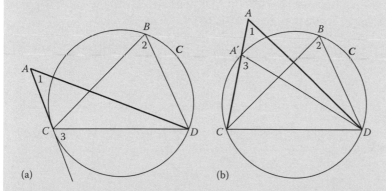

(a)   (b)

Figure 7.18 Concyclic points.

Proof Let $\mathbf{C}$ be the circle passing through B, C, and D. If $A \in \mathbf{C}$, we are finished. Otherwise, consider line CA. If CA is tangent to $\mathbf{C}$ at C, then observe Figure 7.18a; we would have in that case $\angle 1 \cong \angle 2 \cong \angle 3$ (by Corollary B), which contradicts the exterior angle inequality. Hence line CA is a secant of $\mathbf{C}$ and meets $\mathbf{C}$ at some point $A' \neq C$ such that either (CAA') or $(CA'A)$, the case shown in Figure 7.18b. (If (ACA') it can be shown that $m\angle CBD > m\angle CAD$.) Hence, angles $CA'D$ and CBD intercept the same arc and by hypothesis and the inscribed angle theorem,

$$\angle 1 \cong \angle 2 \cong \angle 3$$

which again violates the exterior angle inequality theorem. Therefore, $A \in \mathbf{C}$, as desired. ◣

Corollary D If points A and B lie on opposite sides of line CD, and if $\angle CAD$ is supplementary to $\angle CBD$, the four points A, B, C, and D are **concyclic**.

Proof Left for the reader. (The same method as the above may be used.)

7.4 Using Circles in Geometry

It is interesting to see how the circle theorems of the previous section can be used in geometry to yield additional classical results. One example involves quadrilaterals and whether their vertices lie on a circle. Quadrilaterals with this property are called **cyclic**. Clearly, not all quadrilaterals are cyclic. But, of course, Corollary C of the last section provides a simple answer, which we restate in terms of convex quadrilaterals as the first application of circles to geometry.

> **Theorem 1** A convex quadrilateral can be inscribed in a circle iff the opposite angles are supplementary.

Next is a very old result, first discovered by the Greek geometer C. Ptolemy (c. 150 AD) and elaborated on later by a prominent Hindu mathematician Brahmagupta (c. 600 AD). We have chosen the following proof over more common trigonometric proofs for its elegance.

> **Theorem 2: Ptolemy's Theorem**
>
> If $\lozenge ABCD$ is a cyclic quadrilateral, then
>
> **(1)** $$AB\cdot CD + BC\cdot AD = AC\cdot BD$$
>
> or, equivalently, if a, b, c, and d are the lengths of consecutive sides, and m and n that of the two diagonals,
>
> **(1′)** $$ac + bd = mn$$
>
> *Proof* In Figure 7.19, construct $\angle ABE \cong \angle DBC$, E on diagonal AC. The inscribed angle theorem implies that $\angle BAE \cong \angle BDC$ and $\angle ADB \cong \angle ECB$. Thus, $\triangle ABE \sim \triangle DBC$ and $\triangle ADB \sim \triangle ECB$. Hence, $AB/DB = AE/DC$ and $BD/BC = AD/EC$. That is, letting $x = AE$ and $y = EC$,
>
> $$\frac{a}{n} = \frac{x}{c} \quad \text{and} \quad \frac{n}{b} = \frac{d}{y}$$
>
> or
>
> $$ac = nx \quad \text{and} \quad bd = ny$$
>
> Therefore,
>
> $$ac + bd = nx + ny = n(x+y) = mn$$

(There are two other cases to consider due to betweenness considerations; can you identify them, and does the same idea as the above apply?)

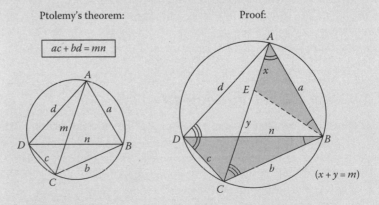

Figure 7.19 Proof of Ptolemy's theorem.

Example 1

Prove that if P is any point on arc AB of the circumcircle of an equilateral triangle ABC, then $PC = PA + PB$. (See Figure 7.20.)

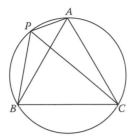

Figure 7.20 Example 1.

Solution

By Ptolemy's theorem, in cyclic quadrilateral $PACB$,

$$PA \cdot CB + AC \cdot PB = PC \cdot AB$$

But $CB = AC = AB$, so these terms divide out in the above expression, leaving the desired result $PA + PB = PC$.

Another application of circles to geometry produces a nice compass-straightedge construction of the tangents to a circle from a given external point.

Construction of External Tangents to a Circle   Suppose P is any point outside circle C and it is required to locate (draw) the two tangent lines to the circle through P. The following construction will suffice (Figure 7.21):

(1) Draw the line PO to the center of C.
(2) Locate the midpoint M of segment PO.
(3) Draw the circle having center M and radius $PM = MO$, intersecting C at T, and T'. Lines PT and PT' are the two tangents to C through P.

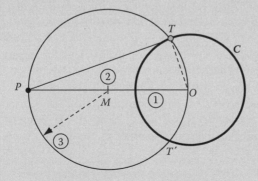

Figure 7.21 Constructing the tangents to circle C.

Proof   Since $\angle PTO$ is inscribed in a semi-circle, $\overleftrightarrow{PT} \perp \overleftrightarrow{OT}$ hence OT and OT' are the tangents to C. ◥

The next theorem involves a circle that is famous for its many unique properties. It is the circle which passes through the midpoints of a triangle. Since it also passes through six other strategic points, it is called the **nine-point** circle.

Theorem 3: The Nine-Point Circle Theorem

In any triangle $\triangle ABC$, consider the following nine points, taken in groups of three (Figure 7.22): L, M, and N—the midpoints of the three sides; D, E, and F—the feet of the altitudes on the three sides; and X, Y, and Z—the midpoints of the segments AH, BH, and CH, where H is the orthocenter—called the **Euler points** of the triangle. These nine points are concyclic.

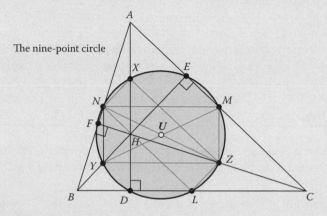

Figure 7.22 The nine-point circle.

Proof   Consider $\triangle HBC$. Since Y and Z are the midpoints of $\overline{HB}$ and $\overline{HC}$ segment YZ is parallel to side BC and equals ½ its length. Similarly, $\overleftrightarrow{NM} \parallel \overleftrightarrow{BC}$ and $NM = \frac{1}{2}BC = YZ$. Hence $\lozenge NYZM$ is a parallelogram. Similarly, observing $\triangle BAD$, $\overleftrightarrow{NY} \parallel \overleftrightarrow{AD}$; since $\overleftrightarrow{AD} \perp \overleftrightarrow{BC}$, then $\overleftrightarrow{NY} \perp \overleftrightarrow{YZ}$, making $\angle NYZ$ a right angle. Thus, $\lozenge NYZM$ is a rectangle. The diagonals NZ and YM meet at a point U, which is the center of a circle $\mathbf{C}$ passing through N, Y, Z, and M. Next we show that $\mathbf{C}$ also passes through points L and X. But it also follows, as in the preceding argument, that $\lozenge XNZL$ is a rectangle, and the midpoint of diagonal NZ is the center of a circle passing through X, N, L, and Z, having diameter NZ, that of $\mathbf{C}$. Hence $\mathbf{C}$ passes through L and X. This leaves the feet of the altitudes. The hypotenuse XL of right triangle XDL is a diameter of $\mathbf{C}$, so $\mathbf{C}$ passes through D (Corollary C, Section **7.2**). Similarly, $\mathbf{C}$ passes through E and F, completing the proof. ◥

Geometric relationships involving the nine-point circle are prolific. We mention three examples, without proof:

- The nine-point circle is exactly ½ the size of the circumcircle of the triangle.

- It is *tangent* to the incircle of the given triangle and to each of the three *excircles* (circles tangent externally to the extended sides).*
- Its center (called the **nine-point center**) lies on the Euler line, the line of collinearity of *H*, *G*, and *O*, as mentioned earlier (see Section 5.11).

The final property of circles to be taken up will provide a unifying principle for the two families of orthogonal circles that will be constructed in Section 7.6. This property is actually rather surprising. It is recommended that the reader explore the ideas in the following project before going on.

Numerical Experiment

Circles [*O*, 10] and [*Q*, 17] intersect at *A* and *B*, with *OQ* = 21 (Figure 7.23). If the common chord *AB* meets the line of centers at *W*, it can be shown that *OW* = 6 and *WQ* = 15 (let the reader verify this using the Pythagorean theorem in $\triangle OAW$ and $\triangle AWQ$, letting *OW* = *u*, with *OA* = 10, *QA* = 17). Now let *P* be any point on line *AB* such that *PW* = *x* (variable) and lying outside both circles (as shown in the figure). Make the following calculations as part of the experiment:

(1) Power *W* (circle *O*) and Power *W* (circle *Q*)
(2) Power *P* (circle *O*) terms of *x*
(3) Power *P* (circle *Q*) terms of *x*

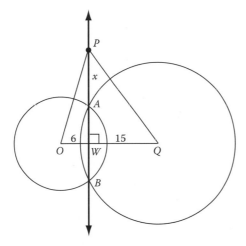

Figure 7.23 Power of *P* with respect to two circles.

As a result of your calculations, do you draw any conclusions? (*Sketchpad* reveals a more general result, which you might want to try.)

The phenomenon evident in the above experiment can be proven without difficulty. It depends on the calculation of the product *PA* · *PB*

---

* This result is due to Karl W. Feuerbach, a German mathematician (1800–1834).

for P on line AB. First, let's experiment with numbers some more. In Figure 7.23 one can determine that $AW = WB = 8$. If $P = W$ then

$$PA \cdot PB = WA \cdot WB = (-8)(8) = -64$$

But observe that

$$-64 = 36 - 100$$

$$= OW^2 - 10^2$$

$$= \text{Power } W \text{ (circle } O)$$

verifying a previous result [$PA \cdot PB = \text{Power } P$ (circle O)]. These ideas can be generalized to establish:

Theorem 4 If two circles [O, r] and [Q, s] intersect at A and B, the locus of points P whose powers with respect to the two circles are equal is line containing the common chord AB.

Proof

1. By **(1)** Section **7.3** we have, as illustrated in Figure 7.24,

$$\text{Power } P \text{ (circle } O) = PA \cdot PB$$

$$\text{Power P (circle } Q) = PA \cdot PB$$

That is, the power of P with respect to the two circles is the same.
2. Conversely, suppose that Power P' (circle O) = Power P' (circle Q) for some point P'. Draw line $P'A$, intersecting circle O at B' and circle Q at C', as in the figure. Then:

$$PA \cdot PB' = PA \cdot PC' \quad \text{or} \quad PB' = PC'$$

Thus $B' = C' = B$, and line $P'A$ coincides with line AB. Therefore, P' lies on line AB, as was to be proved. ◣

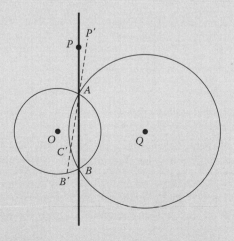

Figure 7.24 Radical axis of two intersecting circles.

Moment for Reflection

If $A = B$ in Theorem 4, the circles are tangent and the line is the common tangent at A (Figure 7.25). Is the power property still valid? See if you can determine the answer to this (with proof).

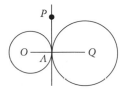

Figure 7.25 Radical axis of tangent circles.

Note: Later the result just established will be obtained for circles *which do not intersect*, and a line (like AB) will exist such that each of its points has equal powers with respect to the two circles. Evidently, then, the locus of points whose powers with respect to any two non-concentric circles are equal is a line. This line is called the **radical axis** of the two circles. ◤

Problems (Section 7.4)

Group A

1. Solve for x in each case, where C is a circle (see Figure P.1a and b).

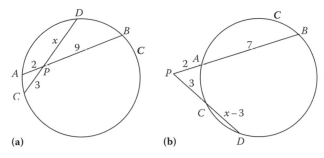

    (a)   (b)

Figure P.1

2. Relative to circle C, solve for x in each case (see Figure P.2).

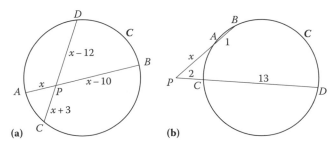

    (a)   (b)

Figure P.2

3. Perpendicular chords *LM* and *PQ* of circle $[O, \sqrt{65}]$ meet at *W* such that *PW* = 6 and *WQ* = 8 (Figure P.3). Find *LW* and *WM*. [**Hint:** Draw radius *OL* and drop perpendiculars from *O* to chords *LM* and *PQ*.]

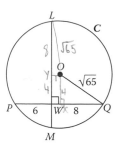

Figure P.3

4. Perpendicular chords *RS* and *PW* of circle *C* meet at point *A* such that *PA* = 6, *AW* = 32, and *AS* = 16, as shown in Figure P.4. Find the radius of *C*.

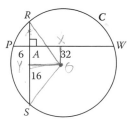

Figure P.4

5. Perpendicular secants *PA* and *PC* of circle *C* meet the circle at the points *A*, *B*, *C*, and *D*, as indicated by Figure P.5, such that *PA* = 5, *PC* = 6, and *CD* = 4. Find the radius of *C*.

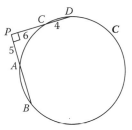

Figure P.5

6. **Proof without Words** Explain how a proof of the following proposition may be obtained by observation (adding one line, as shown in Figure P.6).

Inscribed angles of a circle which subtend the same chord and lie on opposite sides of that chord are supplementary.

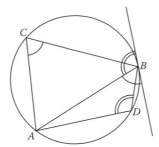

Figure P.6

7. Give a rigorous proof of Corollary **B**, Section **7.3**.

8. **Compass/Straight-Edge Construction for Geometric Mean** Provide a proof for the construction of the geometric mean of a and b (as shown in Figure P.8) indicated in the steps which follow.
   1. Lay off segments AB and BC of length a and b on line l such that (ABC).
   2. Locate the midpoint M of segment AC.
   3. Draw circle C with center M passing through A and C.
   4. Erect the perpendicular to l at B, intersecting the circle at D. The value $c = BD$ is the required geometric mean.

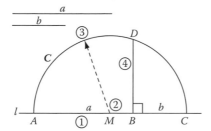

Figure P.8

Group B

9. **Measure for Circular Arcs** The traditional concept for arc measure requires three definitions: one for minor arcs, one for semicircles, and a third one for major arcs. Not only is this cumbersome, it leads to a technical difficulty in justifying the additivity of arc measure, which is needed in order to prove the inscribed angle theorem. It is desirable, therefore, to design a definition that avoids this problem. Our approach is to develop the inscribed angle theorem first (as above), and then use that result in defining arc measure.

Definition Let $A \equiv \overset{\frown}{AB}$ denote the arc of any circle. The measure of arc A is defined to be twice that of any inscribed angle $\angle APB$ which intercepts A (see Figure P.9).

Prove the following results involving this definition.

(a) Prove that arc measure is well-defined, that the choice of point P such that $\angle APB$ intercepts A is immaterial.

(b) Prove that arc measure as defined here is *additive*: If C lies on arc AB then $m\overarc{AB} = m\overarc{AC} + m\overarc{CB}$.

(c) Prove that arc measure as defined here is in agreement with the traditional definition: If an arc has endpoints A and B and O is the center of the circle, then (1) $m\overarc{AB} = m\angle AOB$ if arc AB is a minor arc, (2) $m\overarc{AB} = 180$ if arc AB is a semicircle, and (3) $m\overarc{AB} = 360 - m\angle AOB$ if arc AB is a major arc.

Note: Fundamental results on circles established earlier (Theorem **2** and Corollary **C** of Section **4.5**) play a critical role in the correct handling of these ideas. ◣

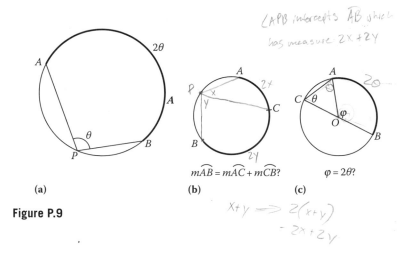

$$m\overarc{AB} = m\overarc{AC} + m\overarc{CB}?$$

$$\varphi = 2\theta?$$

(a)   (b)   (c)

Figure P.9

10. Using the definition for arc measure given in Problem 9, prove each of the relations indicated in Figure P.10. [**Hint:** For (b) and (c) use the (Euclidean) exterior angle theorem.]

$$m\angle ABC = \tfrac{1}{2}x \quad m\angle ABC = \tfrac{1}{2}(x+y) \quad m\angle ABC = \tfrac{1}{2}(x-y)$$

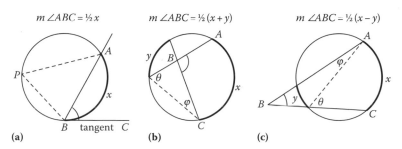

(a)   (b)   (c)

Figure P.10

11. With certain arc and angle measures as indicated in Figure P.11, solve for x, y, and z.

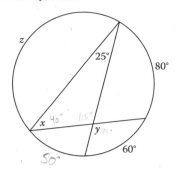

Figure P.11

12. If in Figure P.12 $\overline{MT} \perp \overline{RW}$; $VN = VM = VW = 3$; $RN = 1$, and $m\widehat{MW} = 75$, find

(a) VT (b) NS

(c) $m\widehat{ST}$ (d) $m\widehat{RS}$

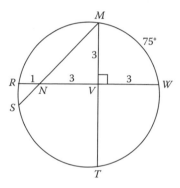

Figure P.12

13. Circle $[P, 10]$ passes through the center of another circle $[Q, r]$, $r > 10$, cutting off an arc of measure 90 on the larger circle. Find r (Figure P.13).

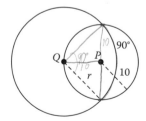

Figure P.13

14. In Figure P.14 for this problem, the center of one circle lies on another circle.

(a) What is the relationship between the two minor arcs A_1 and A_2 determined by the points of intersection A and B of the two circles?

(b) Find $m\widehat{AB}$ if the circles have equal radii.

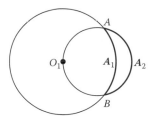

Figure P.14

15.   Two circles intersect at A and B. If $m\overset{\frown}{AB} = 40$ on the larger circle and $m\overset{\frown}{AB} = 120$ on the smaller, and if the radius of the larger circle is 6, find the radius of the smaller circle (Figure P.15). (Use trigonometry and facts about circles and arc measure.)

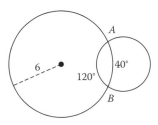

Figure P.15

16.*   **How to Construct the Radius of a Circle**   Suppose it is required to construct the radius of a circle whose center is unavailable. An interesting construction not involving the center follows (illustrated in Figure P.16).
1. Draw any chord AB and construct equilateral triangle ABC interior to the circle.
2. Draw chord BD, whose length equals that of AB.
3. Draw line DC, meeting the circle at E. Segment CE is the desired radius.
Prove this construction by justifying the measurements shown in the figure, and then obtaining the following results:
(a) $m\angle ECW = m\angle UCD = \frac{1}{2}\left(m\overset{\frown}{EW} + m\overset{\frown}{UD}\right)$
Therefore $m\angle UCD = 90 - \frac{1}{2}\varphi + \omega$
(b) $\theta = m\angle EDB = \frac{1}{2}(\varphi + 2\omega) = \frac{1}{2}\varphi + \omega$; $\theta = m\angle BCD = 60 - \frac{1}{2}\varphi + \omega$. Deduce $\varphi = 60$. (This unusual result can be checked using *Sketchpad*.)
(c) $m\angle ECO = m\angle EOC$ (from Figure P.16c)

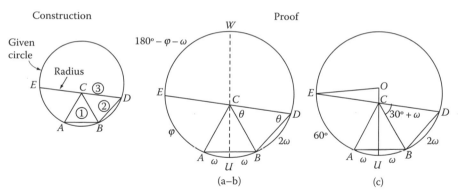

Figure P.16

17. What does the nine-point circle look like for an isosceles triangle? An equilateral triangle? Sketch figures to illustrate.

18. Draw a large scalene triangle (no two sides equal) having an obtuse angle, then sketch its nine-point circle as accurately as possible. Examine the location of U (center of the nine-point circle) relative to the Euler line of the triangle (line of H, G, and O). For more accurate results, carry out this operation as an experiment on *Sketchpad*. Do you observe any significant properties of the nine-point center?

19. **Peculiar Property of Regular Pentagons** If a is a side of any regular pentagon and b the length of a diagonal, show that (Figure P.19)

$$\frac{b}{a} = \frac{a}{b} + 1$$

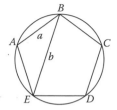

Figure P.19

20. Given any triangle ABC, prove that the power of A with respect to the circle passing through B, C, and G (centroid) equals $\frac{1}{3}(a^2 + b^2 + c^2)$. [**Hint:** Use one of the alternate forms for Power A; see Figure P.20.]

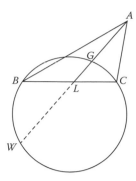

Figure P.20

21.   Two circles C_1 and C_2 are tangent at T, and the common tangent line l is drawn, as shown in Figure P.21. If P is any point on l, show that Power (circle C_1) = Power (circle C_2). [There is a very simple one-line proof.]

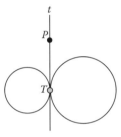

Figure P.21

Group C

22.   Altitude AD of $\triangle ABC$ is extended to point D' on the circumcircle of $\triangle ABC$ (Figure P.22). Prove that D is the midpoint of segment HD'. This result will be used later. [**Hint:** Show that $\angle DAC \cong \angle EBC \cong \angle DBD'$.]

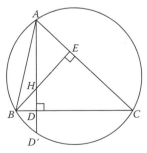

Figure P.22

23.   Diameter AL' of the circumcircle of $\triangle ABC$ is drawn, as shown in Figure P.23. Prove:

(a) ◊*HCL'B* is a parallelogram. (Why are angles ∠*ACL'* and ∠*ABL'* right angles?)

(b) The midpoint *L* of side *BC* is also the midpoint of segment *HL'*. (This result will be used later.)

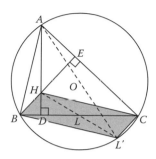

Figure P.23

24. **A Formula for the Area of a Cyclic Quadrilateral** Prove each of the following relations leading to a formula that gives the area *K* of a cyclic quadrilateral in terms of its sides.

(a) $K = $ Area $\triangle ABC + $ Area $\triangle ADC = \frac{1}{2}ab\sin\theta + \frac{1}{2}cd\sin\varphi$. (See Figure P.24.)

(b) $16K^2 = 4(ab + cd)^2\sin^2\theta = 4(ab + cd)^2 - 4(ab + cd)^2 \cdot \cos^2\theta$

(c) $m^2 = a^2 + b^2 - 2ab\cos\theta$ in $\triangle ABD$. A similar expression is derived from $\triangle BDC$.

(d) By subtracting the two equations in **(c)**, one obtains

$$4(ab + cd)^2\cos^2\theta = (a^2 + b^2 - c^2 - d^2)^2$$

Using **(b)**, the final result is

(2) $16K^2 = 4(ab + cd)^2 - (a^2 + b^2 - c^2 - d^2)^2$

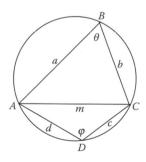

Figure P.24

25. **Brahmagupta's Formula for Area of a Cyclic Quadrilateral** Brahmagupta, who invented negative numbers, was an outstanding mathematician appearing in India around 600 AD. His formulas for quadrilaterals were amazing discoveries for that period. Starting with **(2)** from Problem 24 and using repeated factoring of the form $X^2 - Y^2 = (X + Y)(X - Y)$, rearranging terms when warranted, the result is the relation.

(3) $K = \sqrt{(s-a)(s-b)(s-c)(s-d)}$

where s is the semi-perimeter of the quadrilateral: $s = \frac{1}{2}$ $(a + b + c + d)$. Carry out these details. Note that the special case $d = 0$ yields Heron's formula for a triangle.

26.   **Bretschneider's Formula for the Area of a Cyclic Quadrilateral** The following interesting formula for the area of a cyclic quadrilateral due to Bretschneider will be introduced here without proof. Its form is peculiarly dissimilar to that of (2). Test this relation and Brahmagupta's formula, for the area of the cyclic quadrilateral $ABCD$ shown in Figure P.26 (use K = Area $\triangle ABC$ + Area $\triangle ADC$).

(4) $16K^2 = 4(ac + bd)^2 - (a^2 - b^2 + c^2 - d^2)^2$

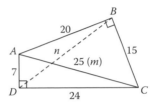

Figure P.26

7.5 Cross Ratio and Harmonic Conjugates

A very prominent and useful concept in geometry, in both theory and applications, is that of the *cross ratio*. The term is suggested by one of its definitions: Given four collinear points A, B, C, and D (Figure 7.26), consider the value

$$\frac{\dfrac{AC}{AD}}{\dfrac{BC}{BD}} = \frac{AC/AD}{BC/BD}$$

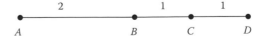

Figure 7.26 Numerical example.

which is a *ratio of ratios* to be termed the **cross ratio**. In the example shown in Figure 7.26, this value is $\frac{3}{4}/\frac{1}{2} = 3/2$; its value for distinct points is almost arbitrary in general. One of the reasons this concept is so important and useful is that, unlike ordinary ratio, it is invariant under arbitrary linear transformations (the topic of the next chapter). A more convenient form is provided in our official definition for it, using directed distance.

Definition 1   Given four distinct collinear points, their **cross ratio** is defined to be the value

$$[AB,CD] \equiv \frac{AC \cdot BD}{AD \cdot BC}$$

Note that the cross ratio is *order dependent*: the order in which the points are taken on the line affects its value. For example, since from the definition, $[AB, DC] = AD \cdot BC/AC \cdot BD$, we observe that this gives the *reciprocal* of the expression in the definition, $[AB, CD]$:

$$[AB,DC] = \frac{1}{[AB,CD]}$$

One might suppose that since there are 24 ways to order the letters A, B, C, and D within the symbol $[AB, CD]$, there should generally be 24 distinct values for the cross ratio of the same four points. But there are only six. The first theorem on cross ratio confirms this, the proof being left for problems (see Problems 6 and 7 that follow).

Theorem 1   Let A, B, C, and D be any four distinct collinear points. Then the 24 permutations of A, B, C, and D result in at most six distinct values for the cross ratio of these four points, as given by the following list, where $\lambda = [AB, CD] \neq 0, 1$:

$$\left\{ \lambda, 1/\lambda, 1-\lambda, 1-1/\lambda, 1/(1-\lambda), \lambda/(\lambda-1) \right\}$$

Corollary   If points A, B, C, and D can be located on a line such that there are fewer than six cross ratios possible, there are then precisely only three values possible, namely -1, $\frac{1}{2}$, and 2.

Example 1

Find the cross ratio $[AB, CD]$ in each of the following cases:

(a) When the points are defined by the coordinates $A[3]$, $B[11]$, $C[8]$, and $D[13]$ on line l, as shown in Figure 7.27.

(b) When the points are $A[a]$, $B[b]$, $C[0]$, and $D[x]$.

(c) Find the value of x in (b) for which the cross ratio is -1.

Figure 7.27 Example 1.

Solution

(a) $[AB,CD] = \dfrac{AC \cdot BD}{AD \cdot BC} = \dfrac{5 \cdot 2}{10 \cdot (-3)} = -\dfrac{1}{3}$

(b) $[AB,CD] = \dfrac{AC \cdot BD}{AD \cdot BC} = \dfrac{(0-a)(x-b)}{(x-a)(0-b)} = \dfrac{a}{b} \cdot \dfrac{x-b}{x-a}$

(c) Directly from the result in **(b)**, if $[AB, CD] = -1$ we obtain

$$a(x-b) = -b(x-a) \quad \text{or} \quad (a+b)x = 2ab$$

Therefore

$$x = \dfrac{2ab}{a+b} = \dfrac{2}{1/a + 1/b}$$

An Experiment in Geometry

In **(c)** of the above example, let $a = 2$ and $b = 6$. Then $x = 3$ and it follows that the four points become $A[2]$, $B[6]$, $C[0]$, and $D[3]$, with $[AB, CD] = -1$. On a sheet of paper, or using *Sketchpad*, draw a line and carefully measure off four such points on that line, using any unit of measure, as in Figure 7.28. Let P be any point in the plane (or on your sheet of paper); draw segments joining P with C, A, and D. Locate Q any point on segment PD. Now draw CQ meeting PA at R, then draw ray DR meeting PC at S. Finally, draw line SQ. Did anything happen? Try this experiment several times with various positions for point P, and also, with A, D and B playing the roles of C, A, and D. *Sketchpad* provides a dynamic demonstration (see Experiment 14, Appendix A).

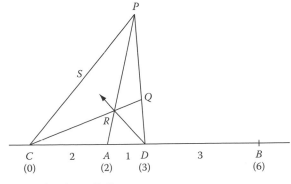

Figure 7.28 Drawing for self discovery.

The fundamental properties of cross ratio will now be established. In all that follows let A, B, C, D, and X represent distinct collinear points.

(1) $$[AB, CD] = [CD, AB]$$

(2) $$[AB, DC] = 1/[AB, CD]$$

(3) $$[AC, BD] = 1 - [AB, CD]$$

(4) If $[AB, CX] = [AB, CD]$ then $X = D$

Proofs For **(1)** write

$$[CD, AB] = \frac{CA \cdot DB}{CB \cdot DA} = \frac{(-AC) \cdot (-BD)}{(-AD) \cdot (-BC)} = \frac{AC \cdot BD}{AD \cdot BC} = [AB, CD]$$

The relation **(2)** was already observed (above). Next, for **(3)**, we use Euler's identity **(1)** of Section **7.1**, with D replacing P throughout:

$$AB \cdot DC + BC \cdot DA + CA \cdot DB = 0$$

Divide through by the term $AD \cdot CB$ to obtain

$$-\frac{AB \cdot CD}{AD \cdot CB} + \frac{CB \cdot AD}{AD \cdot CB} - \frac{AC \cdot BD}{AD \cdot BC} = 0 \quad \text{or}$$

$$-[AC, BD] + 1 - [AB, CD] = 0$$

which is equivalent to **(3)**. Finally, for **(4)**, suppose that $[AB, CX] = [AB, CD]$, or

$$\frac{AC \cdot BX}{AX \cdot BC} = \frac{AC \cdot BD}{AD \cdot BC} \quad \rightarrow \quad \frac{BX}{AX} = \frac{BD}{AD}$$

Thus we have

$$\frac{BA + AX}{AX} = \frac{BA + AD}{AD} \quad \text{or} \quad \frac{BA}{AX} + 1 = \frac{BA}{AD} + 1$$

It follows that $AX = AD$, which implies $X = D$. ∖

Note: The results **(1)** and **(2)** provide the motivation for separating the pairs (A, B) and (C, D) in the notation for cross ratio. ∖

As indicated earlier, the value −1 for cross ratio stands out as a unique special case. This situation occurs whenever points C and D are located on line AB in such a way that (a) C is an interior point of segment AB, (b) D is an exterior point, and (c) the ratios AC/CB and AD/BD are equal, as shown in Figure 7.29. In traditional terms, points C and D are said to divide segment AB *internally and externally in the same ratio*. Indeed, from (c) we obtain

$$\frac{AC}{BC} = -\frac{AD}{BD} \quad \text{or} \quad \frac{AC}{BC} \cdot \frac{BD}{AD} = -1 \quad \text{or} \quad [AB, CD] = -1$$

Definition 2   If the cross ratio of four collinear points A, B, C, and D equals -1, then C and D are said to be **harmonic conjugates** of each other with respect to A and B.

Since $[AB, CD] = [CD, AB]$, it follows from the definition that if C and D are harmonic conjugates with respect to A and B, then A and B are harmonic conjugates with respect to C and D. Figure 7.29a shows the remarkable 4-step straight-edge construction for D, the harmonic conjugate of C with respect to A and B, given A, B, and $C \neq$ midpoint of AB, which is independent of P or Q on line BP. Thus regardless of the choice of points P and Q, the same point D will always be determined. The configuration $ABCDPQRS$ in the figure will be referred to as the **harmonic construction**.

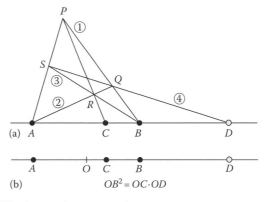

Figure 7.29 The harmonic construction.

Figure 7.29b illustrates an analytic characterization of harmonic conjugates, where O is the midpoint of AB. This result will be used quite frequently, established as the next theorem.

Theorem 2   Consider collinear points A, B, C, and D, with O the midpoint of segment AB, as shown in Figure 7.29b. Then $[AB, CD] = -1$ iff

$$OC \cdot OD = OB^2$$

Proof   The relation $[AB, CD] = -1$ implies, by definition, $AC \cdot BD = -AD \cdot BC$ or

$$AC \cdot BD = AD \cdot CB$$

In this expression, insert O between the pairs (A, C), (B, D), (A, D), and (C, B) to obtain

$$(AO + OC)(BO + OD) = (AO + OD)(CO + OB)$$

$$AO \cdot BO + AO \cdot OD + OC \cdot BO + OC \cdot OD$$

$$= AO \cdot CO + AO \cdot OB + OD \cdot CO + OD \cdot OB$$

Since O is the midpoint of AB, $AO = OB$. Using this, we obtain

$$-OB^2 + AO \cdot OD + (CO \cdot AO) + OC \cdot OD$$

$$= (AO \cdot CO) + OB^2 - OD \cdot OC + OD \cdot AO$$

with the terms in boldface and in boldface parentheses on the two sides of the equation canceling, reducing the expression to

$$2OC \cdot OD = 2OB^2$$

as desired. Since the steps are reversible, the converse holds. ◨

Still another way to characterize harmonic conjugates explains the terminology. Recall that an arithmetic sequence is of the form a, $a + d$, $a + 2d$, $a + 3d, \ldots$, such as the positive integers $1, 2, 3, \ldots$ ($a = d = 1$). A **harmonic sequence** is the sequence of *reciprocals* of an arithmetic sequence, such as $1, \frac{1}{2}, \frac{1}{3}, \ldots$ (in this case, the terms define the well-known *harmonic series* $1 + \frac{1}{2} + \frac{1}{3} + \ldots$). The **harmonic mean** of two numbers a and b is that number c such that a, c, and b is a *harmonic sequence*, that is, $1/a$, $1/c$, $1/b$ is an *arithmetic sequence*. Thus we must have the common difference d equal to

$$1/c - 1/a = 1/b - 1/c.$$

Solving for $1/c$, we obtain the defining relation for the harmonic mean c of a and b.

$$\frac{1}{c} = \frac{1}{2}\left(\frac{1}{a} + \frac{1}{b}\right)$$

Theorem 3   Points C and D are harmonic conjugates of A and B iff AB is the harmonic mean of AC and AD, that is, iff

$$\frac{1}{AB} = \frac{1}{2}\left(\frac{1}{AC} + \frac{1}{AD}\right)$$

Proof   Begin with the relation $[AB, CD] = -1$. Then $AC \cdot BD = -AD \cdot BC = AD \cdot CB$. Also, by **(3)**, $[AC, BD] = 1 - [AB, CD] = 2$. Thus,

$$\frac{AB \cdot CD}{AD \cdot CB} = 2 \quad \text{or} \quad \frac{CD}{AD \cdot CB} = \frac{2}{AB}$$

That is,

$$\frac{2}{AB} = \frac{CB + BD}{AD \cdot CB} = \frac{CB}{AD \cdot CB} + \frac{BD}{AC \cdot BD} = \frac{1}{AD} + \frac{1}{AC}$$

as desired. For the converse, the steps are reversible back to the relation $[AC, BD] = 2$. Therefore, $[AB, CD] = 1 - [AC, BD] = 1 - 2 = -1$. ◨

Problems (Section 7.5)

Group A

1. Find the cross ratios $[KW, TQ]$ and $[WK, TQ]$ of the points $K[3]$, $W[8]$, $T[13]$, and $Q[4]$ as indicated by their coordinates on line l, and verify the relation $[KW, TQ] = [WK, TQ]^{-1}$.

2. Find the cross ratios $[KW, TQ]$ and $[KT, WQ]$ of the points given in Problem 1 and verify the relation $[KT, WQ] = 1 - [KW, TQ]$.

3. What coordinate x will guarantee that $D[x]$ is the harmonic conjugate of $C[4]$ with respect to $A[2]$ and $B[7]$?

4.* If $C[x]$ and $D[y]$ are harmonic conjugates with respect to the pair $A[3]$ and $B[7]$, find a relation between x and y that must hold true.

5. Using the relation you obtained in Problem 4, what happens to x as $y \to \infty$? What geometric fact concerning C must be true in this situation? [**Hint:** Divide the relation mentioned by y throughout, then take the limit.]

Group B

6. Show that the cross ratios of the 24 permutations of A, B, C, and D can be arranged in the following six groups such that the cross ratio in each group is the same, thereby proving Theorem 1:
   (a) $[AB, CD] = [CD, AB] = [DC, BA] = [BA, DC] = \lambda$
   (b) $[AB, DC] = [DC, AB] = [CD, BA] = [BA, CD] = 1/\lambda$
   (c) $[AC, BD] = [BD, AC] = [DB, CA] = [CA, DB] = 1 - \lambda$
   (d) $[AC, DB] = [DB, AC] = [BD, CA] = [CA, BD] = 1/(1 - \lambda)$
   (e) $[AD, CB] = [CB, AD] = [BC, DA] = [DA, BC] = 1 - 1/(1 - \lambda) \equiv \lambda/(\lambda - 1)$
   (f) $[AD, BC] = [BC, AD] = [CB, DA] = [DA, CB] = (\lambda - 1)/\lambda \equiv 1 - 1/\lambda$

7. Prove the following important properties of cross ratio:
   (a) The cross ratio as defined can never equal 0 or 1.
   (b) The set of six possible values for cross ratio has the property of closure under the operations: (1) taking the reciprocal of, and (2) subtracting from 1.

8.* Prove the corollary in this section.

9. Another construction for the harmonic conjugate D given A, B, and C is given by the following (see Figure P.9). You are to show that $[AB, CD] = -1$.

1. Locate the midpoint O of segment AB.
2. Double segment CO to segment CE, making O the midpoint of CE.
3. Construct segment OF congruent to OB and perpendicular to line AB.
4. Construct the perpendicular bisector of EF and determine its intersection G with line AB.
5. The circle $[G, r]$ where $r = GE$ intersects line AB at D.

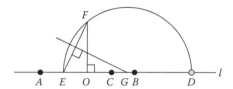

Figure P.9

10. **Line Construction for Tangents to a Circle** Use *Sketchpad* to determine the accuracy of the construction of the tangents from P as shown in Figure P.10. It consists of drawing 7 lines, the first two at random. (A geometric proof involves methods of projective geometry.)

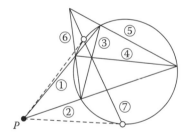

Figure P.10

7.6 The Theorems of Ceva and Menelaus

Many of the results in modern geometry cannot be adequately developed or conveniently proved without the two companion theorems to be introduced here. Curiously, although they are the duals of each other, their discoveries were separated by more than a thousand years. The theorem bearing the name Ceva (from which the term "cevian" originates) was discovered by an Italian mathematician Giovanni Ceva (1647–1736), who noted that it complemented the theorem proved by the Greek astronomer Menelaus of Alexandria discovered around 100 AD. It has bearing on the following problem: Suppose you have a triangle ABC and cevians AD, BE, and CF are drawn as in Figure 7.30.

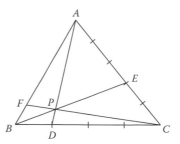

Figure 7.30 Ratios and cevians.

The problem is to determine the ratio AF/FB that guarantees that the three cevians all pass through a common point P. Can you guess the answer? It turns out to be 9/2—a result of Ceva's theorem to be proven momentarily, and difficult to determine without it.

A well-known formula for the area of a *cevian triangle*—the triangle in the interior of another triangle that is determined by its cevians—has bearing on this question. Suppose cevians AD, BE, and CF of $\triangle ABC$ intersect at points P, Q, and R, and suppose $p_1 = BD/BC$, $q_1 = DC/BC$, $p_2 = CE/CA$, $q_2 = EA/CA$, $p_3 = AF/AB$, and $q_3 = FB/AB$. (See Figure 7.31, where the ratios p_i and q_i are marked on each side for convenience.) An intricate formula for the area of $\triangle PQR$ in terms of K, the area of $\triangle ABC$, has been established (stated here without proof):

(1)
$$K' = \frac{(p_1 p_2 p_3 - q_1 q_2 q_3)^2 K}{(p_1 + q_1 q_2)(p_2 + q_2 q_3)(p_3 + q_3 q_1)}$$

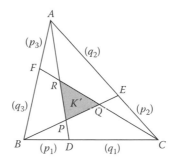

Figure 7.31 A cevian triangle.

For example, if $p_1 = p_2 = p_3 = \frac{1}{3}$ and $q_1 = q_2 = q_3 = \frac{2}{3}$ are the ratios that determine cevian triangle PQR (sometimes referred to as an *aliquot triangle*), then K' turns out to have the value $\frac{1}{7}K$. You might like to experiment with this formula for a few cases using *Sketchpad*. For example, what value for K' results if $p_i = \frac{1}{4}$, for $i = 1, 2, 3$? (See Problem 7.) What happens to this formula if $p_i = \frac{1}{2}$ for $i = 1, 2, 3$, when the cevians are the medians of $\triangle ABC$?

It is evident that $K' = 0$ iff the cevians are concurrent, that is, $p_1 p_2 p_3 = q_1 q_2 q_3$, or

(2)
$$\frac{p_1 p_2 p_3}{q_1 q_2 q_3} = 1$$

Thus is involved the following *ratio product*

(3)
$$\left[\begin{matrix} ABC \\ DEF \end{matrix}\right] = \frac{AF}{FB} \cdot \frac{BD}{DC} \cdot \frac{CE}{EA}$$

which is termed the **linearity number** of D, E, and F with respect to A, B, and C (undefined unless D, E, and F are distinct from A, B, and C). The following diagram provides a useful mnemonic device for the ratios in **(3)**.

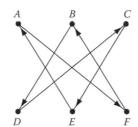

Diagram for the linearity number

The linearity number has two very useful basic properties, easily proven from its definition: Any permutation of the columns in the symbol for the linearity number results in either the original number or its reciprocal. For example:

(4)
$$\left[\begin{matrix} ACB \\ DFE \end{matrix}\right] = \frac{AE}{EC} \cdot \frac{CD}{DB} \cdot \frac{BF}{FA} = \left[\begin{matrix} ABC \\ DEF \end{matrix}\right]^{-1}$$

Linearity numbers over the same five points, in the same order, are equal iff the remaining points are the same. For example:

(5)
$$\left[\begin{matrix} ABC \\ DEX \end{matrix}\right] = \left[\begin{matrix} ABC \\ DEF \end{matrix}\right] \text{ iff } X = F$$

(The proof of **(5)** is to be completed as Problem 9.)

Example 1

An equilateral triangle is shown in Figure 7.32 at right, having sides of length 20 units. Certain segments on those sides are indicated, with their lengths, and cevians are drawn to the interior points thus determined. If the linearity number **(2)** is calculated, the result is

$$\frac{16}{4} \cdot \frac{5}{15} \cdot \frac{9}{11} = \frac{12}{11} \neq 1$$

so we conclude from **(2)** that the cevians are not concurrent. (Note that the figure, although accurately drawn, makes it too close to call without actual calculations!)

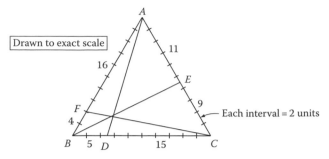

Figure 7.32 Example 4: Are these cevians concurrent?

Example 2

Suppose that cevians *AD*, *BE*, and *CF* of △*ABC* are parallel.

(a) Use parallel projection to show that

$$\left[\frac{ABC}{DEF}\right] = 1$$

(b) Conversely, show that if the above linearity number is 1 and *AD*||*BE*, then the third cevian *CF* is parallel to *AD* and *BE*.

Solution

(a) As shown in Figure 7.33 (only one of three cases depending on betweenness relations), by parallel projection the point triples map as follows, in the order given:

$$BDC \leftrightarrow EAC \leftrightarrow BAF$$

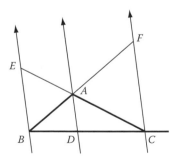

Figure 7.33 Example 2.

[This remains valid whether (*BCD*), (*CBD*), or (*BDC*) holds.] The ratio- and betweenness-preserving property of parallel projection then implies in all cases

$$\left[\frac{ABC}{DEF}\right] = \frac{AF}{FB} \cdot \frac{BD}{DC} \cdot \frac{CE}{EA} = \frac{DC}{CB} \cdot \frac{BD}{DC} \cdot \frac{CB}{BD} = 1$$

(b) By parallel projection we have $BD/DC = EA/AC$. By algebra, the product of the ratios in the linearity number becomes

$$1 = \frac{AF}{FB} \cdot \frac{EA}{AC} \cdot \frac{CE}{EA} = \frac{AF}{FB} \cdot \frac{CE}{AC}$$

or $AF/FB = AC/CE$ (equivalent to $BA/AF = EA/AC$). Therefore, $\overleftrightarrow{CF} \parallel \overleftrightarrow{AD}$.

As it turns out, it is more convenient to prove Menelaus' theorem first, then to prove Ceva's theorem as a corollary. Here is Menelaus' theorem.

> **Menelaus' Theorem** If points D, E, and F lie on the extended sides of $\triangle ABC$ opposite A, B, and C, respectively, these points are collinear iff
>
> $$\begin{bmatrix} ABC \\ DEF \end{bmatrix} = -1$$

Proof The proof of the necessity of the condition (the linearity number = −1) proceeds in two parts (we assume that D, E, and F lie on some line l).

1. First the linearity number is shown to be negative. This follows from the fact that either all three of the points D, E, and F lie exterior to sides BC, AC, and AB, or exactly one of them does so. In the first case, all three ratios in the linearity number are negative, and the product must be negative [recall that $AF/FB > 0$ iff (AFB)]. If one of the points lies in the interior of a side, then by the Postulate of Pasch line l meets precisely one other side of $\triangle ABC$ at an interior point (Figure 7.34a); hence this time we have exactly two of the ratios in the linearity number positive, the other negative, and again the linearity number is negative.

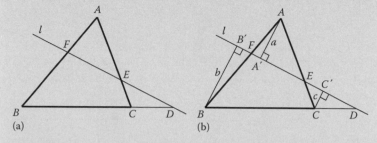

Figure 7.34 Proof that linearity number is −1.

2. Next we show that the linearity number is ±1 (by Part (1) we obtain −1). Here we can ignore signed distance and the ordering of collinear points. Let A', B', and C' be the feet of the perpendiculars from A, B, and C to line l (Figure 7.34b), and let $AA' = a$, $BB' = b$, and $CC' = c$. Then by similar triangles we have

$$\begin{bmatrix} ABC \\ DEF \end{bmatrix} = \frac{AF}{FB} \cdot \frac{BD}{DC} \cdot \frac{CE}{EA} = \pm \frac{a}{b} \cdot \frac{b}{c} \cdot \frac{c}{a} = \pm 1$$

The proof of sufficiency (if the linearity number equals −1, then D, E, and F are collinear) proceeds as follows. First, we must show that line DE cannot be parallel to AB. If it were (Figure 7.35) then by parallel projection, $BD/DC = AE/EC$. That is,

$$\frac{BD}{DC} \cdot \frac{CE}{EA} = 1$$

Since the linearity number equals −1, we obtain

$$\frac{AF}{FB} = -1 \rightarrow AF = -FB = BF$$

or $A = B$, a contradiction. Therefore, DE meets AB at some point X. Then D, E, and X are collinear, and by the first part, and by hypothesis,

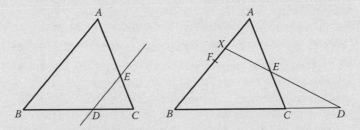

Figure 7.35 Proof of sufficiency.

$$\begin{bmatrix} ABC \\ DEX \end{bmatrix} = -1 = \begin{bmatrix} ABC \\ DEF \end{bmatrix}$$

By property **(5)** of linearity numbers, $X = F$ and D, E, and F are collinear. ◣

Ceva's Theorem   If points D, E, and F lie on the extended sides of $\triangle ABC$ opposite A, B, and C, respectively, then cevians AD, BE, and CF are either parallel, or concurrent, iff

$$\begin{bmatrix} ABC \\ DEF \end{bmatrix} = 1$$

Proof   The case for parallel cevians has already been take care of in Example 1 above, so it remains to be proven that the cevians are concurrent iff the above linearity number equals 1 and no two cevians are parallel. First, assume that the cevians are concurrent at point P (Figure 7.36).

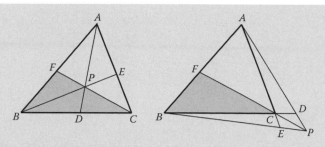

Figure 7.36 Proof that linearity number is 1.

A series of four linearity numbers involving points on the sides of various triangles in the figure can be written down, and because of the collinearity of certain points, each linearity number equals −1 by Menelaus' theorem. For example, in $\triangle FBC$ (shaded in the figure) points A, D, and P, which lie on the sides FB, BC, and FC (extended), are collinear, and by Menelaus' theorem

$$\begin{bmatrix} FBC \\ DPA \end{bmatrix} = -1$$

Since there are an even number of these (four of them), their product is unity. Thus, one obtains

$$\begin{bmatrix} FBC \\ DPA \end{bmatrix} \cdot \begin{bmatrix} PCE \\ ABF \end{bmatrix} \cdot \begin{bmatrix} APE \\ BCD \end{bmatrix} \cdot \begin{bmatrix} AFP \\ CDB \end{bmatrix} = 1$$

and a product of 12 ratios follows. (The reader should verify the validity of the other three linearity numbers in the product.) Most of the terms cancel, such as those in boldface:

$$\left(\frac{FA}{AB} \cdot \frac{BD}{DC} \cdot \frac{CP}{\textbf{\textit{PF}}} \right) \left(\frac{\textbf{\textit{PF}}}{FC} \cdot \frac{CA}{AE} \cdot \frac{EB}{BP} \right) \left(\frac{AD}{DP} \cdot \frac{PB}{BE} \cdot \frac{EC}{CA} \right) \left(\frac{AB}{BF} \cdot \frac{FC}{CP} \cdot \frac{PD}{DA} \right) = 1$$

leaving

$$\frac{FA \cdot BD \cdot EC}{DC \cdot AE \cdot BF} = 1$$

which is equivalent to

$$\frac{AF}{FB} \cdot \frac{BD}{DC} \cdot \frac{CE}{EA} = \begin{bmatrix} ABC \\ DEF \end{bmatrix} = 1$$

Conversely, suppose the linearity number equals 1. Since no two cevians are parallel, let cevians AD and BE meet at P. It follows that line PC (as shown in Figure 7.37a) cannot be parallel to AB using a similar argument that was used in the proof of Menelaus' theorem ($BD/DC = -AB/PC = -EA/CE$ and $AF/FB = -1$), so that PC meets AB at some point X, as in Figure 7.37b. Thus, by the first part just proven and by hypothesis,

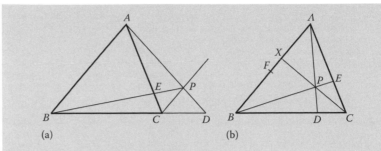

Figure 7.37 Proof of sufficiency.

$$\left[\frac{ABC}{DEX}\right] = 1 = \left[\frac{ABC}{DEF}\right]$$

and $X = F$ by property **(5)**. Therefore, the cevian $CX \equiv CF$ passes through P. ◥

Applications of the theorems of Ceva and Menelaus are prolific, a few examples of which will be included here. A simple proof of the existence of the centroid of a triangle follows immediately from Ceva's theorem: If AL, BM, and CN are the medians of $\triangle ABC$, then since $BL = LC$, $CM = MA$, and $AN = NB$, the linearity number for L, M, and N equals 1 and by Ceva's theorem, the medians are concurrent. A similar analysis for the altitudes is possible, although somewhat more involved. (See Problem 11.) For the angle bisectors AD, BE, and CF, recall the elementary theorem that implies $BD/DC = AB/AC = c/b$. Applied to all three angle bisectors, we obtain

$$\frac{AF}{FB} = \frac{b}{a}, \quad \frac{BD}{DC} = \frac{c}{b}, \quad \frac{CE}{EA} = \frac{a}{c} \rightarrow \left[\frac{ABC}{DEF}\right] = \frac{b}{a} \cdot \frac{c}{b} \cdot \frac{a}{c} = 1$$

By Ceva's theorem, the angle bisectors are concurrent.

Example 3

Figure 7.38a shows parallel lines SQ and AB, with segments AQ and BS intersecting at point R. Finally, line PR is drawn, meeting segment AB at point C. Prove that C is the midpoint of segment AB. (Thus the figure suggests that as the harmonic conjugate of C with respect to A and B recedes to infinity, C approaches the midpoint of AB as limit.)

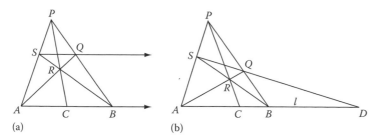

Figure 7.38 Example 3.

Solution

By Ceva's theorem, since *AQ*, *BS*, and *PC* are concurrent

(*)
$$1 = \begin{bmatrix} PAB \\ CQS \end{bmatrix} = \frac{PS}{SA} \cdot \frac{AC}{CB} \cdot \frac{BQ}{QP}$$

Since line *SQ* is parallel to line *AB*, *PS/SA* = *PQ/QB*, which when combined with (*) yields *AC* = *CB*.

Example 4: Harmonic Construction Justified

Figure 7.38b shows the harmonic construction for the harmonic relation involving *A*, *B*, *C*, and *D*. Recall that point *P* is any point not on line *l* and *Q* is any point on line *PB*; *R* and *S* are determined as shown in the figure. Using the theorems of Ceva and Menelaus, show that [*AB, CD*] = −1, and conversely, if [*AB, CD*] = −1 then the point *D* is determined by the harmonic construction in the figure (i.e., line *SQ* will meet line *l* at *D*). Note that this shows that *P* and *Q* can be chosen arbitrarily.

Solution

Since cevians *PC*, *AQ*, and *BS* of △*PAB* are concurrent at *R*, by Ceva's theorem

$$\begin{bmatrix} PAB \\ CQS \end{bmatrix} = 1$$

Collinear points *D*, *Q*, and *S* lie on the sides of △*PAB*, so by Menelaus' theorem

$$\begin{bmatrix} PBA \\ DSQ \end{bmatrix} = -1$$

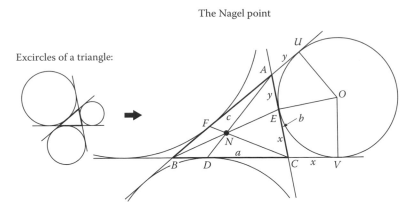

Figure 7.39 The Nagel point.

Therefore, the product of these two equations produces

$$\left[\frac{PAB}{CQS}\right] \cdot \left[\frac{PBA}{DSQ}\right] = \frac{PS}{SA} \cdot \frac{AC}{CB} \cdot \frac{BQ}{QP} \cdot \frac{PQ}{QB} \cdot \frac{BD}{DA} \cdot \frac{AS}{SP} = -1$$

or $\quad \dfrac{AC}{AD} \cdot \dfrac{BD}{BC} = [AB, CD] = -1$

The converse is a corollary of the relation just proved: Suppose that $[AB, CD] = -1$ and points P, Q, R, and S are constructed, as in Figure 7.39. If $SQ\|l$, then by the result in Example 2, C would be the midpoint of segment AB, which would contradict $[AB, CD] = -1$ (since otherwise it would follow that $B = D$). Therefore, let line SQ meet l at X. By the first part proved above, $[AB, CX] = -1 = [AB, CD]$. Thus $X = D$ by **(4)**, Section **7.5**, and SQ passes through D.

Example 5: The Nagel Point of a Triangle

An **excircle** of a triangle is a circle tangent externally to the sides of a triangle, of which there are three. The center of each circle lies on the intersection of the bisectors of the exterior angles. If the points of tangency are D, E, and F (opposite A, B, and C, respectively, as shown in Figure 7.40), prove:

   (a) $CE = s - a$ and $EA = s - c$. Obtain similar expressions for AF, FB, BD, and DC.

   (b) Segments AD, BE, and CF are concurrent at some point N. [The point of concurrency is called the **Nagel point** of the triangle, after C.H. Nagel (1803–1882).]

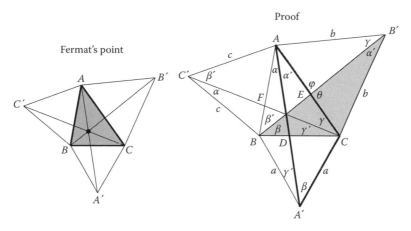

Figure 7.40 The Fermat point.

Solution

(a) Let U and V be the points of tangency of the excircle with extended sides BA and BC. With $x = CE = CV$ and $y = EA = AU$, then $x + y = AC = b$ and $c + y = BU = BV = a + x$.

Thus, we have the system of equations in x and y

$$\begin{cases} x + y = b \\ x - y = c - a \end{cases}$$

which may be easily solved. The result is $x = CE = s - a$ and $y = EA = s - c$.

It follows, by symmetry of notation, that, similar to the result just obtained, $AF = s - b$, $FB = s - a$, $BD = s - c$, and $DC = s - b$.

(b) The linearity number may then be calculated to obtain

$$\left[\frac{ABC}{DEF}\right] = \frac{AF}{FB} \cdot \frac{BD}{DC} \cdot \frac{CE}{EA} = \frac{s-b}{s-a} \cdot \frac{s-c}{s-b} \cdot \frac{s-a}{s-c} = 1$$

and by Ceva's theorem, AD, BE, and CF are concurrent.

Example 6: Fermat's Point

Let equilateral triangles be constructed externally on the sides of a triangle ABC having angles less than 120, and let A', B', and C' be the outer vertices of these equilateral triangles (Figure 7.40). Then segments AA', BB', and CC' are concurrent. (The point of concurrency is called the **Fermat point** of $\triangle ABC$.) Use Ceva's theorem to prove.

Solution

In the shaded and boldfaced triangles we note that $AC = B'C$, $m\angle ACA' = m\angle ACB + 60 = m\angle B'CB$, and $CA' = CB$; by SAS $\triangle ACA' \cong \triangle B'CB$. We then obtain the congruent angles labeled α' and β in the figure. In a similar manner, congruent angles are obtained elsewhere as indicated by the figure. Next, we use the law of sines in $\triangle ECB'$ and $\triangle EB'A$ to obtain the ratio EC/AE:

$$\frac{CE}{b} = \frac{\sin\alpha'}{\sin\theta} \quad \text{and} \quad \frac{b}{EA} = \frac{\sin\varphi}{\sin\gamma}$$

Since $\sin\theta = \sin\varphi$, the product of the above ratios yields the relation

$$\frac{CE}{EA} = \frac{\sin\alpha'}{\sin\gamma}$$

Apply this procedure to the other two triangles on the sides of $\triangle ABC$ to obtain

$$\frac{AF}{FB} = \frac{\sin\beta'}{\sin\alpha} \quad \text{and} \quad \frac{BD}{DC} = \frac{\sin\gamma'}{\sin\beta}$$

Now take the product of the above, rearrange denominators, then apply the law of sines, in turn, to the triangles $\triangle ACC'$, $\triangle BCC'$, and $\triangle CBB'$:

$$\left[\frac{ABC}{DEF}\right] = \frac{\sin\beta'}{\sin\alpha} \cdot \frac{\sin\gamma'}{\sin\beta} \cdot \frac{\sin\alpha'}{\sin\gamma} = \frac{\sin\beta'}{\sin\gamma} \cdot \frac{\sin\gamma'}{\sin\alpha} \cdot \frac{\sin\alpha'}{\sin\beta}$$

$$= \frac{b}{c} \cdot \frac{c}{a} \cdot \frac{a}{b} = 1$$

Therefore, the cevians AD, BE and CF are concurrent.

We close the applications with a singularly interesting result that, although not discovered by Simson, traditionally bears his name; the original discovery was made by William Wallace (1768–1873), who is well-known for his infinite product relations often found in calculus texts.

Example 7: Simson Line

Given any point P on the circumcircle of a triangle ABC, let PD, PE, and PF be the perpendiculars from P to the (extended) sides BC, AC, and AB, respectively (Figure 7.41). Prove that D, E, and F are collinear. (The line of collinearity is the **Simson line** of $\triangle ABC$ corresponding to P.)

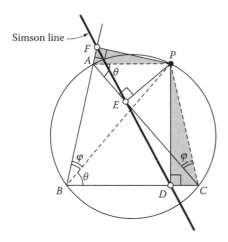

Simson line

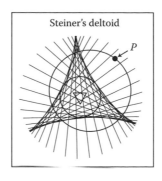

Steiner's deltoid

Figure 7.41 Simson lines of triangles.

Solution

Note that $m\angle PCB = 180 - m\angle PAB$ by a corollary of the inscribed angle theorem. Hence, $\angle PAF \cong \angle PCD$. Since the triangles are right triangles, we obtain $\triangle PAF \sim \triangle PCD$ and $AF/DC = PF/PD$. Since the angles marked θ at B and A inscribe the same arc and are congruent, $\triangle PBD \sim \triangle PAE$ and $BD/EA = PD/PE$. Similarly, the angles marked φ at B and C are congruent and $\triangle PEC \sim \triangle PFB$ and $CE/FB = -PE/PF$. (Here, we have taken PD, PE, PF, AF, DC, and EA as positive.) Thus,

$$\begin{bmatrix} ABC \\ DEF \end{bmatrix} = \frac{AF}{FB} \cdot \frac{BD}{DC} \cdot \frac{CE}{EA} = \frac{AF}{DC} \cdot \frac{BD}{EA} \cdot \frac{CE}{FB}$$

$$= \frac{PF}{PD} \cdot \frac{PD}{PE} \left(-\frac{PE}{PF} \right) = -1$$

By Menelaus' theorem, D, E, and F are collinear.

Note: A dynamic feature of the Simson line is that if P rotates about the circumcircle, as indicated in the inset of Figure 7.41, the Simson line corresponding to P also rotates, but in such a way that it envelops a *hypocycloid of three cusps*, (a *deltoid*, originally known as *Steiner's curve* in deference to this discovery). Moreover, the center of the hypocycloid is the *nine-point center* of the triangle, and the segments joining each two of the three cusp points are parallel to the sides of the *Morley triangle* (see Problem 28, Section 5.12). This remarkable theorem can be animated on *Sketchpad* without difficulty (*Sketchpad Experiment 15*, Appendix A.) This theorem is due to one of the giants of geometry, Jacob Steiner (1796–1863). ◣

Problems (Section 7.6)

Group A

1. Triangle *ABC* has sides *AB* = 13, *BC* = 12, and *AC* = 9. Cevians *AD*, *BE*, and *CF* are drawn such that *BD* = *AE* = 4 and *AF* = 8. Use Ceva's theorem to show that the cevians are concurrent (Figure P.1).

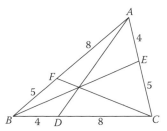

Figure P.1

2. Cevians *AD*, *BE*, and *CF* are drawn in Δ*ABC* passing through point *P* such that *BD* = 3, *DC* = 4, *CE* = 5, and *EA* = 6, as in Figure P.2. If *AB* = 14, find the indicated distances *x* and *y*.

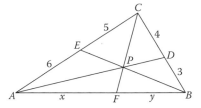

Figure P.2

3. Right triangle *ABC* has sides of length 8, 15, and 17 (a Pythagorean triple), with points *D*, *E*, and *F* on the sides, as shown in Figure P.3. If *AF* = *CD* = 12 and *AE* = *BD* = 3, show from Ceva's theorem that the cevians *AD*, *BE*, and *CF* are concurrent.

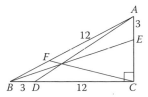

Figure P.3

4. A pennant lies at the top of a flagpole 45′ tall (*AE* in Figure P.4). The pennant is a right triangle having dimensions 5′ × 12′ × 13′ (Δ*ABC* in Figure P.4). If points *D* and *F* are located on two sides of the pennant 4′ from the outer tip, show that *D* and *F* are in line with the base of the flagpole (point *E*).

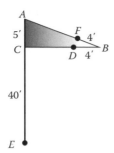

Figure P.4

5. **A Peculiar Property of the 3-4-5 Right Triangle**   Points
    E and F on the sides AC and AB, respectively of right tri-
    angle ABC having sides of length 3, 4, and 5 are such that
    $AE = AF = 1$. Show that line EF meets the third side BC
    (extended) at point D such that C is the midpoint of seg-
    ment BD (Figure P.5).

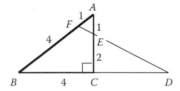

Figure P.5

Group B

6. A 1×2 rectangle $PGCH$ has adjacent sides on the legs of a
    3-4-5 right triangle ABC as shown in Figure P.6. If cevians
    AD, BE, and CF pass through P, show that
    (a) D is the midpoint of BH.   **(c)** $AF = AC = CD$
    (b) E is the midpoint of AG.   **(d)** $CF = \frac{3}{5}\ BE$

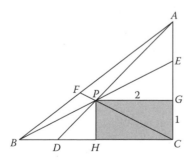

Figure P.6

7. **(a)** Verify the value $1/7K$ if $p_i = \frac{1}{3}$ for $i = 1, 2, 3$, for the area
            K' of the aliquot triangle PQR as mentioned (see for-
            mula for K' introduced above in terms of Figure 7.31).
    (b) Suppose $p_i = 1/p$ and $q_i = 1 - 1/p$ for $i = 1, 2, 3$. Find
            K' if $p = 4$ (see Figure P.7) and $p = 5$.

(c) Using algebra and factoring, derive the general formula for the aliquot triangle in terms of p:

$$K' = \frac{(p-2)^2 K}{p^2 - p + 1}$$

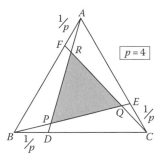

Figure P.7

8. Pythagorean triangles ABC and FDC (sides proportional to those of the 3-4-5 and 5-12-13 right triangles) are placed in the overlapping position shown in Figure P.8. The ratios CE/EA and BD/DC can be obtained directly. For FA/FB, use similar triangles ($\triangle FWD \sim \triangle ECD$ and $\triangle FWB \sim \triangle ACB$) to find x and y. Verify Menelaus' theorem by calculating the appropriate linearity number.

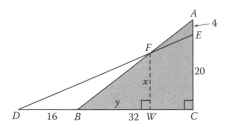

Figure P.8

9. Provide the details for proving (5) by showing that the given condition leads to $XB = FB$ or $X = F$.

10. In $\triangle KMW$, side KM is divided into 3 equal parts and KW into eight equal parts (Figure P.10). Decide whether line EF is parallel to line MW or intersects it at some point X. If the lines intersect, find MX in terms of $a = MW$ using Menelaus' theorem. If this diagram were constructed on *Sketchpad*, and point K were dragged, predict the effect that would be observed.

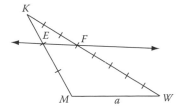

Figure P.10

11. Consider the altitudes AD, BE, and CF of an acute-angled $\triangle ABC$ (Figure P.11). Using similar triangles (e.g., $\triangle BCF \sim \triangle BAD$ which implies that $BD/BF = c/a$), prove that the altitudes of a triangle are concurrent. Include a proof covering the case when $\triangle ABC$ has an obtuse angle. Take care in not using what is to be proved.

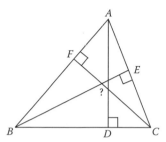

Figure P.11

12. **The Gergonne Point**   The cevians joining the vertices of a triangle with the points of contact of the incircle on the opposite sides (Figure P.12) are concurrent. Prove, using Ceva's theorem. [This point of concurrency is known as the Gergonne point of the triangle, after J.D. Gergonne (1771–1859).] [**Hint:** Show that, for example, $BD = s - b$ and $DC = s - c$, where s is the semi-perimeter.]

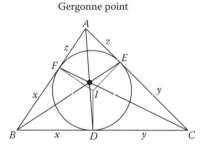

Gergonne point

Figure P.12

Group C

13. **The Theorem of Desargues**   Recall from Chapter 1 the following theorem:

> ***Theorem***   If under some correspondence between the vertices of two triangles, corresponding vertices lie on concurrent lines, then the points of intersection of corresponding sides (if they exist) are collinear.

Prove this for the special case shown in Figure P.13 by working with the linearity numbers

$$\begin{bmatrix} PAB \\ NB'A' \end{bmatrix}, \begin{bmatrix} PBC \\ LC'B' \end{bmatrix}, \begin{bmatrix} PCA \\ MA'C' \end{bmatrix}$$

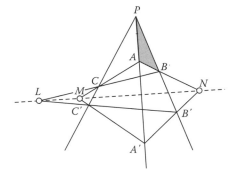

Figure P.13

14.   **The Theorem of Pappus**   The following theorem, mentioned earlier (in Problem 6, Section **1.3**), is due to the Greek astronomer, Pappus of Alexandria (340 AD):

> **Theorem**   Given two collinear point triples (A, B, C) and (A', B', C'), the points of intersection of the three pairs of crossjoins $(AB', A'B)$, $(BC', B'C)$, and $(AC', A'C)$, if they exist, are collinear (see Figure P.14).

Prove this for the special case shown in Figure P.14 by working with the linearity numbers

$$\begin{bmatrix} PQR \\ CBA \end{bmatrix}, \begin{bmatrix} PRQ \\ A'LB \end{bmatrix}, \begin{bmatrix} RQP \\ B'NC \end{bmatrix}, \begin{bmatrix} QPR \\ C'MA \end{bmatrix}, \begin{bmatrix} PQR \\ A'C'B' \end{bmatrix}$$

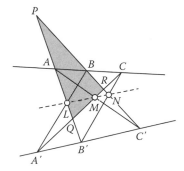

Figure P.14

15.   The following theorem establishes a type of duality between two sets of cevians of any triangle. Use Ceva's theorem and the two-secant theorem to prove it.

Theorem   Suppose a triangle and circle intersect in six distinct points, two on each side of the triangle. If the cevians to three of those points of intersection are concurrent, then so also are the cevians to the other three points of intersection concurrent (see Figure P.15).

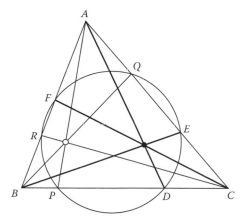

Figure P.15

7.7 Families of Mutually Orthogonal Circles

A well known concept that often appears in engineering mathematics to illustrate the dynamics of flow theory involves *confocal* conics (different conics having the same foci). In coordinate geometry, consider the equation

(1)
$$\frac{x^2}{a^2 - \lambda} + \frac{y^2}{b^2 - \lambda} = 1 \quad (0 < b < a)$$

If $\lambda < b^2$ then $\lambda < a^2$ and we see that **(1)** represents an ellipse, while if $b^2 < \lambda < a^2$ a hyperbola is obtained. By using the usual focus formula for an ellipse, we find that $c^2 = (a^2 - \lambda) - (b^2 - \lambda) = a^2 - b^2$ and $(\pm c, 0)$ are the two foci for **(1)** in the elliptic case. Thus, regardless of the value of λ, the ellipses of **(1)** all have the same two foci. For the hyperbolic case, **(1)** becomes

$$\frac{x^2}{a^2 - \lambda} - \frac{y^2}{\lambda - b^2} = 1$$

and the foci are given by $(\pm c, 0)$ where $c^2 = (a^2 - \lambda) + (\lambda - b^2) = a^2 - b^2$—the same two points as before. Thus the conics represented by **(1)** *are confocal*, regardless of the value of $\lambda < a^2$. The really amazing thing is that each ellipse in **(1)** intersects each hyperbola in **(1)** *at right angles* (see Problem 17). Thus we have two families $\mathcal{F}$ and $\mathcal{G}$ of conics that are *mutually orthogonal*. Figure 7.42 shows an example for the case $a = 5$ and $b = 3$, where the foci are $(\pm 4, 0)$.

If $a^2 = b^2$ then **(1)** represent the family of circles centered at the origin, but the above analysis breaks down since there are no longer *two* families of conics. It is interesting, however, that we can use other means to obtain two families of mutually orthogonal circles.

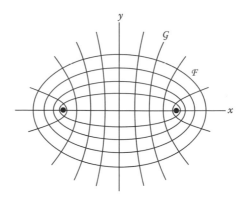

Figure 7.42 Mutually orthonic conics.

Definition 1   Two circles C and D are **orthogonal**, to be denoted by $C \perp D$, iff they intersect and their tangents at the points of intersection are perpendicular (see Figure 7.43).

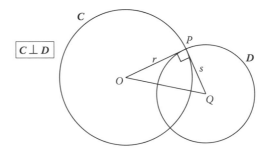

Figure 7.43 Orthogonal circles.

Note: Problem 5 asks the reader to verify that if the tangents at one point of intersection of two circles are perpendicular, then the tangents at the other point of intersection are also perpendicular. Thus the definition presented is unambiguous. ⬆

In view of the definition, if two circles $[O, r]$ and $[Q, s]$ intersect at P they are orthogonal iff any one of the following conditions hold:

(a)   The tangents of $[O, r]$ and $[Q, s]$ at P are perpendicular (definition).
(b)   The radius OP of $[O, r]$ is the tangent to $[Q, s]$ at P, and vice-versa.
(c) $\triangle OPQ$ is a right triangle, and therefore $r^2 + s^2 = OQ^2$.

It is interesting that one can use harmonic conjugates to construct orthogonal circles (other than the obvious way of constructing perpendicular radii). This will eventually lead us to our goal of producing two families of mutually orthogonal circles. Start with any circle and its diameter AB. On line AB, construct harmonic conjugates C and D (C arbitrary) using the harmonic construction, as shown in Figure 7.44. Then draw any circle passing through C and D, having its center on the perpendicular bisector of segment CD. That circle (as well as any other circle through C and D) will be orthogonal to the given circle. The proof is the content of the next theorem.

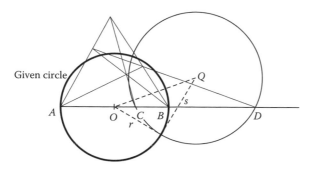

Figure 7.44 Harmonic conjugates and orthogonal circles.

Theorem 1 Suppose that $[O, r]$ is a given circle having diameter AB, and that for two points C and D on line AB $[AB, CD] = -1$. Then any circle $[Q, s]$ passing through C and D is orthogonal to $[O, r]$.

Proof (See Figure 7.44.) Since O is the midpoint of AB, by Theorem **2** in Section **7.5**, we have $OB^2 = OC \cdot OD$. That is,

$$r^2 = OC \cdot OD = \text{Power}\,O\ (\text{circle }[Q, s]) = OQ^2 - s^2$$

or

$$r^2 + s^2 = OQ^2$$

Therefore, by **(c)** above, $[O, r] \perp [Q, s]$. ◣

Thus the circles passing through C and D are each orthogonal to circle O and an entire family $\mathcal{F}$ of circles are orthogonal to circle $[O, r]$, as shown in Figure 7.45. Now one makes a critical observation. There are an infinite number of pairs of points A' and B' on line CD that are harmonic conjugates of C and D. The preceding theorem applies to any circle having $A'B'$ as diameter, and we obtain the result:

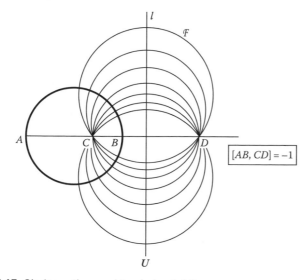

Figure 7.45 Circles orthogonal to circle at AB.

Corollary There exist two families $\mathcal{F}$ and $\mathcal{G}$ of mutually orthogonal circles such that $\mathcal{F}$ consists of all circles passing through two fixed points C and D with their centers lying on the perpendicular bisector l of segment CD, and $\mathcal{G}$ consists of all circles having diameters $A'B'$ such that A' and B' are the harmonic conjugates of C and D, with their centers lying on line CD (see Figure 7.46).

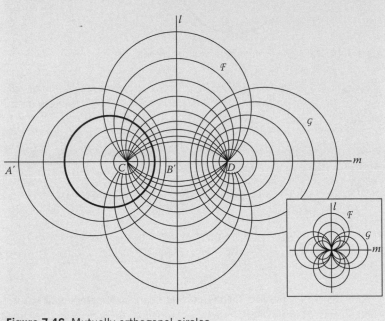

Figure 7.46 Mutually orthogonal circles.

Note: A degenerate case occurs when $C = D$ and the circles of $\mathcal{F}$ and $\mathcal{G}$ all pass through C and are tangent, respectively, to lines l and m, requiring no proof (see inset of Figure 7.46). ⬟

An examination of Figure 7.46 leads to the question of how to characterize the circles in the family $\mathcal{G}$ (those of $\mathcal{F}$ are merely all the circles passing through C and D); is there some simple way to characterize $\mathcal{G}$? It is true that the circles in $\mathcal{G}$ tend to cluster about the points C and D, but not all circles having their centers on line m and having points C (or D) in their interiors are orthogonal to the circles in $\mathcal{F}$.

The answer is found in the concept of the radical axis introduced earlier. Recall that Theorem 4 in Section 7.4 established the fact that given any two intersecting circles, all points on the line containing the common chord have equal powers with respect to the two circles. This line was termed the *radical axis* of the intersecting circles. Now suppose the two circles are gradually "pulled apart;" the question arises as to what becomes of the radical axis. It appears to gradually assume the position of the common tangent when the circles become tangent to each other, but after the circles are too far apart to intersect, the answer is not clear.

Theorem 2   The locus of points whose powers with respect to two non-concentric circles $[O, r]$ and $[Q, s]$ are equal is a line, and this line is perpendicular to the line of centers m at the unique point W on m in that locus. If the circles intersect, the locus is the line containing the common chord or the common tangent, whichever applies.

Proof   The existence of W follows by showing, by algebra, that the condition $WO^2 - r^2 = WQ^2 - s^2$ (powers of W are equal) leads to $WO = (QO^2 + r^2 - s^2)/2QO$ (left to the reader). Since WO is directed distance, W is unique on m. Now let l be the perpendicular to line m at W.

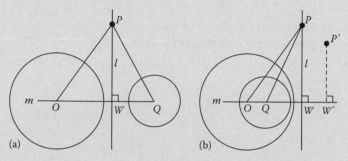

Figure 7.47 Radical axis for non-intersecting circles.

1. Suppose $P \in l$. (Figure 7.47a shows the two cases possible, assuming the circles do not intersect.) By definition, Power P (circle O) = $PO^2 - r^2$, and Power P (circle Q) = $PQ^2 - s^2$. Also, $WO^2 - r^2 = WQ^2 - s^2$. By the Pythagorean theorem in right triangles OPW and QPW, $PO^2 - r^2 = PW^2 + WO^2 - r^2 = PW^2 + WQ^2 - s^2 = PQ^2 - s^2$ and Power P (circle O) = Power P (circle Q).
2. Suppose that Power P'(circle O) = Power P'(circle Q), to show that $P' \in l$. Let W' be the foot of the perpendicular from P' on line m. Thus, again by the Pythagorean theorem,

$$P'O^2 - r^2 = P'W'^2 + W'O^2 - r^2$$

$$P'Q^2 - s^2 = P'W'^2 + W'Q^2 - s^2$$

Since $P'O^2 - r^2 = P'Q^2 - s^2$, it follows that $W'O^2 - r^2 = W'Q^2 - s^2$. But there can be only one point W on m whose powers with respect to circles O and Q are the same, hence $W' = W$ and $P \in l$. ◣

Definition 2   The line l of Theorem 2 is called the **radical axis** of the two circles.

Since the radical axis of two intersecting circles is the line containing their common chord, the circles in $\mathcal{F}$ are those which have their centers on l and whose radical axis is m. It turns out that, reciprocally, the circles in $\mathcal{G}$ are precisely those which have their centers on m and whose radical axis is l.

(The two families of circles are thus said to be **coaxial**.) To see this, let C_1 and C_2 be any two circles in G (Figure 7.48). By properties already established,

$$[AB,CD] = -1 = [CD,AB] \quad \text{and} \quad [EF,CD] = -1 = [CD,EF]$$

Let l intersect m at W, the midpoint of segment CD

$$\text{Power } W(\text{circle } C_1) = WA \cdot WB = WC^2$$

$$= WD^2 = WE \cdot WF = \text{Power } W \ (\text{Circle } C_2)$$

where we have used Theorem **2**, Section **7.5**. Therefore, by Theorem 2 above, l is the radical axis of C_1 and C_2.

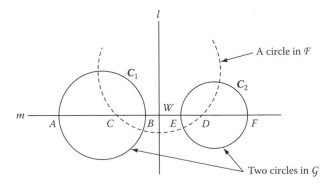

Figure 7.48 Radical axis for G.

This section will be concluded with a coordinate characterization for the families F and G, which will provide a direct analogy to the mutually orthogonal conics discussed at the outset. First, we need to find a coordinate criterion for orthogonal circles in terms of their equations. Recall that the circle $[O, r]$ has the standard equation

(2) $$(x - h)^2 + (y - k)^2 = r^2$$

where the center is the point $O(h, k)$ and the radius is r. By algebra, this equation transforms into the following form (by squaring out the two binomials)

$$x^2 + y^2 - 2hx - 2ky + h^2 + k^2 = r^2$$

or the general form

(3) $$x^2 + y^2 + ax + by + c = 0$$

where $a = -2h$, $b = -2k$, and $c = h^2 + k^2 - r^2$. Thus, if we are given the equation for a circle in the general form **(3)**, we can identify its center as the point $O(-\frac{1}{2}a, -\frac{1}{2}b)$ and its radius as $r = \sqrt{h^2 + k^2 - c} = \frac{1}{2}\sqrt{a^2 + b^2 - 4c}$. The result we seek is the following.

Theorem 3 Two circles C_1 and C_2 represented by the respective general equations $x^2 + y^2 + a_ix + b_iy + c_i = 0$ $(i = 1, 2)$ are orthogonal iff

(4) $a_1a_2 + b_1b_2 = 2c_1 + 2c_2$

Proof Using condition **(C)** on orthogonal circles, if O_1 and O_2 are the centers of C_1 and C_2 respectively, and r_1 and r_2 their radii, then $C_1 \perp C_2$ iff $(O_1O_2)^2 = r_1^2 + r_2^2 = \frac{1}{4}(a_1^2 + b_1^2 - 4c_1) + \frac{1}{4}(a_2^2 + b_2^2 - 4c_2)$. Using the distance formula for $(O_1O_2)^2$, where $O_1 = (-\frac{1}{2}a_1, -\frac{1}{2}b_1)$ and $O_2 = (-\frac{1}{2}a_2, -\frac{1}{2}b_2)$ we obtain the equation

$$(-\tfrac{1}{2}a_1 + \tfrac{1}{2}a_2)^2 + (-\tfrac{1}{2}b_1 + \tfrac{1}{2}b_2)^2$$

$$= \tfrac{1}{4}(a_1^2 + b_1^2 - 4c_1) + \tfrac{1}{4}(a_2^2 + b_2^2 - 4c_2)$$

which reduces to **(4)**, by algebra (details left to the reader). ◢

Corollary A circle $x^2 + y^2 + ax + by + c = 0$ is orthogonal to the unit circle $x^2 + y^2 = 1$ iff $c = 1$.

Proof With $a_2 = b_2 = 0$ and $c_2 = -1$ in the equation of the unit circle, **(4)** becomes $0 = 2c_1 + 2c_2 = 2c_1 - 2$, or $c_1 = 1$. ◢

Note: Since a and b are arbitrary reals for the circles orthogonal to the unit circle, we find that all circles orthogonal to any given circle is a two-parameter family of circles. [The family of *all* circles in the plane is a three-parameter family as revealed by **(3)**.] ◢

Now we are ready to tackle the coordinate equations for the mutually orthogonal families $\mathcal{F}$ and $\mathcal{G}$ mentioned above. For convenience, take the lines l and m of Figure 7.46 as coordinate axes, and let C and D have coordinates $(\pm 1, 0)$, as shown in Figure 7.49. Any circle in $\mathcal{F}$ has its center on the y-axis, so it will have the general equation

$$x^2 + y^2 + by + c = 0$$

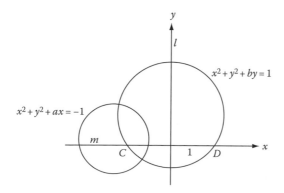

Figure 7.49 Equations for $\mathcal{F}$ and $\mathcal{G}$.

This circle also passes through (1, 0) so by substitution, $1^2 + 0^2 + b \cdot 0 + c = 0$ or $c = -1$. Thus, any circle in $\mathcal{F}$ has equation $x^2 + y^2 + by = 1$, b arbitrary—a one-parameter family of circles. The circles in $\mathcal{G}$ have their centers on the x-axis, having general equation $x^2 + y^2 + ax + c = 0$. Using the orthogonality condition for this circle and any circle in $\mathcal{F}$, we obtain from the orthogonality condition **(4)**,

$$a \cdot 0 + 0 \cdot b = 2(-1) + 2c \quad \text{or} \quad c = 1$$

Thus the desired equation for a circle in $\mathcal{G}$ is $x^2 + y^2 + ax + 1 = 0$ (also a one-parameter family, in agreement with $\mathcal{F}$). The resulting families $\mathcal{F}$ and $\mathcal{G}$ then take on the coordinate form

(5)
$$\begin{cases} \mathcal{F}: x^2 + y^2 + \lambda y = 1 \\ \mathcal{G}: x^2 + y^2 + \mu x = -1 \end{cases}$$

where λ and μ are arbitrary real numbers, with $\mu > 2$. (Note that $\mu \le 2$ for $\mathcal{G}$ leads to $x^2 + (y + \frac{1}{2}\mu)^2 = \frac{1}{4}\mu^2 - 1 \le 0$ which is either a point circle at $(0, \pm 1)$ when $\mu = \pm 2$, or an imaginary circle; for $\mu > 2$, the circles in the family $\mathcal{G}$ cluster around the points $(0, \pm 1)$ in agreement with the behavior of the original family pictured in Figure 7.46.)

Problems (Section 7.7)

Group A

1. Under what circumstances will the radical axis between two non-intersecting circles be equidistant from the circles?

2. What is the radical axis of two circles that have a common tangent l?

3. Prove that the power of the point $P(x_0, y_0)$ with respect to the circle $x^2 + y^2 + ax + by + c = 0$ is the value given by

   (6) $x_0^2 + y_0^2 + ax_0 + by_0 + c = 0$

4.* Show that the radical axis of two circles

   $$C_1: x^2 + y^2 + a_1 x + b_1 y + c_1 = 0$$

   $$C_2: x^2 + y^2 + a_2 x + b_2 y + c_2 = 0$$

   has a coordinate equation of the form

   (7) $(a_1 - a_2)x + (b_1 - b_2)y + (c_1 - c_2) = 0$

   (the result of merely subtracting the equations of C_1 and C_2).

5. Prove that the angle between the tangents to two intersecting circles at one point of intersection is congruent to the angle at the other point of intersection.

Group B

6. In Figure P.6, circles C_1, C_2, and C_3 intersect each other at the points A and B, C and D, and at E and F, respectively. Prove that chords AB, CD and EF are concurrent.

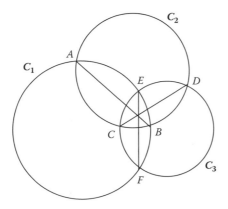

Figure P.6

7. Generalize the result in Problem 6 by showing that the three radical axes of the pairs of circles C_1, C_2, and C_3 are concurrent (Figure P.6).

8. Find the radical axes of each pair of the following circles using (7) (Problem 4), and find the coordinates (x_0, y_0) of the point of concurrency of those radical axes:

$$C_1: x^2 + y^2 + 3x + 4y + 2 = 0$$

$$C_2: x^2 + y^2 + 2x + 4y + 4 = 0$$

$$C_3: x^2 + y^2 + 4x + 3y - 1 = 0$$

Sketch the graph of the three circles and their radical axes. [Note that C_2 does not meet either C_1 or C_3 but C_1 and C_3 intersect at the rational points $A(-1, -4)$ and $B(\frac{1}{2}, -5/2)$.]

9. Show the validity of the following three-step construction for the radical axis of two non-concentric, non-intersecting circles C and D (Figure P.9).
   1. Draw any circle E intersecting C and D at the respective points C, C' and D, D'.
   2. Draw lines CC' and DD', intersecting at point P.

3. Construct the perpendicular from P to the line of centers OQ. This perpendicular is the required radical
axis.

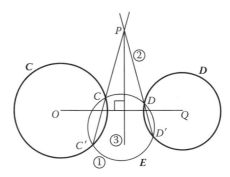

Figure P.9

10.* Using the concepts involving the families $\mathcal{F}$ and $\mathcal{G}$, find
a construction that will produce the unique circle that is
orthogonal to each of two non-concentric, non-intersecting
circles C and D and passing through a given point P not
lying on the line of centers (see Figure P.10).

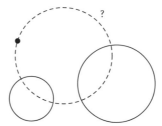

Figure P.10

11. Consider the points $C(0, 0)$ and $D(2, 1)$ in the coordinate
plane. Find the parametric equations of the family $\mathcal{F}$ of all
circles passing through C and D.

12. Using the orthogonality condition **(4)** above, find the equation of the member of $\mathcal{G}$ passing through $V(2, 2)$ that is
orthogonal to all the circles in $\mathcal{F}$ of Problem 11.

13. The unit circle centered at the origin has equation C: $x^2 +
y^2 = 1$.
   (a) Find the general two-parameter equation of all circles
   orthogonal to C.
   (b) Find the general one-parameter equation of all circles
   orthogonal to C passing through $A(2, 1)$.
   (c) The circles of **(b)** must also pass through a second
   point B (why?). Find its coordinates.

Group C

14. If you are familiar with determinants, carry out the algebraic manipulations that prove in general that the coordinates of the point of concurrency of the radical axes of each pair of non-concentric circles whose equations are

$$C_1: x^2 + y^2 + a_1 x + b_1 y + c_1 = 0$$

$$C_2: x^2 + y^2 + a_2 x + b_2 y + c_2 = 0$$

$$C_3: x^2 + y^2 + a_3 x + b_3 y + c_3 = 0$$

are given by $x_0 = M/D$, $y_0 = N/D$ where M, N, and D are the determinants.

$$M = \begin{vmatrix} 1 & b_1 & c_1 \\ 1 & b_2 & c_2 \\ 1 & b_3 & c_3 \end{vmatrix}, \quad N = \begin{vmatrix} 1 & a_1 & c_1 \\ 1 & a_2 & c_2 \\ 1 & a_3 & c_3 \end{vmatrix}, \quad D = \begin{vmatrix} 1 & a_1 & b_1 \\ 1 & a_2 & b_2 \\ 1 & a_3 & b_3 \end{vmatrix}$$

(Use **(7)** in Problem 4. Note: $D = 0$ iff any two of the circles are concentric.)

15. Using **(7)**, Problem 4, verify that if a and b are not both zero, the radical axis of the one-parameter family of circles

$$C_\lambda: x^2 + y^2 + \lambda(ax + by) = a^2 + b^2$$

has an equation of the form

$$l: ax + by = 0$$

Note: Let the reader also verify that any two circles in this family intersect at the same two points on l, and thus represent the family $\mathcal{F}$ of the corollary of Theorem 1. Show that the corresponding family $\mathcal{G}$ in this case has the parametric form $C_\mu: x^2 + y^2 + \mu(bx - ay) = -a^2 - b^2 = 0$. ◣

16.* **Analytic Characterization of Coaxial Systems** The most general one-parameter family of circles is given by

(8) $C_t: x^2 + y^2 + a(t)x + b(t)y + c(t) = 0$ (t real)

where $a(t)$, $b(t)$, and $c(t)$ are (continuous) functions of the real variable t.

(a) Show this family represents a family of circles of type $\mathcal{F}$ or $\mathcal{G}$ iff $a(t)$, $b(t)$, and $c(t)$ are linear equations (but for a change of parameter). [Use the fact that the radical axis of such circles must be some line $ax + by + c = 0$; apply **(7)** to the equation of one particular circle and the general equation.] Thus, **(8)** becomes, for certain constants a, a', b, b', c, and c',

(8′) $C_t: x^2 + y^2 + (at + a')x + (bt + b')y + ct + c' = 0$

(b) Define a discriminant of the form

(9) $\Delta = (a^2 + b^2)(a'^2 + b'^2 - 4c') - (aa' + bb' - 2c)^2$

Show that **(8′)** represents a family of type $\mathcal{F}, \mathcal{G}$, or the degenerate case iff $\Delta > 0$, $\Delta < 0$, or $\Delta = 0$ respectively.

17.   Show that the one-parameter family $\mathcal{F}$ of ellipses having foci at $(\pm 1, 0)$ is mutually orthogonal to the one-parameter family $\mathcal{G}$ of hyperbolas having the same two foci. (Use calculus to find the slope dy/dx of a curve using implicit differentiation.)

18.   Find the locus of a point P such that the difference of its powers with respect to two circles is a constant.

8

Transformations in Modern Geometry

A difficult problem in geometry can often be transformed into a much simpler one by use of a geometry-preserving mapping in the plane. When this is possible, conclusions are made for the new simplified problem that are invariant under the mapping, and these conclusions are then transformed back to the original problem using the inverse mapping, thus solving the problem. For a nongeometric example, suppose we wanted to find the product of 358 and 762 in base 9. Since multiplication in base 9 is rather nebulous, we first change the numbers to base 10, find the product (which can be done on a pocket calculator), then transform the answer back to base 9. (In case you want to try this, the answer is 310857, base 9.)

The general class of transformations most useful in geometry would presumably be those which preserve collinearity (thus mapping lines to lines). Special subclasses would include mappings which preserve angle measure or distance. A systematic study of these transformations will be undertaken here. Although this theory is interesting in itself, its applications in solving geometry problems is our main goal.

8.1 Projective Transformations

Early studies in geometry in the seventeenth and eighteenth centuries were devoted to an understanding of the problem the artist faces in painting a three-dimensional scene on a two-dimensional canvass. This study resulted in a major area of geometry known as *projective geometry*. If we analyze what an artist must do to provide a faithful rendition of a particular landscape, the result is a *projective map*, defined by taking a point P in the scene to be depicted, and intersecting the line of sight EP with the plane of the canvass (see Figure 8.1). This determines point P', the *image* of P. Thus, as P traces out the rails of a railroad track, for example, P' traces out its image on the canvass. This is why parallel lines sometimes map to intersecting lines—the parallel rails of a railroad track extending into the distance become lines which intersect on the horizon. And circles often map to ellipses—the circular rim of a drinking glass held upright is drawn as an oval (not a circle).

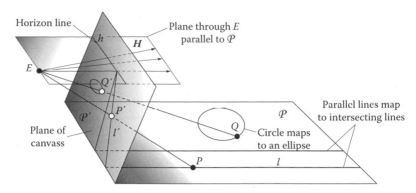

Figure 8.1 A central projection.

In order to pursue this type of analysis further, a basic knowledge of three-dimensional geometry is needed. We assume these four fundamental principles:

(a) Three noncollinear points determine a unique plane.
(b) If two planes intersect, they do so in a line.
(c) A line containing two points of a plane lies entirely in that plane.
(d) *Parallel Postulate*: Given a plane and a point not on it, there is exactly one plane passing through the given point and parallel to (does not intersect) the given plane.

These principles will get us through the initial theory adequately. For a coordinate-based treatment justifying these concepts, see Problems 17–23, which introduce a three-dimensional coordinate system and show how to work with it.

We formally define a **central projection** with **center** E from one plane $\mathcal{P}$ to another plane $\mathcal{P}'$ to be the mapping that associates each point of $\mathcal{P}$ with the intersection of line $\overleftrightarrow{EP}$ and the plane $\mathcal{P}'$, as illustrated in Figure 8.1. Note that lines in $\mathcal{P}$ map to lines in $\mathcal{P}'$: if P traces out a line l in $\mathcal{P}$, the line EP generates a plane, whose intersection with $\mathcal{P}'$ is a line—the *image* of l—denoted l'. Thus, *a central projection preserves collinearity*. We define the **horizon** of the projection to be the line h of intersection of $\mathcal{P}'$ with the plane H passing through E that is parallel to $\mathcal{P}$ (as in Figure 8.1).

Note that, generally, parallel lines in $\mathcal{P}$ map to lines which intersect on the horizon. Different families of parallel lines L, M,… map to lines that intersect on h; each family corresponds to a unique point of intersection L', M',… on h, as shown in Figure 8.2. A central projection may thus be used to define, and to study, the so-called *region about a point at infinity* in $\mathcal{P}$. Each such region maps to a finite region about a point on h in $\mathcal{P}'$. It may also be used to provide an elegant proof of Desargues' theorem. (A rather involved proof appeared previously in Problem 13, Section **7.6**, using linearity numbers and Menelaus' theorem.) A properly chosen central projection reduces this problem to an elementary exercise on ratios and parallel projection as in the analysis which follows.

Consider Figure 8.3 where the diagram depicting the hypotheses of Desargues' theorem in plane $\mathcal{P}'$ is drawn, so positioned in order that the horizon passes through points L' and M', with N' the intersection of lines

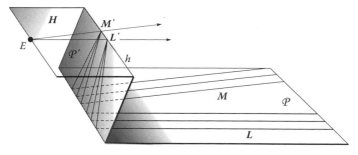

Figure 8.2 Parallel lines mapping to intersecting lines.

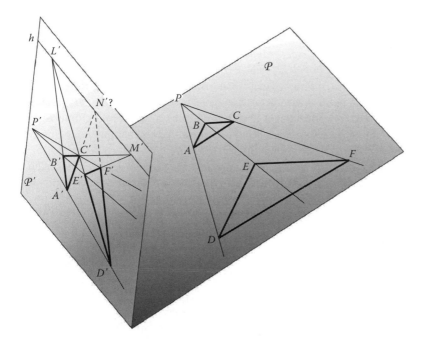

Figure 8.3 Projective proof of Desargues' theorem.

$A'C'$ and $D'F'$. We are now reversing the mapping, going from $\mathcal{P}'$ to $\mathcal{P}$ (the *inverse map*). Thus, the reverse central projection from E maps the lines through L and M in $\mathcal{P}'$ to parallel lines in $\mathcal{P}$. Accordingly, the image lines AB and DE are parallel, as are BC and EF. Since we want to show that the lines $A'C'$ and $D'F'$ meet on h, it suffices to show that lines AC and DF are parallel in $\mathcal{P}$. But this is a comparatively easy task. Observe that by parallel projection $PA/PD = PB/PE$ and $PB/PE = PC/PF$. Thus, $PA/PD = PC/PF$ and lines AC and DF must be parallel. Hence lines $A'C'$ and $D'F'$ meet on h and therefore L', M', and N' are collinear. ◣

Since we are interested mainly in transformations taking place in just the Euclidean plane $\mathcal{P}$ (having domain and range in $\mathcal{P}$), one can map the plane $\mathcal{P}'$ back to $\mathcal{P}$ by choosing a second center F for a central projection, as shown in Figure 8.4. Thus, a point P in $\mathcal{P}$ projects to Q in $\mathcal{P}'$ via E,

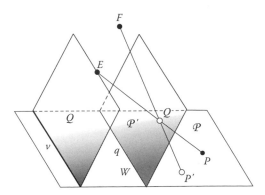

Figure 8.4 Product of two central projections.

and then Q is projected to P' in $\mathcal{P}$ via F. This defines the **composition** or **product** f of the *two central projections* f_1 and f_2 (one followed directly by the other, i.e., $f = f_2 \circ f_1$). (The centers of central projections are allowed to be points at infinity, which render such mappings as **parallel projections**. That is, P' is the intersection of $\mathcal{P}'$ with the line through P parallel to some fixed line e in space.) In this particular case, the line of intersection of $\mathcal{P}$ and $\mathcal{P}'$ (line q in the figure) is a line of points that map to themselves (points like W shown in the figure); these are called *fixed points*.

The composition of a finite number of such projections is called a **projectivity**, or **projective transformation**, on $\mathcal{P}$. Since each individual central projection evidently preserves collinearity, then the composite map also preserves collinearity, and thus a *projectivity maps lines into lines*. Transformations of the plane having this property are thus said to be **linear**.

Note that in this example f is defined for all points in $\mathcal{P}$ except those lying on the line of intersection v of $\mathcal{P}$ with the plane parallel to $\mathcal{P}'$ passing through E. The domain of f in this case is the set of all points in $\mathcal{P}$ not on line v, denoted $\mathcal{P} \backslash v$ for convenience. The line v defined in this manner for such a projectivity is called the **vanishing line**. Not all projectivities have vanishing lines; for example, when $\mathcal{P}'$ is parallel to $\mathcal{P}$, as shown in Figure 8.5, the projectivity via centers E and F is defined for every point in $\mathcal{P}$. In general, the domain of a projective transformation on $\mathcal{P}$ is either the set $\mathcal{P} \backslash v$, where v is the vanishing line of f, or the entire plane $\mathcal{P}$ if the vanishing line does not exist (as in the example in Figure 8.5).

It can be shown by extensive use of algebra in three-dimensional coordinate space that any projectivity f on $\mathcal{P}$ has the coordinate form, for certain constants a_k, b_k, and c_k ($k = 1, 2, 3$),

(1)
$$\begin{cases} x' = \dfrac{a_1 x + b_1 y + c_1}{a_3 x + b_3 y + c_3} \\[2mm] y' = \dfrac{a_2 x + b_2 y + c_2}{a_3 x + b_3 y + c_3} \end{cases}$$

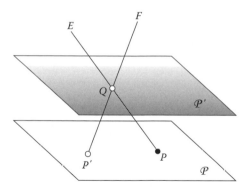

Figure 8.5 A projective map having no vanishing line.

such that the third-order determinant of the matrix of coefficients $[a_k\ b_k\ c_k]$ is nonzero (x' and y' denote the coordinates of the image point of $P(x, y)$ under f, which will be made clear as we proceed). The coordinate form of a projectivity thus makes it much easier to study such transformations.

Example 1

Consider the projectivity f given in coordinate form by

$$x' = \frac{x}{x - y + 1} \qquad y' = \frac{y}{x - y + 1}$$

(a) Find the coordinates of the images A', B', C' of $A(-1, 1)$, $B(1, 1)$, and $C(2, 1)$. Note the betweenness relation (ABC), and find whether $(A'B'C')$ holds in this case. (Are A', B', and C' collinear?)

(b) Show this mapping is one-to-one by finding its inverse f^{-1}. (Solve for the variables x and y in terms of x' and y').

(c) Find the vanishing line v of f in order to identify the domain of f.

(d) Show that $y = 2x$ and $x + y = 3$, which intersect on v, map to parallel lines (as in Figure 8.6).

Solution

(a) By direct substitution, $x = -1$ and $y = 1$ produce $x' = 1$, $y' = -1$. Thus, $A' = (1, -1)$. Similarly, $B' = (1, 1)$ and $C' = (1, \frac{1}{2})$. Thus A', B', and C' are collinear, but $(A'C'B')$ holds instead of $(A'B'C')$.

(b) Clearing fractions, we have

$$x'(x - y + 1) = x \quad \text{or} \quad (x' - 1)x - x'y = -x'$$

$$y'(x - y + 1) = y \quad \text{or} \quad y'x - (y' + 1)y = -y'$$

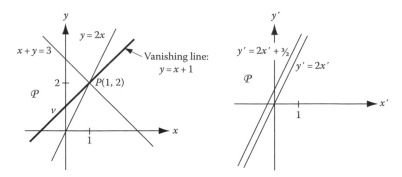

Figure 8.6 Example 1.

This is a system of linear equations in x and y with coefficients $(x' - 1)$, $-x'$, y', and $-(y' + 1)$, respectively. In order to solve this system for x and y, one can use the method of elimination, or Cramer's rule. We prefer the latter:

$$x = \frac{\begin{vmatrix} -x' & -x' \\ -y' & -(y'+1) \end{vmatrix}}{\begin{vmatrix} (x'-1) & -x' \\ y' & -(y'+1) \end{vmatrix}} = \frac{x'y' + x' - x'y'}{-x'y' - x' + y' + 1 + x'y'} = \frac{x'}{-x' + y' + 1}$$

$$y = \frac{\begin{vmatrix} (x'-1) & -x' \\ y' & y' \end{vmatrix}}{-x' + y' + 1} = \frac{-x'y' + y' + x'y'}{-x' + y' + 1} = \frac{y'}{-x' + y' + 1}$$

Therefore, f^{-1} is given by

$$x = \frac{x'}{-x' + y' + 1} \qquad y = \frac{y'}{-x' + y' + 1}$$

Since this form yields unique values x, y for given values x', y' for which x and y are defined, f cannot map two distinct points into the same point, and is therefore one-to-one.

(c) The vanishing line v is given by setting the denominators of x' and y' to zero: $x - y + 1 = 0$ or $y = x + 1$. Thus, the domain of f is the xy-plane with $y = x + 1$ deleted.

(d) As in (b), substituting the equations for f^{-1} into the equations $y = 2x$ and $x + y = 3$ (which intersect at (1, 2) on v) we obtain

$$\frac{y'}{-x' + y' + 1} = 2\left(\frac{x'}{-x' + y' + 1}\right)$$

and

$$\frac{y'}{-x' + y' + 1} + \frac{x'}{-x' + y' + 1} = 3$$

which simplify to

$$y' = 2x' \quad \text{and} \quad y' = 2x' + \tfrac{3}{2}$$

These are lines having slope $m = 2$, thus are parallel.

In the theory of transformations, the concept of *fixed points* (mentioned above) plays an important role.

Definition   Given a projective mapping f on $\mathcal{P}$, a **fixed point** of f is any point P in the plane for which $f(P) = P$ (or, in terms of coordinates, points $P(x, y)$ such that $x' = x$ and $y' = y$).

Example 2

Find the fixed points of the projective transformation of Example 1, that is, for the projectivity

$$x' = \frac{x}{x - y + 1} \quad y' = \frac{y}{x - y + 1}$$

Solution

To find the fixed points, we set $x' = x$ and $y' = y$ in the above equations and proceed to solve for x and y. (Note that this only gives us necessary conditions for fixed points; these solutions must be checked to verify them as actual fixed points.) Clearing fractions, we obtain

(∗) $x(x - y + 1) = x, \quad x^2 - xy + x = x, \quad x(x - y) = 0$

and

(∗∗) $y(x - y + 1) = y, \quad xy - y^2 + y = y, \quad y(x - y) = 0$

Thus, we obtain from (∗) the two cases $x = 0$ or $x - y = 0$ (i.e., $y = x$), and from (∗∗) $y = 0$ or $y = x$. The candidates for fixed points must then lie on the line $y = x$. Indeed, substituting (a, a) into the transformation equations results in the image point (a, a), the result being that every point on the line $y = x$ is a fixed point. (Note that $(0, 0)$ is a special fixed point. It has the property that every line passing through it, thus having equation $y = mx$ for some constant m, are *fixed lines*, mapping to the line $y' = mx'$.)

It can be shown that although ratios such as AB/BC for collinear points A, B, and C are distorted, the cross ratio of four collinear points is left unchanged (**invariant**) under projective transformations. That is, if A, B, C, and D are distinct collinear points in the domain of a projective transformation f, then $[AB, CD] = [A'B', C'D']$. Other than lines, this is practically the only important invariant of projective transformations. (Certainly, parallelism is not an invariant since parallel lines can map to intersecting lines.) Problem 7 reveals that betweenness is also not preserved. Thus, the list of invariants for projectivities is very short.

Invariants for Projective Transformations

1. Lines and collinearity
2. Cross ratio

Problems (Section 8.1)

Group A

1. Consider the projective mapping f given in coordinate form by

    $$x' = \frac{x}{y-1} \quad y' = \frac{y}{y-1}$$

    (a) Find the inverse of f by solving the system for x and y in terms of x' and y'.
    (b) What is v, the vanishing line? Give its equation.
    (c) Find, by substitution of the inverse equations you found in (a), the images under f of the two lines l: $y = 2x - 5$ and m: $y = -x + 4$. Show that l and m intersect on v and that their images are parallel lines.

2. Repeat the instructions (a), (b), (c) of Problem 1 for the mapping

    $$x' = \frac{-x+2y}{x-y+1} \quad y' = \frac{y}{x-y+1}$$

    For (c), use $y = 2x - 2$ and $y = 3x - 5$.

3. Consider the two projectivities given by

    $$f: x' = \frac{x}{y}, \quad y' = \frac{y+2}{y}$$

    $$g: x' = \frac{x+1}{x+3y-2}, \quad y' = \frac{y}{x+3y-2}$$

Find the coordinate form of the products *fg* and *gf*, and show that these products are also projective transformations. Are they the same? [**Hint:** To find *fg*, raise the primes in *f* so that x' is replaced by x'', y' by y'', x by x', and y by y', then use direct substitution of the equations for *g* into those of *f*, and simplify by algebra.]

4. The projectivity *f* given below clearly has the origin as a fixed point (substitute (0, 0) for (*x*, *y*)). Show that there exists only one other fixed point, by solving the system of equations resulting from setting $x' = x$ and $y' = y$.

Note: The two fixed points *A* and *B* of *f* imply that the line $l \equiv \overrightarrow{AB}$ is a fixed line. However, *f* moves the individual points of *l*, except for *A* and *B*. This is in stark contrast with the behavior of affine mappings to be studied next (see Corollary C below).

$$x' = \frac{3x - y}{y + 1} \quad y' = \frac{x + y}{y + 1}$$

5. The projective mapping given below has three isolated fixed points. Find them.

$$x' = \frac{x + y}{x + 1} \quad y' = \frac{4x + 4y - 12}{x + 1}$$

6. Find all fixed points of the projective mapping given in Problem 2, that is

$$x' = \frac{-x + 2y}{x - y + 1} \quad y' = \frac{y}{x - y + 1}$$

and describe them geometrically (are they isolated, lie on a line, etc.?)

Group B

7. Verify that under the mapping of Problem 2 the points *A*(−2, 0), *B*(0, 0), and *C*(2, 0) map to points *A′*, *B′*, and *C′* such that *B′* does not lie between *A′* and *C′*, even though *B* is the midpoint of segment *AC*. (This shows that projectivities in the plane need not preserve betweenness nor midpoints.) Also show that [*AB, CD*] = [*A′B′, C′D′*], where *D* = (7, 0).

8. Using **(1)**, show that a projective mapping that fixes the points *A*(0, 0), *B*(1, 0), and *C*(0, 1) must have the coordinate form

$$x' = \frac{(a + 1)x}{ax + by + 1} \quad y' = \frac{(b + 1)y}{ax + by + 1}$$

for some *a* and *b*, constants. [**Hint:** Substitute 0 for *x*, *y*, *x′*, and *y′* for *A*, and similarly for *B* and *C*.]

9. Show that a projective mapping that fixes the points $A(0, 0)$, $B(1, 0)$, $C(0, 1)$, and $D(1, 1)$ is the identity mapping $x' = x$, $y' = y$. (See Problem 8.)

10.* Consider the mapping f of Problem 1 and the three circles

(a) $x^2 + y^2 = \frac{1}{2}$ (b) $x^2 + y^2 = 1$ (c) $x^2 + y^2 = 2$

(see Figure P.10). Find the image of each circle under f. by substituting the inverse relations into these equations. Note that the circle in (a) does not intersect the vanishing line, the circle in (b) is tangent to it, and the circle in (c) intersects it in two distinct points.

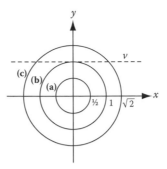

Figure P.10

11. In Problem 10, the projective transformation of Problem 1,

$$x' = \frac{x}{y-1} \quad y' = \frac{y}{y-1}$$

was shown to map the circle $x^2 + y^2 = 2$ into a hyperbola. This circle intersects the vanishing line $y = 1$ at the points $(\pm 1, 1)$, and the tangent lines to the circle at those two points are, respectively, $y = x + 2$ and $y = -x + 2$. Show that these tangents map into the asymptotes of the hyperbola. Sketch the graph.

12. **Pappus' Theorem** Use a projective mapping to prove the following: If points A, B, and C lie on one line, and A', B', and C' on another, then the intersection of the cross-joins $L = \overleftrightarrow{AB'} \cap \overleftrightarrow{BA'}$, $M = \overleftrightarrow{A'C} \cap \overleftrightarrow{CA'}$, and $N = \overleftrightarrow{BC'} \cap \overleftrightarrow{CB'}$ are collinear. (See Figure P.12.)

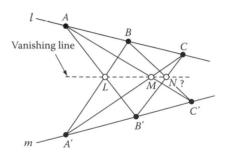

Figure P.12

13. **The Harmonic Construction**   Using a central projection, prove that the harmonic construction is independent of the choice of points P and Q on line PB. Figure P.13 shows the case when P and Q are replaced by T and U; you must show that lines SQ and WU meet on line AB. (Your proof should lead to parallelograms $P'Q'R'S'$ and $T'U'V'W'$ having corresponding sides parallel, and one pair of diagonals parallel. Translate so that the parallelograms have a vertex in common in order to reach the desired conclusion.)

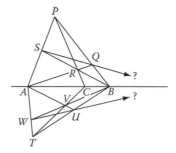

Figure P.13

Group C

14. The third-order determinant of coefficients of the projectivity of Problem 5 is given by

$$\begin{vmatrix} 1 & 1 & 0 \\ 4 & 4 & -12 \\ 1 & 0 & 1 \end{vmatrix}$$

If you are familiar with determinants, calculate the value of this determinant to show it is nonzero.

15. The third-order determinant of the coefficients of the mapping on $\mathcal{P}$ given below equals zero. Find something wrong with the mapping. Is it a projectivity?

$$x' = \frac{x+y}{4x+y+1} \qquad y' = \frac{2x-y+1}{4x+y+1},$$

16. **Fundamental Theorem for Projective Transformations** It can be proven that given two sets of four points (A, B, C, D) and (P, Q, R, S), such that no three points are collinear within each group, there exists a unique projective transformation which maps the first set onto the second set. Illustrate this theorem by finding the coordinate form (1) of the transformation that takes $A(0, 0)$, $B(1, 0)$, $C(0, 1)$, and $D(1, 1)$ to $P(3, -2)$, $Q(3, -5)$, $R(7, -5)$, and $S(1, -2)$, respectively.

17. **Coordinates for Three-dimensional Euclidean Geometry (E^3)**   Starting with the x, y coordinate plane $\mathcal{P}$, erect a coordinated line l perpendicular to $\mathcal{P}$ at the origin $O(0, 0)$, and let O be the origin of the one-dimensional coordinate system for l (see Figure P.17). Given any point P in space, let $Q(x, y)$ be the foot of P on $\mathcal{P}$, and let the plane passing through P parallel to $\mathcal{P}$ intersect l at $R[z]$. The coordinates of P are then defined to be (x, y, z), and we write $P(x, y, z)$. (At this point line l is called the *z-axis*.)*

Evidently, the points of $\mathcal{P}$ have coordinates $(x, y, 0)$. Thus, we agree that the "equation" of the plane $\mathcal{P}$ is $z = 0$. The equation of the plane $\mathcal{P}'$ parallel to $\mathcal{P}$ and at a distance 1 above would be $z = 1$. In general, a single linear equation $ax + by + cz = d$ for constants a, b, c, d represents a plane (not a line). More on this appears in Problem 20, not yet established.

   (a) Show that the equation $y = x$ represents the plane perpendicular to $\mathcal{P}$ passing through the line given by the *two* equations $y = x, z = 0$ in $\mathcal{P}$.

   (b) Find a, b, c, and d for which the plane $ax + by + cy = d$ passes through $A(1, -1, 0)$, $B(2, 0, 1)$, and $C(3, 8, -1)$. Is the equation uniquely determined? This indicates the coordinate version of the geometric principle that *three noncollinear points determine a plane*.

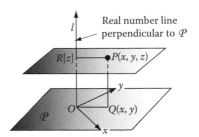

Figure P.17

18. **Distance Formula for E^3**   Suppose $P(x_1, y_1, z_1)$ and $Q(x_2, y_2, z_2)$ are any two points in space. To find the distance PQ, drop the perpendiculars from P and Q to the xy-plane $\mathcal{P}$ at the points $P'(x_1, y_1, 0)$ and $Q'(x_2, y_2, 0)$, as shown in Figure P.18. By the distance formula for $\mathcal{P}$ we have $P'Q'^2 = (x_1 - x_2)^2 + (y_1 - y_2)^2$. The planes parallel to $\mathcal{P}$ passing through P and Q meet the z-axis at z_1 and z_2, and the point $R(x_2, y_2, z_1)$ evidently forms a right triangle with P and Q such that $\lozenge PRQ'P'$ is a parallelogram. Also, $RQ = |z_1 - z_2|$. Then $PQ^2 = PR^2 + RQ^2 = P'Q'^2 + (z_1 - z_2)^2$ and the distance formula for PQ emerges:

(2) $$PQ = \sqrt{(x_1 - x_2)^2 + (y_1 - y_2)^2 + (z_1 - z_2)^2}$$

---

* Orientation of the axes is also involved; without elaborating further, the system we construct is a *right-hand* system.

(a) Use the Pythagorean theorem and the distance formula to show that $A(1, 12, 12)$, $B(7, 9, 6)$, and $C(11, 1, 14)$ are the vertices of a right triangle.

(b) Project the triangle ABC orthogonally into the xy-plane to $\Delta A'B'C'$ and determine whether $\Delta A'B'C'$ is a right triangle.

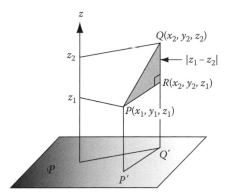

Figure P.18

19. **Orthogonality Condition in E^3**   For any three points $P_i(x_i, y_i, z_i)$, $i = 0, 1, 2$, use the Pythagorean theorem and the distance formula to derive the following necessary and sufficient condition for lines P_0P_1 and P_0P_2 to be perpendicular (Figure P.19):

(3) $$(x_1 - x_0)(x_2 - x_0) + (y_1 - y_0)(y_2 - y_0)$$
$$+ (z_1 - z_0)(z_2 - z_0) = 0$$

Use this and the slope formula in $\mathcal{P}$ to prove that lines in the xy plane having slopes m_1 and m_2 are perpendicular iff $m_1m_2 = -1$.

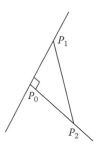

Figure P.19

Note: The numbers $x_1 - x_0$, $y_1 - y_0$, and $z_1 - z_0$ are called **direction numbers** for the line through P_0 and P_1, since, as may be proven, these numbers are independent of the points P_0 and P_1 chosen on that line. Thus, $x_2 - x_0$, $y_2 - y_0$, and $z_2 - z_0$ are direction numbers for line P_0P_2, and one concludes that lines $P_0P_1 \perp P_0P_2$ iff the direction numbers of the two lines have the property that the *sum of the products of corresponding direction numbers equals zero* (known as the *inner product* in vector terminology). ◥

20. **Equations for Planes in E^3**   A theorem in axiomatic three-dimensional geometry states that the lines perpendicular to a fixed line l at some point P_0 on that line generate (lie in) a unique plane $\mathcal{P}'$ perpendicular to l at P_0. (See Figure P.20 for this problem.) Let $Q(p, q, r)$ be any point on l, $Q \neq P_0$. Use **(3)** to show that if $P(x, y, z)$ is any point in $\mathcal{P}'$ and if $a = p - x_0$, $b = q - y_0$, and $c = r - z_0$, then

(4) $a(x - x_0) + b(y - y_0) + c(z - z_0) = 0$

This equation then becomes the *equation of $\mathcal{P}'$*, where the coefficients a, b, and c are the *direction numbers* of the **normal** to $\mathcal{P}'$ (line l), as mentioned in Problem 19. Removing parentheses in the above equation produces the form

$$ax + by + cz = d$$

where $d = ax_0 + by_0 + cz_0$. Verify the calculations and conclusions indicated in this problem.

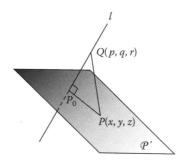

Figure P.20

21. Prove that if $P(x, y, z)$ lies on the intersection of planes $\mathcal{P}_1$: $3x + y - z = 2$ and $\mathcal{P}_2$: $x + y + z = 0$, there exists a real number t such that $x = 2t$, $y = -4t + 1$, and $z = 2t - 1$. Conversely, show that for any real t the coordinates of the point $(2t, -4t + 1, 2t - 1)$ satisfy the equations for both $\mathcal{P}_1$ and $\mathcal{P}_2$. Thus, we have **parametric equations** in t for the line of intersection $\mathcal{P}_1 \cap \mathcal{P}_2$.

22. **Equations for Lines in E^3**   Let line l be given, with $P_0(x_0, y_0, z_0)$ any point on l. Let $Q(x_1, y_1, z_1)$ and $Q'(x_2, y_2, z_2)$ be two further points on l, and let $\mathcal{P}'$ be the plane perpendicular to l at P_0, as shown in Figure P.22. Then the equation of $\mathcal{P}'$, as shown in Problem 20, is

$$a(x - x_0) + b(y - y_0) + c(z - z_0) = 0$$

where $a = x_1 - x_0$, $b = y_1 - y_0$, and $c = z_1 - z_0$. But the equation of $\mathcal{P}'$ is also given by

$$(x_2 - x_0)(x - x_0) + (y_2 - y_0)(y - y_0) + (z_2 - z_0)(z - z_0) = 0$$

(since $Q(p, q, r)$ was an arbitrary point on l in Problem 20). So the numbers $(x_2 - x_0)$, $(y_2 - y_0)$, $(z_2 - z_0)$ must be proportional to a, b, c. Thus, there exists a real number t such that $x_2 - x_0 = at$, $y_2 - y_0 = bt$, and $z_2 - z_0 = ct$. Now let $Q'(x_2, y_2, z_2) = P(x, y, z)$ be any point on l; show that the **parametric equations**

$$x = (x_1 - x_0)t + x_0$$
$$(5) \qquad y = (y_1 - y_0)t + y_0$$
$$z = (z_1 - z_0)t + z_0.$$

are the equations for line l—the line passing through (x_1, y_1, z_1) and (x_0, y_0, z_0). The numbers

$$a = x_1 - x_0, \quad b = y_1 - y_0, \quad c = z_1 - z_0$$

are the direction numbers for l as previously defined (Problem 19). They are evidently proportional to those determined by any other pair of points on l. (Why?)

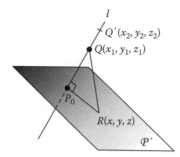

Figure P.22

23. Use the results of the preceding problem to solve each of the following:

(a) Find a proof showing that two lines whose direction numbers are a, b, c and a', b', c', respectively, are perpendicular iff $aa' + bb' + cc' = 0$. (See Problem 19 in this connection.)

(b) Find the parametric equations for the particular line passing through $A(3, 1, 0)$ and $B(1, -2, 4)$.

(c) Determine whether the lines

$$l_1: \{x = 2t + 3, \ y = -t + 1, \ z = t + 4\}$$

$$l_2: \{x = 3t - 2, \ y = 4t - 2, \ z = -2t + 5\}$$

intersect. (Use s for t in the equations for l_2 since the parameters for l_1 and l_2 are independent and could be different.)

(d) Show that $l_1 \perp l_2$.

8.2 Affine Transformations

Projective transformations preserve collinearity, but do not preserve betweenness as we have seen. We now consider projective transformations which do. Since they evidently cannot have a vanishing line, they map the entire plane $\mathcal{P}$ onto itself, and are linear, thus the term *affine transformation*. The formal definition follows.

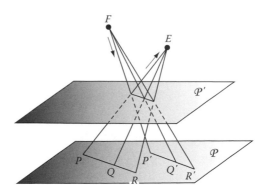

Figure 8.7 Betweenness-preserving property of an affine map.

> **Definition 1**   An **affine transformation** is a betweenness-preserving projective transformation whose domain and range is the entire plane $\mathcal{P}$.

Recall that if a function f is one-to-one, it has an inverse denoted f^{-1}, where $f^{-1} \circ f = f \circ f^{-1} =$ the identity function. The same is true for transformations in the plane. The betweenness-preserving property of an affine mapping makes it one-to-one (for if $A \neq B$ and $A' = B'$, then the betweenness relation (ACB) is not preserved under the mapping). Thus, an affine mapping has an inverse that is defined on the entire plane.

A simple example will illustrate these ideas. As in a previous discussion, suppose the centers of two central projections are ordinary points, E and F, and plane $\mathcal{P}'$ is parallel to $\mathcal{P}$. One can then observe that betweenness is preserved under this mapping, as illustrated in Figure 8.7. Here, the segment PR is mapped by the product of these two central projections onto segment $P'R'$, with the betweenness relation (PQR) corresponding to $(P'Q'R')$. Also, note that by merely reversing the order of the two central projections, using center F first, then center E, point P' is mapped to P, providing the *inverse of the original map*. (The above is true also if $\mathcal{P}'$ is not parallel to $\mathcal{P}$ and the central projections are parallel projections, with E and F points at infinity.) Thus, such a product of projections not only preserves betweenness, but also possesses an inverse, with the range also the entire plane $\mathcal{P}$.

An affine transformation maps lines into lines, as noted above. A technicality to be cleared up is whether an affine transformation can map a line into just part of a line, or, for that matter, a segment into just part of another segment. The following argument shows how to reach the answer one expects: Suppose line l is mapped into line m, and that there is a point $W \in m$ whose pre-image C does not lie on l (Figure 8.8). Thus, if A and B lie on l, the three noncollinear points A, B, and C map to points A', B', and W lying on m. It now follows that since any point X in the plane lies on a line q that intersects two sides of $\triangle ABC$, then along with A, B, and C, and all points on lines AB, BC, and AC, thus line q, X is mapped into line m. That is, the entire plane $\mathcal{P}$ is mapped into line m, contradicting the existence of the inverse map, which was just shown. A similar argument applies to segments.

Our discussion has proved an important theorem about affine transformations.

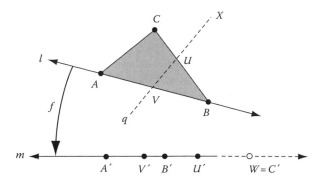

Figure 8.8 Can a line map to part of a line?

Theorem 1   An affine transformation is a one-to-one mapping, possesses an inverse, and maps lines onto lines, and segments onto segments.

Corollary A   An affine transformation maps parallel lines to parallel lines, and thus any parallelogram is mapped onto another parallelogram.

Proof   Let $l \parallel m$ and suppose the image of l and m under an affine map f is l' and m', respectively. If l' is not parallel to m', let W be the point of intersection. Then if P is the pre-image of W under the given, affine mapping (thus $P' = W$), by Theorem, 1 P lies on both l and m, contradicting $l \parallel m$. ◥

An important feature of affine transformations is a direct result of the preceding corollary. Since the diagonals of a parallelogram bisect each other, the midpoint of diagonals AC and BD in parallelogram $ABCD$ map to the midpoint of the corresponding diagonals $A'C'$ and $B'D'$ of parallelogram $A'B'C'D'$. If we start with a given segment AC and its midpoint M, a parallelogram ◊$ABCD$ can be constructed having AC as diagonal, and M will also be the midpoint of diagonal BC. Hence, if f maps AC to $A'C'$ and M to M', M' must lie on both diagonals of the image parallelogram, hence M' is the midpoint of segment $A'C'$. This proves the important

Corollary B   An affine transformation preserves midpoints of segments.

A characteristic of affine transformations, in addition to preserving midpoints of segments (and also the cross ratio of collinear points as any projective transformation does), also preserves ratios of *two collinear segments* as well.

Theorem 2   An affine transformation f preserves ratios of collinear or parallel segments. That is, if AB and CD are either collinear or parallel segments, then $AB/CD = A'B'/C'D'$.

Proof   It suffices to show that if C lies between A and B, then $AC/AB = A'C'/A'B'$. (One can see that this implies the general case by the use of parallelograms.) Consider (ACB), as shown in Figure 8.9. Suppose that $AC/AB \neq A'C'/A'B'$, and for sake of argument, let $AC/AB > A'C'/A'B'$.

Construct point D on segment AB such that $AD/AB = A'C'/A'B' < AC/AB$, and thus (ADC), as shown in the figure. Begin a bisection process on segment AB, obtaining the midpoint of AB, then determining the midpoints of the subsegments thus determined, and so on. At some point, one of the midpoints P will lie on segment DC. Let f map the midpoints on AB to midpoints on $A'B'$. Since f preserves betweenness, P' will lie on segment $D'C'$ and there will be the same number n of midpoints on AB between A and P as there are on $A'B'$ between A' and P'. Let the resulting congruent subsegments on AB and $A'B'$ have lengths a and b. Thus, $AP = na$, and if $AB = ma$, we obtain the contradiction

$$\frac{AD}{AB} < \frac{AP}{AB} = \frac{na}{ma} = \frac{nb}{mb} = \frac{A'P'}{A'B'} < \frac{A'C'}{A'B'} = \frac{AD}{AB}$$

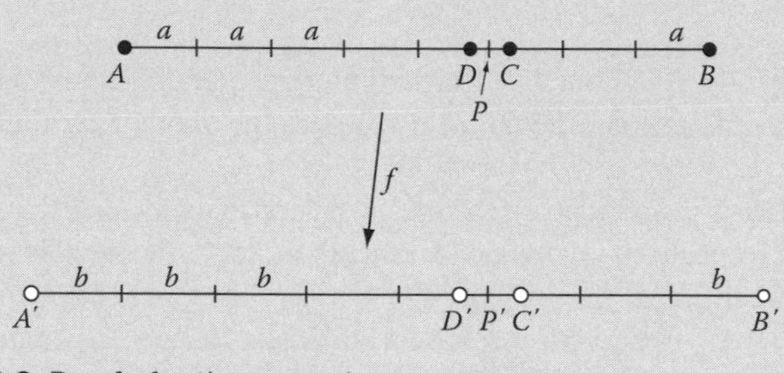

Figure 8.9 Proof of ratio-preserving property.

Corollary C If an affine transformation fixes two points A and B on a line, then it fixes all other points on that line.

Corollary D A nontrivial affine map having an *axis* (a line of fixed points) cannot have a fixed point not on its axis. Consequently, an affine mapping that fixes the vertices of a triangle is trivial (the identity).

Proof The argument chosen for this will closely resemble a previous one, illustrated in Figure 8.8. Using that same figure, suppose the vertices of triangle ABC are fixed points under the affine mapping f. If X is any point in the plane, there exists a line q passing through X and meeting two sides of $\triangle ABC$ at, say U and V. By Corollary C, all points of the sides of $\triangle ABC$ are fixed points, hence so are U and V. Then corollary C requires that X be fixed as well. Thus, every point in the plane is a fixed point of f, hence f is the identity mapping.

In spite of the preceding results, affine transformations are quite general, and geometric figures in the plane can be badly deformed. For example, an equilateral triangle can be mapped to *any other triangle* by an affine transformation. In order to aid an understanding of the nature of affine transformations, as opposed to isometries and angle-preserving transformations discussed later, we introduce a general kind of line reflection that is appropriate for affine mappings.

Definition 2 An **affine reflection** in line l in the direction of line m not parallel to l, and having parameter $k \neq 0$, is the mapping defined as follows: For each point $P \notin l$, determine the line through P parallel to m and the point Q of its intersection with l. Let P' be that point on line PQ such that $QP' = k \cdot PQ$, directed distance understood. (Thus if $k > 0$, P' lies on the opposite ray of QP, and if $k < 0$, on ray QP itself.) If $P \in l$, let $P' = P$. Line l is called the **axis** of reflection.

For an example of this mapping, see Figure 8.10, where $\triangle ABC$ is mapped to $\triangle A'B'C'$ under the affine reflection shown, with $k = 2$. In this example, none of the angles of $\triangle ABC$ are preserved. An ordinary reflection in line l is a special case of an affine reflection, when $m \perp l$ and $k = 1$.

It is rather clear that an affine reflection is a projective map via parallel projections in space, with the intermediate plane $\mathcal{P}'$ passing through the axis (see Problem 14). Thus, an affine reflection is an affine transformation, and products of such reflections are also affine transformations. A sequence of results proves that a product of three or fewer affine reflections can be constructed, which will map any given triangle to another given triangle.

We start by showing that a product of at most two affine reflections f_1 and f_2 can be constructed mapping any given segment AB onto any other segment DE. Because of later considerations, this will be done in such a manner that B' does not lie on line DE. If $A = D$ and $B \neq E$, as shown in Figure 8.11a, let H be any point in the plane not on lines DE or AB, and let M be the midpoint of segment BH. The affine reflection f_1 with axis $l_1 = $ line DM, direction BH and $k = 1$, will map B to $H \equiv B'$. Then with L any point of HE, the affine reflection f_2 with axis $l_2 = $ line DL, direction $HE \equiv B'E$ and $k = HL/LE$, will map B' to E. Thus, $f_2 f_1$ maps A to D (=$A' = A''$) and B to $B' \notin$ line DE, then to $B'' = E$. If $A \neq D$ and $B \neq E$, as in Figure 8.11b, we proceed as follows: Choose H (which is to become B') any point not on line DE such that line $BH \parallel$ line AD. Let l_1 be any line passing through the midpoints M and N of segments BH and AD; the affine reflection f_1 with axis l_1, direction AD and $k = 1$, will map A to D and B to $H = B'$. Now with L any point of

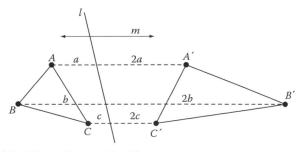

Figure 8.10 Affine reflection ($k = 2$).

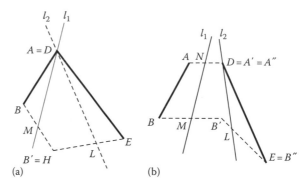

Figure 8.11 Reflecting *AB* to *DE*.

BE and l_2 = line *DL*, let f_2 be the affine reflection with axis l_2, direction *B'E*, and $k = LE/B'L$, which maps *D* to itself and *B'* to *E*. Thus, the product $f_1 f_2$ maps *A* to *D* and *B* to *E*, with $B' \notin$ line *DE*. Since betweenness is preserved, segment *AB* is mapped onto segment *DE* in either case.

Theorem 3   Given any two triangles *ABC* and *DEF*, there exists an affine transformation mapping $\triangle ABC$ onto $\triangle DEF$ consisting of the product of three, or fewer, affine reflections.

Proof   We have already proven that a product of two or fewer affine reflections f_1 and f_2 will map *AB* to *DE*, with $B' \notin$ line *DE*. If $C'' = F$, we are finished. Otherwise, consider line $C''F$. If $C''F$ intersects line *DE* at *Q* (Figure 8.12a), define $k = QF/C''Q$ (directed distance understood), and consider the affine mapping f_3 having axis $\overleftrightarrow{DE} = l_3$ and parameter *k*. This will map C'' to *F* since $QF = kC''Q$. (If $C''F$ is parallel to line *DE*, as in Figure 8.12b, or if $C'' \in$ line *DE*, then an adjustment of line l_2 through *D* will move C'' on line $C'C$ away from *n*, the parallel to *DE* through *F*, or away from line *DE*—possible, since $B' \notin$ line *DE*). Thus, the product $f_3 f_2 f_1$ will map *A* to *D*, *B* to *E* (since *D* and *E* are fixed points of f_3), and *C* to *F*, as desired. ◣

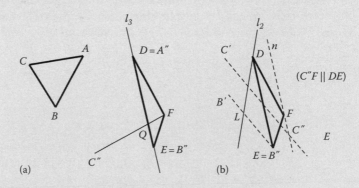

Figure 8.12 Reflecting $\triangle ABC$ to $\triangle DEF$.

Corollary E   Every affine mapping is a product of three, or fewer, affine reflections.

Proof   Let f be any affine mapping, and let $\triangle ABC$ be any triangle, with A', B', and C' the images of A, B, and C under f. Theorem 3 established the existence of a product of three or fewer affine reflections mapping A, B, and C to A', B', and C'. Let g be that product, which is also an affine mapping. Now consider the product $g^{-1}f$. The mapping $g^{-1}f$ maps A to A', then A' back to A, so A is a fixed point. Similarly, B and C are also fixed points, and by Corollary D of Theorem 2 $g^{-1}f$ fixes every point in the plane. Hence, if P is any point, evidently $f(P) = g(P)$ since $g^{-1}f$ maps P to itself. Thus, the two mappings f and g agree at every point, hence $f = g$, and f is a product of three or fewer affine reflections. ◣

The following theorem is now self-evident, normally proven with coordinates requiring a certain amount of algebra. The proof of uniqueness closely resembles the argument given in the preceding corollary for $f = g$.

Fundamental Theorem for Affine Transformations   Given any two triads of noncollinear points (A, B, C) and (D, E, F), there exists a unique affine transformation that maps the first triad onto the second.

An illustration of how to use affine mappings to solve a geometry problem appears in the next example. Other applications appear in the problem section.

Example 1

The familiar theorem concerning the medians of a triangle can be proven with an affine transformation that maps a given triangle *ABC* to an equilateral triangle *A′B′C′*. By Corollary B, the medians *AL*, *BM*, and *CN* of $\triangle ABC$ map to those of $\triangle A'B'C'$, as shown in Figure 8.13. Using properties of 30°–60° right triangles, medians *A′L′*, *B′M′*, and *C′N′* must meet at *P*, the circumcenter of $\triangle A'B'C'$. Its inverse image *G* is the point of concurrency of segments *AL*, *BM*, and *CN*. Also, the proportionality property of *P*, being true for $\triangle A'B'C'$, is also true for *G* (by Theorem 2) proving that *G* is the point on each median that is two-thirds the distance from a vertex of the triangle to the midpoint of the opposite side.

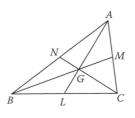 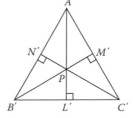

Figure 8.13 Proof of a familiar theorem.

The final concept to be considered is the coordinate form for an affine transformation. Since there can be no vanishing line, it must be concluded from the form **(1)**, Section **8.1** of a projective transformation that $a_3 = b_3 = 0$ and $c_3 \neq 0$. Thus, $x' = (a_1 x + b_1 y + d_1)/c_3$ and $y' = (a_2 x + b_2 y + d_2)/c_3$. Hence, an arbitrary affine transformation can be represented in coordinate form by the following equations (where a, b, c, d, x_0, and y_0 are constants):

(1)
$$\begin{cases} x' = ax + by + x_0 \\ y' = cx + dy + y_0 \end{cases} \quad (ad \neq bc)$$

Example 2

Consider the linear transformation f given by

$$\begin{cases} x' = -x + 3y \\ y' = x + 2y \end{cases}$$

Find the images A', B', and C' of the points $A(-1, 1)$, $B(1, 1)$, and $C(1, 3)$, graph, and compare $\triangle ABC$ with the image triangle $\triangle A'B'C'$.

Solution

By substitution, we find $A' = (4, 1)$, $B' = (2, 3)$, and $C' = (8, 7)$. The graph is shown in Figure 8.14. Thus $\triangle A'B'C'$ is a distortion of $\triangle ABC$, an elongated isosceles triangle. Although it is an isosceles triangle like $\triangle ABC$, with $A'C' = B'C'$, its base is $A'B'$ instead of AC as in $\triangle ABC$.

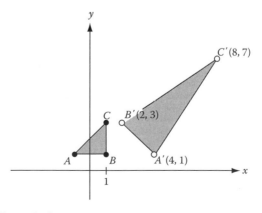

Figure 8.14 Example 2.

Example 3

For each of the following affine transformations, find all fixed points and describe them geometrically:

(a) $\begin{cases} x' = 3x - y - 5 \\ y' = 2x - y - 4 \end{cases}$ (b) $\begin{cases} x' = 3x - y \\ y' = 4x - y \end{cases}$

Solution

Set $x' = x$ and $y' = y$ in each case. For **(a)** we obtain

$$x = 3x - y - 5 \quad \text{or} \quad 2x - y = 5$$

$$y = 2x - y - 4 \quad \text{or} \quad 2x - 2y = 4$$

Subtracting produces $y = 1$, and $x = 3$. Thus, $(3, 1)$ is the only fixed point. For **(b)** we have

$$x = 3x - y \quad \text{or} \quad y = 2x$$

$$y = 4x - y \quad \text{or} \quad y = 2x$$

In this case, there is a line of fixed points, namely, $y = 2x$. (This can easily be checked by substituting $(t, 2t)$ for real t, representing any point on $y = 2x$.) There are no other fixed points.

Example 4

Consider the general form **(1)** of an affine transformation f.

(a) Find the form of an affine transformation f (conditions on the coefficients) that fixes the points $A(0, 1)$ and $B(0, 2)$ (two points on the y-axis).
(b) Show that f fixes all points on the y-axis. Can f have a fixed point that is not on the y-axis? Use algebra.

Solution

(a) Start with

$$\begin{cases} x' = ax + by + x_0 \\ y' = cx + dy + y_0 \end{cases}$$

and substitute the coordinates of the given fixed points A and B ($x' = x$ and $y' = y$):

$$0 = b + x_0, \quad 1 = d + y_0 \quad \text{(from point } A\text{)}$$

$$0 = 2b + x_0, \quad 2 = 2d + y_0 \quad \text{(from point } B\text{)}$$

Thus, eliminating x_0, $b = 0$ and $x_0 = 0$, and eliminating y_0, $d = 1$ and $y_0 = 0$. Therefore, the required form of f is

$$\begin{cases} x' = ax \\ y' = cx + y \end{cases}$$

where a and c are arbitrary, $a \neq 0$.

(b) A point on the y-axis must have coordinates $(0, t)$ for arbitrary t. Substituting these coordinates into the equations just found for f we obtain $x' = a \cdot 0 = 0$ and $y' = c \cdot 0 + t = t$, hence $(0, t)$ is a fixed point. Suppose some other point (p, q) for $p \neq 0$ were a fixed point. Then

$$p = ap \quad \text{or} \quad a = 1 \quad \text{and} \quad q = cp + q \quad \text{or} \quad c = 0$$

Thus f reduces to $x' = x$ and $y' = y$—the equations for the identity mapping. Thus, unless f is the trivial identity, f cannot fix a point not on the y-axis—an illustration of Corollary D above.

The invariants for affine transformations include those of the more general projective transformations, and, as we have seen, include a few more. As we continue to specialize projective transformations, the list of invariants will continue to grow.

Invariants for Affine Transformations

1. Lines
2. Betweenness
3. Parallelism
4. Ratios of collinear segments
5. Cross ratio

Problems (Section 8.2)

Group A

1. Consider the affine transformation given by

$$\begin{cases} x' = 10x \\ y' = y \end{cases}$$

   (a) Find the images A', B', and C' of the vertices $A(-1, 1)$, $B(1, 1)$, and $C(1, 3)$ of a right triangle, graph, and compare $\triangle ABC$ with $\triangle A'B'C'$.

   (b) Is $\triangle A'B'C'$ a right triangle? Is it similar to $\triangle ABC$?

2. Consider the affine transformation

$$\begin{cases} x' = 3x - 6y \\ y' = 3x - 2y \end{cases}$$

(a) Find the images A', B', and C' of the vertices $A(0, 0)$, $B(4, 0)$, and $C(4, 3)$ of a right triangle, graph, and compare $\triangle ABC$ with $\triangle A'B'C'$

(b) Is $\triangle A'B'C'$ a right triangle? Is it similar to $\triangle ABC$?

3. Find the images of the points $P(3, 0)$, $Q(4, 1)$, $R(2, 3)$, and $S(1,2)$ under the affine mapping given by

$$\begin{cases} x' = 3x - 6y \\ y' = 3x - 2y \end{cases}$$

and carefully graph both the original points, and their images. Make any significant observations you can.

4.* For each of the following affine transformations, find all the fixed points, and describe them geometrically.

(a) $\begin{cases} x' = x + y + 2 \\ y' = 3y + 4 \end{cases}$ **(b)** $\begin{cases} x' = 3x + y + 2 \\ y' = 2x - y + 8 \end{cases}$

5. Can an affine transformation have exactly three fixed points?

6. Do you notice any particular attributes of the affine transformation:

$$\begin{cases} x' = 0.6x - 0.8y \\ y' = 0.8x + 0.6y \end{cases}$$

Answer this by experimenting with the mapping, finding the image of $\triangle ABC$ of Problem 1, for starters.

Group B

7. Use affine transformations to prove that the segment joining the midpoints of two sides of a triangle is parallel to the third side and has one-half its length.

8. Given point P in the interior of $\angle ABC$, construct a transversal DE of $\angle ABC$ such that P is the midpoint of segment DE. [**Hint:** Map $\angle ABC$ to a right angle $\angle A'B'C'$.]

9. Prove that a trapezoid can be mapped by an affine transformation onto one having two right angles.

10. Obtain the following formula for the transformed slope m' of the line l', the image of the line l having slope m, under an affine transformation **(1)**.

(2) $m' = \dfrac{c + dm}{a + bm}$

(To save work, it suffices to find the coordinates of the images of $(0, 0)$ and $(1, m)$ in **(1)** and to use the slope formula for m'. Explain why this is sufficient.)

11.   Consider the affine transformation given here that fixes the origin.

$$\begin{cases} x' = ax + by \\ y' = cx + dy \end{cases}$$

If this transformation is also slope-preserving (each line maps to one that is parallel to it), what conditions on a, b, c, and d must be satisfied? [Using **(2)** write down the resulting equations for such a mapping, and make a cursory examination of the properties it must satisfy.]

12.   Show that the transformation on the coordinate plane given by

$$x' = 2x - 3y + 1, \quad y' = 6x - 9y - 5$$

maps the entire plane $\mathcal{P}$ to the line $3x' - y' + 2 = 0$, so is not an affine transformation. What condition in **(1)** does this violate?

13.   Give a coordinate proof of the fundamental theorem (above), using **(1)**. [The algebra is made easier by first showing that the mapping that takes $O(0, 0)$, $U(0, 1)$, and $V(1, 0)$ onto $D(p, q)$, $E(r, s)$, and $F(u, v)$ is uniquely determined, and the one taking O, U, and V onto $A(a, b)$, $B(c, d)$, and $C(e, f)$ is also determined. How will this produce the required affine mapping? Does it produce a unique mapping?]

Group C

14.* Show that the product of two parallel projections, where the first map takes points in $\mathcal{P}$ to points of a plane $\mathcal{P}'$ perpendicular to $\mathcal{P}$ via lines parallel to some line e not perpendicular to $\mathcal{P}$, followed by a second projection from $\mathcal{P}'$ back to $\mathcal{P}$ via lines parallel to another line f not perpendicular to $\mathcal{P}$, is an affine reflection with $\mathcal{P} \cap \mathcal{P}'$ as axis.

15.   **A Proposition on Real Numbers**   Prove the following:

> **Theorem**   Suppose that $a, b, c,$ and d are real numbers such that
>
> (a) $ad \neq bc$
> (b) $a^2 + c^2 = b^2 + d^2$
> (c) $ab = -cd$
>
> Then either $b = -c$ and $a = d$ or $b = c$ and $a = -d$.

[**Hint:** Using (c) and assuming $c \neq 0$, $b \neq 0$, let $a/c = -d/b = x$, and substitute into (b). You must also examine the cases $c = 0$ or $b = 0$.]

16. **Coordinate Form of Conformal Affine Transformations**
A *conformal* transformation is one that preserves angles,
and thus perpendicularity. Using the theorem in Problem
15 and the relation **(2)** (Problem 10), show that any angle-
measure preserving affine transformation has one of the
two forms

$$\begin{cases} x' = ax - by + x_0 \\ y' = bx + ay + y_0 \end{cases} \qquad \begin{cases} x' = ax + by + x_0 \\ y' - bx - ay + y_0 \end{cases}$$

(Preserving perpendicularity means that if $mn = -1$ then
$m'n' = -1$ in **(2)** for variable m.)

17. **Direct, Opposite Affine Transformations** An interesting phenomena is how the orientation of points is
affected by an affine mapping. Given three points A, B,
and C oriented in the counterclockwise (positive) direction, it is evident that an *affine reflection* will map these
points to those oriented in the clockwise direction and
thus reverses the orientation. Such a mapping is called
opposite; a mapping which preserves orientation is
called **direct**. A fundamental result for affine mappings
in general, is that an affine transformation f: $\{x' = ax +$
$by, y' = cx + dy\}$ is direct iff the determinant of its
coefficients $(ad - bc)$ is positive. Verify this result by
finding the images of the positively ordered points
$A(0, 0)$, $B(1, 0)$, and $C(0, 1)$ under f. (It can be shown that
the orientation behavior of f on just one triangle implies
that same behavior for all triangles.)

18. **Area-Preserving Affine Transformations** Another
interesting phenomena is the connection between the
determinant d of coefficients of an affine mapping f and
the area of a polygon and its image under f: *the area
of the mapped polygon equals the area of the original
polygon times d.* Thus, f preserves area iff $d = 1$. Verify
this for the particular case of $\triangle ABC$ and its image
$\triangle A'B'C'$ under the mapping f: $\{x' = 3x + 5y, y' = x + 2y\}$
(verify that $d = 1$ here) where A is the origin, $B = (2, -1)$,
and $C = (2, 4)$.

The following problems require a knowledge of matrices and
matrix products, but you might want to try these even without prior
knowledge of matrices, and figure out these concepts for yourself.

19. Find the images P', Q', R', and S' under the affine map f:
$\{x' = 3x - 2y, y' = 2x + 4y\}$ if $P = (3, 0)$, $Q = (4, 1)$, $R = (2, 3)$,
and $S = (1, 2)$ (as in Problem 3 above) by calculating the
matrix product

$$\begin{bmatrix} 3 & -6 \\ 3 & -2 \end{bmatrix}\begin{bmatrix} 3 & 4 & 2 & 1 \\ 0 & 1 & 3 & 2 \end{bmatrix} = \begin{bmatrix} 9 & \cdots \\ 6 & \end{bmatrix}$$

20. Consider the points $U(3, 4)$ and $V(2, -1)$. By direct substitution (without matrices), find the images U', V', U'', and V'' under the affine maps g and $fg \equiv f \circ g$ if

$$f: \begin{cases} x' = 3x + 5y \\ y' = 2x + 4y \end{cases} \qquad g: \begin{cases} x' = 2x - 2y \\ y' = x + 3y \end{cases}$$

[**Hint:** For fg, apply g first, then f.]

21. Suppose in Problem 20 you wanted to find, in addition to those of U and V, the images of $W(-2, 4)$ and $T(-1, 5)$ under the product fg. In view of Problem 19, this suggests the matrix product

$$\begin{bmatrix} 3 & 5 \\ 2 & 4 \end{bmatrix} \left(\begin{bmatrix} 2 & -2 \\ 1 & 3 \end{bmatrix} \begin{bmatrix} 3 & 4 & 2 & 1 \\ 0 & 1 & 3 & 2 \end{bmatrix} \right)$$

What is the most efficient way to proceed? (The direct product shown here repeats the calculations made in Problem 20.) [**Hint:** Use the associate law for matrix products $(AB)C = A(BC)$.]

22. In working Problem 21 you might have observed the following connection between a matrix product and the product of the affine maps f and g:

$$f \leftrightarrow \begin{bmatrix} 3 & 5 \\ 2 & 4 \end{bmatrix} \quad \text{and} \quad g \leftrightarrow \begin{bmatrix} 2 & -2 \\ 1 & 3 \end{bmatrix} \quad \text{implies}$$

$$fg \leftrightarrow \begin{bmatrix} 3 & 5 \\ 2 & 4 \end{bmatrix} \begin{bmatrix} 2 & -2 \\ 1 & 3 \end{bmatrix} = \begin{bmatrix} 11 & 9 \\ 8 & 8 \end{bmatrix}$$

Verify this for the particular point $U = (3, 4)$, whose image under fg was calculated in Problem 20. Note that this involves the associative law of matrix products.

8.3 Similitudes and Isometries

An important special case of affine transformations is the family of angle-preserving and distance-preserving mappings.

Definition 1   A **similitude** is any affine mapping of the plane for which there is a positive constant k, called the **dilation factor**, such that $P'Q' = k \cdot PQ$ for all points P, Q. If $k = 1$, the mapping preserves distances, and is an **isometry**.

It is easy to see that a similitude maps distinct points to distinct points, thus is one-to-one, and maps any triangle to one that is similar to it by the SSS similarity criterion. Thus, a similitude is angle preserving. It will eventually be shown that given any two similar triangles, a unique similitude exists which maps the first triangle onto the second. An analogous result is true for isometries. This shows that similitudes and isometries exist in abundance.

By the result of Problem 16 above, the most general similitude has the coordinate form

(1)
$$\begin{cases} x' = ax - \varepsilon by + x_0 \\ y' = bx + \varepsilon ay + y_0 \end{cases} \quad a^2 + b^2 > 0, \varepsilon = \pm 1$$

where a and b are constants. (The mapping is direct iff $\varepsilon = 1$.) To find the dilation factor, consider $A(0, 0)$ and $B(1, 0)$ and their images $A'(x_0, y_0)$ and $B'(a + x_0, b + y_0)$. Then $A'B' = kAB = k \cdot 1$ implies $k = A'B'$, or, by the distance formula,

(2)
$$k = \sqrt{a^2 + b^2}$$

If $k = 1$ then **(1)** is an isometry, so the coordinate form of an isometry is given by **(1)**, with $a^2 + b^2 = 1$.

Example 1

Consider the transformation f given by

$$\begin{cases} x' = 3x - 4y + 5 \\ y' = 4x + 3y - 2 \end{cases}$$

- **(a)** Find the coordinates of the images A', B', and C' for the points $A(0, 1)$, $B(0, 2)$, and $C(1, 1)$ and sketch the graph showing $\triangle ABC$ and its image $\triangle A'B'C'$.
- **(b)** Find the dilation factor k, and show that the lengths of the sides of $\triangle A'B'C'$ are k times those of $\triangle ABC$, thus showing that the triangles are similar.

Solution

- **(a)** Substituting the coordinates of A, B, and C into the equations for f, we find $A' = (1, 1)$, $B' = (-3, 4)$, and $C' = (4, 5)$. The graph is shown in Figure 8.15.
- **(b)** From **(2)** we find $k = \sqrt{3^2 + 4^2} = \sqrt{25} = 5$. The distance formula produces $A'B' = \sqrt{(1 + 3)^2 + (1 - 4)^2} = \sqrt{16 + 9} = 5 = 5AB$. Similar calculations show that $B'C' = 5 \cdot BC$ and $A'C' = 5 \cdot AC$.

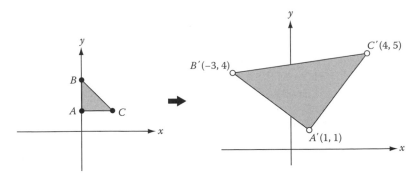

Figure 8.15 Example 1.

Example 2

Repeat the instructions of Example 1 for the transformation f given by

$$\begin{cases} x' = \tfrac{3}{5}x - \tfrac{4}{5}y + 5 \\ y' = \tfrac{4}{5}x + \tfrac{3}{5}y - 2 \end{cases}$$

Solution

(a) The images of $A(0, 1)$, $B(0, 2)$, and $C(1, 1)$ are as follows (found by substituting coordinates): $A'(4.2, -1.4)$, $B'(3.4, -0.8)$, and $C'(4.8, -0.6)$ (exact decimals). The graph is shown in Figure 8.16.

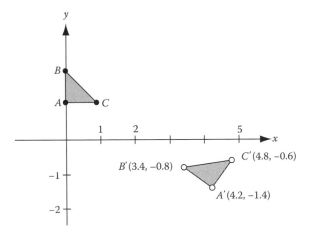

Figure 8.16 Example 2.

(b) $k = \sqrt{(\tfrac{3}{5})^2 + (\tfrac{4}{5})^2} = 1$. Thus, the mapping is an isometry, and the triangles are congruent, as is evident from the figure.

It is apparent from Figure 8.15 that the image of $\triangle ABC$ under f in Example 1 was an enlarged copy of $\triangle ABC$, rotated by a certain amount. This is a generally true behavior for similitudes: any similitude simply enlarges, or shrinks (the technical term is *dilates*) any given figure, then translates or rotates it, and reflects it in some line for the final result. The actual proof of this will appear later (Corollary **B** in Section **8.4**). Isometries have a similar property, but there is no enlarging or shrinking. Another result is that if an affine transformation maps a single triangle to one similar to it, then it must be a similitude. An example will illustrate this result (which we do not prove here).

Example 3

An affine mapping f takes the points $A(2, 4)$, $B(1, 1)$, and $C(3, 1)$ to $A'(11, 1)$, $B'(6, 6)$, and $C'(10, 8)$, respectively. Graph these points to verify that $\triangle ABC \sim \triangle A'B'C'$, then use the coordinate form of an affine transformation to determine the correct coefficients from the conditions imposed on it, in order to verify that f is a similitude.

Solution

The graph is shown in Figure 8.17; it is apparent that the triangles are similar, isosceles triangles. Using the coordinate form of an affine transformation, with its 6 coefficients a, b, c, d, x_0, and y_0, we proceed to substitute the coordinates of corresponding points for x, y, x', and y' into **(1)**, Section **8.2** in order to determine conditions on the coefficients. For A and A', we obtain

$$11 = 2a + 4b + x_0, \quad 1 = 2c + 4d + y_0$$

Continuing this process, we obtain the further conditions

$$6 = a + b + x_0, \quad 6 = c + d + y_0$$

$$10 = 3a + b + x_0, \quad 8 = 3c + d + y_0$$

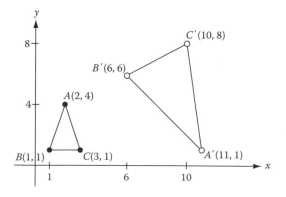

Figure 8.17 Example 3.

The result is a system of six linear equations in six unknowns. However, the system divides itself into two sets of three equations in three unknowns (the three equations on the left, and those on the right), which are easily handled. Thus

$$11 = 2a + 4b + x_0$$

$$6 = a + b + x_0$$

$$10 = 3a + b + x_0$$

By elimination, the second and third of the above three equations yield

$$2a = 4 \quad \text{or} \quad a = 2$$

The remaining variables are easily calculated, yielding the results $b = 1$, $x_0 = 3$. Taking the system of three equations in the right column above, the results are $2c = 2$ or $c = 1$, $d = -2$ and $y_0 = 7$. Hence, going back to the original affine transformation **(1)** with the values for the coefficients just found, we obtain the correct form for a similitude (with $\varepsilon = -1$):

$$\begin{cases} x' = 2x + y + 3 \\ y' = x - 2y + 7 \end{cases}$$

An important special case of a similitude is the transformation described next. It is a prominent tool for solving geometry problems.

Definition 2   A **dilation** (also, **homothetic**) mapping is a transformation for which there exists a point C, called its **center**, and a real number $k \neq 0$, called the **dilation factor**, such that a given point P is mapped to P' lying on ray CP with $CP' = k \cdot CP$ if $k > 0$, or lying on the ray opposite CP with $CP' = -k \cdot CP$ if $k < 0$. Such a dilation will be denoted by $\mathcal{D}[C, k]$.

This mapping can be realized as a projectivity by observing the composition of the two central projections, the first from $\mathcal{P}$ to a plane $\mathcal{P}'$ parallel to $\mathcal{P}$ from center A, the second from $\mathcal{P}'$ back to $\mathcal{P}$ from center B, where line AB is perpendicular to $\mathcal{P}$ and neither A nor B lie in $\mathcal{P}$ (see Figure 8.18). The mapping is defined in general as the product of the central projections having centers A and B, as in the previous definition of affine mappings, where A and B play the role of E and F in the previous construction. Note that in this case, point C (the foot of the perpendicular of A on plane $\mathcal{P}$) is a fixed point of the mapping. The dilation factor turns out to be a certain cross ratio.

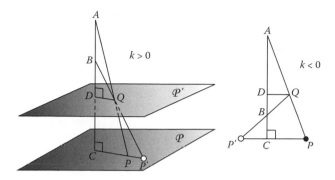

Figure 8.18 Dilation as a projective transformation.

Theorem   The projective mapping described above is a dilation having center C with dilation factor $k = [AB, DC]$.

Proof   Since $\overrightarrow{DQ} \parallel \overrightarrow{CP}$, we have both $DQ/CP = AD/AC$ and $CP'/DQ = BC/BD$. Multiplying these two equations yields the expression

$$\frac{DQ \cdot CP'}{CP \cdot DQ} = \frac{AD \cdot BC}{AC \cdot BD} = [AB, DC]$$

or $CP' = k \cdot CP$ where $k = [AB, DC]$. Note that if (ABD) then $k > 0$ and P' lies on ray CP, while if (ADB) then $k < 0$ and P' lies on the ray opposite CP. Thus, the product of the two central projections is the dilation $\mathcal{D}[C, k]$. ∖

The coordinate form for a dilation $\mathcal{D}[O, a]$ having center at the origin and dilation factor a is predictably

(3) $$\begin{cases} x' = ax \\ y' = ay \end{cases} \quad a \neq 0$$

An illustration showing the use of a dilation to solve a geometry problem appears in the next example.

Example 4

Problem: Given an acute-angled a triangle, $\triangle ABC$, the object is to find a construction that will inscribe a square $PQRS$ inside it such that one side of the square QR lies on the base of the triangle and the other two vertices P and S lie on the other two sides of the triangle.

Solution

Construct *any* square whose base lies on the extended side BC with one vertex on extended side BA, as shown in Figure 8.19. Now use a dilation $\mathcal{D}[B, k]$ centered at B to pull the square to the proper position. Since squares map to squares under dilations, the image is the square desired. This leads to the following compass/straight-edge construction: After (1) constructing the arbitrary square, (2) draw a line through the opposite vertex of that square, and (3) intersect this line with side *AC* to determine a vertex *P* of the required square. From point *P*, the rest of the square can be easily constructed, dropping the perpendicular from *P* to side *BC*, and so on. This is called the method of *false position*.

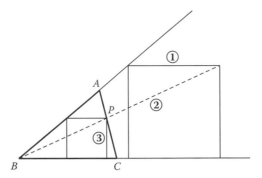

Figure 8.19 Inscribing a square in a triangle.

A more significant application for dilations involves the nine-point circle. Recall that it is the circle passing through the midpoints of the sides of a triangle and six other points.

Example 5: Properties of the Nine-Point Circle

Show that the nine-point circle of $\triangle ABC$ can be mapped by a dilation with center *H* to the circumcircle. Find further results of this mapping. Note the relevance of Problems 22–23, Section **7.4**.

Solution

In Figure 8.20 is shown the nine-point circle and its nine points, including the three Euler points—the midpoints *X*, *Y*, and *Z* of the segments *HA*, *HB*, and *HC*, where *H* is the orthocenter. Consider the dilation $\mathcal{D}[H, 2]$. This maps *X*, *Y*, and *Z* on the nine-point circle to *A*, *B*, and *C* on the circumcircle, which, because the dilation factor is 2, is twice the size of the nine-point circle. Thus, we have the following properties, listed in order of logical dependence:

(a) The nine-point circle has a radius one-half that of the circumcircle.
(b) Since centers of circles are preserved under dilations, the center of circle ***C*** maps to the center of the image circle ***C′***, therefore *U* maps to *O*.
(c) The nine-point center lies on the Euler line and is the midpoint of segment *HO* joining the orthocenter and circumcenter.

(d) The nine-point center and circumcenter are harmonic conjugates of the orthocenter and centroid. That is, $[HG, UO] = -1$ (see Problem 12).

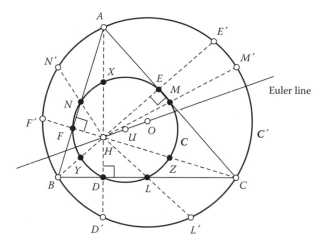

Figure 8.20 Transforming the nine-point circle.

The list of invariants for similitudes follows. (To give a list for isometries, one can simply add the item *distance*—an invariant that obviously implies all the others.)

Invariants for Similitudes

1. Lines
2. Betweenness
3. Parallelism
4. Ratios of collinear segments
5. Cross ratio
6. Angle measure
7. Circles

Problems (Section 8.3)

Group A

1. Identify each of the following affine transformations by their coordinate forms

(a) $\begin{cases} x' = 2x \\ y' = 3y \end{cases}$ **(b)** $\begin{cases} x' = 3x \\ y' = 3y \end{cases}$ **(c)** $\begin{cases} x' = 2x + y \\ y' = -x + 2y \end{cases}$

2. Identify each of the following affine transformations by their coordinate forms.

(a) $\begin{cases} x' = -x + y \\ y' = x + y \end{cases}$ (b) $\begin{cases} x' = \frac{5}{13}x - \frac{12}{13}y \\ y' = \frac{12}{13}x + \frac{5}{13}y \end{cases}$ (c) $\begin{cases} x' = x + 2y \\ y' = x + 3y \end{cases}$

3.   Find the images of the points $A(0, 1)$, $B(1, 1)$, and $C(1, 2)$, and graph, under each of the transformations

(a) $\begin{cases} x' = 5x \\ y' = 5y \end{cases}$ (b) $\begin{cases} x' = 3x - 2y \\ y' = 2x + 3y \end{cases}$

4.   Find the images of the points $P(3, 0)$, $Q(4, 1)$, $R(2, 3)$, and $S(1, 2)$ under the affine mapping given by

$$\begin{cases} x' = 3x - 2y \\ y' = 2x + 4y \end{cases}$$

and carefully graph the results. Do you find anything significant? (Compare with Problem **3**, Section **8.2**.)

5.   Repeat the directions for Problem 4 using the different affine transformation

$$\begin{cases} x' = 3x \\ y' = 3y \end{cases}$$

Do you find anything significant?

6.*   Find the dilation factor k for each of the following similitudes to determine which one is an isometry.

(a) $\begin{cases} x' = 12x + 5y \\ y' = -5x + 12y \end{cases}$ (b) $\begin{cases} x' = -\frac{2}{3}x + \frac{\sqrt{5}}{3}y \\ y' = \frac{\sqrt{5}}{3}x + \frac{2}{3}y \end{cases}$

Group B

7.   Use the method of false position to find a compass/straight-edge construction for the circle which is tangent to the sides of a given angle and passing through a given point in the interior of that angle. (See Figure P.7.)

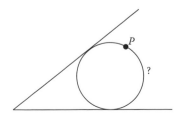

Figure P.7

8. Each of the following four matrices provide the coefficients $a, b, c,$ and d of an affine mapping $\{x' = ax + by, y' = cx + dy\}$ in the obvious manner. Three of the mappings take squares into squares, but have decidedly different ways of doing so. Without plotting points, find which of the three mappings do so, and tell what the different ways are.

(a) $\begin{bmatrix} 5 & 2 \\ 2 & 5 \end{bmatrix}$ (b) $\begin{bmatrix} -4 & 3 \\ 3 & 4 \end{bmatrix}$ (c) $\begin{bmatrix} \frac{1}{3} & -\frac{1}{4} \\ \frac{1}{4} & \frac{1}{3} \end{bmatrix}$ (d) $\begin{bmatrix} \frac{4}{5} & -\frac{3}{5} \\ \frac{3}{5} & \frac{4}{5} \end{bmatrix}$

9. If a similitude preserves distance for just one pair of points A and B, then it must do so for every pair, and is an isometry. Prove this
   (a) Synthetically
   (b) Using coordinates

10. If a similitude has a line of fixed points, is it a reflection in that line? Prove, or disprove, giving a counterexample if false.

11. If a similitude has exactly one fixed point, is it a rotation about that point? Prove, or disprove.

12. Letting $HO = 6a$, show that $GO = 2a$ and $GU = -a$. Thus, it follows that $[HG, UO] = -1$ in any triangle ABC with standard notation. Prove.

13. **Scheiner's Pantograph** Versions of an instrument invented in 1630 by Christolph Scheiner are in use to this day. Called a *pantograph*, it can be used to trace a copy of a given pattern and either uniformly decrease, or increase its size, in effect, performing a similitude. The mechanism is fastened at a pivot point O with the remaining rods allowed to move freely in the manner illustrated (Figure P.13). The main component is parallelogram $ABCD$. Show that the mapping $P \rightarrow P'$ is a similitude with a fixed ratio k, and find its value.

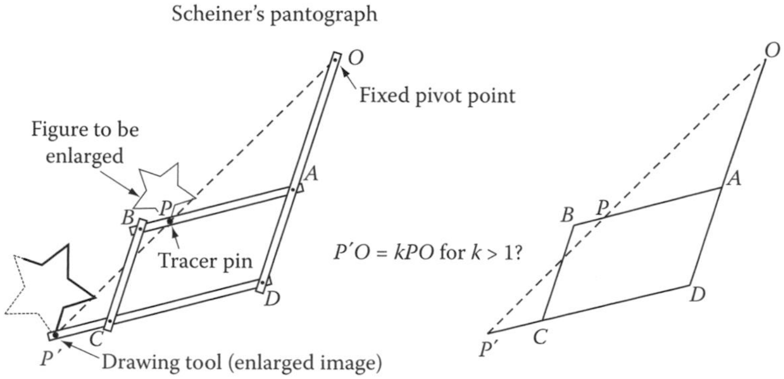

Scheiner's pantograph

Figure P.13

Group C

14. Consider triangles *ABC* and *DEF* having vertices $A(-3, 2)$, $B(3, 4)$, $C(1, -10)$, $D(10, -9)$, $E(28, 5)$, and $F(-18, 27)$. Carefully graph these triangles, and using the formula $\tan \theta = (m_2 - m_1)/(1 + m_1 m_2)$ for the angle between lines having slopes m_1 and m_2, show that the angles of the two triangles are congruent, and thus $\triangle ABC \sim \triangle DEF$.

15. Referring to Problem 14, there exists a unique affine transformation *f* which maps $\triangle ABC$ to $\triangle DEF$. Using the coordinate form for an affine transformation, use the information given to find the coefficients and thus verify that *f* is a similitude.

8.4 Line Reflections: Building Blocks for Isometries and Similitudes

It is perhaps surprising that all isometries in the plane are simply products of reflections in lines—the simplest type of isometry. Line reflections have a multitude of applications in mathematics and physics, as well as in the real world. It is fitting that a section be devoted to this important area of geometry. For convenience, we repeat the definition from a previous section where reflections were originally introduced (in particular, see Section **3.5**).

Definition A **reflection** in the line *l* is the mapping which takes each point *P* in $\mathcal{P}$ to the unique point *P'* also in $\mathcal{P}$ such that *l* is the perpendicular bisector of the segment *PP'*. (If $P \in l$, we take $P' = P$; thus *l* is a line of fixed points for the reflection.)

It is evident that one may construct the reflection *P'* for each point *P* not on *l* by dropping the perpendicular *PM* to point *M* on *l*, then doubling the segment *PM* to segment *PP'* (segment-doubling theorem). It should be noticed that reflections are special cases of affine reflections. It is almost obvious that reflections preserve distance, and consequently, reflections are isometries. In Figure 8.21 is shown two cases, (1) when *A* and *B* lie on the same side of *l*, and (2) when they do not. Thus, we see that in case (1) $\lozenge ABNM \cong \lozenge A'B'NM$ (by SASAS), so that $AB = A'B'$ by CPCPC, and in case (2) $AB' = A'B$ by case (1) so that trapezoid $AA'BB'$ is isosceles, with congruent diagonals, *AB* and *A'B'*.

We include one property of reflections that has practical applications. If a ray of light strikes a mirror or other reflective surface, the angle at which the light ray meets the mirror always equals the angle at which it is reflected off the mirror. The angle between the normal to the mirror and the incoming light ray is called the **angle of incidence**, labeled θ_I, while that between the normal and the reflected ray is the **angle of**

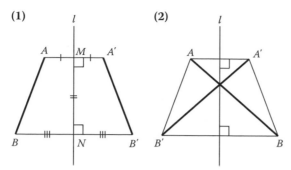

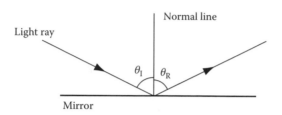

Figure 8.21 Distance-preserving property.

Figure 8.22 Law of physics: $\theta_I = \theta_R$.

reflection, labeled θ_R (see Figure 8.22). The law of physics governing this phenomenon then becomes

$$\theta_I = \theta_R$$

(i.e., *the angle of reflection equals the angle of incidence*).

The geometric principle of line reflection enters the physics law in the following way. Let line l represent the surface of the mirror, as in Figure 8.23. If point E represents the eye of the observer and the ray of light from an object at P is reflected off the mirror (or other reflecting surface) at point R toward point E, then geometrically, the two angles at R are equal, having the value θ, as indicated (complementary to θ_I and θ_R). But the virtual image that is actually seen by the observer at E is point P'—the (geometric) reflection in l of P. Why should this be? The following argument gives an answer. First, the actual total distance traveled by the light ray from P to the observer equals the distance from the virtual image P' to the observer, which can be verified by experiment (the size of the image is identical to the object reflected). Thus we obtain from Figure 8.23

$$PR + RE = P'E = P'R + RE \quad \text{so} \quad PR = P'R$$

Since $\angle MRP'$ and $\angle ERW$ are vertical angles, $\angle MRP' \cong \angle ERW \cong \angle MRP$ hence $\triangle MRP' \cong \triangle MRP$. It follows that l is the perpendicular bisector of segment PP', and therefore, P' is the reflection of P in line l.

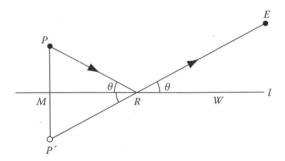

Figure 8.23 Image of point *P* from observer *E*.

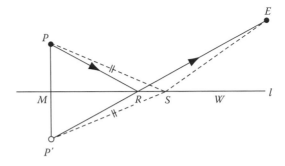

Figure 8.24 Least distance property of light rays.

A mathematical reason for the physical law $\theta_1 = \theta_R$ has been proposed: The ray of light could be assumed to seek the least total distance $PR + RE$ to reach point *E*. If there were some other point on *l*, say *S*, for which the reflected ray to *E* yields a smaller total distance from *P* to *E*, then we would have $PS + SE < PR + RE$ (Figure 8.24). But on the other hand, by the triangle inequality, and with *l* the perpendicular bisector of segment *PP′*,

$$PR + RE = P'R + RE = P'E \leq P'S + SE = PS + SE$$

a contradiction. Therefore, point *R* is the optimal point, thus $\angle ERW \cong \angle MRP' \cong \angle PRM$, showing that the angle of incidence equals the angle of reflection.

Note that the drawing in Figure 8.23 shows how to locate the point *R* of reflection in any application of the principle. If we want to find point *R* on line *l* that minimizes the total distance $PR + RQ$ for any two points *P* and *Q*, we merely reflect *P* in line *l* to *P′*, then draw line *P′E*, whose intersection with *l* is the desired point. This principle is often overlooked in the calculus where, as you might recall, in order to solve the "river" problem, you are supposed to use differentiation (which can be difficult to handle). An example will illustrate.

Example 1: The River Problem

Variations of this problem often appears in calculus texts. One version is that a person walking from A to B wants to stop by the riverside to fill a bucket with water to take to point B (Figure 8.25). We assume that the x-axis represents the riverbank and, for example, $A = (1, 3)$, $B = (5, -1)$. The problem is to find what point $P(t, 0)$ along the river bank minimizes the total walking distance, $AP + PB$. Note that the solution can be provided geometrically by simply drawing line AB' where B' is the reflection of point B in line l (the x-axis), and finding its intersection with the x-axis, since, as analyzed above, that point will minimize the value $AP + PB$. In order to contrast this with the calculus method, you are to solve this problem in two ways:

(a) By using calculus in the traditional manner
(b) By using geometry

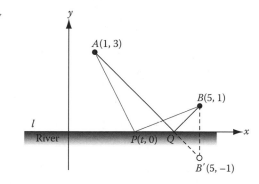

Figure 8.25 Example 1.

Solution

(a) Let $P(t, 0)$ represent a variable point on the riverbank. By the distance formula,

$$AP = \sqrt{(t-1)^2 + (0-3)^2} = \sqrt{t^2 - 2t + 10}$$

$$BP = \sqrt{(t-5)^2 + (0-1)^2} = \sqrt{t^2 - 10t + 26}$$

We want to minimize the function

$$F(t) = AP + BP = \sqrt{t^2 - 2t + 10} + \sqrt{t^2 - 10t + 26}$$

so we take its derivative. After simplifying, we find

$$F'(t) = \frac{t-1}{\sqrt{t^2 - 2t + 10}} + \frac{t-5}{\sqrt{t^2 - 10t + 26}}$$

Setting $F'(t) = 0$, squaring to eliminate the radicals, and using a little algebraic manipulation, we can write

$$\frac{t^2 - 2t + 1}{t^2 - 2t + 10} = \frac{t^2 - 10t + 25}{t^2 - 10t + 26}$$

or

$$1 - \frac{9}{t^2 - 2t + 10} = 1 - \frac{1}{t^2 - 10t + 26}$$

Some more algebra gives us

$$9(t^2 - 10t + 26) = t^2 - 2t + 10$$

$$9t^2 - 90t + 234 = t^2 - 2t + 10$$

or $t^2 - 11t + 28 = 0$. Thus, we finally arrive at the required condition (which may not be sufficient)

$$(t - 4)(t - 7) = 0$$

The only solution lying on [1, 5] is $t = 4$. Therefore, $Q = (4, 0)$ is the optimal point.

(b) We find that $B' = (5, -1)$. The equation of line AB' is given by

$$\frac{y + 1}{x - 5} = \frac{3 + 1}{1 - 5}$$

which simplifies to $y = 4 - x$. This line cuts the x-axis at $y = 0$ or $x = 4$, so that $Q = (4, 0)$ is the optimal point. (*Geometry is better!*)

A second example shows how to solve a problem involving *three* mirrors.

Example 2

Consider the lines l, m, and n in Figure 8.26 representing three reflecting mirrors. What path is taken by the light ray AB originating from point A and reflected by the three mirrors?

Solution

Reflect point A first in line l to point A', and let A'' and A''' be the reflections of A' and A'' in lines m and n, respectively. (Thus, A''' is the image of A under the product of three line reflections.) The ray from A will then take the path $ABCD$, determining point D on n, the intersection of line EA''' with n the reflected ray takes the path $A''C$ to point D, then C on line m, and finally B on line l. The result is the broken line $ABCDE$. The reader should check to see that the reflection law of physics is satisfied at all reflection points. An equivalent solution involving points B' and C' is shown in the figure.

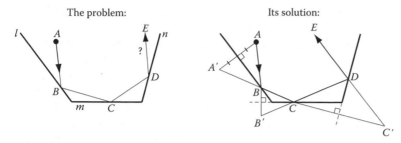

Figure 8.26 Three mirror problem.

The next example is another maxima–minima problem suggesting calculus once again, but for which calculus is not the appropriate method for solution. The problem is to find the triangle having minimal perimeter that is inscribed within another triangle, assumed to be acute-angled. The solution was originally discovered by J. F. Fagnano in 1775; the reflection proof we use is due to L. Fejr.

Example 3

Consider an acute-angled triangle ABC and an inscribed triangle PQR. We want to vary the vertices P, Q, and R so as to decrease the perimeter of $\triangle PQR$ [we let $p(\triangle PQR) \equiv PQ + QR + RP$ denote the perimeter]. As shown in Figure 8.27a, reflect point P in line AC to P', and P in line AB to P''. By properties of reflections, $AP' = AP = AP''$, with $\triangle AP'Q \cong \triangle APQ$ and $\triangle AP''R \cong \triangle APR$. Thus, since $AP' = AP''$, $\triangle AP'P''$ is isosceles, with base $P'P''$ and legs AP', AP''. Suppose that Q^* and R^* denote the points of intersection of line $P'P''$ with lines AC and AB, respectively. Again by properties of reflections, $P'Q^* = Q^*P$ and $P''R^* = R^*P$. By the polygonal inequality, we observe the inequality

$$p(\triangle PQ^*R^*) = PQ^* + Q^*R^* + R^*P$$

$$= P'Q^* + Q^*R^* + R^*P'' = P'P''$$

$$\leq P'Q + QR + RP''$$

$$= PQ + QR + RP = p(\triangle PQR)$$

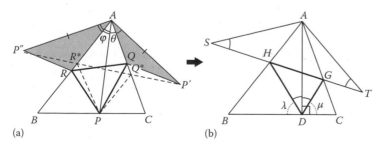

(a)   (b)

Figure 8.27 Proving Fagnano's theorem.

with equality only when $Q = Q^*$ and $R = R^*$. Thus, $\triangle PQ^*R^*$ is an inscribed triangle having strictly less perimeter than $\triangle PQR$ if either $Q \neq Q^*$ or $R \neq R^*$. This is true regardless of the position of P on side BC. Thus, if we consider only triangles PQ^*R^* where Q^* and R^* are constructed as above (whose perimeters are less than or equal to those of all other inscribed triangles), the task remains to minimize $p(\triangle PQ^*R^*)$ as P varies on BC. As we noted above, however, the perimeter of $\triangle PQ^*R^*$ is precisely the length of base $P'P''$ of isosceles triangle $AP'P''$. Thus, to minimize $p(\triangle PQ^*R^*)$, we must minimize the length of segment $P'P''$.

Note that the vertex angle of $\triangle P'AP''$ has constant angle measure: By congruent triangles, $m\angle P'AP = m\angle P'AQ^* + m\angle Q^*AP = 2\theta$, where $\theta = m\angle PAC$; similarly, $m\angle PAP'' = 2\varphi$ where $\varphi = m\angle PAB$, and it follows that

$$m\angle P'AP'' = 2\theta + 2\varphi = 2A$$

where A denotes the measure of $\angle BAC$ (standard notation). Thus, since the vertex angle of the isosceles triangle $AP'P''$ remains constant, the base $P'P''$ can only depend on the length of each leg, $AP' = AP'' = AP$ (isosceles triangles with congruent vertex angles are similar triangles). Consequently, the length of $P'P''$ is minimal when the legs are minimal. Segment AP is minimal precisely when $AP \perp BC$ and $P = D$, where AD is the altitude from vertex A. Let the corresponding position for segment $P'P''$ be $\overleftrightarrow{ST}$ and those of Q^* and R^*, G and H, as shown in Figure 8.27b. Thus, $p(\triangle PQR)$ takes on its minimum when $P = D$, $Q = Q^* = G$, and $R = R^* = H$. [Proof: $p(\triangle PQR) \geq P'P'' \geq ST = p(\triangle DGH)$.] This argument then proves that an inscribed triangle PQR has minimum perimeter when, and only when, $\overleftrightarrow{AP} \perp \overleftrightarrow{BC}$ and Q and R lie on line $P'P''$. Note for later reference that the angles marked in the figure at D are congruent to the congruent base angles of $\triangle AP'P''$, and thus are congruent. (See Problem 10 for an elegant formula for the perimeter of $\triangle PGH$.)

The argument in the previous example leads to the classic result:

Theorem 1: Fagnano's Theorem

The orthic triangle DEF of an acute-angled triangle ABC is the inscribed triangle having least perimeter.

Proof The last statement in L. Fejr's argument in Example 3 shows that $\lambda = \mu = 90 - m\angle AST = 90 - \frac{1}{2}(180 - m\angle SAT) = 90 - \frac{1}{2}(180 - 2A) = A$. Recall the orthic triangle theorem of Section **5.11**, where it was shown that $m\angle EDC = A = m\angle FDB$. This shows that G coincides with E, and H coincides with F, proving that $\triangle DGH$ is the orthic triangle. Fejr's argument also showed that $\triangle DGH$ (i.e., the orthic triangle) has minimal perimeter. ◣

The result alluded to earlier will now be proven. It is relevant to the following question: Is it possible to vary an isometry by adding more reflections? In particular, is an isometry that is made up of the product of a dozen or more reflections, for example, significantly different from one made up of three or four? A similar question arises with respect to affine mappings and affine reflections. The answer, of course, came in the fundamental theorem for affine transformations appearing above: An affine transformation is the product of three or fewer affine reflections. So we anticipate a similar result for isometrics. Two preliminary results are observed in order to reach our goal: the fundamental theorem for isometries. (Note that here, mere affine reflections are not involved; we are using *Euclidean* reflections exclusively, as defined above.)

(1)   Given $\triangle ABC \cong \triangle DEF$, there exist three, or fewer,
        reflections whose product maps $\triangle ABC$ onto $\triangle DEF$.

Proof   This already appeared in a slightly different form in Section **3.4** (the SAS theorem derived from **Axiom M** using three or fewer reflections). We present essentially the same argument, slightly revised to fit the current situation. Let two distinct congruent triangles be given, say $\triangle ABC \cong \triangle DEF$; thus we may assume that $A \neq D$. Let l_1 be the perpendicular bisector of segment AD. Reflect $\triangle ABC$ to $\triangle A'B'C'$ in line l_1, as shown in Figure 8.28. (The shaded triangles indicate the different positions of $\triangle ABC$ under subsequent line reflections in this argument.) It follows that $A' = D$ by definition of reflection.

Construct the perpendicular bisector l_2 of segment $B'E$ (if $B' = E$, label the vertices of $\triangle A'B'C'$ by A'', B'', C'' respectively). Line l_2 must pass through D because $DB' = A'B' = AB = DE$ and D is equidistant from B' and E. Now reflect $\triangle A'B'C'$ to $\triangle A''B''C''$ in line l_2, with B' mapping to E, hence $B'' = E$. Also, since $A' = D$ lies on l_2, D is a fixed point and $A'' = A' = D$. Thus, segment $A'B'$ maps to segment $A''B''$, which coincides with segment DE. If C'' coincides with F, we are finished, for then $\triangle A''B''C'' \equiv \triangle DEF$ is the image of $\triangle ABC$ under either one, or two line reflections. Otherwise, $C'' \neq F$ and we use a third reflection in line $l_3 \equiv$ line DE to map $\triangle A''B''C''$ to $\triangle DEF$: Since D and E are each equidistant from F and C'' (Why?), line DE is the perpendicular bisector of segment FC'' and C'' maps to F under reflection in line DE. Thus, $\triangle DEF$ is the image of $\triangle A''B''C''$ under the reflection in line l_3, and it is therefore the image of $\triangle ABC$ under the product of either one, two, or three line reflections. ◣

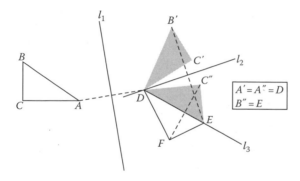

Figure 8.28 Reflecting $\triangle ABC$ to $\triangle DEF$.

(2)   At most one isometry can map a given triangle to
one that is congruent to it.

Proof   This result is a corollary to the fundamental theorem for affine
transformations. Suppose there were two isometries f and g mapping
$\triangle ABC$ to $\triangle DEF$. Since isometries are affine transformations, by the fundamental theorem there can be only one, so $f = g$. ⬟

The direct result of combining **(1)** and **(2)** is

Theorem 2: Fundamental Theorem for Isometries

Given $\triangle ABC \cong \triangle DEF$, there exists a unique isometry that maps $\triangle ABC$ to $\triangle DEF$.

Corollary A   An isometry is the product of three or fewer line reflections.

The analogous theorem for similitudes follows easily. One first observes
that if $\triangle ABC \sim \triangle DEF$, then a dilation $\mathcal{D}[A, k]$ centered at A maps segment BC
to PQ such that $AP = DE$ and $AQ = DF$, so that $\triangle APQ \cong \triangle DEF$ (see Figure
8.29). (The proper value for k is $DE/AB = AP/AB$.) There exists an isometry f
that maps $\triangle APQ$ to $\triangle DEF$. Thus fd maps $\triangle ABC$ to $\triangle DEF$, proving:

Fundamental Theorem for Similitudes   Given two similar triangles,
there exists a unique similitude which maps the first triangle to the
second.

Corollary B   A similitude is the product of a dilation and three or
fewer line reflections.

The coordinate form for a reflection will now be derived, in keeping
with our procedure for previous transformations. The reader is no doubt
already familiar with the reflections in the two coordinate axes, given by

$$\begin{cases} x' = x \\ y' = -y \end{cases} \text{(reflection in the x-axis)} \qquad \begin{cases} x' = -x \\ y' = y \end{cases} \text{(reflection in the y-axis)}$$

The coordinate form for an arbitrary reflection is surprisingly complicated in view of its elementary nature. But in spite of this its derivation is
straightforward. Suppose that line l is given by the equation $ax + by = c$,

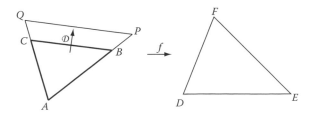

Figure 8.29 Similitude mapping $\triangle ABC$ to $\triangle DEF$.

(a, b, not both zero) and that $P(x, y)$ is reflected to $P'(x', y')$. First, assume that $a \neq 0$ and $b \neq 0$, and observe that the slope of the line of reflection is $-a/b$, so the slope of the perpendicular PP' is b/a. By the slope formula applied to $P(x, y)$ and $P'(x', y')$,

$$\frac{b}{a} = \frac{y' - y}{x' - x}$$

Also, the midpoint of segment PP' lies on l and satisfies the equation $ax + by = c$, where x is replaced by $(x' + x)/2$ and y by $(y' + y)/2$, so a second relation emerges:

$$a\left(\frac{x' + x}{2}\right) + b\left(\frac{y' + y}{2}\right) = c$$

These two relations yield the following system of equations in x' and y'

$$\begin{cases} bx' - ay' = bx - ay \\ ax' + by' = 2c - ax - by \end{cases}$$

This system can be solved by Cramer's rule (which will be left to the reader) with the result:

(3) Reflection in $ax + by = c$ $\begin{cases} kx' = (b^2 - a^2)x - 2aby + 2ac \\ ky' = -2abx - (b^2 - a^2)y + 2bc \end{cases}$ $(k = a^2 + b^2)$

(This relation is observed to be valid also in the cases a or $b = 0$.)

Example 4

According to **(3)**, the reflection in the line $y = 2x$, (with $a = -2$, $b = 1$, $c = 0$, and $k = 5$) is given by the equations

$$\begin{cases} 5x' = -3x + 4y \\ 5y' = 4x + 3y \end{cases}$$

Verify that this is indeed a reflection in $y = 2x$ by performing the following operations.

(a) Calculate the images of the points $A(2, 4)$, $B(5, 0)$, and $C(3, 1)$ and draw the graphs of $\triangle ABC$ and $\triangle A'B'C'$.
(b) Show that the points of the line $y = 2x$ are all fixed points by calculating the image of the points $(t, 2t)$ for real t.

Solution

(a) Substitution yields, for A', $5x' = -3 \cdot 2 + 4 \cdot 4 = -6 + 16 = 10 \rightarrow x' = 2$, and $5y' = 4 \cdot 2 + 3 \cdot 4 = 8 + 12 = 20 \rightarrow y' = 4$; thus $A'(2, 4)$. Similarly, the images of B and C are $B'(-3, 4)$ and $C'(-1, 3)$. The graph is shown in Figure 8.30, showing clearly the reflection expected.

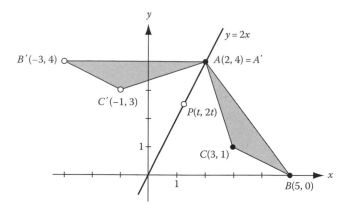

Figure 8.30 Example 4.

(b) Substitution of t and $2t$ for x and y produces

$$5x' = -3 \cdot t + 4 \cdot 2t = -3t + 8t = 5t \rightarrow x' = t$$

$$5y' = 4 \cdot t + 3 \cdot 2t = 4t + 6t = 10t \rightarrow y' = 2t$$

An important alternate form for **(3)** *is obtained by assuming the equation of l ($ax + by = c$) is written in the* normal form*: $x \sin\theta - y\cos\theta = p$, where $0 \le \theta < \pi$. Any line may be put into normal form by simply dividing by the nonzero quantity $\sqrt{a^2 + b^2}$ and setting*

$$\sin\theta = \frac{a}{\sqrt{a^2 + b^2}} \quad \text{and} \quad \cos\theta = \frac{b}{\sqrt{a^2 + b^2}}$$

(Problem 13 asks you to show that θ measures the angle between the positive x-axis and l, and that $c(a^2 + b^2)^{-1/2} = P$ is the distance from $O(0, 0)$ to l.) Thus, a new form for the reflection in l is obtained by replacing a, b, and c by the parameters $\sin\theta$, $-\cos\theta$, and p in **(3)**, with $k = (\sin\theta)^2 + (-\cos\theta)^2 = 1$. We obtain

$$x' = x(\cos^2\theta - \sin^2\theta) - y(-2\sin\theta\,\cos\theta) + 2p\sin\theta$$

$$= x\cos 2\theta + y\sin 2\theta + 2p\sin\theta$$

making use of double-angle formulas from trigonometry. A similar form follows for y'. Thus one obtains:

(4) Reflection in $x\sin\theta - y\cos\theta = p$
$\begin{cases} x' = x\cos 2\theta + y\sin 2\theta + 2p\sin\theta \\ y' = x\sin 2\theta - y\cos 2\theta - 2p\cos\theta \end{cases}$

Problems (Section 8.4)

Group A

1. Consider the arrangement of balls shown in Figure P.1. In a game of billiards, the best shot to sink the ball is the side pocket nearest the cue ball because of the position of other balls blocking easier shots. Provide, by sketch, the geometric construction that locates the point on AB for a trick bank shot that will sink the ball.

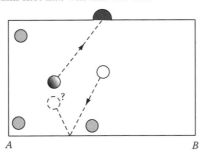

Figure P.1

2. With the coordinate system shown in Figure P.2, where units are in inches, the 7-ball is blocking the cue ball at $A(5, 20)$ from the 4-ball at $B(20,10)$. In the game of snooker, the balls must be pocketed in numerical order. In order to make a trick shot that lands the 4-ball in the side pocket at C, at what point on the opposite side should the cue ball be hit, assuming that an angle trajectory requires hitting the 4-ball? (You are to give the x-coordinate of this point; it will only be an approximation of the exact point, which is influenced by the angle of impact at which the cue ball must strike the 4-ball.)

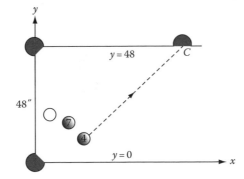

Figure P.2

3. Show that a light ray from P that is reflected from a mirrored surface t_1 at Q at an angle of 45° will be reflected from a second mirrored surface t_2 at right angles to t_1 in a direction parallel to the original direction PQ. (See Figure P.3.)

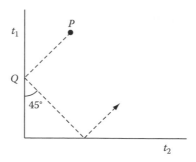

Figure P.3

4.   What happens in Problem 3 if t_2 is tilted at an angle of 60°
     instead of being perpendicular to t_1, and PQ is reflected
     from t_1.
     (a) At an angle of 60°?
     (b) At an angle of 30°?

5.   Generalize the outcome of Problem 3 to show that a light
     ray reflected at any angle from t_1 will be reflected from t_2
     in a direction parallel to the original direction PQ. Do you
     think it necessary for $t_1 \perp t_2$ to achieve this phenomenon?
     If you can, give proofs (Figure P.5).

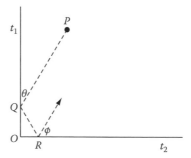

Figure P.5

6.   Give a construction that will inscribe a parallelogram in
     a given rectangle. Discuss uniqueness of such parallelo-
     grams (can a vertex of the parallelogram be chosen arbi-
     trarily on one side of the rectangle?).

7.   Use the result of Problem 6 to show that one can hit a ball
     on a rectangular pool table in such a way that it will return
     to its original position after it caroms off the four sides of
     the table. (If not for friction, the ball would theoretically
     continue on the same path indefinitely.) Prove your con-
     clusions (Figure P.7).

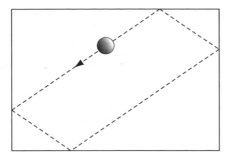

Figure P.7

8. Establish a result similar to that obtained in Problem 7 if the pool table were triangular instead of rectangular. Show that there is only one trajectory possible (the ball cannot be placed arbitrarily).

Group B

9. A triangle *ABC* and its orthic triangle *DEF* were drawn. Someone erased the sides of △*ABC* in the drawing. Show how one might recover the original triangle *ABC*.

10.* Prove that the perimeter *p* of the orthic triangle (*ST* in Example 3) of any acute-angled triangle *ABC* in standard notation is given by $p = 2h_a \sin A$.

11. At what point $P(t, 0)$ along the *x*-axis (modeled as a river-bank) should pipe be laid to join each of two towns at $A(1, 5)$ and $B(8, 3)$ to a pumping station at *P* so as to minimize the length of pipe needed? (Units are in miles.) Work this problem in two ways (see Figure P.11):
    (a) By calculus
    (b) By geometry

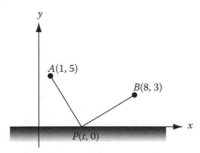

Figure P.11

12. **The Bridge Problem** A road and bridge combination is being planned joining two towns, *A* and *B*, separated by

a river (see Figure P.12). Assume that the river banks are straight, parallel lines, that the bridge will be built perpendicular to the river banks, and that the roads are to be straight. Where should the bridge be built so that the total distance across the bridge and along the roadway between the two towns and bridge is minimized?

Note: The seemingly logical solution depicted in the figure, where the midpoint of the bridge is the midpoint between the points C and D lying on the straight line joining A and B does *not* give the correct answer. (It turns out that the solution does not involve reflections at all.)

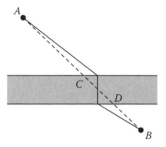

Figure P.12

13.   **The Normal Form for the Equation of a Line**   As shown in Figure P.13 at right, let φ $(0 \le \varphi < \pi)$ be the angle between the positive x-axis and the line l: $ax + by = p$, where we assume that $a^2 + b^2 = 1$. Then set $a = \sin\theta$ and $b = -\cos\theta$, $0 \le \theta < \pi$, as in (4). Provide details for the following.

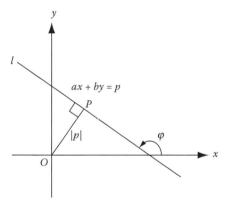

Figure P.13

(a)   Show that for $\varphi \ne \pi/2$, $\tan\varphi$ = slope of l = $\tan\theta$. Thus $\varphi = \theta$ and θ is the counter clockwise angle between the positive x-axis and l (also true if $\varphi = \pi/2$). The equation $x\sin\theta - y\cos\theta = p$ is called the **normal form**.

(b)   Show that p is the signed distance from the origin $O(0, 0)$ to l where $P > 0$ if P (the foot of O on l) lies in quadrants I or IV, or on the positive x on y axis, and $p < 0$ otherwise. (If l passes through O then $p = 0$.) [**Hint:** Find the equation of line OP and the coordinates of P, using $a^2 + b^2 = 1$; note that $a = \sin\theta \ge 0$.]

14. **Distance from a Point to a Line** Prove that if the equation $ax + by + c = 0$ of line l is written in normal form, then the distance from $P(x_0, y_0)$ to l is simply $|ax_0 + by_0 + c|$. [**Hint:** Let the line l' parallel to l passing through (x_0, y_0) have normal equation $ax + by = q$; the required distance is $|q - p|$. Consider all cases.]

Group C

15. In Example 3, what if the angle at A were obtuse in triangle ABC? Discuss whether an inscribed triangle with minimal perimeter exists for this case. What bearing does the orthic triangle have?

16. Find a construction that shows the exact direction a laser beam at A should be aimed in order to remain within the restricted area shown in Figure P.16 and to hit the target at B by reflection in the mirrored surface M. (**Note:** The path shown in the figure is not the correct answer.)

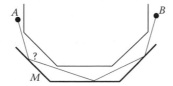

Figure P.16

17. **Construction of Tangent to an Ellipse and Reflective Property** The ellipse has the property that, in the context of the surface of revolution which it defines, light or sound waves from one focus will be reflected to the other. This may be proven geometrically, without calculus. Assume the characteristic property for an ellipse $PF_1 + PF_2 = 2a$ (constant) for all points P on the ellipse ($=E$), where F_1, F_2 are the two foci (Figure P.17), and that a line is tangent to E iff it has exactly one point in common with E (the point of contact). Let P be any point on the ellipse and construct line t perpendicular to the bisector of $\angle F_1PF_2$ at P. Let W be constructed so that (F_2PW) and $PW = PF_1$, with G the intersection of t with segment F_1W.

    (a) Prove that W is the reflection of F_1 in t and that t is tangent to E at P. (Consider $Q \neq P$ on t; use the triangle inequality for $\triangle QWF_2$ to show that $QF_1 + QF_2 > 2a$.)

    (b) Conclude that a ray originating at F_1 will be reflected to F_2.

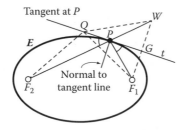

Figure P.17

18.   Give a tangent construction for the hyperbola and parabola analogous to that of Problem 17, and establish the correct reflection phenomena for these two conic sections in terms of their foci (without calculus).

8.5 Translations and Rotations

Two fundamental transformations for coordinate geometry are the translations and rotations, much used in the calculus. Although their properties are easy to describe, they seem to resist a clear geometric definition. We know that a translation should be a kind of "sliding" action, without rotation, but how can this be defined geometrically, without using coordinates?

In synthetic (non-coordinate) geometry, consider the phenomena illustrated in Figure 8.31. In the diagram at left we have the product of two reflections over *parallel* lines l and m at a distance d apart. The image of the original configuration (the block letter "F") looks as though it could be obtained by directly sliding the figure to the new position, a total distance of $a + d + (d - a) = 2d$ (see figure). In the diagram to the right, the same configuration ("F") has been reflected in two *intersecting lines l* and m, and here it appears that the letter "F" has been rotated about point O, the point of intersection of the two lines. A close look also reveals that the angle of rotation is twice that of the angle between l and m (here, one obtains $\varphi + \theta + (\theta - \varphi) = 2\theta$). These diagrams can be easily created on *Sketchpad* to illustrate the results mentioned here.

Definition   A **translation** is the product of two reflections in parallel lines. A **rotation** is the product of two reflections in intersecting lines. In the latter case, the point of intersection C of the two lines is a fixed point, called the **center of rotation**.

An almost trivial fact, but of great importance, is that, as a result of the definition, *translations and rotations are isometries in the plane* $\mathcal{P}$. This is a much used concept in all that follows. We consider first the case of translations, making some pertinent observations.

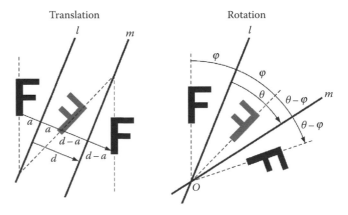

Figure 8.31 Products of two reflections.

You no doubt have been introduced to vectors. A translation *defines* a vector in the following sense. A vector is the entire set of directed line segments having the same length and the same direction. A translation can be defined so that the initial points of all such segments map to the terminal points of those same segments. The basis for this, and for vectors as well, is the following important, characteristic feature of any translation.

Theorem If P and Q are any two points, and P' and Q' represent their images under a translation, the quadrilateral $\lozenge PQQ'P'$ is a parallelogram.

Proof First, we prove that the product of the reflections in parallel lines l and m maps a given line n to one n' parallel to n. There are two cases: (1) $n \parallel l$, and (2) $n \nparallel l$ for this part.

1. $(n \parallel l)$ This case follows from the evident fact that a line parallel to the line of reflection is mapped to a line that is also parallel to the line of reflection (use parallelograms). Thus, in Figure 8.32(1) line n, which is parallel to l, maps to a line q parallel to l, hence to m. In turn, q is reflected in m to $q' \equiv n'$ parallel to q. Hence, $n \parallel n'$.

2. $(n \nparallel l)$ Choose any line $k \parallel l$, meeting n at some point A (since k cannot be parallel to n), as shown in Figure 8.32(2). Let k' be the image of line k under the product of the two reflections, as n' is the image of n; since reflections are angle-preserving, $\theta = \varphi$. But by (1) just established, $k \parallel k'$ and n meets k' at some point B, with $\theta = \omega$ (corresponding angles), hence $\varphi = \omega$ and it follows that $n' \parallel n$.

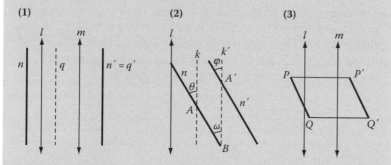

Figure 8.32 Translation of a line segment.

3. Thus, we have established the fact that a translation maps line PQ to line $P'Q'$ parallel to PQ [Figure 8.33(3)]. Since distances are preserved under translations, $PQ = P'Q'$, and thus it follows that a pair of opposite sides of $\lozenge PQQ'P'$ are parallel and congruent, hence $\lozenge PQQ'P'$ is a parallelogram, as desired. ◣

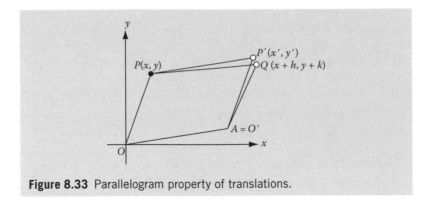

Figure 8.33 Parallelogram property of translations.

The coordinate form for a translation can now be established from its definition. Suppose the point $P(x, y)$ is mapped to $P'(x', y')$ in the coordinate plane, and suppose that the origin $O(0, 0)$ maps to $O' \equiv A(h, k)$, as in Figure 8.33. By the above theorem, $\Diamond POAP'$ is a parallelogram. Consider the point $Q(x + h, y + k)$. By calculating the slopes of the lines involved, we conclude that $\Diamond POAQ$ is also a parallelogram: Slope $(PQ) = k/h = $ Slope (OA) and Slope $(AQ) = y/x = $ Slope (OP). Hence $Q = P'$ and we arrive at the familiar coordinate form for a translation:

(1)   **Translation from** O **to** (h, k) $\begin{cases} x' = x + h \\ y' = y + k \end{cases}$

A numerical example will verify **(1)** directly.

Example 1

Consider parallel lines l: $x + y = 1$ and m: $x + y = 3$. Use the coordinate form **(3)**, Section **8.4**, to find the product of the reflections in l and m (in that order). This is a translation that maps (0, 0) first to (1, 1) under reflection in l, then to (2, 2) under reflection in m, as indicated in Figure 8.34.

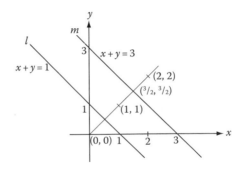

Figure 8.34 Example 1.

Solution

Using the coordinate form for a reflection, we have for l the parameters $a = b = c = 1$ so that $b^2 - a^2 = 0$, $k = a^2 + b^2 = 2$, and $ac = bc = 1$. Thus, the reflection takes on the coordinate form $2x' = 0 \cdot x - 2 \cdot y + 2 = -2y + 2$, $2y' = -2 \cdot x - 0 \cdot y + 2 = -2x + 2$ or

(∗)
$$\begin{cases} x' = -y + 1 \\ y' = -x + 1 \end{cases} \quad \text{(Reflection in } l\text{)}$$

The reflection in m likewise takes on the coordinate form (with $c = 3$) $2x' = 0 \cdot x - 2 \cdot y + 6 = -2y + 6$, $2y' = -2 \cdot x - 0 \cdot y + 6 = -2x + 6$, or

(∗∗)
$$\begin{cases} x'' = -y' + 3 \\ y'' = -x' + 3 \end{cases} \quad \text{(Reflection in } m\text{)}$$

(using double primes in order to set this up for substitution of (∗) into (∗∗), thus computing the composition of the two reflections.) By substitution,

$$x'' = -(-x + 1) + 3 = x + 2 \quad \text{and} \quad y'' = -(-y + 1) + 3 = y + 2$$

and the final result is the transformation given by $x' = x + 2$, $y' = y + 2$, agreeing with **(1)**.

We now turn our attention to rotations. The coordinate form of a rotation can be obtained by applying the form **(4)** in Section **8.4** for a line reflection. For convenience, at first we consider the special case when the center of rotation is the origin. Since the reflections in l and m defining the rotation must each pass through the origin (Figure 8.35), **(4)** reduces to the following:

$$l \colon x \sin\theta - y \cos\theta = 0 \quad \begin{cases} x' = x \cos 2\theta + y \sin 2\theta \\ y' = x \sin 2\theta - y \cos 2\theta \end{cases}$$

$$m \colon x \sin\varphi - y \cos\varphi = 0 \quad \begin{cases} x'' = x' \cos 2\varphi + y' \sin 2\varphi \\ y'' = x' \sin 2\varphi - y' \cos 2\varphi \end{cases}$$

As before, the form for the product of these two reflections is obtained by substitution. Thus, making use of the addition identities for sine and cosine as appropriate, we obtain

$$x'' = (x \cos 2\theta + y \sin 2\theta)\cos 2\varphi + (x \sin 2\theta - y \cos 2\theta)\sin 2\varphi$$

$$= x(\cos 2\theta \cos 2\varphi + \sin 2\theta \sin 2\varphi) + y(\sin 2\theta \cos 2\varphi - \cos 2\theta \sin 2\varphi)$$

$$= x \cos(2\varphi - 2\theta) - y \sin(2\varphi - 2\theta)$$

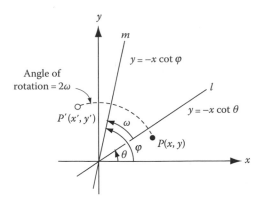

Figure 8.35 Proving the coordinate form of a rotation.

Similarly, one obtains for y'' (letting the reader provide the details)

$$y'' = x\cos(2\varphi - 2\theta) + y\sin(2\varphi - 2\theta)$$

Recall that θ and φ represent the positive angles between the positive x-axis and lines l and m, so their difference is the angle between lines l and m, as in Figure 8.35. If ω represents that angle, the product of the two reflections takes on the form

$$x' = x\cos 2\omega - y\sin 2\omega \quad \text{and} \quad y' = x\sin 2\omega + y\cos 2\omega$$

But the angle of rotation is twice the angle between l and m, that is, 2ω. If we change notation and let θ be the angle of rotation, the coordinate equations become

(2)
$$\begin{cases} x' = x\cos\theta - y\sin\theta \\ y' = x\sin\theta + y\cos\theta \end{cases}$$

(This form should be familiar to most readers.) The most general form for a rotation when the center of rotation $C(h, k)$ is different from the origin can be obtained by translating to the origin O, rotating about an angle θ using **(2)**, then translating back to C. The final result is

(3) Rotation about $C(h, k)$ through an angle θ:

$$\begin{cases} x' - h = (x - h)\cos\theta - (y - k)\sin\theta \\ y' - k = (x - h)\sin\theta + (y - k)\cos\theta \end{cases}$$

A numerical example follows.

Example 2

Consider the rotation defined by reflection in the lines l: $y = 3x - 3$ and m: $y = -2x + 7$, which intersect at $C(2, 3)$. The slopes of the lines are $m_1 = 3$ and $m_2 = -2$, so the formula $\tan\omega = (m_2 - m_1)/(1 + m_1 m_2)$ gives the angle ω between l and m.

(a) Show that under the product of the reflections in l and m an arbitrary point $P(x, y)$ will map to the point $P'(x', y')$ according to **(3)**, a rotation about C through an angle 2ω.

(b) Using the formula for $\tan \omega$, verify the angle of rotation as given by **(3)**.

Solution

(a) With l: $3x - y = 3$ and thus $a = 3$, $b = -1$, $c = 3$, and $k = 3^2 + (-1)^2 = 10$, the reflection in l for x' has the coordinate form $10x' = -8x + 6y + 18$. Similarly, $10y' = 6x + 8y - 6$. Hence the reflection in l is given by

$$(*) \qquad \begin{aligned} 5x' &= -4x + 3y + 9 \\ 5y' &= 3x + 4y - 3 \end{aligned} \qquad \text{(Reflection in } l\text{)}$$

The reader can work out the similar form for the reflection in m: $2x + y = 7$ to obtain

$$(**) \qquad \begin{aligned} 25x'' &= -3(5x') - 4(5y') + 140 \\ 25y'' &= -4(5x') + 3(5y') + 70 \end{aligned} \qquad \text{(Reflection in } m\text{)}$$

where the equations were adjusted in order to avoid fractions. To find the equations for the product of the two reflections, we substitute $(*)$ into $(**)$:

$$25x'' = -3(-4x + 3y + 9) - 4(3x + 4y - 3) + 140 =$$
$$- 25y + 125$$
$$25y'' = -4(-4x + 3y + 9) + 3(3x + 4y - 3) + 70 =$$
$$25x + 25$$

which reduces to

$$\begin{cases} x'' = -y + 5 \\ y'' = x + 1 \end{cases}$$

Note that this can be put into the form

$$\begin{cases} x'' - 2 = -(y - 3) = (\cos \pi/2)(x - 2) - (\sin \pi/2)(y - 3) \\ y'' - 3 = x - 2 = (\sin \pi/2)(x - 2) + (\cos \pi/2)(y - 3) \end{cases}$$

which is equivalent to **(3)**, with $\theta = \pi/2$.

(b) $\tan \omega = (-2 - 3)/(1 + (-6)) = -5/-5 = 1$ so that $\omega = \pi/4$. The angle of rotation is therefore $2\omega = \pi/2$, in agreement with (a).

The use of translations and rotations to solve problems in geometry is illustrated in the examples and problems which follow. The first example shows how to solve a problem that is analogous to a now classical problem (appearing here as Problem 12). The example follows.

Example 3

Let *l*, *m*, and *n* be any three parallel lines (Figure 8.36). Find a construction that will produce an isosceles right triangle with its three vertices on the three parallel lines.

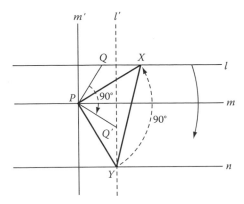

Figure 8.36 Placing a right triangle on three parallel lines.

Solution

Draw any line segment *PQ* extending from line *m* to line *l*, as shown. Using *P* as center, rotate segment *PQ* and line *l* about a 90° angle to the position shown by the dotted line. The image *l'* of line *l* will meet line *n* at some point *Y*; use the reverse rotation about *P* through an angle of 90° to obtain the pre-image *X* of point $Y \equiv X'$ and returning *l* to its original position. Since rotations are isometries, $PY = PX$, and angle *XPY* is a right angle, so $\triangle XPY$ is a solution to the problem. [Since *Q* was arbitrary, do you think there are other (noncongruent) solutions?]

Note: Starting out with segment *PQ* as in the above solution, which was just a guess for the solution, but likely incorrect, is another example of the method of *false position*, mentioned previously. As you might expect, it is a prominent feature in classical geometry problems and their solutions. It is interesting to create your own procedure for doing such constructions on *Sketchpad*. ◼

Example 4: Fermat's Point

Recall that Fermat's point was defined previously as the point of concurrency of the segments *AA**, *BB**, and *CC** where *A**, *B**, and *C** are vertices of the equilateral triangles constructed externally on the sides of $\triangle ABC$. (See Figure 7.40, Section **7.6**.) A problem that often appears in calculus texts (but for which calculus is an inappropriate method of solution) goes back to Fermat (1601–1665). The solution given here was proposed by J. E. Hoffman in 1929. The problem is to locate a point *P* inside a triangle *ABC* (assumed to have angles less than 120°), which minimizes the sum $PA + PB + PC$. (This problem is sometimes called the *warehouse problem*—find a location for a manufacturing plant using items from three warehouses that minimizes the total distance of travel from the warehouses to the plant.) The geometric solution not only reveals the minimum value, it also shows where to locate the point, not immediately evident from the calculus solution.

Start with any point P inside $\triangle ABC$, as shown in Figure 8.37. Perform the rotation about the vertex A through an angle of measure 60°, mapping $\triangle APB$ to $\triangle AP'B'$. Since $AP' = AP$ and $m\angle P'AP = 60$, $\triangle AP'P$ is equilateral. Also note that $PB = P'B'$. It follows that $\triangle ABB'$ is also equilateral, with $B' = C^*$. Hence, the sum whose minimum we seek is

$$x = PA + PB + PC = PP' + P'B' + PC$$

which, by the triangle inequality, yields

$$x \geq C^*C = c$$

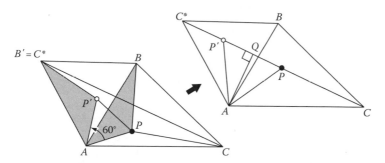

Figure 8.37 Using a rotation to construct the Fermat point.

Thus, we have found an absolute minimum (namely c) for all such sums x as P varies throughout the interior of $\triangle ABC$. (C^*C is constant as P varies.) If we can now prove that the sum actually equals c for some position of P, we will have established the existence of the minimum we are seeking.

Drop the perpendicular AQ to segment C^*C and locate P on segment QC such that $m\angle QAP = 30°$, defining P' as before. Again the sum x will equal the quantity mentioned above, but this time both P and P' will lie on segment C^*C, with $(C^*P'PC)$. Hence

$$x = PA + PB + PC = P'P + C^*P' + PC = C^*C = c$$

and the proof is complete, including the correct position for point P on segment CC^*. Since there was nothing special about vertices A and B, the same optimal position for P would have placed it on segments AA^* and BB^* as well. Hence, *P is the Fermat point of* $\triangle ABC$, introduced in Section **7.6**, Example 6.

Moment for Reflection

Consider the very last statement of the solution to Example 4. Do you think the argument given would provide an independent proof that segments *AA**, *BB**, and *CC** are concurrent, without using Ceva's theorem (as was used previously in Example 6, Section **7.6**)? **Note:** You might find it interesting to see what *Sketchpad* has to offer relative to the ideas involved here.

Example 5: Napoleon's Theorem

A theorem credited to the famous general is the following: Let triangle *ABC* be given, with equilateral triangles *BCF*, *ACG*, and *ABH* constructed externally on the three sides, with *P*, *Q*, and *R* their respective centers (centroids), as illustrated in Figure 8.38. Then △*PQR* is equilateral. A coordinate proof seems at first to be the only feasible method, but this turns out to be all but intractable. It turns out that products of rotations can be used to construct an elegant proof.

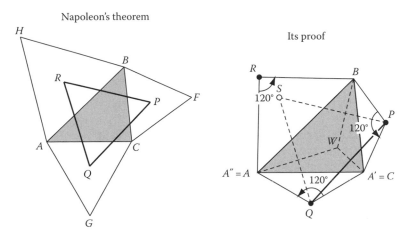

Figure 8.38 Proving Napoleon's theorem.

To begin, note that *m∠BPC* = 120°, with *PB* = *PC*. Thus, point *C* is the rotation of point *B* about center *P* through a positive angle of 120°. Similarly, *A* is the rotation of *C* about center *Q*, and *B* is the rotation of *A* about center *R*, each through a positive angle of 120°.

Construct equilateral triangle *PQS* with *S* on the same side of line *PQ* as *R*; the goal is to prove *R* = *S*. Let *W* be the reflection of *B* in line *PS*, *A′* the reflection of *W* in line *PQ*, and *A″* the reflection of *W* in line *QS*. Thus, *A′* is the image of *B* under the product of reflections in lines *PS* and *PQ*, thus a rotation about *P* through an angle 2·60° = 120°. Therefore, *A′* = *C*. Similarly, *A″* is the image of *C* under the product of reflections in lines

PQ and *QS*, thus a rotation about *Q* through an angle of 120°, hence *A″* = *A*. Note that the product of the reflections in lines *QS* and *PS* is the rotation about *S* through an angle of 120°, and that, under this rotation, *A* maps to *B*. But *B* is also the rotation about *R* through an angle of 120°, hence triangles *RAB* and *SAB* are isosceles triangles having equal vertex angles and a common base *AB*. It follows that *S* = *R*, as desired.

Problems (Section 8.5)

Group A

1. Show that a translation has no fixed points unless it is the identity.

2. Show that a rotation has exactly one fixed point unless it is the identity.

3. Using Corollary **A**, Section **8.4**, prove that a direct isometry is either a translation or a rotation.

4. Find the determinant of coefficients of **(1)** in Section **8.3** and show that $\varepsilon = 1$ iff the isometry is a direct mapping. (Refer to Problem 17, Section **8.2** for the result needed)

5. Using **(1)** in Section **8.3**, give a coordinate proof that an isometry that has exactly one fixed point is a rotation. (It may be assumed that the fixed point is the origin.)

6. Without plotting points, find the angle of rotation θ for the mapping given in coordinate form by

$$\begin{cases} x' = \tfrac{1}{2}x - \tfrac{\sqrt{3}}{2}y \\ y' = \tfrac{\sqrt{3}}{2}x + \tfrac{1}{2}y \end{cases}$$

Verify your answer by plotting the point $A(1, 0)$ and its image A' under this mapping.

7. Find the angle of rotation θ to the nearest one-tenth degree for the mapping whose coordinate form is

$$\begin{cases} x' = \tfrac{12}{13}x - \tfrac{5}{13}y \\ y' = \tfrac{5}{13}x + \tfrac{12}{13}y \end{cases}$$

Find the image of $A(13, 0)$ by substitution, draw an accurate graph of A and A', then measure $\angle AOA'$ with a protractor as a check. (This may be accomplished more satisfactorily using *Sketchpad*.)

8. Find the images of $A(0, 0)$, $B(2, 0)$, and $C(0, 3)$ under the rotation

$$\begin{cases} x' = 0.6(x-2) - 0.8(y-3) + 2 \\ y' = 0.8(x-2) + 0.6(y-3) + 3 \end{cases}$$

and graph the triangles ABC and $A'B'C'$ in order to observe a rotation about $C(2, 3)$. Find the angle of rotation to the nearest one-tenth degree and verify using your graph.

9. Isosceles triangles ABC and DEF have bases BC and EF lying on line l (Figure P.9), with $\angle A$ less than $\angle D$ and the altitude from A of $\triangle ABC$ greater than that from D of $\triangle DEF$. The problem is to find a construction for a line m parallel to l, which intersects the interiors of the two triangles in segments having equal lengths. Explain the details of the following solution to the problem.
   (1) Draw line n through D perpendicular to line EF, and translate $\triangle ABC$ to $\triangle A'B'C'$ so that A' falls on line n.
   (2) By hypothesis, the legs of $\triangle A'B'C'$ will intersect those of $\triangle DEF$ at certain points G and H.
   (3) Line GH is the solution to the problem.

The problem   The solution

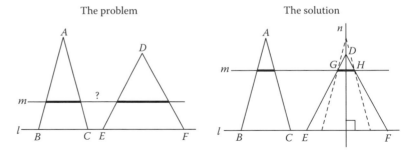

Figure P.9

Group B

10. Find a construction that will inscribe an equilateral triangle inside a square, with one vertex of the triangle coinciding with a vertex of the square (Figure P.10). (There is a direct construction not involving transformation theory.)

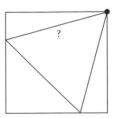

Figure P.10

11. Use a rotation to inscribe an equilateral triangle in a square with one vertex specified on one side of the square (see Figure P.11). Can this vertex be given arbitrarily? [**Hint:** Rotate the square in the false position method.]

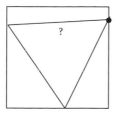

Figure P.11

12. **Classical Construction Problem**   Give a construction that will inscribe the vertices of an equilateral triangle onto each of three parallel lines. First solve if the lines are spaced equally (not requiring transformation theory), then when they are not.

13. Given two circles C and D, and line l outside both, construct a line $m \parallel l$ which cuts the two circles in chords of equal lengths. Identify a few situations in which there will be no solution (Figure P.13).

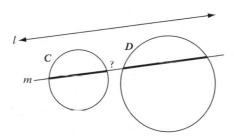

Figure P.13

14. **Half-Turns** A rotation about an angle of 180° is called a **half-turn** in geometry. Using **(3)**, find the coordinate form of a half-turn
     (a) About the point $C(h, k)$
     (b) About the origin $O(0, 0)$. (This mapping is the familiar **central reflection** through the origin O. The graph $y = f(x)$ is said to be *symmetric about O* iff $f(-x) = -f(x)$ for all x (such as $y = x^3$). Such graphs are invariant under central reflection through O.

15. **An Application of Half-Turns** Given two circles C and D intersecting at A as in Figure P.15, find a line through A which cuts the two circles in chords of equal lengths (give a construction).

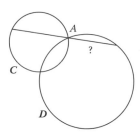

Figure P.15

Group C

16. **The Linear Group** The set of linear (affine) transformations in the plane forms what is called a *group* under the composition product rule, having an identity *e* (the identity mapping), and an inverse for each element. A *subgroup S* is a subset of a group having the property that the inverse of each element in *S* and the product of any two elements of *S* belong to *S* (i.e., *S* is a subset of a group that is *closed* under products and inverses). For example, the set of all integer powers of 2 (the set $\{2^n: n = 0, \pm 1, \pm 2, \ldots\}$) is a subgroup of the real numbers $x \neq 0$ under ordinary multiplication, but the subset $\{2^n: n = 0, 1, 2, \ldots\}$ is not, since it does not contain the inverse of any of its elements. The subgroup $\{2^n: n = \text{an integer}\}$ is an example of a **cyclic group** since it has a single generator (in this case, the element 2) all of whose elements is either a positive, zero, or negative power of that generator.

    The set of all affine transformations is called the **linear group** of dimension 2, denoted *GL*(2). It has a rich complex system of subgroups, whose theory is still in progress. We consider a few results here and in the next few problems.

    Prove that the set *T* of all area-preserving affine mappings is a subgroup of *GL*(2), and that, in turn, the set of all isometries *S* is a subgroup of *T*. (Is the subset of similitudes a subgroup of *GL*(2)? Of *T*?) We write this relationship as $S < T < GL(2)$. Can matrix products be used for problems of this sort?

17. Show that if *D* is the set of all dilations about the origin and *R* is the set of all similitudes, then $D < R < GL(2)$.

18. Show that the set of all line reflections about a single line is a two-element subgroup of *GL*(2) (the identity *e* is considered to be a line reflection, although trivial).

19. The following matrices define a four-element subgroup of *GL*(2) under matrix multiplication. If you are familiar with matrices, show that they form a four-element group, closed under matrix multiplication, and such that the inverse of each element is present.

$$E = \begin{bmatrix} 1 & 0 \\ 0 & 1 \end{bmatrix}, \quad S = \begin{bmatrix} -1 & 0 \\ 0 & 1 \end{bmatrix}, \quad U = \begin{bmatrix} 1 & 0 \\ 0 & -1 \end{bmatrix}, \quad B = \begin{bmatrix} -1 & 0 \\ 0 & -1 \end{bmatrix}$$

Note: The inverse of a 2 × 2 matrix is given by

$$\begin{bmatrix} a & b \\ c & d \end{bmatrix}^{-1} = k^{-1} \begin{bmatrix} d & -b \\ -c & a \end{bmatrix} \equiv \begin{bmatrix} d/k & -b/k \\ -c/k & a/k \end{bmatrix} \quad \text{where } k = ad - bc$$

20. Prove that all matrices of the form $\begin{bmatrix} 1 & a \\ 0 & b \end{bmatrix}$, a and b real

numbers define a subgroup of **GL** (2) by showing closure under matrix multiplication, and showing that the inverse of each such matrix exists and is a matrix of this form.

21. Consider the numbers four numbers $1, -1, i, -i$ in the complex field, where $i^2 = -1$. Show that these numbers constitute a four-element group under ordinary multiplication. Is it a cyclic group?

8.6 Circular Inversion

Up to now, the focus of attention has been on *linear* transformations. However, there is a very important type of transformation of the plane that does not preserve collinearity or betweenness. Under this particular transformation, lines can map into circles and circles into lines. It is called an *inversion*, or *circular inversion*, for reasons that will become clear momentarily. Since this work is needed in the presentation of the model for hyperbolic geometry in the next chapter, we include a brief development.

To *invert* a nonzero number means to *take its reciprocal*. Thus, we can define the *inversion mapping* $f(x) = 1/x$ over the nonzero reals. By analogy, the inversion mapping in the plane works the same way: If O is the origin and P is any point different from O, one defines P' as that point such that $OP' = 1/OP$ (for convenience, we take P' on ray OP). Thus, if $OP = 3$ then $OP' = 1/3$, for example. Since we need a slightly more general mapping, we state

---

Definition 1 Let C be a circle having center at O and radius $r > 0$, as shown in Figure 8.39. The **inverse** of any point $P \neq O$ relative to circle C is that point P' on ray OP such that

$$OP' = \frac{r^2}{OP}$$

(See Figure 8.39.) The mapping thus defined is a one-to-one transformation on $\mathcal{P}$ whose domain is the plane $\mathcal{P}$ with point O deleted (known as the *punctured plane*) and is called an **inversion** (or **inverse mapping**) with respect to C. The circle C is called the **circle of inversion**, its center O, the **center of inversion**, and its radius r, the **radius of inversion**.

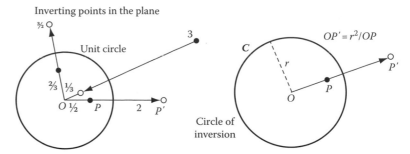

Figure 8.39 Circular inversion.

Note: Observe that f^2 (i.e., f followed by f) is the identity mapping, e. Any transformation having this property (like line reflections) is called **involutory**. Such mappings are their own inverses due to the relation $f^2 = e$. It is easy to see that circular inversion obeys this property. ◖

We now observe a few key geometric properties of circular inversion, along with their "one-line" proofs.

(1)   The circle of inversion maps to itself; every point on C is a fixed point.

Proof   If $P \in C$, then $OP = r$ then $OP' = r^2/OP = r$, and $P' = P$. ◖

(2)   A line passing through O maps to itself.

Proof   For each P on line l passing through O, P' lies on ray $OP \subset l$. ◖

(3)   If P' and Q' are the inverses of the points P and Q, then $\Delta POQ \sim \Delta Q'OP'$.

Proof   (See Figure 8.40.) Since $OP' = r^2/OP$ and $OQ' = r^2/OQ$, we have $OP'/OQ' = OQ/OP$. Since the angle at O is the same for both triangles, $\Delta POQ \sim \Delta Q'OP'$ by SAS. ◖

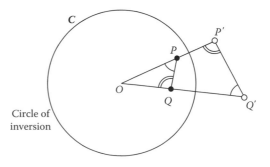

Figure 8.40 Property of similar triangles.

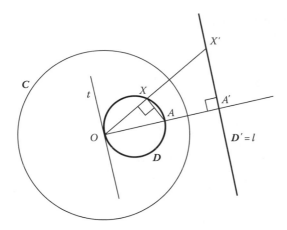

Figure 8.41 A circle through O inverts to a line.

(4)   If P' and Q' are the inverses of the points P and Q, then $\angle OPQ \cong \angle OQ'P'$, and $\angle OQP \cong \angle OP'Q'$.

Proof   (See Figure 8.40.) By **(3)** $\triangle POQ \sim \triangle Q'OP'$ ⧄

(5)   Let D be a circle passing through O, the center of inversion. Then D maps to line l that is parallel to the tangent t to D at O.

Proof   (See Figure 8.41.) If segment OA is a diameter of D, then for each point X in D, $m\angle OXA = 90 = m\angle OA'X'$. Hence, X' lies on the unique perpendicular l to line OA at A'. ⧄

(6)   A circle D not through O maps to a circle D' not through O.

Proof   Left to the reader as Problem 11.

Note: It might be helpful to observe with regard to **(6)** that when D does not contain O in its interior, the two common tangents to D and D' are lines passing through O, as shown in Figure 8.42. ⧄

Many of the applications of circular inversion depend on those properties that remain invariant under inversion. We have already encountered two invariants, namely, the circle of inversion itself, which remains unchanged under inversion, and any line passing through O, the center of inversion

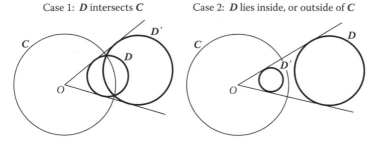

Figure 8.42 Inversion of circles not passing through O.

(although the individual points of such a line may change positions). A third invariant is the object of the next theorem.

Theorem 1   If a circle D is orthogonal to the circle of inversion, it maps to itself. That is, $D' = D$.

Proof   Let $D \equiv [D, s]$ be orthogonal to $C \equiv [O, r]$, the circle of inversion, as shown in Figure 8.43. We must show that the image D' of D under inversion coincides with D. If P is any point on D not on C, and Q is the other intersection of line OP with D, then by the power relation for circles (Definition 1 in Section **7.3**), and using the Pythagorean theorem,

$$OP \cdot OQ = \text{Power } O \text{ (circle } D) = OD^2 - s^2 = r^2$$

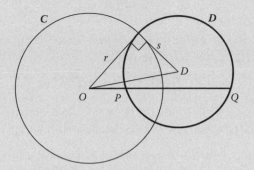

Figure 8.43 Invariance of circles orthogonal to C.

That is, $OP \cdot OQ = r^2$ and P and Q are inverse points. It follows that $D' \subseteq D$, and conversely $D \subseteq D'$. Hence, $D' = D$, as desired. ◤

Note: It is important to remember that the individual points of D are not fixed points, unlike the points of the circle of inversion, which are. ◤

Theorem 2   The inversion mapping is **conformal**, that is, the angle measure between any two curves remains unchanged under inversion.

Proof   We must show that if C_1 and C_2 are any two curves intersecting at P, and t_1 and t_2 are their tangent lines at P, respectively, the angle θ between t_1 and t_2 equals that of the corresponding tangents t_1' and t_2' at P' to the image curves C_1', C_2', as indicated by θ' in Figure 8.44. Since P is not the center of inversion, by **(5)** t_1 and t_2 invert to circles t_1' and t_2' through O and the angle θ' between the tangent lines to C_1' and C_2' equals the angle between the circles t_1' and t_2' at P', which, in turn, equals the angle between them at O (φ in the figure). Also by **(5)**, t_1 and t_2 are parallel, respectively, to the tangents to t_1' and t_2' at O. Thus, $\theta = \varphi = \theta'$, as desired. (It was tacitly assumed that O was not on either t_1 or t_2, but those cases are covered by assuming that a line is parallel to itself.) ◤

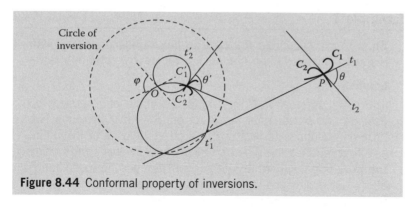

Figure 8.44 Conformal property of inversions.

For the next property, a modification of the definition of cross ratio is needed.

Definition 2   If A, B, C, and D are any four distinct points in the plane, the **generalized cross ratio** is defined to be the number

$$R[AB, CD] = \frac{AC \cdot BD}{AD \cdot BC}$$

where directed distance is understood if any three of the points are collinear.

Observe that when all four points are collinear, $R[AB, CD]$ agrees with the previously defined cross ratio, $[AB, CD]$.

Theorem 3   The generalized cross ratio of four points is an invariant under a circular inversion.

Proof   Let the images of A, B, C, and D be, respectively, A', B', C', and D'. By **(3)**, $\triangle OAC \sim \triangle OC'A'$ (Figure 8.40) and

$$\frac{AC}{A'C'} = \frac{OC}{OA'}$$

Similarly,

$$\frac{BD}{B'D'} = \frac{OD}{OB'}, \quad \frac{A'D'}{AD} = \frac{OA'}{OD}, \quad \frac{B'C'}{BC} = \frac{OB'}{OC}$$

The product of these ratios yields the expression

$$\frac{AC \cdot BD}{A'C' \cdot B'D'} \cdot \frac{A'D' \cdot B'C'}{AD \cdot BC} = \frac{OC}{OA'} \cdot \frac{OD}{OB'} \cdot \frac{OA'}{OD} \cdot \frac{OB'}{OC}$$

that is,

$$\frac{AC \cdot BD}{AD \cdot BC} \cdot \frac{A'D' \cdot B'C'}{A'C' \cdot B'D'} = 1 \quad \text{or} \quad \frac{AC \cdot BD}{AD \cdot BC} = \frac{A'C' \cdot B'D'}{A'D' \cdot B'C'}$$

which is the desired result, $R[AB, CD] = R[A'B', C'D']$. ◣

Example 1

Show that if the circle **D** inverts to line *l* under inversion with respect to circle **C**, then line *l* is the radical axis of **C** and **D**.

Solution

The result is clear if **D** intersects the circle of inversion, so assume that **D** lies either inside, or outside, of **C**. We show two different solutions to illustrate the different uses of inversion. The first requires the direct calculation of inverse points using the power concept, while the second requires no calculations at all, but makes use of orthogonal systems of circles $\mathcal{F}$ and $\mathcal{G}$ which appeared in Section 7.7.

1. Let the diameter *OP* of circle **D** meet *l* at *Q*, as shown in Figure 8.45. Then *P* and *Q* are inverse points and, since orthogonality is preserved under inversion, line $OQ \perp l$. If we show that the power of *Q* with respect to circles **C** and **D** are equal, this will prove that *l* is the radical axis of **C** and **D**. We have $OQ \cdot OP = r^2$ and, by definition of power,

$$\text{Power } P(\text{circle } \mathbf{C}) = OQ^2 - r^2 = OQ^2 - OQ \cdot OP =$$
$$OQ(OQ - OP) = -OQ(QO + OP) =$$
$$QO \cdot QP = \text{Power } P(\text{circle } \mathbf{D}).$$

2. The line *m* of centers of **C** and **D** is perpendicular to *l* at *Q*; let *A* and *B* be the intersections of *m* with **C** (Figure 8.46). Construct circle **E** centered at *Q* orthogonal to **C**, meeting line *m* at *C* and *D*. Then *C* and *D* are harmonic conjugates of *A* and *B*, hence all circles through *C* and *D* (including **E**) are orthogonal to circle **C** and belong to the family $\mathcal{F}$ having their centers on line *l*. Under inversion, *l* and **E**, which are orthogonal, map to **D** and **E** (since **E** maps to itself). Hence **D** is orthogonal to **E** and, with **C**, belongs to the family $\mathcal{G}$. But *l* is the radical axis for $\mathcal{G}$, hence is the radical axis of **C** and **D**.

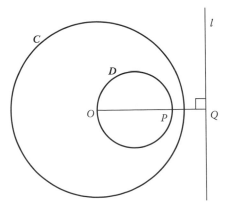

Figure 8.45 Example 1.

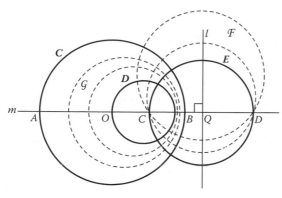

Figure 8.46 Proof by picture.

Example 2

Let P, Q, and R be the points of contact of the incircle of $\triangle ABC$ with its sides, and let T, V, and W be the midpoints of the sides of $\triangle PQR$ (Figure 8.47). Show that the circumcircle $[O, R]$ of $\triangle ABC$ inverts to the nine-point circle of $\triangle PQR$ under inversion with respect to the incircle of $[I, r]$ of $\triangle ABC$.

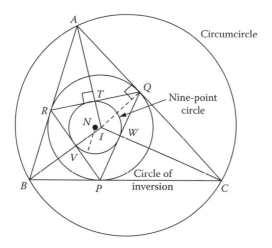

Figure 8.47 Example 2 (used for Euler's formula).

Solution

Observe that line AI is the perpendicular bisector of QR, hence passes through T the midpoint of segment QR. Similarly, lines BI and CI pass through points V and W. Thus, we need to show that A and T are inverse points with respect to the incircle. (A similar proof would apply to B and V, and to C and W.) Since OT is the altitude of right triangle AOI to hypotenuse AI, we have

$$QT^2 = AT \cdot TI = (AI - TI)TI = AT \cdot TI - TI^2$$

or

$$AI \cdot TI = QT^2 + TI^2 = r^2$$

proving that $TI = r^2/AI$. Thus, T is the image of A under inversion with respect to circle $[I, r]$. Similarly, V is the inverse of B and W is the inverse of C. Therefore, the circumcircle (passing through A, B, and C) maps to the circle passing through T, V, and W—the incircle $[N, s]$.

Note: This result leads to a simple proof of Euler's formula for the distance d between O and I in terms of the two radii: $d^2 = R^2 - 2rR$. (See Problem 12.) ⬗

Example 3

If circle $\boldsymbol{A} \equiv [A, a]$ maps to circle $\boldsymbol{B} \equiv [B, b]$ under inversion with respect to circle $\boldsymbol{C} \equiv [C, c]$, and if $d = CA > 0$, prove the following conversion formula that gives the radius of $\boldsymbol{B}$ in terms of c, d, and the radius of $\boldsymbol{A}$

(7)
$$b = \frac{ac^2}{|a^2 - d^2|}$$

Solution

Let $\boldsymbol{A}$ and $\boldsymbol{B}$ cut the line of center of the three circles in points D, E and E', D', respectively. The condition $CA > 0$ determines the direction for the line of centers. Figure 8.48 shows just one case, which will suffice for reference, but one observes that other cases are possible, such as the ordering $(D'DCE')$. Thus, $2b = \pm E'D' = \pm(CD' - CE') = \pm(c^2/CD + c^2/EC)$ (using directed distance on line CD). Hence,

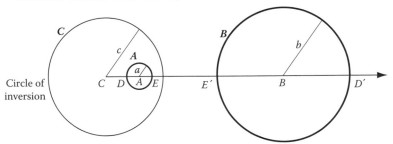

Figure 8.48 Proving identity **(7)**.

$$2b = \left| \frac{c^2}{EC} + \frac{c^2}{CD} \right| = \left| \frac{c^2(EC + CD)}{EC(CD)} \right| =$$

$$= \left| \frac{c^2 ED}{(CA + AE)(DA + AC)} \right| = \left| \frac{c^2(-2a)}{(a + d)(a - d)} \right| = \frac{2ac^2}{|a^2 - d^2|}$$

which is equivalent to the desired result.

Example 4: The Ptolemaic Inequality

Using circular inversion, one can establish an interesting generalization of Ptolemy's theorem (Theorem 2 in Section **7.4**), which stated that when four points A, B, C, and D lie on a circle, in that order, then $AB \cdot CD + BC \cdot AD = AC \cdot BD$. But when the four points are not so arranged, then the strict inequality $AB \cdot CD + BC \cdot AD > AC \cdot BD$ holds—a "triangle inequality" on the three products for any four points in the plane.

Start with three distinct noncollinear points A, B, and C, and let D be any other point taken as the center of inversion. (Figure 8.49 shows the case when A, B, and C lie on a circle not containing D, but this assumption is not necessary for what follows.) For convenience, let the radius of inversion be 1 (not shown in the figure). Then A, B, and C map to points A', B', and C'. Now by similar triangles, and since $A'D = 1/AD$,

$$\frac{A'B'}{AB} = \frac{A'D}{BD} = \frac{1}{AD \cdot BD}$$

that is,

$$A'B' = \frac{AB}{AD \cdot BD}$$

Similarly,

$$B'C' = \frac{BC}{BD \cdot CD} \quad \text{and} \quad A'C' = \frac{AC}{AD \cdot CD}$$

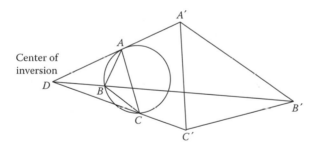

Figure 8.49 Proving the Ptolemaic inequality.

Since $A'B' + B'C' \geq A'C'$, with equality only when $(A'B'C')$ (which occurs only when the points are concyclic in the order A–B–C–D, and property **(5)** applies), we obtain

$$\frac{AB}{AD \cdot BD} + \frac{BC}{BD \cdot CD} \geq \frac{AC}{AD \cdot CD}$$

which converts by simple algebra to the following:

(8)   **Ptolemaic Inequality** $AB \cdot CD + BC \cdot AD \geq AC \cdot BD$ (with equality iff A, B, C, and D are concyclic).

It should be noticed that this argument proves Ptolemy's theorem as a special case.

As with previous transformations of the plane, we list the invariants of inversions.

Invariants for Inversions

1. Points on the circle of inversion
2. Circles not through the center of inversion
3. Lines through the center of inversion
4. Curvilinear angle measure
5. Generalized cross ratio

Problems (Section 8.6)

Group A

1. Using a unit radius of inversion, give an example to show that the center of a circle need not map to the center of the inverted circle under inversion.

2. Using your example in Problem 1, test the validity of **(7)**.

3. If the center of inversion O is the origin in coordinate geometry, and the radius of inversion is 2, find the coordinates of the images A', B', and C' under inversion of $A(1, 1)$, $B(2, 0)$, and $C(1, -1)$. Graph these points, along with the circle of inversion, and the circles ABC and $A'B'C'$. Draw a conclusion.

4.* Draw a diagram that shows the image under inversion of the configuration shown in Figure P.4 (orthogonal circles D, E, and line l) if the center of inversion is point A and the centers of the two circles lie on l.

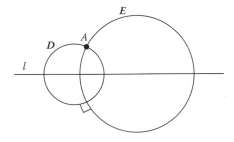

Figure P.4

5. Draw a diagram that shows the image under inversion of the configuration shown in Figure P.5 (four concurrent lines l, m, n, and k, with $l \perp n$ and $m \perp k$) if the center of inversion is point C.

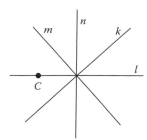

Figure P.5

Group B

6. Consider Figure 7.46 illustrating the mutually orthogonal families $\mathcal{F}$ and $\mathcal{G}$. What is the inversion of these two families if the center of inversion is the point C through which all the circles in $\mathcal{F}$ pass? Make a sketch. Show that this proves that if $\mathcal{F}$ is the family of circles passing through two fixed points C and D, there exists a family of circles orthogonal to every circle in $\mathcal{F}$.

7. Using the fact that the points $P(x, y)$ and $P'(x', y')$ on ray OP satisfy the relation $x' = \lambda x$, $y' = \lambda y$ where $\lambda = PO/P'O$, find the following coordinate form for an inversion with respect to the circle $x^2 + y^2 = r^2$

    (9) $$x' = \frac{r^2 x}{x^2 + y^2}, \quad y' = \frac{r^2 y}{x^2 + y^2}$$

8. Use the coordinate form **(9)** to prove that, under inversion with respect to $[O, r]$, the circle $x^2 + y^2 + ax = 0$ $(a \neq 0)$, inverts to a line parallel to the tangent to this circle at $(0, 0)$.

9. Use the coordinate form **(9)** to prove that a circle orthogonal to $[O, r]$ inverts to itself. (Use the condition **(4)** in Section **7.7** for orthogonal circles in coordinate geometry.)

10. Let D_1 and D_2 be any two circles, and l their radical axis. Show that if inversion leaves l fixed, and the center of inversion does not lie on D_1 and D_2, l is the radical axis of the image circles D_1' and D_2'.

11. Prove relation **(6)** above.

12.* **Proof of Euler's Formula** Apply the formula **(7)** of Example 3 to the circles $[O, R]$, $[N, s]$, and $[I, r]$ of Example 2 in order to derive Euler's formula for the distance between the circumcenter and incenter of a triangle in terms of their radii R and r:

    (10) $$d^2 = R^2 - 2rR$$

Group C

13. **Peaucellier's Cell: How to Draw a Straight Line**   A device for drawing a perfectly straight line was invented by a French army officer, A. Peacellier (1832–1913). Depicted in Figure P.13, it is an application of the inversion mapping. Show how the device will generate a line by a pencil or pen placed at P', assuming $PAP'B$ is a rhombus. (Assume that the device is free to pivot about the fixed points O and C indicated in the figure.)

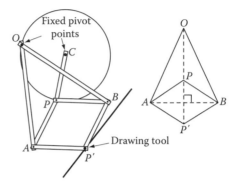

Figure P.13

14. **Steiner's Porism**   Jacob Steiner was a gifted and prolific geometer who contributed many amazing geometric results. One of these is the solution to the problem of how to complete a chain of circles within two fixed circles. In Figure P.14, the boundary circles, C and D, create an uneven annular ring. Suppose one begins constructing circles inside the ring, mutually tangent to one another: $A_1, A_2, A_3, \ldots$, and tangent to C and D. The question is whether A_n for some integer $n > 0$ large enough, will be tangent to A_1, thereby completing the chain. A second related question is whether the position of the first circle A_1 affects the conclusion, and what determines the value of n. An inversion maps the system of circles to an arrangement that makes the answers to these two questions almost trivial, certainly quite elementary. Find this inversion, and discuss the results.

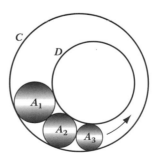

Figure P.14

9

Non-Euclidean Geometry: Analytical Approach

A deeper study of elliptic and hyperbolic geometry will be undertaken here to explore a theory largely unattainable by synthetic methods. The emphasis, with the exception of the very last section, will be on exploring further identities in unified trigonometry and their use in establishing theorems analogous to classical Euclidean geometry. For example, analogies to the theorems of Menelaus and Ceva are established for the two non-Euclidean geometries, which, in turn, provide proofs of the concurrency of the medians and altitudes of a triangle, as well as Desargues' theorem, and other results that are synthetically inaccessible.

The term *unified geometry* now applies to just the two non-Euclidean geometries since the identities derived here are obviously not valid in Euclidean geometry. The prominent feature of these formulas is their uncanny resemblance, even though there is an insurmountable chasm between elliptic and hyperbolic geometry, and the seeming impossibility of connecting trigonometric sine and cosine with the hyperbolic sine and cosine. This makes it all the more interesting that a common and quite natural axiom system can be formulated for these two geometries simultaneously. Of course, it is well known in the literature that a common system is possible; Felix Klien (1849–1925) discovered a universal model for them using methods of projective geometry. And as far as the trigonometric and hyperbolic functions are concerned, they are indeed connected in the complex number field, where formulas for $\sinh z$ and $\cosh z$ can be established in terms of $\sin z$ and $\cos z$ for complex numbers $z = x + iy$. [They are $\sinh z = -i \sin iz$, and $\cosh z = \cos iz$, derived basically from the well-known Euler relation $e^{i\theta} = \cos \theta + i \sin \theta$.]

In our development, the trigonometry of the right triangle as exhibited in the tables given in Section **6.12** will be the starting point. At times, a separate development will be indicated for hyperbolic and elliptic geometry in order to present an analysis that is more familiar. Since the methods of proof are virtually identical, we shall show the development for only one geometry, leaving the other one for the reader. The separate formulas will be routinely set side by side, with those of the opposite geometry in brackets. It should be noted, however, that separate proofs are

489

unnecessary due to the unified treatment presented in Chapter 6 (Section 6.12), made possible by the use of the functions $S(x)$, $C(x)$, and $T(x)$ (sine or hyperbolic sine, cosine or hyperbolic cosine, and tangent or hyperbolic tangent, respectively) and by applying the unified sign symbol $\sigma = \pm 1$. For the identities proved later in the chapter that are more involved, a unified proof will be used in order to save work.

9.1 Law of Sines and Cosines for Unified Geometry

As mentioned, since the starting point is the list of trigonometric forms, given earlier in the last section of Chapter 6, the reader needs to review those at this time. Also, for convenience, we assume the value $\kappa = 1$ throughout this section.

Unlike the law of sines that appeared in Chapter 6 and involved limits, essentially applying only to triangles having "zero length," we develop here a general identity valid for all triangles (including the limit relation of Chapter 6). Begin with any triangle ABC, and let CD be the altitude on side AB, with D on line AB (Figure 9.1). We have the cases (ADB), (DAB), (ABD), or, in elliptic geometry, $(AD*B)$. We give the proof for elliptic geometry, leaving hyperbolic geometry for the reader. We also will be using the fact that the trigonometry established for right triangles in Chapter 6, and the table of relations in Section 6.12 applies to all right triangles. (See Problem 19, Section **6.12**.)

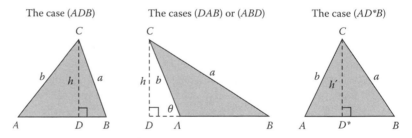

Figure 9.1 Law of sines.

If (ADB), consider the trigonometric relations for right triangles CAD and CBD:

$$\sin A = \frac{\sin h}{\sin b} \quad \text{and} \quad \sin B = \frac{\sin h}{\sin a}$$

Invert the second equation (at right) and multiply the two resulting equations to obtain

$$\frac{\sin A}{\sin B} = \frac{\sin h}{\sin b} \cdot \frac{\sin a}{\sin h} = \frac{\sin a}{\sin b}$$

The cases (DAB) and (ABD), which are logically equivalent, can be handled similarly, as follows. Referring to Figure 9.1 for the case (DAB), in right triangles CAD and CBD, we have

$$\sin A = \sin(\pi - A) = \sin\theta = \frac{\sin h}{\sin b} \quad \text{and} \quad \sin B = \frac{\sin h}{\sin a}$$

As before, multiply the first equation in sin A by the inverted second equation in sin B to obtain the same relation as before. Finally, in the case (AD^*B), we find no change in the essential steps since $\angle CAB$ is identical to $\angle CAD^*$ [since (AD^*B)] and $\angle CBA \equiv \angle CBD^*$, so that sin A and sin B are the same as before, with $h' = CD^*$ replacing h, which is eventually eliminated.

Since the labeling of these results is immaterial, we obtain for any triangle ABC in standard notation the following (with the result for hyperbolic geometry shown in brackets):

(1)   **Law of Sines for Non-Euclidean Geometry**

$$\frac{\sin a}{\sin A} = \frac{\sin b}{\sin B} = \frac{\sin c}{\sin C} \qquad \left[\frac{\sinh a}{\sin A} = \frac{\sinh b}{\sin B} = \frac{\sinh c}{\sin C} \right]$$

For the law of cosines, a little more work is necessary. Construct the same altitude of $\triangle ABC$ as before, with the same cases (Figure 9.2) and with $c_1 = AD$ and $c_2 = DB$. Again we give the proof for elliptic geometry. Consider the case (ADB). In right triangle ADC, the initial result to be used later comes from the trigonometric relation $\cos A = \tan c_1/\tan b$:

$$\tan c_1 = \tan b \cos A$$

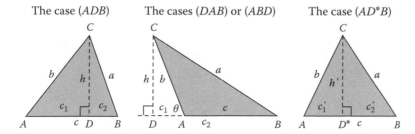

Figure 9.2 Law of cosines.

Now using right triangles BCD and ADC (where $\cos h = \cos b/\cos c_1$), we have

$$\cos a = \cos h \cos c_2 = \cos h \cos(c - c_1) = \cos h (\cos c \cos c_1 + \sin c \sin c_1)$$

$$= \cos c(\cos h \cos c_1) + \frac{\cos b}{\cos c_1} \sin c \sin c_1$$

$$= \cos c(\cos b) + \cos b \sin c \tan c_1$$

$$= \cos b \cos c + \cos b \sin c \tan b \cos A$$

$$= \cos b \cos c + \sin b \sin c \cos A$$

That is, in $\triangle ABC$,

$$\cos a = \cos b \, \cos c + \sin b \, \sin c \, \cos A$$

The cases (DAB) and (ABD) are handled in much the same way, again left to the reader as an interesting exercise for deriving this same result. Finally, consider the case $(AD*B)$. As in the case for the law of sines, we have $\angle CAB \equiv \angle CAD*$, $\angle CBD \equiv \angle CBD*$, and $c = AB = AD* + BD* = c_1' + c_2'$. Thus, $\tan b \cos A = \tan c_1'$, and the steps in the analysis are the same, with c_1', c_2', and h' replacing c_1, c_2, and h throughout.

Since the corresponding identity for hyperbolic geometry differs from that of elliptic geometry, we show one case for $\triangle ABC$: when the altitude from C has its foot D outside segment AB. Since $\alpha = \infty$, the cases are either (DAB) or (ABD), and we consider the case (DAB) (Figure 9.3). In right triangle ACD, we have $\cos \theta = \tanh c_1/\tanh b$ or $\tanh c_1 = \tanh b \cos \theta = \tanh b \cos(\pi - A)$. That is,

$$\tanh c_1 = -\tanh b \, \cos A$$

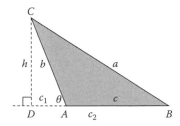

Figure 9.3 Proof for hyperbolic geometry.

Using right triangles CBD and CAD (where $\cosh h = \cosh b/\cosh c_1$), we can write

$$\cosh a = \cosh h \, \cosh c_2 = \cosh h \, \cosh(c + c_1)$$

$$= \cosh h(\cosh c \, \cosh c_1 + \sinh c \, \sinh c_1)$$

$$= \cosh c(\cosh h \, \cosh c_1) + \frac{\cosh b}{\cosh c_1} \sinh c \sinh c_1$$

$$= \cosh c \, \cosh b + \cosh b \, \sinh c \, \tanh c_1$$

$$= \cosh b \, \cosh c + \cosh b \, \sinh c(-\tanh b \, \cos A)$$

$$= \cosh b \, \cosh c - \sinh b \, \sinh c \, \cos A$$

The other cases for hyperbolic geometry will be left to the reader. Since the notation is immaterial, the following relations have thus been established for any non-Euclidean triangle ABC, where the results for hyperbolic geometry appear in brackets.

(2)   **Law of Cosines for Non-Euclidean Geometry**

$$\cos a = \cos b \cos c + \sin b \sin c \cos A$$
$$\cos b = \cos a \cos c + \sin a \sin c \cos B$$
$$\cos c = \cos a \cos b + \sin a \sin b \cos C$$

$$\begin{bmatrix} \cosh a = \cosh b \cosh c - \sinh b \sinh c \cos A \\ \cosh b = \cosh a \cosh c - \sinh a \sinh c \cos B \\ \cosh c = \cosh a \cosh c - \sinh a \sinh c \cos C \end{bmatrix}$$

Example 1

Solve $\triangle ABC$ in elliptic geometry if $a = 2$, $b = 1.8$, and $C = 20° \approx 0.349$ rad. That is, find (to three-decimal accuracy) the values for c, A, and B (in radians). Note that $A + B + C > \pi$. (The conversion factor from radians to degrees is $180°/\pi$.)

Solution

We begin with the appropriate identity from the law of cosines **(2)**:

$$\cos c = \cos 2 \cos 1.8 + \sin 2 \sin 1.8 \cos 0.349$$

$$= 0.09455 + 0.83213$$

$$= 0.92668$$

Therefore,

$$c = 0.38531 \quad \text{and} \quad c \approx 0.385$$

(In order to achieve three-place accuracy, and since there is considerable round-off error, all intermediate calculations are being executed to five-place accuracy.)

By the elliptic law of sines **(1)**,

$$\frac{\sin B}{\sin C} = \frac{\sin b}{\sin c} \quad \rightarrow \quad \sin B = \frac{\sin 1.8 \sin 0.349}{\sin 0.38531}$$

Having determined that $\angle B$ cannot be obtuse from the information given $(a > b)$,

$$\sin B = 0.88609 \rightarrow B = 1.08884 \text{ rad} \rightarrow B \approx 1.089 \approx 62.4°$$

Finally, in order to determine A, we use the law of sines:

$$\frac{\sin A}{\sin B} = \frac{\sin a}{\sin b} \quad \rightarrow \quad \sin A = \frac{\sin 2 \sin 1.08884}{\sin 1.8}$$

$$\sin A = 0.82736 \quad \rightarrow \quad A = 0.97438 \approx 55.8°$$

or

$$A = \pi - 0.97438 = 2.16721 \approx 124.2°\,*$$

Since $\angle A$ is obtuse, $A \approx 2.167$ rad $\approx 124.2°$.

Note: Unless your pocket calculator provides access to the hyperbolic functions, in working with numerical problems such as those in Example 1, it is convenient to write a short program for the formulas of hyperbolic trigonometry, as mentioned in Chapter 6, Section 6.12. These forms are being repeated here for the convenience of the reader. Using INPUT X, the following programs should be entered:

$$\sinh X = (e^X - e^{-X})/2, \qquad \sinh^{-1} X = \ln(X + \sqrt{X^2 + 1}),$$

$$\cosh X = (e^X + e^{-X})/2, \qquad \cosh^{-1} X = \ln(X + \sqrt{X^2 - 1}),$$

$$\tanh X = \sinh X / \cosh X \qquad \tanh^{-1} X = \frac{1}{2}\ln\left(\frac{1+X}{1-X}\right)$$

Problems (Section 9.1)

In all problems in this section, assume the value $\kappa = 1$ unless indicated otherwise.

Group A

1. Solve $\triangle ABC$ in hyperbolic geometry to three-decimal accuracy if $a = 1.5$, $b = 2$, and $C = 20° \approx 0.349$ rad.

2. If $A = 30°$, $B = 120°$, and $a = 0.5$, solve $\triangle ABC$
    (a) In Euclidean geometry
    (b) In hyperbolic geometry

3. Solve $\triangle ABC$ in elliptic geometry if $a = \pi/2$, $b = \pi/9$, and $B = 20°$. Interpret the result on the unit sphere.

4. Let $\triangle ABC$ be an equilateral triangle in elliptic geometry having equal sides of length a and equal angles of measure A (Figure 9.4). Find a formula for a in terms

---

* The TI-83 does not produce the two possible solutions that must be examined in order to obtain the correct answer. In effect, every problem of this type is the "ambiguous case."

of A (analogous to Problem 9, Section **6.12**). Note the impossibility of doing this in Euclidean geometry.

    (a)   Use the formula in **(a)** to prove that $A > 60°$.

    (b)   Use the formula in **(a)** to find a when $A = 61°$.

    (c)   Use the formula in **(a)** to find a when $A = 60.001°$. Discuss this result as it compares with that of **(b)**.

5.   In view of the AAA congruence criterion for non-Euclidean geometry, there is only one triangle up to isometry for which $A = 30°$, $B = 120°$, and $C = 60°$. Find the sides using the *law of cosines for angles* for elliptic geometry (see Problem 11 below): $\cos C = -\cos A \cos B + \sin A \sin B \cos c$.

6.   In hyperbolic geometry, solve for $\triangle ABC$ given its three angles $A = 30°$, $B = 120°$, and $C = 20°$ using $\cos C = -\cos A \cdot \cos B + \sin A \sin B \cosh c$.

Group B

7.   Give the proof of the law of sines for hyperbolic geometry.

8.   Give the proof of the law of cosines for hyperbolic geometry for the remaining case (*ADB*).

9.   **Cevian Formula**   Derive the formula for a Cevian triangle in hyperbolic geometry, illustrated in Figure P.9, using the addition identity for $\sinh(c_1 + c_2) = \sinh c$ and using the law of cosines for $\triangle BDC$ and $\triangle ACD$ to obtain expressions for $\cosh a$ and $\cosh b$:

    (3) $\cosh d = p \cosh a + q \cosh b$

where $p = \sinh c_1 / \sinh c$ and $q = \sinh c_2 / \sinh c$.

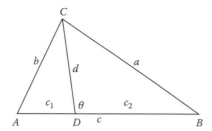

Figure P.9

10.   Prove the elliptic version of **(3)**:

    (4) $\cos d = p \cos a + q \cos b$

where $p = \sin c_1 / \sin c$ and $q = \sin c_2 / \sin c$. Write down Cevian formula for unified geometry using the functions $C(x)$ and $S(x)$ introduced in Section 6.12.

Group C

11.   **Law of Cosines for Angles**   The dual of the law of cosines (for sides) is the *law of cosines* for angles:

(5) $\cos C = -\cos A \cos B + (\sin A \sin B) \cos c$

$[\cos C = -\cos A \cos B + (\sin A \sin B) \cosh c]$

Prove this for elliptic geometry for the case (ADB) in Figure 9.2. [**Hint:** Start with the right side of the desired equation: $Q \equiv -\cos A \cos B + (\sin A \sin B)\cos c = -\tan c_1 \cdot \tan c_2/\tan b \tan a + \sin^2 h \sin c/\sin b \sin a$. Thus, $(\sin a \cdot \sin b) Q = -\tan c_1 \tan c_2 \cos a \cos b + \sin^2 h \cos c$; use law of cosines to show that $Q = \cos C$.]

12.   Give the unified proof for **(3)**, as defined in Section 6.12 (also reviewed in the next section).

9.2 Unifying Identities for Unified Trigonometry

The law of sines and cosines can be given a form more appropriate to unified geometry, where a single form suffices for both geometries. This is achieved by using the special functions introduced in Section 6.12. For convenience, we repeat definitions here.

(1) $S(x) = \begin{cases} \sin \kappa x & \text{in elliptic geometry,} \\ \sinh \kappa x & \text{in hyperbolic geometry} \end{cases}$

(2) $C(x) = \begin{cases} \cos \kappa x & \text{in elliptic geometry,} \\ \cosh \kappa x & \text{in hyperbolic geometry} \end{cases}$

(3) $T(x) = \begin{cases} \tan \kappa x & \text{in elliptic geometry,} \\ \tanh \kappa x & \text{in hyperbolic geometry} \end{cases}$

Also, for convenience, we list the major identities involving these functions, where we recall that $\sigma = +1$ for elliptic geometry and $\sigma = -1$ for hyperbolic geometry.

(4) $C^2(x) + \sigma S^2(x) = 1$

(5) $T(x) = S(x)/C(x)$

(6) $S(x \pm y) = S(x)C(y) \pm C(x)S(y)$

(7) $C(x \pm y) = C(x)C(y) \mp \sigma S(x)S(y)$

(8) $S(2x) = 2S(x)C(x)$

The tables for the right triangle relations for the two geometries in Chapter 6, Section 6.12, can be combined into a single one, as shown here.

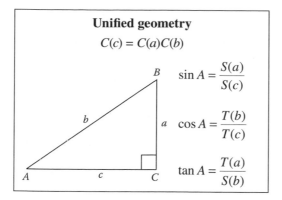

The identities **(1)** and **(2)** then take on the following appearance:

(9)   **Unified Law of Sines** $\dfrac{S(a)}{\sin A} = \dfrac{S(b)}{\sin B} = \dfrac{S(c)}{\sin C}$

(10) Unified Law of Cosines
$$\begin{cases} C(a) = C(b)C(c) + \sigma S(b)S(c)\cos A \\ C(b) = C(a)C(c) + \sigma S(a)S(c)\cos B \\ C(c) = C(a)C(b) + \sigma S(a)S(b)\cos C \end{cases}$$

Note that these identities are more general that those of **(1)** and **(2)** in the previous section since they include the arbitrary constant κ.

A unified proof can be given for each of these forms as well. Since we intend to use this method later, we present the proof of the first identity in **(10)** for the special case (ADB), where D is the foot of A on side AB. (See Figure 9.4.)

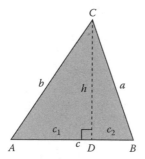

Figure 9.4 Unified proof for law of cosines.

First, note that in right triangle ACD, $\cos A = T(c_1)/T(b)$, and hence $T(c_1) = T(b) \cos A$, which will be used in one of the steps below. Starting with the Pythagorean relation in $\triangle CDB$,

$$C(a) = C(h)C(c_2) = C(h)C(c - c_1) = C(h)[C(c)C(c_1) + \sigma S(c)S(c_1)]$$
$$= C(c)[C(h)C(c_1)] + \sigma C(h)S(c)S(c_1)$$
$$= C(c)[C(h)C(c_1)] + \frac{C(b)}{C(c_1)} \sigma S(c)S(c_1)$$
$$= C(c)C(b) + \sigma C(b)S(c)T(c_1)$$
$$= C(c)C(b) + \sigma C(b)S(b)T(b)\cos A$$
$$= C(b)C(c) + \sigma S(b)S(c)\cos A,$$

which is the first equation in (10). Let the reader try proving another identity in either (9) or (10) for another case, such as (*DAB*), in order to gain a first-hand knowledge of this method.

9.3 Half-Angle Identities for Unified Geometry

A familiar theme running through the three geometries involves the incenter of a triangle, the half-angle formulas in terms of the sides, and terminating in Heron's formula for area. For motivation, and since it is instructive, we carry this out first for Euclidean geometry, which will have analogies to the two non-Euclidean geometries.

Begin with ΔABC in standard notation, with $s = \frac{1}{2}(a + b + c)$. Since $\frac{1}{2}A < \pi/2$, we can assume that $\sin\frac{1}{2}A > 0$ and $\cos\frac{1}{2}A > 0$. Using the half-angle formula for the sine function,

$$\sin^2 \tfrac{1}{2}A = \frac{1 - \cos A}{2} \cdot \frac{2bc}{2bc} = \frac{2bc - 2bc\cos A}{4bc}$$
$$= \frac{2bc - (b^2 + c^2 - a^2)}{4bc} = \frac{a^2 - (b^2 - 2bc + c^2)}{4bc}$$
$$= \frac{a^2 - (b - c)^2}{4bc} = \frac{[a - (b - c)][a + (b - c)]}{4bc}$$

Now, using the relation $2s = a + b + c$, we obtain for the first expression in brackets above, $a - b + c = 2(s - b)$. Similarly, $a + b - c = 2(s - c)$. Thus, the first half-angle relation emerges:

(1)
$$\sin\tfrac{1}{2}A = \sqrt{\frac{(s - b)(s - c)}{bc}}$$

In the same manner, we obtain

$$\cos\tfrac{1}{2}A = \frac{2bc + (b^2 + c^2 - a^2)}{4bc} = \frac{(b + c)^2 - a^2}{4bc} = \frac{[(b + c) + a][(b + c) - a]}{4bc}$$

Thus, the half-angle identity for cosine becomes

(2)
$$\cos \tfrac{1}{2} A = \sqrt{\frac{s(s-a)}{bc}}$$

To obtain the area K of $\triangle ABC$ from the above, recall a familiar area formula for K (Figure 9.5 shows its simple derivation) and a double-angle identity ($\sin 2x = 2 \sin x \cos x$ with $x = \tfrac{1}{2}A$). Thus,

$$K = \tfrac{1}{2} bc \sin A = bc \sin \tfrac{1}{2} A \cos \tfrac{1}{2} A$$

$$= bc \sqrt{\frac{(s-b)(s-c)}{bc}} \sqrt{\frac{s(s-a)}{bc}}$$

$$= \sqrt{s(s-a)(s-b)(s-c)},$$

which is the familiar Heron's formula. (The reader should compare this derivation with the one given earlier in Section **7.2** using the Cevian method.)

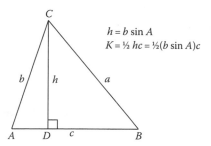

$$h = b \sin A$$
$$K = \tfrac{1}{2} hc = \tfrac{1}{2}(b \sin A)c$$

Figure 9.5 Proving Heron's formula from half-angle identities.

Analogously, in hyperbolic geometry, the law of cosines gives us

$$\sinh b \sinh c \cos A = \cosh b \cosh c - \cosh a$$

and we can write

$$\sin^2 \tfrac{1}{2} A = \frac{1 - \cos A}{2} \cdot \frac{\sinh b \sinh c}{\sinh b \sinh c} = \frac{\sinh b \sinh c - \sinh b \sinh c \cos A}{2 \sinh b \sinh c}$$

$$= \frac{\sinh b \sinh c - (\cosh b \cosh c - \cosh a)}{2 \sinh b \sinh c}$$

$$= \frac{-(\cosh b \cosh c - \sinh b \sinh c) + \cosh a}{2 \sinh b \sinh c}$$

$$= \frac{\cosh a - \cosh(b-c)}{2 \sinh b \sinh c}$$

where we have used the addition identity **(7)** in the previous section for hyperbolic trigonometry. At this point, we make use of the following sum-to-product identity, which can be derived directly from **(7)** by merely summing. [Let $u = x + y$, $v = x - y$, which leads to $x = \tfrac{1}{2}(u + v)$ and $y = \tfrac{1}{2}(u - v)$.] We then have

(3) $\cosh u - \cosh v = 2 \sinh \tfrac{1}{2}(u + v) \sinh \tfrac{1}{2}(u - v)$

Continuing where we left off, with $u = a$ and $v = b - c$,

$$\sin^2 \tfrac{1}{2} A = \frac{2 \sinh \tfrac{1}{2}(a+b-c) \, \sinh \tfrac{1}{2}(a-b+c)}{2 \sinh b \, \sinh c}$$

$$= \frac{\sinh(s-c) \sinh(s-b)}{\sinh b \, \sinh c}$$

and the first of the half-angle identity for hyperbolic geometry emerges. The details for the cosine identity are completely analogous, and so it will be left to the reader to derive the companion identity

$$\cos^2 \tfrac{1}{2} A = \frac{\sinh s \, \sinh(s-a)}{\sinh b \, \sinh c}$$

With a similar analysis for elliptic geometry, where the law of cosines is:

$$\sin b \sin c \cos A = \cos a - \cos b \cos c$$

we can derive analogous identities. Thus, emerge the **half-angle identities** for non-Euclidean geometry,

(4)
$$\begin{cases} \sin \tfrac{1}{2} A = \sqrt{\dfrac{\sin(s-b)\sin(s-c)}{\sin b \, \sin c}} & \left[\sin \tfrac{1}{2}A = \sqrt{\dfrac{\sinh(s-b)\sinh(s-c)}{\sinh b \, \sinh c}} \right] \\[3ex] \cos \tfrac{1}{2} A = \sqrt{\dfrac{\sin s \, \sin(s-a)}{\sin b \, \sin c}} & \left[\cos \tfrac{1}{2}A = \sqrt{\dfrac{\sinh s \, \sinh(s-a)}{\sinh b \, \sinh c}} \right] \end{cases}$$

The same identities can be combined for a much simpler unified version (the constant κ can clearly be inserted without affecting the rest of the identity):

(5) $\sin^2 \tfrac{1}{2} A = \sqrt{\dfrac{S(s-b)S(s-c)}{S(b)S(c)}}$ $\cos^2 \tfrac{1}{2} A = \sqrt{\dfrac{S(s)S(s-a)}{S(b)S(c)}}$

Unfortunately, although a Heron-like formula is valid for non-Euclidean geometry, its derivation is much more involved, as might be expected. Simply multiplying two equations as was done for Euclidean geometry fails. Since area in non-Euclidean geometry involves angle measure, a more complicated process is to be expected. The identities known as the *equations of Gauss* are needed in the derivation. So, we postpone this until those identities are established.

9.4 The Shape of a Triangle in Unified Geometry: Cosine Inequality

An important inequality for non-Euclidean geometry is based on the following proposition:

Theorem 1   If a triangle is constructed in the Euclidean plane having sides of length equal to those of a triangle in elliptic geometry, then each angle of the elliptic triangle is greater than the corresponding angle of the Euclidean triangle. (See Figure 9.6.)

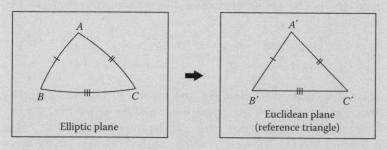

Figure 9.6 An elliptic triangle and its reference triangle.

Suppose we are given a triangle ABC in the elliptic plane, having sides of length a, b, and c and angles of measure A, B, and C. Then, the triangle inequality holds, and we can construct a triangle in the Euclidean plane having sides of length a, b, and c, respectively. The angles of the Euclidean triangle may well be different from those of the elliptic triangle, so we use A', B', and C' for those measures. It is to be proved that

$$A > A', \quad B > B', \quad \text{and} \quad C > C'$$

Using the half-angle identities of the previous section,

$$\sin\tfrac{1}{2}A = \sqrt{\frac{\sin\kappa(s-b)\sin\kappa(s-c)}{\sin\kappa b\,\sin\kappa c}}, \quad \sin\tfrac{1}{2}A' = \sqrt{\frac{(s-b)(s-c)}{bc}}$$

Hence, in order to prove that $A > A'$, we need to prove that $\sin \tfrac{1}{2}A > \sin \tfrac{1}{2}A'$, or that

$$\frac{\sin\kappa(s-b)\sin\kappa(s-c)}{\sin\kappa b\,\sin\kappa c} > \frac{(s-b)(s-c)}{bc}$$

or that

$$\frac{\sin\kappa(s-b)}{\kappa(s-b)} \cdot \frac{\sin\kappa(s-c)}{\kappa(s-c)} > \frac{\sin\kappa b}{\kappa b} \cdot \frac{\sin\kappa c}{\kappa c}$$

It thus suffices to prove

(1) $$\frac{\sin\kappa(s-b)}{\kappa(s-b)} > \frac{\sin\kappa c}{\kappa c} \quad \text{and} \quad \frac{\sin\kappa(s-c)}{\kappa(s-c)} > \frac{\sin\kappa b}{\kappa b}$$

But this merely involves the increasing/decreasing behavior of the function

$$(2) \qquad f(x) = \frac{\sin x}{x}, \quad 0 < x < \pi$$

It is an elementary exercise in calculus to prove that the derivative f' is negative on this range,* and thus that f is decreasing: $f(x) > f(y)$ for $x < y$.

Observe that by the triangle inequality:

$$s - b = \tfrac{1}{2}(a - b + c) < \tfrac{1}{2}(c + c) = c$$
$$s - c = \tfrac{1}{2}(a + b - c) < \tfrac{1}{2}(b + b) = b$$

Hence $\kappa(s - b) < \kappa c$ and therefore, $f(\kappa(s - b)) > f(\kappa c)$; similarly, $f(\kappa(s - c)) > f(\kappa b)$. Thus

$$\frac{\sin \kappa(s-b)}{\kappa(s-b)} > \frac{\sin \kappa c}{\kappa c} \quad \text{and} \quad \frac{\sin \kappa(s-c)}{\kappa(s-c)} > \frac{\sin \kappa b}{\kappa b}$$

and it follows that $\sin \tfrac{1}{2}A > \sin \tfrac{1}{2}A'$ and $A > A'$. In the same manner, it is proven that $B > B'$ and $C > C'$, thereby establishing Theorem 1.

An analogous theorem may be derived for hyperbolic geometry. Here, the conclusion is that the corresponding angles A, B, and C in the hyperbolic plane are less than A', B', and C', respectively. Figure 9.7 illustrates this relationship. The proof is basically the same, except that this time we must show that the function $f(x) = (\sinh x)/x$ is increasing instead of decreasing. The details will be left as a problem.

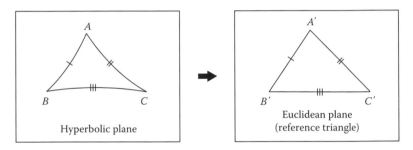

Figure 9.7 A hyperbolic triangle and its reference triangle.

The "shape" of a triangle in non-Euclidean geometry may thus be imagined in terms of the physical renditions of these requirements. The drawings in Figures 9.6 and 9.7 are suggestive of the actual appearance of a triangle in each model, if certain conditions are met. On the sphere, for example, such a drawing is basically accurate only if the point of the observer is directly above the center of a spherical triangle, looking directly down on a small portion of the sphere. We may recall, however, that drawings in models cannot be taken literally, and they cannot go much beyond mere suggestions of theoretical results.

If one makes use of the inequality concept defined in Chapter 6, the above results can be combined into one comprehensive result. For

---

* This involves showing that $g(x) \equiv x \cos x - \sin x < 0$; the derivative of g is given by $g'(x) = -x \sin x$, which is negative on $0 < x < \pi$, which shows that for $0 < x < \pi$, $g(x) < g(0) = 0$.

convenience, we call $\Delta A'B'C'$ the **reference triangle** corresponding to ΔABC, where the corresponding sides of the two triangles are congruent.

Theorem 2 Let ΔABC be any triangle in non-Euclidean geometry, with $\Delta A'B'C'$ the reference triangle whose sides have the same lengths as those of ΔABC. Then, in standard notation, $A < A'$, $B < B'$, and $C < C'$.

An immediate corollary to this result is an important inequality, which follows. If ΔABC is any triangle in non-Euclidean geometry, then, in standard notation,

(3)   **Cosine Inequality** $a^2 > b^2 + c^2 - 2bc \cos A$

Proof Let $\Delta A'B'C'$ be the reference triangle. Then, the Euclidean law of cosines is valid for $\Delta A'B'C'$, whose sides are a, b, and c. By Theorem 2, and since cosine is decreasing on $[0, \pi]$,

$$\cos A > \cos A' = \frac{b^2 + c^2 - a^2}{2bc}$$

which is equivalent to the desired result. ⟍

An interesting related result immediately follows; it also has implications involving the imagined shape of a triangle in non-Euclidean geometry.

Median Inequality The length m_a of median AL of ΔABC in non-Euclidean geometry satisfies the inequality

(4) $m_a^2 < \tfrac{1}{2}b^2 + \tfrac{1}{2}c^2 - \tfrac{1}{4}a^2$

Proof Observe from applications of the cosine inequality, first for ΔABL (Figure 9.8),

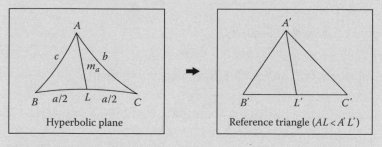

Figure 9.8 Comparing medians.

$$c^2 > (\tfrac{1}{2}a)^2 + (m_a)^2 - 2(\tfrac{1}{2}a)(m_a)\cos\theta \quad \text{and} \quad c^2 > \tfrac{1}{4}a^2 + m_a^2 - am_a \cos\theta$$

Similarly, in ΔACL, we have

$$b^2 > \tfrac{1}{4}a^2 + m_a^2 - am_a \cos(\pi - \theta) = \tfrac{1}{4}a^2 + m_a^2 + am_a \cos\theta$$

so that, upon the addition of the two inequalities, we obtain

$$b^2 + c^2 > \tfrac{1}{2}a^2 + 2m_a^2,$$

which is equivalent to the desired result. ⟍

The implication of (4) is that in hyperbolic geometry, for example, the median of $\triangle ABC$ is less than the corresponding median of the reference triangle in Euclidean geometry. This immediately follows from the Euclidean formula (Section 7.2) for the median of a triangle: $m_a'^2 = \frac{1}{2}b^2 + \frac{1}{2}c^2 - \frac{1}{4}a^2$. Figure 9.8 is suggestive of this result.

Example 1

Use (4) to prove that in any non-Euclidean triangle ABC, if M and N are the midpoints of sides AC and AB as in Figure 9.9, then

(5) $MN < \frac{1}{2}BC$

Solution

This amounts to proving $m < \frac{1}{2}a$ in Figure 9.15. To that end, using the median inequality (4) in $\triangle ABC$, and with $MB = d$, we have

$$d^2 < \frac{1}{2}a^2 + \frac{1}{2}c^2 - \frac{1}{4}b^2$$

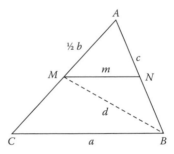

Figure 9.9 A median inequality.

Similarly, in $\triangle AMB$ with MN as median, we have

$$m^2 < \frac{1}{2}d^2 + \frac{1}{2}(\frac{1}{2}b)^2 - \frac{1}{4}c^2 = \frac{1}{2}d^2 + \frac{1}{8}b^2 - \frac{1}{4}c^2$$

Thus

$$m^2 < \frac{1}{2}(\frac{1}{2}a^2 + \frac{1}{2}c^2 - \frac{1}{4}b^2) + \frac{1}{8}b^2 - \frac{1}{4}c^2$$

$$= \frac{1}{4}a^2 + \frac{1}{4}c^2 - \frac{1}{8}b^2 + \frac{1}{8}b^2 - \frac{1}{4}c^2$$

$$= \frac{1}{4}a^2$$

or, $m < \frac{1}{2}a$, as desired.

Problems (Section 9.4)

In all problems in this section, assume the value $\kappa = 1$ unless indicated otherwise.

Group A

1. Using one of the half-angle formulas, derive the following relation for any angle of an equilateral triangle having side a in length:

    (6) $$\sin \tfrac{1}{2} A = \frac{\sin \tfrac{1}{2} a}{\sin a} \qquad \left[\sin \tfrac{1}{2} A = \frac{\sinh \tfrac{1}{2} a}{\sinh a} \right]$$

2. Using the result from Problem 1, show that $A = 0.919$ rad $= 52.6°$ for each angle of a unit equilateral triangle in hyperbolic geometry, then use this to verify the cosine inequality **(3)**.

3. Verify the cosine inequality **(3)** for each base angle of the isosceles triangle ABC shown in Figure P.3, where $a = c = 2$, and $b = 1$. (Use right triangle trigonometry to find A for each geometry.)

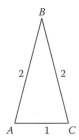

Figure P.3

4. In $\triangle ABC$ suppose that $a = 2$, $b = 1.8$, $c = 0.385$, and $C = 20° \approx 0.349$ (Example 1, Section 9.1). Calculate the length of the median from point A by the law of cosines in $\triangle ALC$. Then, use this calculation to verify the median inequality **(4)** (Figure P.4).

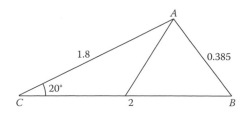

Figure P.4

Group B

5. **Cevian Inequality**   A generalization of the median inequality is easily derived. Let d be the length of Cevian AD in $\triangle ABC$ as shown, and let $p = BD/BC$, $q = DC/DC$, or $BD = pa$, and $DC = qa$, with $p + q = 1$. Use the law of cosines for triangles ABD and ADC to prove for $p > 0$ and $q > 0$ in general

    (7) $$d^2 < pb^2 + qc^2 - pqa^2$$

6. Prove Theorem 1 explicitly for hyperbolic geometry using the same method of proof, including the proof that $f(x) \equiv$ $(\sinh x)/x$ is increasing.

7. In $\triangle ABC$, points D and E are located on sides AB and AC, such that $AD/AB = AE/AC = \tfrac{2}{3}$. Show that $DE < \tfrac{2}{3} BC$ (Figure P.7). [**Hint:** Draw segment DC and make two applications of **(7)**, Problem 5.]

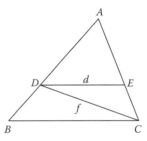

Figure P.7

Group C

8. Generalize the result of Problem 7 using an undetermined ratio p, $0 < p < 1$, in place of $\tfrac{2}{3}$.

9. Using the half-angle forms **(4)** Section **9.3**, establish the following formula for the inradius of an elliptic triangle (with $\kappa = 1$) in terms of its sides (Figure P.9):

    (8) $$\tan r = \sqrt{\dfrac{\sin(s-a)\sin(s-b)\sin(s-c)}{\sin s}}$$

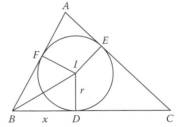

Figure P.9

9.5 The Equations of Gauss: Area of a Triangle

The identities known as the *equations of Gauss* can be used to prove a Heron-style formula for the area of a non-Euclidean triangle in terms of its sides, thereby completing the project begun in Section **9.3**. These identities first became well known for spherical trigonometry (developed from spherical coordinates independently) and became useful for problems of navigation. But Gauss observed that spherical trigonometry was closely parallel to that of hyperbolic trigonometry, the formulas having virtually the same form. (These he was able to prove along the line of reasoning used by Bolyai.) He became convinced that the age-old quest for finding a logical flaw in hyperbolic geometry was futile. Actual models proving this would come later.

In this section, we are going to need the following sum-to-product identity (which appeared in Chapter 6, with $u = x + y$, $v = x - y$):

$$C(u) + C(v) = 2C\left(\tfrac{1}{2}(u+v)\right)C\left(\tfrac{1}{2}(u-v)\right)$$

Three analogous ones are obtained from the addition identities **(6)** Section **9.2** (again with $u = x + y$, $v = x - y$):

$$C(u) - C(v) = 2\sigma S\left(\tfrac{1}{2}(u+v)\right)S\left(\tfrac{1}{2}(v-u)\right)$$

$$S(u) + S(v) = 2S\left(\tfrac{1}{2}(u+v)\right)C\left(\tfrac{1}{2}(u-v)\right)$$

$$S(u) - S(v) = 2C\left(\tfrac{1}{2}(u+v)\right)S\left(\tfrac{1}{2}(u-v)\right)$$

In order to establish the equations of Gauss, we start out with the ordinary addition identity for the sine function, and continue on, using the half-angle formulas **(5)**, Section **9.3**:

$$\sin\tfrac{1}{2}(A+B) = \sin\tfrac{1}{2}A\,\cos\tfrac{1}{2}B + \cos\tfrac{1}{2}A\,\sin\tfrac{1}{2}B$$

$$= \sqrt{\frac{S(s-b)S(s-c)}{S(b)S(c)}\cdot\frac{S(s)S(s-b)}{S(a)S(c)}} + \sqrt{\frac{S(s)S(s-a)}{S(b)S(c)}\cdot\frac{S(s-a)S(s-c)}{S(a)S(c)}}$$

$$= \sqrt{\frac{S^2(s-b)S(s-c)S(s)}{S(a)S(b)S^2(c)}} + \sqrt{\frac{S(s)S^2(s-a)S(s-c)}{S(a)S(b)S^2(c)}}$$

$$= \frac{S(s-b)}{S(c)}\sqrt{\frac{S(s)S(s-c)}{S(a)S(b)}} + \frac{S(s-a)}{S(c)}\sqrt{\frac{S(s)S(s-c)}{S(a)S(b)}}$$

$$= \left(\frac{S(s-b) + S(s-a)}{S(c)}\right)\sqrt{\frac{S(s)S(s-c)}{S(a)S(b)}}$$

$$= \left(\frac{2S\left[\tfrac{1}{2}(s-b+s-a)\right]C\left[\tfrac{1}{2}(s-b-s+a)\right]}{2S(\tfrac{1}{2}c)C(\tfrac{1}{2}c)}\right)\cos\tfrac{1}{2}C$$

$$= \frac{2S(\tfrac{1}{2}c)C(\tfrac{1}{2}(a-b))}{2S(\tfrac{1}{2}c)C(\tfrac{1}{2}(c))}\cos\tfrac{1}{2}C$$

Dividing both sides by cos ½C and simplifying the fraction, the first of the equations of Gauss emerges:

(1)
$$\frac{\sin \tfrac{1}{2}(A+B)}{\cos \tfrac{1}{2}C} = \frac{C(\tfrac{1}{2}(a-b))}{C(\tfrac{1}{2}c)}$$

In a similar manner, the addition identity for $\sin\tfrac{1}{2}(A-B)$ leads to the same exact radical expressions as before, which, after common factoring, gives us

$$\sin \tfrac{1}{2}(A-B) = \left(\frac{S(s-b)-S(s-a)}{S(c)} \right) \sqrt{\frac{S(s)S(s-c)}{S(a)S(b)}}$$

Using another sum-to-product identity, the final result emerges:

$$\sin\tfrac{1}{2}(A-B) = \frac{2C(\tfrac{1}{2}(2s-a-b))S(\tfrac{1}{2}(a-b))}{2S(\tfrac{1}{2}c)C(\tfrac{1}{2}c)} \cos\tfrac{1}{2}C$$

$$= \frac{S(\tfrac{1}{2}(a-b))}{S(\tfrac{1}{2}c)} \cos\tfrac{1}{2}C$$

or the second equation of Gauss,

(2)
$$\frac{\sin \tfrac{1}{2}(A-B)}{\cos \tfrac{1}{2}C} = \frac{S\left(\tfrac{1}{2}(a-b)\right)}{S(\tfrac{1}{2}c)}$$

Finally, the expansion of the two cosine functions cos ½(A + B) and cos ½(A − B) yields the last two formulas. (The reader is encouraged to try one of these, using the above method as a model for proof.) The complete list is as follows.

(3) Equations of Gauss

$$\frac{\sin \tfrac{1}{2}(A+B)}{\cos \tfrac{1}{2}C} = \frac{C\left(\tfrac{1}{2}(a-b)\right)}{C(\tfrac{1}{2}c)} \qquad \frac{\sin \tfrac{1}{2}(A-B)}{\cos \tfrac{1}{2}C} = \frac{S\left(\tfrac{1}{2}(a-b)\right)}{S(\tfrac{1}{2}c)}$$

$$\frac{\cos \tfrac{1}{2}(A+B)}{\sin \tfrac{1}{2}C} = \frac{C\left(\tfrac{1}{2}(a+b)\right)}{C(\tfrac{1}{2}c)} \qquad \frac{\cos \tfrac{1}{2}(A-B)}{\sin \tfrac{1}{2}C} = \frac{S\left(\tfrac{1}{2}(a+b)\right)}{S(\tfrac{1}{2}c)}$$

The main goal of this section is to derive Heron-style formulas for the area of a triangle in terms of its sides. Recall that the area of $\triangle ABC$ was previously defined as a constant times either the excess or defect of the triangle. The two cases can be combined by the use of the sign symbol σ:

$$K = k\sigma(A+B+C-\pi)$$

We start with the following half-angle identity for the tangent function:

(4)
$$\tan\tfrac{1}{2}(x+y) = \frac{\sin x + \sin y}{\cos x + \cos y}$$

The proof is immediate upon using the sum-to-product identities as follows:

$$\frac{\sin x + \sin y}{\cos x + \cos y} = \frac{2\sin\frac{1}{2}(x+y)\cos\frac{1}{2}(x-y)}{2\cos\frac{1}{2}(x+y)\cos\frac{1}{2}(x-y)} = \tan\frac{1}{2}(x+y)$$

Thus, writing K' for K/k for convenience, and noting that $\tan^2(\sigma x) = (\sigma \tan x)^2 = \tan^2 x$, we have

$$\tan^2(\tfrac{1}{4}K') = \tan^2\tfrac{1}{4}[(A+B+C-\pi)] = \left(\frac{\sin\frac{1}{2}(A+B)+\sin\frac{1}{2}(C-\pi)}{\cos\frac{1}{2}(A+B)+\cos\frac{1}{2}(C-\pi)}\right)^2$$

$$= \left(\frac{\sin\frac{1}{2}(A+B)-\cos\frac{1}{2}C}{\cos\frac{1}{2}(A+B)+\sin\frac{1}{2}C}\right)^2$$

To set this up for the use of Gauss's equations (where the first and third formulas are actually used), one modifies the expression inside the parentheses to obtain

$$\frac{C(\frac{1}{2}c)\sin\frac{1}{2}(A+B)-C(\frac{1}{2}c)\cos\frac{1}{2}C}{C(\frac{1}{2}c)\cos\frac{1}{2}(A+B)+C(\frac{1}{2}c)\sin\frac{1}{2}C}$$

$$= \frac{C(\frac{1}{2}(a-b)\cos\frac{1}{2}C - C(\frac{1}{2}c)\cos\frac{1}{2}C}{C\left(\frac{1}{2}(a+b)\right)\sin\frac{1}{2}C + C(\frac{1}{2}c)\sin\frac{1}{2}C}$$

$$= \frac{\cos\frac{1}{2}C}{\sin\frac{1}{2}C}\cdot\frac{C\left(\frac{1}{2}(a-b)\right)-C(\frac{1}{2}c)}{C\left(\frac{1}{2}(a+b)\right)+C(\frac{1}{2}c)}$$

$$= \frac{\cos\frac{1}{2}C}{\sin\frac{1}{2}C}\cdot\frac{2\sigma S\left(\frac{1}{4}(a-b+c)\right)S\left(\frac{1}{4}(a-b-c)\right)}{2C\left(\frac{1}{4}(c+a+b)\right)C\left(\frac{1}{4}(a+b-c)\right)}$$

$$= \frac{\cos\frac{1}{2}C}{\sin\frac{1}{2}C}\cdot\frac{-\sigma S\left(\frac{1}{2}(s-b)\right)S\left(\frac{1}{2}(s-a)\right)}{C(\frac{1}{2}s)C\left(\frac{1}{2}(s-c)\right)}$$

Thus, we obtain from the half-angle identities for hyperbolic geometry,

$$\tan^2\tfrac{1}{4}K' = \frac{S(s)S(s-c)}{S(s-a)S(s-b)}\cdot\frac{S^2\left(\frac{1}{2}(s-b)\right)S^2\left(\frac{1}{2}(s-a)\right)}{C^2(\frac{1}{2}s)C^2\left(\frac{1}{2}(s-c)\right)}$$

Now apply the identity $S(x) = 2S(\frac{1}{2}x)C(\frac{1}{2}x)$ to the entries in the fraction on the left. The result is the expression

$$\frac{4S(\frac{1}{2}s)C(\frac{1}{2}s)S\left(\frac{1}{2}(s-c)\right)C\left(\frac{1}{2}(s-c)\right)S^2\left(\frac{1}{2}(s-b)\right)S^2\left(\frac{1}{2}(s-a)\right)}{4S\left(\frac{1}{2}(s-a)\right)C\left(\frac{1}{2}(s-a)\right)S\left(\frac{1}{2}(s-b)\right)C\left(\frac{1}{2}(s-b)\right)C^2(\frac{1}{2}s)C^2\left(\frac{1}{2}(s-c)\right)}$$

$$= \frac{S(\frac{1}{2}s)S\left(\frac{1}{2}(s-c)\right)S\left(\frac{1}{2}(s-b)\right)S\left(\frac{1}{2}(s-a)\right)}{C\left(\frac{1}{2}(s-a)\right)C\left(\frac{1}{2}(s-b)\right)C(\frac{1}{2}s)C\left(\frac{1}{2}(s-c)\right)}$$

$$= T(\frac{1}{2}s)T\left(\frac{1}{2}(s-a)\right)T\left(\frac{1}{2}(s-b)\right)T\left(\frac{1}{2}(s-c)\right)$$

Taking square roots, the result is a Heron-style identity in unified geom-etry for the area of any triangle in terms of its sides. To be more explicit, we have established the two area formulas

(5) $\tan \tfrac{1}{4} K' = \sqrt{\tanh \tfrac{1}{2}\kappa s \tanh \tfrac{1}{2}\kappa(s-a)\tanh \tfrac{1}{2}\kappa(s-b)\tanh \tfrac{1}{2}\kappa(s-c)}$

(hyperbolic geometry)

(6) $\tan \tfrac{1}{4} K = \sqrt{\tan \tfrac{1}{2}\kappa s \tan \tfrac{1}{2}\kappa(s-a)\tan \tfrac{1}{2}\kappa(s-b)\tfrac{1}{2}\tan \tfrac{1}{2}\kappa(s-c)}$

(elliptic geometry)

The identity for hyperbolic geometry was established using what we called the *equations of Gauss*, which were apparently established by him only for spherical geometry. The area formula for elliptic geometry is known as *L'Huillier's formula*, due to Simon L'Huillier (1760–1810), which was well known in spherical geometry by Gauss's time.

The similarity of the two formulas for area for geometries which have almost nothing in common globally, is by itself suggestive of the con-sistency of hyperbolic geometry. Of course, for small triangles, general properties are the same or analogous (such as congruence and inequali-ties)—the content of unified geometry. However, hyperbolic geometry is suspect, while elliptic geometry has a Euclidean model. Gauss could see from such similarities that hyperbolic geometry must be as valid as spherical geometry.

Example 1

In a previous example, calculations were made that show an elliptic triangle ABC has the following dimensions (where $\kappa = 1$):

$a = 2$	$b = 1.8$	$c = 0.385$
$A = 124.2° = 2.167$	$B = 62.4° = 1.089$	$C = 20° = 0.349$

Use this information to check the validity of **(6)**.

Solution

The area defined by excess is $K' = A + B + C - \pi = 3.605 - \pi = 0.463$. The area as given by **(6)** requires finding $s = \tfrac{1}{2}(a + b + c)$: $s = \tfrac{1}{2}(4.185) = 2.0925$. Furthermore,

$$s - a = 0.0925 \quad s - b = 0.2925 \quad s - c = 1.7075$$

Thus, from **(6)**, $\tan \tfrac{1}{4} K'$ is given by

$$\sqrt{\tan(1.04625)\tan(0.04625)\tan(0.14625)\tan(0.85375)}$$

$$= 0.11625$$

and

$$\tfrac{1}{4} K' = \tan^{-1} 0.11625 \rightarrow K' = 0.46292 \approx 0.463$$

in agreement (up to at least three decimals).

Example 2

Non-Euclidean geometry is "locally Euclidean" in the sense that for small triangles, the formulas for hyperbolic and elliptic geometry are approximately the same as the corresponding formulas for Euclidean triangles, and the smaller the triangle, the smaller the error involved. This means that the simpler Euclidean formulas make good approximations in small regions of the unified plane. Recall, for example, that if $\triangle ABC$ is a non-Euclidean triangle, the angle sum $A + B + C$ approaches $\pi = 180°$ as limit as a, b, $c \rightarrow 0$.

Test this principle by showing in an example that for a triangle having sides $< \frac{1}{2}$ the discrepancy between the following two area formulas is less than 0.005:

(a) $K = \sqrt{s(s-a)(s-b)(s-c)}$

(b) $\tan \frac{1}{4}K' = \sqrt{\tanh\frac{1}{2}s \, \tanh\frac{1}{2}(s-a)\frac{1}{2}\tanh(s-b)\frac{1}{2}\tanh(s-c)}$

Solution

Consider a triangle in each geometry having sides $a = 0.2$, $b = 0.3$, and $c = 0.4$. Then, $s = 0.45$ and

(a) $K = \sqrt{(0.45)(0.25)(0.15)(0.05)} = 0.02905$

(b) $\tan\frac{1}{4}K' = \sqrt{\tanh(0.225)\tanh(0.125)\tanh(0.075)\tanh(0.025)}$

$= 0.0071754$

$\frac{1}{4}K' = \tan^{-1}(0.0071754) = 0.0071753 \rightarrow K' = 0.02870$

The difference between the two results is $0.00035 < 0.005$.

Example 3

Use two of the formulas of Gauss to prove that for an isosceles right triangle $(C = \pi/2)$

(a) $\tan A = \operatorname{sech} a$

(b) Using hyperbolic trigonometry for a right triangle with right angle at C, then $\tan A = \tanh a/\sinh b$. Simplify this for the special case $a = b$ and show that the identity in (a) agrees with to this simplified identity, thereby providing a check on **(3)**.

Solution

(a) Dividing equations in the first and third of Gauss's formulas, we have

$$\frac{\dfrac{\sin\frac{1}{2}(A+B)}{\cos\frac{1}{2}C}}{\dfrac{\cos\frac{1}{2}(A+B)}{\sin\frac{1}{2}C}} = \frac{\dfrac{\cosh\frac{1}{2}(a-b)}{\cosh\frac{1}{2}c}}{\dfrac{\cosh\frac{1}{2}(a+b)}{\cosh\frac{1}{2}c}} \quad \text{or}$$

$$\tan\frac{1}{2}(A+B)\tan\frac{1}{2}C = \frac{\cosh\frac{1}{2}(a-b)}{\cosh\frac{1}{2}(a+b)}$$

with $C = \pi/2$ and $a = b$ (hence $A = B$), the above identity reduces to

$$\tan A = 1/\cosh a = \operatorname{sech} a$$

(b) Setting $a = b$, the right triangle relation reduces to

$$\tan A = \tanh a/\sinh a = 1/\cosh a = \operatorname{sech} a.$$

Problems (Section 9.5)

In all problems in this section, assume the value $\kappa = 1$ unless indicated otherwise.

Group A

1. Test the formula for area in hyperbolic geometry by using an equilateral triangle having sides of length 4 units, and performing the following calculations using 5 decimal accuracy:
   (a) Use hyperbolic trigonometry of the right triangle to find the measure A of each angle.
   (b) Find the area $K' = \pi - 3A$ and calculate $\tan \frac{1}{4} K'$.
   (c) Find s and $s - a$ and then calculate $\tan \frac{1}{4} K'$ using **(5)**.

2. Repeat Problem 1 for elliptic geometry for an equilateral triangle having sides of length 2 $(< \pi)$ units.

3. Choose any of the equations of Gauss **(3)** to obtain a contradiction if one assumes there exists a right triangle having acute angles of measure 45°.

4. Write a formula (simplified) for the area of an isosceles triangle having legs $2p$ and base $2q$ for positive p and q.

Group B

5. Test the last of the equations of Gauss in elliptic geometry by deriving the following results.
   (a) Simplify the last equation in **(3)** for the case of an equilateral triangle having side a and angle A.
   (b) Use one of the relations for a right triangle with sides a and $\frac{1}{2}a$, and show it coincides with your result in (a).

6. Choose any one of the equations of Gauss **(3)** for elliptic geometry and, following the procedure given for deriving those in the above arguments, prove it.

Group C

7. **Formula for the Area of a Right Triangle**   Starting with the relation for area of a right triangle $K' = \pi - A - B - C = \pi/2 - A - B$ for the area of $\triangle ABC$, use the identities involving $C(u) \pm C(v)$ and Gauss's equations to derive the relation for elliptic geometry

(7)
$$\tan \tfrac{1}{2}K' = \tan \tfrac{1}{2}\kappa a \, \tan \tfrac{1}{2}\kappa b$$
$$[\tan \tfrac{1}{2}K' = \tanh \tfrac{1}{2}\kappa a \, \tanh \tfrac{1}{2}\kappa b]$$

8. **Boundedness of Area of Right Triangles**   The maximal area for all right triangles in hyperbolic geometry is $\frac{1}{2}k\pi$. Prove this using the formula **(7)** and the limit relations $e^x \to \infty$ and $\tanh x \to 1$ as $x \to \infty$. (This may also be easily derived synthetically using asymptotic parallelism.)

Note: Regarding this phenomenon of hyperbolic geometry Gauss once wrote in a letter to a friend: "I have arrived at much which most people would consider sufficient for proof, but which proves nothing from my viewpoint. For example, if it could be proved that a rectilinear triangle is possible with an area exceeding any given area, I would be in a position to prove rigorously the whole of [Euclidean] geometry." ⟍

9. **Area and Perimeter of a Regular Polygon**   Using **(7)** and elliptic trigonometry for right triangle OMB in Figure P.9, derive the following relations in elliptic geometry for the perimeter P and area K' of a regular polygon having n sides, apothem a (distance from center to a side) and radius r (distance from center to a vertex):

(8)
$$\begin{cases} \sin \dfrac{\kappa P}{2n} = \sin \kappa r \sin \dfrac{\pi}{n} & \left[\sinh \dfrac{\kappa P}{2n} = \sinh \kappa r \sin \dfrac{\pi}{n} \right] \\[2mm] \tan \dfrac{K'}{4n} = \tan \dfrac{\kappa a}{2} \tan \dfrac{\kappa P}{4n} & \left[\tan \dfrac{K'}{4n} = \tanh \dfrac{\kappa a}{2} \tanh \dfrac{\kappa P}{4n} \right] \end{cases}$$

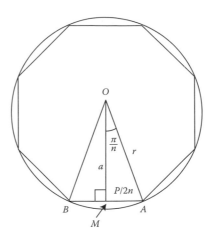

Figure P.9

10.* **Area and Circumference of a Circle** Assume that a sequence of inscribed regular polygons of a circle C having perimeters P_n and areas K_n exist, such that as $n \to \infty$, P_n and K_n approach the circumference C and area A of the circle. Using **(8)**, prove:

(9)
$$\begin{cases} \kappa C = 2\pi \sin \kappa r & [\kappa C = 2\pi \sinh \kappa r \\ \\ A = 4k\pi \sin^2 \tfrac{1}{2}\kappa r & [A = 4k\pi \sinh^2 \tfrac{1}{2}\kappa r \end{cases}$$

Note: Observe that these formulas approximate the Euclidean formulas for small r ($C = 2\pi r$ and $K = \pi r^2$), provided $k\kappa^2 = 1$. This provides evidence that there is a link between the two constants k for area and κ for distance. Indeed, the formula $k = 1/\kappa^2$ is valid, obtained by both Bolyai and Lobachevski. It can be established for spherical geometry as indicated in Problem 12 below. ◥

11. **Unboundedness of Area of Hyperbolic Plane** Use **(9)** to show that the area of the hyperbolic plane is unbounded. Note that this seems to contradict the result in Problem 8; how can this be explained?

12. **Relationship Between k and κ** Use a sphere of radius R as a model to show that the constant k for area satisfies the relation $k = R^2$, so that $k = 1/\kappa^2$. (Recall the relation established earlier $\kappa = \pi/\alpha$.) The corresponding result for hyperbolic geometry is also $k = 1/\kappa^2$ (see Note in Problem 10). [**Hint:** Let $r \to \pi/2\kappa$ in **(9)** and compare the resulting expression with the area of a hemisphere, using the standard area formula for a sphere of radius r: $K = 4\pi r^2$.]

9.6 Directed Distance: Theorems of Menelaus and Ceva

A bit of groundwork needs to be laid before we can give valid proofs of the non-Euclidean version of the theorems of Ceva and Menelaus. Again, it will be convenient to prove Menelaus' theorem first and use this result to prove the first half of Ceva's theorem. All this work is worth the effort because it enables us to prove in non-Euclidean geometry, for example, that the medians, and altitudes, of a triangle are concurrent. For convenience, we take $\kappa = 1$ throughout this section; it can be seen that the conclusions obtained in this special case extend to arbitrary values of κ.

It is first necessary to introduce directed distance for non-Euclidean geometry. The same approach that was used in Euclidean geometry is valid. Let l be any line and take a coordinate system for points on l as generated by the Ruler Postulate. Then, define for any two points $A[a]$ and $B[b]$ on l, with their coordinates, the **directed distance**

$$\overline{AB} = b - a$$

(where we abandon the use of the bar to denote segments in this section).

Properties of Directed Distance for Non-Euclidean Geometry
For any three distinct collinear points $A[a]$, $B[b]$, and $C[c]$,

(1) $$\overline{AB} + \overline{BA} = 0$$

(2) $$\overline{AB} + \overline{BC} + \overline{CA} = 0$$

(3) $$\overline{AB} = \overline{AC} \text{ iff } B = C$$

(4) (ABC) iff $\overline{AB}$ and $\overline{BC}$ are either both positive or both negative
[In elliptic geometry, one must include the condition $|AC| < \pi$.]

We note that in hyperbolic geometry, $AB = |\overline{AB}|$ (as in Euclidean geometry), but in elliptic geometry, $AB = |\overline{AB}|$ if $|\overline{AB}| \leq \pi$ and $AB = 2\pi - |\overline{AB}|$ if $|\overline{AB}| > \pi$ (coordinatization theorem for lines). The bars will now be removed to eliminate cumbersome notation; however, as in modern Euclidean geometry (Chapter 7 in particular), the reader is advised that AB does not always represent a positive real number.

A non-Euclidean version of Stewart's theorem can be derived: For any four points A, B, C, and P, with A, B, and C collinear,

(5) $$\begin{cases} \sin AB \cos PC + \sin BC \cos PA + \sin CA \cos PB = 0 \\ [\sinh AB \cosh PC + \sinh BC \cosh PA + \sinh CA \cosh PB = 0] \end{cases}$$

As in Euclidean geometry, we follow the convention of taking a given directed distance between two points to be positive whenever possible.

The proof of **(5)** is virtually the same for both geometries, given here only for elliptic geometry. This will be proven first for the case when P lies on the line l of A, B, and C. Making use of **(2)**, which allows one to write $AB = AC + CB$, for example, one obtains

$$\sin AB \cos PC + \sin BC \cos PA + \sin CA \cos PB =$$

$$\sin(AC + CB)\cos PC + \sin BC \cos(PC + CA) + \sin CA \cos(PC + CB)$$

The expansion of this last expression leads to six terms as follows:

$$\sin AC \cos CB \cos PC + \cos AC \sin CB \cos PC +$$

$$\sin BC \cos PC \cos CA - \sin BC \sin PC \sin CA +$$

$$\sin CA \cos PC \cos CB - \sin CA \sin PC \sin CB$$

An examination of this expression, and using the identities $\sin(-x) = -\sin x$ and $\cos(-x) = \cos x$, reveals that the first and fifth terms cancel (due to $AC = -CA$), as do the second and third terms and the fourth and sixth terms—proving the original expression equals zero, as desired. When P does not lie on line l, take Q as the foot of P on l (if the distance from P to l is $\pi/2$ then P is a pole of l and $PA = PB = PC = \pi/2$ making all terms zero, thus proving the theorem in this case). As shown in Figure 9.10, this forms a number of right triangles. First, note that $\cos XY = \cos YX$ for all X, Y, which will be used in what follows. By the Pythagorean relation for elliptic geometry, $\cos PA = \cos h \cos QA$, and similarly for PB and PC. Thus,

$$\sin AB \cos PC + \sin BC \cos PA + \cos CA \cos PB =$$

$$\sin AB \cos h \cos QC + \sin BC \cos h \cos QA + \sin CA \cos h \cos QB =$$

$$\cos h \left(\sin AB \cos QC + \sin BC \cos QA + \sin CA \cos QB \right)$$

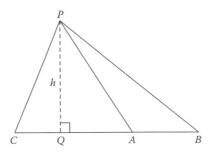

Figure 9.10 Proving Stewart's theorem.

But, by the first case, since Q lies on l, the expression inside the parentheses is zero. This finally proves **(5)** in general for elliptic geometry. The proof of **(5)** for hyperbolic geometry is evidently very similar.

Note: It is interesting that, just as in Euclidean geometry, the previous result leads to a generalized Cevian formula, where the Cevian is not required to intersect the opposite side at an interior point. This formula agrees with the one derived in Problem 9, Section **9.1**. ⬟

Following the agenda for Euclidean geometry, the **cross-ratio** is defined for any four distinct collinear points A, B, C, and D for non-Euclidean geometry, as follows:

$$\mathcal{R}[AB,CD] = \begin{cases} \dfrac{\sin AC \sin BD}{\sin AD \sin BC}, & \text{for elliptic geometry} \\[2em] \dfrac{\sinh AC \sinh BD}{\sinh AD \sinh BC}, & \text{for hyperbolic geometry} \end{cases}$$

(In elliptic geometry, it must be required, in addition to A, B, C, and D being distinct points, that no two of the points A, B, C, and D are extreme opposites.) This cross-ratio has essentially the same properties as those for Euclidean geometry. For example,

$$\mathcal{R}[AB,DC] = 1/\mathcal{R}[AB,CD] \quad \text{and} \quad \mathcal{R}[CD,AB] = \mathcal{R}[AB,CD]$$

In order to prove the property

$$\mathcal{R}[AB,CX] = \mathcal{R}[AB,CD] \quad \text{implies} \quad X = D \text{ or } X = D^*,$$

a special lemma is needed (this result is also important in working with the theorems of Menelaus and Ceva).

Lemma 1   Let A, B, D, and X be distinct, collinear points. In elliptic geometry, if

$$\frac{\sin AX}{\sin BX} = \frac{\sin AD}{\sin BD}$$

and no two of A, B, and D are extremal opposites, then either $X = D$ or $X = D^*$. In hyperbolic geometry, if

$$\frac{\sinh AX}{\sinh BX} = \frac{\sinh AD}{\sinh BC}$$

then $X = D$.

Proof (for elliptic geometry)   For convenience, let $x = AX$, $y = BX$, $a = AD$, and $b = BD$. The hypothesis then reads $\sin x/\sin y = \sin a/\sin b$. Since $AB = AD + DB = AX + XB$, then $x - y = AB = a - b$. Thus, if $x - y = \theta$, then $y = x - \theta$ and $b = a - \theta$. Therefore,

$$\sin y = \sin(x - \theta) = \sin x \cos \theta - \cos x \sin \theta$$

and, dividing by sin x yields

$$\frac{\sin y}{\sin x} = \cos\theta - \cot x \sin\theta$$

(If sin $x = 0$ then sin $a = 0$ and $\overline{AD} = 0$ or π, contrary to hypothesis.) Similarly, considering sin $b = \sin(a - \theta)$, it follows that

$$\frac{\sin b}{\sin a} = \cos\theta - \cot a \sin\theta$$

By hypothesis, $\cos\theta - \cot x \sin\theta = \cos\theta - \cot a \sin\theta$. Since sin $\theta = \sin AB \neq 0$ ($0 < AB < \pi$), then we obtain

$$\cot x = \cot a$$

It now follows that either $x = a$ (which yields $AX = AD$ or $X = D$) or $|x - a| = \pi$. (See the graph of the cotangent function provided in Figure 9.11.) In the latter case, $|XD| = |x - a| = \pi$ and $X = D^*$, finishing the proof for elliptic geometry. The simpler proof for hyperbolic geometry is left as a problem (Problem 6). ◤

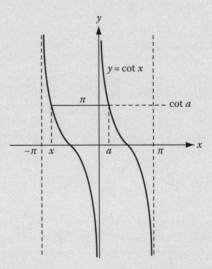

Figure 9.11 Proving Lemma 1.

Finally, the linearity number for non-Euclidean geometry may be defined. Given any triangle ABC and points D, E, and F on the extended sides, distinct from the vertices of the triangle and their extremal opposites, do not coincide with the vertices of the triangle let the number

$$
\mathcal{L}\begin{bmatrix} ABC \\ DEF \end{bmatrix} = \begin{cases} \dfrac{\sin AF \sin BD \sin CE}{\sin FB \sin DC \sin EA} & \text{for elliptic geometry} \\[3mm] \dfrac{\sinh AF \sinh BD \sinh CE}{\sinh BF \sinh DC \sinh EA} & \text{for hyperbolic geometry} \end{cases}
$$

be defined, called the **(unified) linearity number** of D, E, and F with respect to $\triangle ABC$. To save writing and to speed up the process of writing linearity numbers, the following notation will be introduced: Let sPQ denote $\sin PQ$ in elliptic geometry, or $\sinh PQ$ in hyperbolic geometry.

$$
\mathcal{L}\begin{bmatrix} ABC \\ DEF \end{bmatrix} = \frac{sAF}{sFB} \cdot \frac{sBD}{sDC} \cdot \frac{sCE}{sEA}
$$

Its properties mirror those of the Euclidean linearity number, two of which are noted below, the proofs of which are identical to those given in Euclidean geometry (Chapter 7), where it is necessary to use Lemma 1 for property **(7)**.

(6)
$$
\mathcal{L}\begin{bmatrix} ACB \\ DFE \end{bmatrix} = \mathcal{L}\begin{bmatrix} ABC \\ DEF \end{bmatrix}^{-1}
$$

(7)   If $\mathcal{L}\begin{bmatrix} ABC \\ DEF \end{bmatrix} = \mathcal{L}\begin{bmatrix} ABC \\ DEF \end{bmatrix}$ then either $X = F$ or $X = F^*$.

Since the method of proofs for the theorems of Ceva and Menelaus are basically the same as those used in Euclidean geometry (with some modifications necessary), these two theorems will be left as problems (see Problems 10–12). As in Euclidean geometry, Menelaus' theorem is proved first, which is used for the first half of Ceva's theorem. (You should at this point review the proofs of these two theorems for Euclidean geometry.) Two technical details needed in hyperbolic geometry will be provided here because they require special proofs. This is necessary because parallel Cevians in hyperbolic geometry do not behave the same way as they do in Euclidean geometry.

For Menelaus' theorem, after proving that collinear "menelaus" points D, E, and F lead to a value of -1 for the linearity number, the converse follows this line: If the linearity number equals -1 then consider the intersection X of the line determined by two of the menelaus points of two sides of the triangle with the remaining side; by the first part, the linearity numbers involving X and the appropriate menelaus point F both equal -1, and we conclude that $X = F$ or $X = F^*$, and thus the menelaus points are collinear (recall that if a line in elliptic geometry passes through F^*, it must also pass through F). In hyperbolic geometry, it needs to be proven that point X exists. The following result takes care of this detail.

Lemma 2 If D, E, and F are points of triangle ABC lying on the extended sides $\overleftrightarrow{BC}$, $\overleftrightarrow{CA}$ and $\overleftrightarrow{AB}$ respectively, and the linearity number of D, E, and F with respect to $\triangle ABC$ equals -1, then two of these points determine a line that meets the third side at some point X. Explicitly, either $\overleftrightarrow{DE} \cap \overleftrightarrow{AB} \neq \emptyset$, $\overleftrightarrow{EF} \cap \overleftrightarrow{BC} \neq \emptyset$, or $\overleftrightarrow{DF} \cap \overleftrightarrow{AC} \neq \emptyset$.

Proof There is nothing to prove in elliptic geometry, so we turn to hyperbolic geometry. Since the linearity number is negative, and, for example, (BDC) iff BD and DC have the same sign (and the same for sinh BD and sinh DC), either (1) two of the points lie interior to the sides of $\triangle ABC$ and one lies exterior or (2) all three points lie exterior to the sides.

1. Suppose D and E lie interior to sides BC and CA, as shown in Figure 9.12a, and F lies exterior to side AB. In this case, we show that line DF meets line AC. Since either (BAF) (the case shown in the figure) or (ABF), line DF cannot meet side AB. Thus, it must meet side AC of $\triangle ABC$ (postulate of Pasch) at some point X. The proof is similar for the other case.

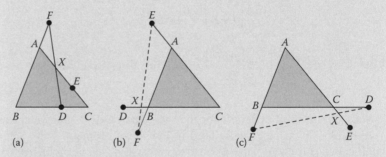

Figure 9.12 Proving Lemma 2.

2. Next, if none of the points D, E, and F lie interior to the sides of $\triangle ABC$, suppose that all the intersections mentioned in the desired conclusion fail. That is, (D, E) lie on the same side of line AB, (E, F) lie on the same side of line BC, and (D, F) lie on the same side of line AC. The first of these assumptions forces a diagram like those of Figure 9.12b or 9.12c, with either (DBC) and (EAC) or (BCD) and (ACE). Suppose (DBC) and (EAC). Since F and D lie on the same side of line AC, we must have (ABF). This forces E and F to lie on opposite sides of line BC, a contradiction. The last case is (BCD) and (ACE). Again, (ABF) must hold, and this forces D and F to lie on opposite sides of line AC, a contradiction. ◣

As you can see the last result required some rather tedious betweeen-ness arguments. The next result is similar, unfortunately. It is needed in Ceva's theorem, also in the converse part, where we must prove that if the linearity number equals 1, the cevians are concurrent. In hyperbolic geometry, we must make the assumption that any two cevians meet, not implied by the unit value of the linearity number (just as it is not in Euclidean geometry). We must be able to conclude that, for example, if cevians $\overleftrightarrow{AD}$ and $\overleftrightarrow{BE}$ intersect at P, then we can draw line $\overleftrightarrow{CP}$, which will intersect $\overleftrightarrow{AB}$ at some point X. The linearity numbers involving X and F are then both 1, so that $X = F$ and the cevians are concurrent. The following theorem will validate this.

Lemma 3   Suppose that lines $\overleftrightarrow{AD}, \overleftrightarrow{BE}$ and $\overleftrightarrow{CF}$ are cevians of $\triangle ABC$, such that $\overleftrightarrow{AD} \cap \overleftrightarrow{BE} = P, \overleftrightarrow{BE} \cap \overleftrightarrow{CF} = Q$, and $\overleftrightarrow{CF} \cap \overleftrightarrow{AD} = R$. Then one of the following must be true:

$$\overleftrightarrow{CP} \cap \overleftrightarrow{AD} \neq \emptyset, \quad \overleftrightarrow{AQ} \cap \overleftrightarrow{BC} \neq \emptyset, \quad \overleftrightarrow{BR} \cap \overleftrightarrow{AC} \neq \emptyset$$

Proof   Suppose neither of the desired intersections exist. That is,

$$\overleftrightarrow{CP} \cap \overleftrightarrow{AB} = \emptyset, \quad \overleftrightarrow{AQ} \cap \overleftrightarrow{BC} = \emptyset, \quad \text{and} \quad \overleftrightarrow{BR} \cap \overleftrightarrow{AC} = \emptyset$$

Hence neither P, Q, nor R can lie in the interior of $\triangle ABC$ or else either $\overleftrightarrow{CP}$ meets $\overleftrightarrow{AB}$, $\overleftrightarrow{AQ}$ meets $\overleftrightarrow{BC}$ or $\overleftrightarrow{BR}$ meets $\overleftrightarrow{AC}$ as illustrated in Figure 9.13a. Also, by assumption, P lies on the same side of line AB as C, that is, in the half-plane $H(C, \overleftrightarrow{AB})$. Furthermore, P cannot lie in the interior of $\angle A'CB'$ (as shown in Figure 9.13b, or else ray PC enters the interior of $\triangle ABC$ and then meets side AB, again denying $\overleftrightarrow{CP} \cap \overleftrightarrow{AB} = \emptyset$. Hence P lies in one of the two shaded regions of Figure 9.13b, and either (1) $P \in \text{Int} \angle ABC$ or (2) $P \in \text{Int} \angle BAC$.

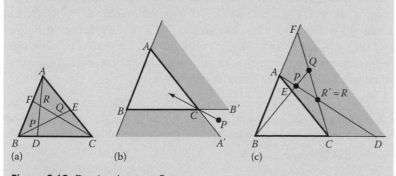

Figure 9.13 Proving Lemma 3.

1. $P \in \text{Int}\angle ABC$. Hence (BCD) follows. Also (AEC) and (BEP) hold. Now Q lies on line BP and must lie on ray BP since (QBP) would put Q and A on opposite sides of line BC, denying $\overrightarrow{AQ} \cap \overleftrightarrow{BC} = \emptyset$, and (BQP) would imply that ray AQ meets side BD or line BC, a contradiction. Therefore, (BPQ) and $(BEPQ)$, as in Figure 9.13c. Since therefore $(\overrightarrow{CB}\,\overleftrightarrow{CE}\,\overrightarrow{CQ})$, and since F lies on line CQ, (BAF) must hold. Thus, $(\overrightarrow{DB}\,\overleftrightarrow{DA}\,\overrightarrow{DF})$ implies by the crossbar principle that $\overrightarrow{DA}$ meets segment CF at some point R' which must coincide with R. This means that ray BR meets segment AC, again by the crossbar principle, contradicting $\overleftrightarrow{BR} \cap \overleftrightarrow{AC} = \emptyset$.

2. $P \in \text{Int}\angle BAC$. Using the same argument as in (1), only with R playing the role of Q, one obtains a contradiction in this case also. This proves that our original assumption was false, hence the theorem follows. ⟋

Menelaus' Theorem for Non-Euclidean Geometry Consider any triangle ABC, with menelaus points D, E, and F on its sides. Then, D, E, and F are collinear iff

$$\mathcal{L}\begin{bmatrix} ABC \\ DEF \end{bmatrix} = -1.$$

Ceva's Theorem for Non-Euclidean Geometry Consider any triangle ABC, with cevians AD, BE, and CF. If each pair of cevians intersect, they are concurrent iff

$$\mathcal{L}\begin{bmatrix} ABC \\ DEF \end{bmatrix} = 1.$$

Note: In elliptic geometry, the directed distances on lines AB, BC, and AC used to define the linearity number must be chosen correctly. It suffices to take as the origin of the coordinate system on these lines to be the points A, B, or C. Thus, it will follow that $\sin BD / \sin DC > 0$ iff (BDC), and similarly for the other Menelaus points: For if B is the origin, it follows that $|BC| < \pi$ and by **(4)**, $(BDC) > 0$ iff BD and DC have the same sign. Such properties are crucial in the proof of these theorems because they determine the sign of the linearity number. ⟋

Ceva's theorem helps to complete the theory for the concurrence of special segments begun in Chapter 3, where it was shown that the angle bisectors and perpendicular bisectors (if any two of them intersect) of any triangle are concurrent. The following result will be left as an interesting problem.

Corollary The medians of any triangle and the altitudes of an acute-angled triangle are concurrent.

Problems (Section 9.6)

Group A

1. Prove properties **(1)**, **(2)**, and **(3)** for directed distance.

2. Prove for hyperbolic geometry the following unified version of Euler's identity (for A, B, C, and D collinear):

(8) $S(AB) \, S(CD) + S(AC) \, S(AD) + S(CA) \, S(BD) = 0$

3. Show **(4)** is not valid in elliptic geometry if $|AD| > \pi$.

4.* **General Cevian Formula for Non-Euclidean Geometry**
Starting with Stewart's theorem **(5)** applied to $\triangle ABC$ with Cevian AD, establish the general Cevian formula below by solving the equation for $\cosh d$ [$\cos d$]. Note that the assumption (BDC) made for the restricted Cevian formulas **(5)** and **(6)** derived in Problems 9 and 10, Section **9.1**, has been eliminated here, and p and q are allowed to be negative.

> **Theorem**   In any triangle ABC, if D lies on line BC and $d = AD$, then, in standard notation,
>
> $$\cosh d = p\cosh b + q\cosh a \quad [\cos d = p\cos b + q\cos a]$$
>
> where $p = \sinh DB / \sinh AB$, $q = \sinh AD / \sinh AB$.

5. Prove property **(6)** for the linearity number stated above.

Group B

6. Prove Lemma 1 for hyperbolic geometry.

7. Using Lemma 1 prove that if $\mathcal{R}[AB,CX] = \mathcal{R}[AB,CD]$, then either $X = D$ or $X = D^*$.

8. Prove property **(7)** for the linearity number using Lemma 1.

9. **Prelude (1) to Menelaus' Theorem**   In the figure for this problem, lines AA' and BB' are perpendicular to line BD, with $a = AA'$ and $b = BB'$. Prove in either of the two cases shown, $(DA'B')$ or $(A'DB')$, that

$$\frac{sAD}{sDB} = \pm \frac{S(a)}{S(b)}$$

10. **Prelude (2) to Menelaus' Theorem**   In Figure P.10, perpendicular lines AA', BB', and CC' have been drawn to a line l intersecting the three sides of $\triangle ABC$ at D, E, and F. Using the result of Problem 9, prove that the linearity number of $\triangle ABC$ with respect to D, E, and F is ± 1.

Then, use the postulate of Pasch to show that either exactly two of the menelaus points lie interior to the sides of $\triangle ABC$, or none do so, and that, therefore the linearity number is -1.

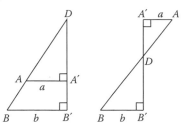

Figure P.10

11.   **Proof of Menelaus' Theorem**   Using Problem 10 and Lemma 2, prove Menelaus' theorem (Figure P.11).

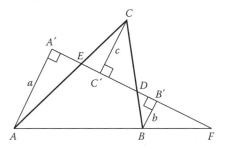

Figure P.11

Group C

12.   **Proof of Ceva's Theorem**   Use the same method as was used in Euclidean geometry (Section **7.5**) in order to prove the theorem of Ceva for non-Euclidean geometry. (Use Menelaus' theorem to prove that the linearity number equals $+1$ if the Cevians are concurrent by showing that

$$\mathcal{L}\begin{bmatrix} FBC \\ DPA \end{bmatrix} \mathcal{L}\begin{bmatrix} PCE \\ ABF \end{bmatrix} \mathcal{L}\begin{bmatrix} APE \\ BCD \end{bmatrix} \mathcal{L}\begin{bmatrix} AFP \\ CDB \end{bmatrix} = 1$$

where the left side reduces to the desired expression for $\mathcal{L}$, then use Lemma 3 for the converse.)

13.   Given an acute-angled triangle ABC in hyperbolic geometry (Figure P.13), let altitudes AD, BE, and CF be drawn, having lengths $AD = d$, $BE = e$, and $CF = f$. Derive the following:

$$\frac{sCE}{sDC} \equiv \frac{\sinh CE}{\sinh DC} = \frac{\tanh e}{\tanh d},$$

$$\frac{sAF}{sEA} = \frac{\tanh f}{\tanh e}, \quad \frac{sBD}{sFB} = \frac{\tanh d}{\tanh f}$$

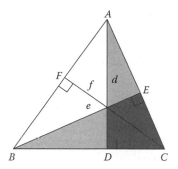

Figure P.13

14. **Medians and Altitudes are Concurrent**   Prove each of the following propositions:
    (a) The medians of any triangle in unified geometry are concurrent.
    (b) The altitudes of any triangle in elliptic geometry, and of an acute-angled triangle in hyperbolic geometry, are concurrent. (See Problem 13.)

15. **The Harmonic Construction in Non-Euclidean Geometry**   Using cross-ratio and Lemma 1, prove that the harmonic construction $PQRS$ produces four collinear points A, B, C, and D such that $\mathcal{R}[AB, CD] = -1$, and that it is independent of the choice of points P and Q. (See Example **4**, Section **7.6**.)

16. **Desargues' Theorem in Non-Euclidean Geometry**   Prove the theorem of Desargues using Menelaus' theorem and linearity numbers. (It must be assumed in hyperbolic geometry that certain points of intersection needed in the proof exist, but that this assumption can possibly be removed by the method of false position and the continuity of linear points in taking limits.)

17. **Pappus' Theorem in Non-Euclidean Geometry**   Prove the theorem of Pappus using Menelaus' theorem and linearity numbers. (See comment concerning hyperbolic geometry in Problem 16.)

9.7 Poincarè's Model for Hyperbolic Geometry

The great endeavor in mathematics beginning with Euclid's treatment of parallelism was proving Euclid's elaborate parallel axiom (the Fifth Postulate) as a theorem. Euclid postponed the insertion of the axiom until the last moment (fully 28 propositions were demonstrated prior to this—the content of which is the extensive material known as *absolute geometry*). Was this a matter of convenience, or did Euclid actually anticipate a logical difficulty inherent with the parallel axiom? Whatever the reason, in view of his decision to delay the introduction of parallelism into his system, Euclid has been called the *first non-Euclidean geometer*! Thus, the tools available for proving the parallel postulate included the large body of material that preceded the Fifth Postulate and logic. The possibility for success certainly seemed feasible.

A further encouragement for mathematicians was also psychological. The *converse* of Proposition 27 *implies the Fifth Postulate*. For virtually all the basic theorems of geometry, the converse of a given theorem is usually valid and can be easily demonstrated, often by a simple indirect proof. It is then highly probable, so it seems, that one could construct a proof of the converse of Proposition 27, a theorem provable in absolute geometry, thereby proving the Fifth Postulate. These matters present a strong incentive for trying to find an indirect proof of the parallel axiom, which, in turn, would imply that hyperbolic geometry does not exist. The stage was set for a long, 2000-year struggle of attempts by mathematicians to prove the parallel axiom, some of whom, like Legendre, devoted their entire lives to the project (calling to mind the four-color conjecture of modern times).

Few realized that what they were doing was to no avail. As we mentioned earlier, Gauss had strong suspicions that the effort was futile, but he did not succeed in producing an actual model for hyperbolic geometry. Although the strong parallel between the formulas derivable for hyperbolic geometry and those of spherical geometry is strong evidence, this by itself does not constitute a proof. Bolyai and Lobachevski rediscovered the work of Gauss in this area and were the first to publish their findings (see Historic Note on the lives of these two in Section **5.5**). The research that Gauss had done were unknown to both of them, for Gauss had withheld his discoveries due to his stated fear of ridicule. "I fear the scream of dullards if I make my views explicit"—so wrote Gauss in 1829 in a letter to a friend. Their developments consisted of an elaborate structure, including all the formulas we have derived for hyperbolic geometry, and a surface in three-dimensional hyperbolic geometry—analogous to a sphere in Euclidean geometry which Bolyai called a *horosphere*—whose intrinsic geometry was *Euclidean*! These strange results led Bolyai to proclaim his discovery of a "new world," but was deeply discouraged by the seeming lack of approval by Gauss. (It should be pointed out that the method we used in Chapter 6 is not the same as Bolyai's development, where he used the theory of horospheres.)

Thus, mathematicians in the eighteenth and nineteenth centuries were working in an antischolarly atmosphere, fraught with strong emotional feelings and major philosophical beliefs about the physical world—Immanuel Kant advocated the philosophy that only one valid perception of the universe existed. This meant that Euclidean geometry, as a "perception" of the universe, was the only valid geometry—all others must lead to contradictions and foolishness. Thus, the published work of Bolyai and Lobachevski was not seriously regarded until 1868 when E. Beltrami proposed, with proof, the pseudosphere as a model for hyperbolic geometry (locally). He later discovered, with Klein, a circular model Euclidean plane for hyperbolic geometry in the large. Later, in 1882, H. Poincaré reformulated the Beltrami–Klein model to create another circular model—the one we now consider.

Start with a unit circle C, traditionally called the **absolute**, with interior S. Take as a *point* any member of S, and take as a *line* any diameter of C, or that portion of any circle orthogonal to C that lies in S, as illustrated in Figure 9.14. In order to distinguish the objects of the model from Euclidean objects, the prefix h- will be used from now on in order to designate any such object or concept. Thus, so far, we have introduced *h-points* and *h-lines*, and it is clear that there exist infinitely many h-lines, and an h-line contains infinitely many points. So far, **Axiom 1** holds.

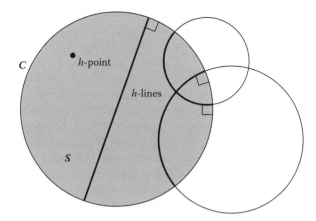

Figure 9.14 Poincarè's disk.

The orthogonality condition for circles introduced earlier shows that if C is the circle $x^2 + y^2 = 1$ in the xy-plane and $x^2 + y^2 + ax + by = c$ is orthogonal to C, then $c = -1$ and an h-line has the form for certain constants a and b, as illustrated in Figure 9.15,

$$L: x^2 + y^2 + ax + by + 1 = 0, \quad x^2 + y^2 < 1$$

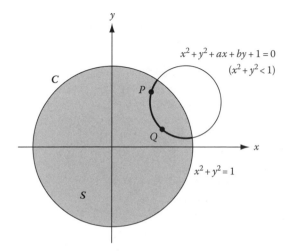

Figure 9.15 Equation for h-lines.

Now consider two h-points, $P(x_1, y_1)$ and $Q(x_2, y_2)$. In order for P and Q to lie on an h-line whose equation is given by the above, its coordinates must satisfy this equation, and thus we obtain, after rearranging,

$$ax_1 + by_1 = -x_1^2 - y_1^2 - 1$$

$$ax_2 + by_2 = -x_2^2 - y_2^2 - 1$$

This is a system of linear equations in a and b (since x_i and y_i for $i = 1, 2$ represent specific real numbers). This system has a unique solution provided

$$\det \begin{bmatrix} x_1 & y_1 \\ x_2 & y_2 \end{bmatrix} \neq 0$$

that is, provided $x_1 y_2 \neq x_2 y_1$. But if $x_1 y_2 = x_2 y_1$, then P and Q are observed to be collinear with the origin $O(0, 0)$, lying on a line having slope y_1/x_1 if $x_1 \neq 0$. Thus, we see that if P and Q are in line with the origin, the h-line containing P and Q is a diameter of C, and if P and Q are not, then there is a unique arc of a circle orthogonal to C passing through P and Q. In either case, there is a unique h-line containing P and Q, and each two h-points in S lie on a unique h-line (**Axiom 2** and **Axiom 4**).

An appropriate concept for distance is somewhat involved. Let two h-points be given, A and B; we aim to define the h-distance (AB) from A to B that will satisfy **Axiom 3**. This must be done in such a manner that (a) as A and B vary in S, (AB) range over the positive real numbers from 0 to ∞ (thus, the ordinary Euclidean distance AB will not suffice), and (b) if points A, B, and C lie on a line, with B on the arc or diameter AC then $(AB) + (BC) = (AC)$. The definition that has been found to work is to let M and N be the endpoints of the unique h-line AB determined by A and B, as illustrated in Figure 9.16 (either a segment or circular arc), and to take as the **h-distance** the number

(1) $$(AB) = \left| \ln R[AB, MN] \right| \equiv \left| \ln \left(\frac{AM \cdot BN}{AN \cdot BM} \right) \right|$$

where "ln" denotes the natural logarithm and $R[AB, MN]$ is the generalized cross-ratio introduced in Section **8.6**.

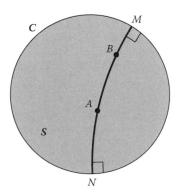

Figure 9.16 Distance in the Poincarè model.

(You may want to review this, but its properties will be apparent as we proceed.) By definition, $(AB) \geq 0$. Furthermore, if $A = B$, then $R[AB, MN]$ reduces to $AM \cdot AN/AN \cdot AM = 1$ and $(AB) = \ln 1 = 0$, and conversely. Finally, note that as $A \to M$, then $R[AB, MN] \to 0$, and therefore, $\ln R[AB, MN] \to \infty$. Thus, **Axiom 3** holds, with $\alpha = \infty$.

The axiom on betweenness, **Axiom 5**, follows from a perhaps surprising (but certainly indispensable) property of h-distance, in view of its complicated definition. Let A, B, and C be any three points on a circular arc (or segment) orthogonal to C, thus collinear, with B lying on subarc $\overset{\frown}{AC}$, as shown in Figure 9.17. First, it is apparent that by choosing the endpoints M and N judiciously, one can eliminate the absolute values in **(1)**. Thus,

$$(AB) = \ln R[AB, MN] = \ln \frac{AM \cdot BN}{AN \cdot BM} \quad \text{and}$$

$$(BC) = \ln R[BC, MN] = \ln \frac{BM \cdot CN}{BN \cdot CM}$$

and, by a property of logarithms,

$$(AB) + (BC) = \ln \frac{AM \cdot BN}{AN \cdot BM} + \ln \frac{BM \cdot CN}{BN \cdot CM}$$

$$= \ln \frac{AM \cdot BN}{AN \cdot BM} \cdot \frac{BM \cdot CN}{BN \cdot CM} = \ln \frac{AM \cdot CN}{AN \cdot CM}$$

This last expression is observed to be $\ln R[AC, MN] = (AC)$. Hence, this proves

(2) $(AB) + (BC) = (AC),$

which exhibits the h-betweenness relation (ABC) in terms of h-distance. Thus, if D is any other point on h-line AB (circular arc $\overset{\frown}{AB}$) distinct from A, B, and C, it is apparent that either (DAB), (ADB), (BDC), or (BCD), thus **Axiom 5**. At this point, **h-segments**, **h-rays**, and **h-angles** become meaningful, where one uses the definitions stated earlier, modified for h-betweenness.

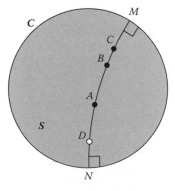

Figure 9.17 Betweenness in the Poincarè model.

The ruler postulate for rays is a direct result of **(2)**. Consider h-ray AB (circular arc ABM in Figure 9.17). For any point X on h-ray, AB define its *coordinate* to be $x = \ln R[AX, MN] \equiv (AX)$. Thus, the coordinate of A is $\ln R[AA, MN] = 0$ and that of B is $\ln R[AB, MN] > 0$. For any two points X and Y on h-ray AB, we evidently have, by **(2)**,

$$(XY) = |(AX) - (AY)| = |x - y|,$$

which is the content of **Axiom 6**.

The concept for angle measure is more transparent. One takes the ordinary Euclidean angle measure between the tangents to the h-rays (which are either segments or circular arcs in the model), which define the angle, as illustrated for $\angle ABC$ in Figure 9.18, where

(3) $m\angle ABC = \theta = m\angle UBV$

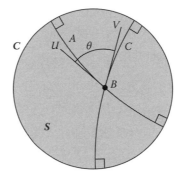

Figure 9.18 Angle measure in the Poincarè model.

This will make evident all the basic properties of angle measure (at least) **Axioms 7** and **8** since at each point these are merely statements about Euclidean angle measure. The remaining axioms about angle measure, **Axiom 10** and **Axiom 11**, and the plane separation axiom, **Axiom 9**, require further analysis.

In order to work a little with h-distance, we obtain a useful formula for points on any diameter of C. Let A any point on diameter MN and O the center of C (Figure 9.19). Since C is the unit circle, $OM = ON = 1$, and the definition of h-distance yields

$$(OA) = \ln R[OA, MN] = \ln \frac{OM \cdot AN}{ON \cdot AM} = \ln \frac{AN}{AM}$$

That is, with $x = (OA)$, $y = AN$ and $z = AM$, $x = \ln (y/z)$, or

$$e^x = \frac{y}{z} \quad \text{and} \quad e^{-x} = \frac{z}{y}$$

By definition of sinh x, it follows that

$$\sinh x = \tfrac{1}{2}(e^x - e^{-x}) = \frac{1}{2}\left(\frac{y}{z} - \frac{z}{y}\right) = \frac{y^2 - z^2}{2yz} = \frac{(y+z)(y-z)}{2yz}$$

To relate this to the Euclidean distance $OA \equiv a$, note that $y = AN = 1 + a$ and $z = AM = 1 - a$. Thus,

$$\sinh(OA) = \frac{(1+a+1-a)(1+a-1+a)}{2(1+a)(1-a)} = \frac{2a}{(1-a^2)}$$

showing the relationship between the h-distance and ordinary Euclidean distance from O to A:

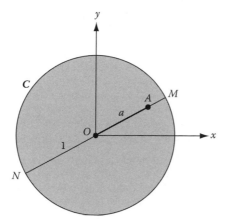

Figure 9.19 Euclidean distance and h-distance related.

(4)
$$\sinh(OA) = \frac{2a}{1-a^2}$$

One can obtain from the identity $\cosh^2 x = 1 + \sinh^2 x$, the additional relationship

(5)
$$\cosh(OA) = \frac{1+a^2}{1-a^2}$$

These formulas can be used for various calculations, such as that presented in the following example.

Example 1

In Figure 9.20 is shown an h-right-triangle OAB, where $A = (\frac{1}{2}, \frac{1}{2})$ and $B = (b, 0)$. Obtain calculations for (OA), (OB), and (AB), then verify the two right triangle relations for hyperbolic geometry (showing that $\kappa = 1$ for the Poincarè model):

(a) $\cosh(OA) = \cosh(OB)\cosh(AB)$

(b) $\dfrac{\sinh(AB)}{\sinh(OA)} = \sin\theta$

where $\theta = 45°$. (Note that one can readily observe that the angle sum of $\triangle OAB$ is $<180°$.)

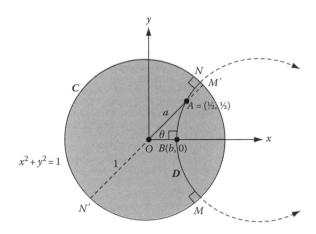

Figure 9.20 Example 1.

Solution

(a) Note that circle **D** is orthogonal to **C** and to the *x*-axis. Thus, it has an equation of the form $x^2 + y^2 + px + 1 = 0$ since its center lies on the *x*-axis. This circle also passes through *A*, so the coordinates of *A* satisfy this equation and we obtain, by substitution,

$$(\tfrac{1}{2})^2 + (\tfrac{1}{2})^2 + p(\tfrac{1}{2}) + 1 = 0 \rightarrow \tfrac{3}{2} + \tfrac{1}{2}p = 0 \rightarrow p = -3$$

Hence **D**: $x^2 + y^2 - 3x + 1 = 0$. To find the coordinates of $B(b, 0)$, set $y = 0$ and we find

$$x^2 - 3x + 1 = 0 \rightarrow x = \frac{3 - \sqrt{5}}{2}$$

Thus, $b = (3 - \sqrt{5})/2, 0)$. To find cosh (OA) and cosh (OB), we use the relation **(5)**. We find that, with $a = \sqrt{\tfrac{1}{2}}$,

$$\cosh(OA) = \frac{1 + \tfrac{1}{2}}{1 - \tfrac{1}{2}} = 3 \quad \text{and}$$

$$\cosh(OB) = \frac{1 + [\tfrac{1}{2}(3 - \sqrt{5})]^2}{1 - [\tfrac{1}{2}(3 - \sqrt{5})]^2} = 3\sqrt{5}/5$$

In order to find (AB), the coordinates of *M* and *N* (the set **D** ∩ **C**) must be found. Thus, we solve the system

$$x^2 + y^2 - 3x = -1$$

$$x^2 + y^2 = 1$$

Placing 1 for $x^2 + y^2$ in the first equation, the solution is $x = \tfrac{2}{3}$, $y = \pm\sqrt{1 - x^2} = \pm\tfrac{1}{3}\sqrt{5}$. Thus, $M(\tfrac{2}{3}, -\tfrac{1}{3}\sqrt{5})$, $N(\tfrac{2}{3}, \tfrac{1}{3}\sqrt{5})$. We can now find (AB). By definition,

$$(AB) = \ln R[AB, MN] = \ln \frac{AM \cdot BN}{AN \cdot BM} = \ln \frac{AM}{AN}$$

By the distance formula for Euclidean geometry,

$$AM^2 = (\tfrac{1}{2} - 2/3)^2 + (\tfrac{1}{2} + 1/3\sqrt{5})^2 = (5 + 2\sqrt{5})/6$$

$$AN^2 = (\tfrac{1}{2} - 2/3)^2 + (\tfrac{1}{2} - 1/3\sqrt{5})^2 = (5 - 2\sqrt{5})/6$$

and

$$(AB) = \tfrac{1}{2}\ln \frac{AM^2}{AN^2} = \tfrac{1}{2}\ln \frac{5 + 2\sqrt{5}}{5 - 2\sqrt{5}} = \tfrac{1}{2}\ln(9 + 4\sqrt{5})$$

$$= \tfrac{1}{2}\ln(2 + \sqrt{5})^2 = \ln(2 + \sqrt{5})$$

Hence

$$\cosh(AB) = \tfrac{1}{2}\left[2 + \sqrt{5} + \frac{1}{2 + \sqrt{5}}\right]$$

$$= \sqrt{5}$$

Thus,

$$\cosh(OB)\cosh(AB) = (3\sqrt{5}/5) \cdot \sqrt{5} = 3$$

verifying (a).

(b) To calculate sinh (AB) and sinh (OA), merely use the identity $\sinh^2 x = \cosh^2 x - 1$ together with the calculations already made in (a). We find

$$\sinh^2(AB) = (\sqrt{5})^2 - 1 = 4 \quad \text{or} \quad \sinh(AB) = 2$$

$$\sinh^2(OA) = 3^2 - 1 = 8 \quad \text{or} \quad \sinh(OA) = \sqrt{8} = 2\sqrt{2}$$

Thus, $\sinh(AB)/\sinh(OA) = 2/2\sqrt{2} = \sqrt{2}/2$, which is sin 45°.

The program of verifying Axioms 1–12 in the Poincarè model continues, having done so for **Axioms 1–8**. Consider **Axiom 9**, the plane separation axiom. Let h-line l be given. The segment or arc determined by l obviously splits S into two disjoint sets H_1 and H_2, each of which is the intersection of S with a either a Euclidean half plane or the interior, or exterior, of a circle orthogonal to C. (Figure 9.21 illustrates the latter case.) One can see that for each point $A \in H_1$ and $B \in H_2$ the h-segment joining A and B (a circular arc or segment joining A and B) must cross l at some point.

To prove the convexity of H_1 (or H_2), which is clear if l is a diameter of C, consider two points A and B in H_1 when l is a circular arc. If the arc or segment AB contains a point not in H_1 then it meets l at some point C and therefore must contain a point D in H_2 (since two circles orthogonal to C cannot be tangent). But from the above, since $B \in H_1$ and $D \in H_2$, this means that h-segment DB crosses l at a second point E, contradicting the property that two h-points determine a unique h-line. Thus, h-segment AB cannot meet l and thus must belong completely to H_1. Therefore, H_1 is convex as well as H_2 by the same argument. This completes the verification of **Axiom 9**. **Axioms 10** and **11** can be verified by observation.

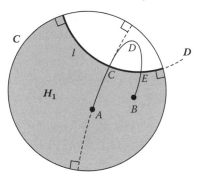

Figure 9.21 Plane separation postulate verified.

This brings us to **Axiom 12**, the SAS postulate. For the proof of this crucial axiom, we use the alternate approach that appeared in Chapter 3, Section 3.4. We have only to exhibit an h-reflection in a given line l (a mapping that is h-isometric and h-angle preserving on S) in order to guarantee **Axiom M**, the motion postulate (which implies the SAS postulate, as shown in Section 3.4). This will involve circular inversions, introduced in Section 7.6. We already know that such mappings preserve generalized cross-ratio (thus, h-distance preserving) and curvilinear angle measure (thus, h-angle preserving). The only question is whether such mappings can leave the proposed line of reflection fixed and can map the points of S into itself.

If l is a diameter, the ordinary Euclidean reflection in line l will suffice. In the case when l is a circular arc contained by the circle D having center D and radius r (Figure 9.22), take D as the circle of inversion. In this case, since D—the center of inversion—lies outside C, the mapping is defined for all points of S. The points of D are fixed points of the inversion, so the points of l are fixed. What we must be sure of, however, is that points of S map into points of S. To see this, let P and Q be inverse points with respect to D, with $P \in S \cap \text{Int } D \equiv H_1$, C_1 and C_2 the intersection points of ray DP and C, and E the intersection of ray DP and D. Since $E \in D$, E maps to itself, and since C is orthogonal to D, C_1 maps to C_2. Thus

$$DC_1 < DP < DE$$

leads to

$$\frac{r^2}{DC_2} < \frac{r^2}{DQ} < \frac{r^2}{DE}$$

and therefore

$$DE < DQ < DC_2$$

Hence P maps to a point of S on the opposite side of h-line l. Thus, as is characteristic of reflections, points on one side of a line are mapped to points on the opposite side. We leave it as a problem to show that the circular inversion in D also maps lines to lines, segments to segments, and angles to angles. We have already noted that circular inversion also preserves generalized cross-ratio and is conformal, and h-distance and h-angle measure are preserved. Such a mapping will be called an **h-reflection** in line l. Thus **Axiom 12** holds, and this completes the proof of:

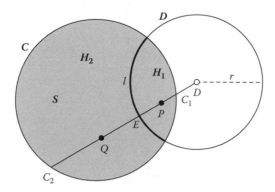

Figure 9.22 h-reflection in l.

> ***Theorem 1***   The Poincarè disk S is a model for unified geometry, with $\alpha = \infty$.

Problem 2 will ask you to decide what type of parallel postulate holds for the geometry of S. Your answer, if correct, will prove

> ***Corollary***   The Poincarè disk S is a model for hyperbolic geometry.

We might note that the above corollary does not necessarily mean that the axioms of hyperbolic geometry are consistent in any absolute sense; only relative consistency has been shown: Hyperbolic geometry has been shown to be *as consistent as Euclidean geometry*. There is no known proof of the absolute consistency of Euclidean (coordinate) geometry, nor of the real number system. In fact, the now famous theorem by Kurt Godel (1906–1978) proves that within any system of axioms as involved as the axiomatic treatment of the natural numbers, there exist *undecidable* propositions, those which cannot be proven within the system and for which no counterexample can be constructed, while staying within the system. Furthermore, the proposition that the system is consistent is *one of those undecidable issues*. So it is impossible to prove that the axioms for the natural numbers are consistent and to stay within the system of natural numbers. One must go outside the system, using models, as has been an ongoing concept in our study of geometry. This is a very unsettling state of affairs for mathematics in general. Much controversy has resulted and continues to this day, largely unknown to the public. But for us at the very least, this points up the importance of examples and models, and shows their limitations as well.

Example 2

Consider any point M distinct from O (center of the absolute) and at a Euclidean distance a from O. Locate point D on ray OM at a Euclidean distance $1/a$ from O and draw a circle **D** centered at D having radius $r = \sqrt{1-a^2}/a$. Show that **D** is a circle of inversion orthogonal to **C**, which maps M to O. Thus, **S** $\cap$ **D** is an h-line which reflects M to O. (See Figure 9.23.)

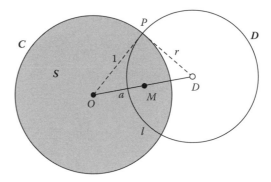

Figure 9.23 Example 2.

Solution

If we show that the Pythagorean relation holds for $\triangle OPD$, this will prove **D**$\perp$**C**. Observe

$$OP^2 + r^2 = 1 + \frac{1-a^2}{a^2} = \frac{a^2+1-a^2}{a^2} = \frac{1}{a^2} = OD^2$$

as desired. To show that M and O are inverse points with respect to **D**, it suffices to prove that $DO \cdot DM = r^2$:

$$DO \cdot DM = DO(DO - a) = \frac{1}{a}\left(\frac{1}{a} - a\right) = \frac{1}{a^2} - 1 = \frac{1-a^2}{a^2} = r^2$$

Example 3: The h-Midpoint Construction

Let AB be any h-segment lying on a circular arc m orthogonal to **C**. Construct the tangents to m at A and B, intersecting at point P (Figure 9.24), then draw line OP meeting arc AB at M. Show that M is the h-midpoint of h-segment AB.

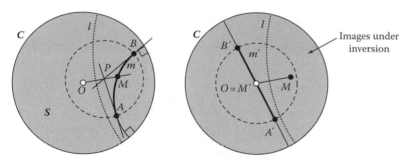

Figure 9.24 Constructing the midpoint of $\overline{AB}$.

Solution

Construct h-line l as in Example 3 that will reflect point M to point O. Since h-line AB ($\equiv m$) maps to h-line $A'B'$ under this inversion, and this line passes through $M' = O$, h-line $A'B'$ is a diameter of C. (The diagram on the right in Figure 9.24 shows these images.) Now the circle that we drew with center P is orthogonal to m, hence its image is a circle orthogonal to h-line $A'B'$, and its center must lie on the invariant line OM (dashed circle in the diagram). That is, the dashed circle has a diameter on line $A'B'$, so its center is the intersection of lines OM and $A'B'$, or O (= M'). Therefore, $M'A' = M'B'$ implies $(M'A') = (M'B')$, thus $(MA) = (MB)$.

Problems (Section 9.7)

Group A

1. What does the Poincarè model show about the angle sum of h-triangles?

2. Do parallel lines exist in the Poincarè model? If so, what is their behavior? Given a point not on a given line, how many lines in the model can be drawn through the given point and parallel to the given line? What does Figure 5.31 from Section **5.9** look like in the model, where $A = O$?

3. What do Saccheri quadrilaterals look like in the Poincarè model? [**Hint:** Take as base an h-segment lying on a diameter of C.]

4. A realization of the equilateral triangle of Problem 4, Section **5.8** appears in the Poincarè model as shown in Figure P.4. Explain why each shaded h-triangle must be equilateral *by observation*. Show that the eight equilateral

triangles in the figure making up the octagon shown are congruent. What is the Euclidean equivalent of this figure, and how does it compare?

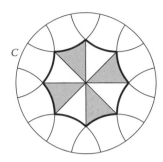

Figure P.4

Group B

5. Show that the h-octagon of Problem 4 constructed from tha eight congruent equilateral triangles is regular. Could such an octagon be used to tile the huperbolic plane (using octagons congruent to this one)? Explain why, or why not.

6. In $\triangle OAB$ of Example 1 (Figure 9.20), the h-lengths of sides OA and AB were calculated. Verify the right triangle relation

$$\tan\theta = \frac{\tanh(AB)}{\sinh(OB)}$$

7. Using **(4)** and **(5)**, derive the following forms for the h-distance (AB) for points A and B lying on a diameter of C, as in Figure P.7, where $a = OA$ and $b = OB$.

(a) $\sinh(AB) = \dfrac{2(a+b)(1+ab)}{(1-a^2)(1-b^2)}$ if (AOB)

(b) $\sinh(AB) = \dfrac{2(a-b)(1-ab)}{(1-a^2)(1-b^2)}$ if (OAB)

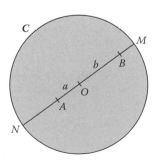

Figure P.7

8. Make a sketch using the Poincarè model showing that in hyperbolic geometry, there exists for each angle ∠ABC a line that is entirely contained by the interior of the angle, yet does not intersect its sides.

Group C

9. Suppose that *l* and *m* are perpendicular h-lines, meeting at point *M*, such that *l* passes through *O*, the center of *C*. If *m* has endpoints *U* and *V* on *C* and the chord *UV* meets *l* at *A*, show that *M* is the h-midpoint of h-segment *OA*. (See Figure P.9.)

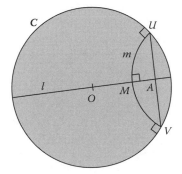

Figure P.9

9.8 Other Models: Surface Theory

Poincarè's model is only one of many that have been developed over the years. One that resembles Poincarè's model most closely is due to Felix Klein (1849–1929). This model also consists of a circle (or ellipse) as absolute, with the interior points as the points of the geometry, but with ordinary *chords* (Euclidean segments) as the lines. The distance concept is the same as that defined for the Poincarè model, but angle measure must be redefined. Another model is due to E. Beltrami (1835–1900), which consists of the upper half plane of coordinate geometry [all points (x, y), such that $y > 0$], where the x-axis plays the role of the absolute. The lines of the geometry are defined to be either vertical lines $x = a$ (a constant), $y > 0$, or semicircles having diameters along the x-axis. The metric is defined on semicircles much as in the Poincarè model, and angle measure is Euclidean, as in the Poincarè model. (The Poincarè model can be mapped onto the upper-plane model by a circular inversion.) It should be mentioned that Poincarè first introduced his model by working in the complex plane and observing the fact that any fractional linear transformation of the form $z' = (az + b)/(cz + d)$ where a, b, c, and d are real, and $ad \neq bc$ over the unit disk $|z| < 1$ is conformal and distance preserving.

Another model that preceded all the others was discovered by Beltrami, involving surface theory, an area also studied quite intensely by Carl Gauss in his search for various non-Euclidean geometries. Gauss was the first to fully develop the study of surfaces in three-dimensional Euclidean geometry. He pursued it to great lengths because he realized that this study would shed light on the existence of hyperbolic geometry, and it had other applications as well. The ultimate result later led to Riemann's theory of generalized metrics in R^n (n-dimensional affine space), which provided the exact tools Einstein needed for his theories. Gauss set up a parametric representation for surfaces in E^3, which today often appears in the study of multidimensional calculus: $x = f(u, v)$, $y = g(u, v)$, $z = h(u, v)$ for u and v real. For example, the unit sphere $x^2 + y^2 + z^2 = 1$ can be represented by

$$x = \cos u \sin v, \quad y = \sin u \sin v, \quad z = \cos v \, (v \neq 0, \pi)$$

for u, v real numbers. (Try squaring the coordinates x, y, and z, summing, and see what you get.) The geometry of the unit sphere can then be studied more conveniently by using the parameters u, v (which are actually *spherical coordinates*, closely associated with latitude, longitude measurements on the earth's surface). Thus, if $u = 0$, then $x = \sin v$, $y = 0$, and $z = \cos v$, which yield the great circle lying in the xz-plane. (If you need more information about three-dimensional coordinates, see Problems 17–23 in Section **8.1**, where a detailed presentation is outlined.) Thus, by restricting u and v in certain ways, we obtain the "lines" of a sphere. These, of course, were used as a model for the lines we developed in axiomatic elliptic geometry in Chapter 1.

You can see how this idea can be generalized to any surface in E^3, where the above functions f, g, and h are modified to arbitrary two-variable functions. Thus, suppose S represents some such surface, as shown in Figure 9.25. Geometry on S is defined by taking as "points" the ordinary points on that surface, as "distance" the shortest arc distance possible on S between two points, and as "lines" the *geodesics* of S (i.e., those curves on which the arc length coincides locally with distance). This enables us to define betweenness, segments, and rays. Finally, one uses ordinary angle measure for the "angle" formed by two rays. Although our discussion is merely descriptive, these concepts do have exact analytical definitions, many of which are encountered in multidimensional calculus; we do not pursue that here.

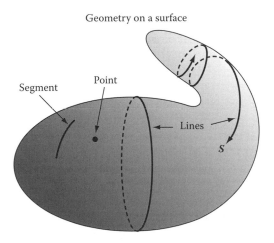

Figure 9.25 Geometry on a surface.

Theoretically, this procedure describes the **plane geometry associated with** S, or, as it is usually called, the **intrinsic geometry** of S. It is conceivable that one can produce as many plane geometries as there are surfaces, with the shape of the surface controlling the geometric theory. Those surfaces would be *models* for various axiomatic systems, just as the sphere is a model for elliptic geometry.

Two fundamental problems suggest themselves at this point: (1) Given a surface S, find a characteristic maximal set of independent axioms $\mathcal{E}$ that are valid for the intrinsic geometry of S, and (2) given an axiomatic system $\mathcal{E}$, find a surface S whose intrinsic geometry is a realization of $\mathcal{E}$. More briefly stated, the problems are as follows: Given a surface as model find the set of axioms which fit it, and, given a set of axioms, find a surface that serves as a model. As far as our axiomatic system is concerned (Axioms 1–12), a sphere serves as a model for elliptic geometry, and the coordinate plane is a model for parabolic geometry. However, hyperbolic geometry *does not have a global realization on a surface in Euclidean space.**

Of particular interest in classical differential geometry, however, is the study of *local* properties of surfaces, as opposed to the global properties we have been discussing. To illustrate, consider an entire sphere S_1 and just part of another sphere S_2 having the same radius as S_1 (Figure 9.26). While the global intrinsic geometries of these two surfaces are indeed different (the maximal distance α would be different, for example), all previously proven theorems regarding betweenness, angle measure, congruence, and inequalities—if restricted to small enough regions—would be identical in the two geometries. To be more specific, each point P of S_1 corresponds to a point Q of S_2, such that for some $r > 0$ the intrinsic geometry of S_1 confined to points within a distance r of P coincides with that of S_2 confined to points within a distance r of Q. To put it still another way, elliptic geometry can be realized *locally*, but not *globally*, on S_2.

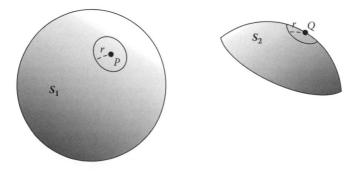

Figure 9.26 Local isometry between two surfaces.

---

* This theorem was discovered by G. Lütkemeyer in 1902, also proved by Hilbert. It is known as *Hilbert's theorem*, not to be confused with the theorem concerning hyperparallel lines, also known by that name. The theorem does not say *there is no model for hyperbolic geometry*, just that one does not exist as the intrinsic geometry on a surface in E^3.

If we compromise our previous goal of seeking global interpretations of axiomatic systems on surfaces in $\mathbf{E}^3$ to that of finding only local representations, then the tools of classical differential geometry are available, and much is known on the subject. Under this less restrictive requirement then, hyperbolic geometry has a local realization on a hornlike surface of revolution—the surface discovered by Beltrami—called a **pseudosphere** (illustrated in Figure 9.27). It is generated by a special curve, called the **tractrix**. This curve can be described informally as the path taken by a weight attached to a rope and dragged in a direction orthogonal to its initial position, as in Figure 9.27. Its precise x–y equation appears in the figure. The pseudosphere is then obtained by revolving the tractrix about the x-axis. This surface has the rare property of possessing *constant negative curvature*, as studied in differential geometry.

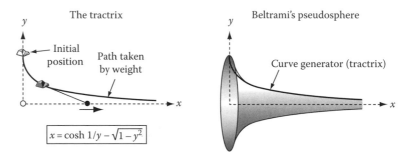

The tractrix

y

Initial position

Path taken by weight

x

$$x = \cosh 1/y - \sqrt{1 - y^2}$$

Beltrami's pseudosphere

y

Curve generator (tractrix)

x

Figure 9.27 Hyperbolic geometry realized locally.

Further examples are given by the cylinder and cone. The intrinsic geometry of the cylinder is a local realization of the Euclidean plane. To see this, suppose a cylinder is rolled on a plane surface, and various configurations in the plane are inked onto the cylinder, as illustrated in Figure 9.28. This inking process intuitively defines a mapping of the plane onto the cylinder.

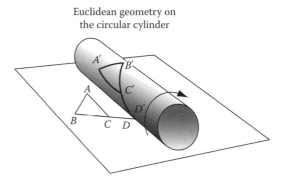

Euclidean geometry on the circular cylinder

Figure 9.28 Euclidean geometry on a cylinder.

Thus, $\triangle ABC$ and extended side BD have an image $A'B'C'D'$, as shown in the figure. Clearly, this mapping is both distance preserving and angle preserving. For example, for small enough triangles, the characteristically Euclidean property concerning the angle sum of a triangle is true on the cylinder, as well as any other property, including the Pythagorean theorem for small enough right triangles. This "inking-and-rolling" procedure can be repeated for the cone, as shown in Figure 9.29. This discussion leads to the interesting conjecture that there must be a locally distance-preserving mapping from the cylinder to the cone. This conjecture is true. In fact, a more general theorem from differential geometry states that *any two developables are locally isometric* (developables being surfaces that are generated by tangent planes).

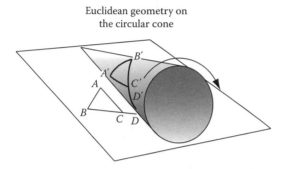

Figure 9.29 Euclidean geometry on a cone.

In this vein, one might toy with the idea of possibly producing a locally distance-preserving mapping from the plane to the sphere by the inking-and-rolling procedure, as depicted in Figure 9.30. Of course, in this case, one would have to roll the sphere in more than one direction, back and forth. Possibly contrary to our intuitive feelings on the matter, however, this mapping process does not succeed, as asserted in the theorem which follows. It is the formal confirmation of the ancient problem of creating an accurate planar map of the sphere: even the smallest portion of the earth's surface represented by such a map will always introduce some error.

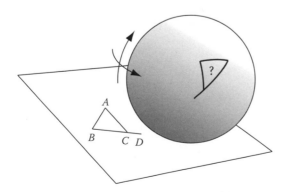

Figure 9.30 Euclidean geometry on a sphere?

> **Theorem** There exists no locally distance-preserving map from the plane to the sphere.
>
> *Proof* Recall that if *MN* is the segment joining the midpoints *M* and *N* of the sides *AB* and *AC* of certain triangles *ABC* in elliptic geometry, then $MN > \frac{1}{2}BC$, contradicting equality for the corresponding triangle in the Euclidean plane. (Proof of this inequality occurred previously in Example 1, Section **9.4**.) ◣

Problems (Section 9.8)

Group A

1. Let *S* be the intrinsic geometry on a cube 10 cm on a side, shown in Figure P.1. (This will be the topic of the next several problems.) If *A* is the point in the front face of the cube that makes a 3×3 cm square with the vertex closest to it, *B* and *D* are points on the side and bottom located similarly, and *C* is a vertex of the cube, as shown in Figure P.1, find the (intrinsic) distances *AB*, *AC*, and *AD*. **Note:** It helps to lay out the faces of the cube on a flat Euclidean plane in order to determine shortest paths.

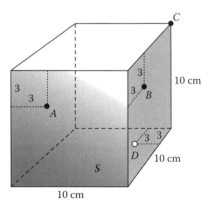

Figure P.1

2. We would expect the geometry of *S* in Problem 1 to resemble elliptic geometry since it has a bounded metric, and the cube is perfectly symmetrical, like the sphere. Find α, the maximal distance for this cube.

3. Again referring to Problem 1, in your own drawing sketch, a few paths on the cube that would represent:
    (a) Segments in *S*
    (b) Lines in *S*
    (c) Do two points determine a line in the geometry of *S*?

4. Examine angle measure and the angle sum of triangles in
    S of Problem 1. Is it $180°$ for all triangles?

5. Explain why the cylinder can only be a local realization of
    the Euclidean plane.

6. What portion of the plane is covered by one revolution
    of a
    (a) Cylinder of radius 1?
    (b) Cone of radius 1, height 2?

7. If you are familiar with three-dimensional coordinates
    x, y, z, consider the unit sphere $x^2 + y^2 + z^2 = 1$ given by the
    Gaussian representation (Figure P.7)

    $$x = \cos u \sin v \quad y = \sin u \sin v \quad z = \cos v \ (v \neq 0, \pi)$$

    (a) Find the parametric equations for the meridian line
        whose plane makes a $45°$ angle with the x-axis, as
        shown.
    (b) Show that the relation $\tan v = \sec u$ represents a great
        circle by finding its $x-y-z$ equation.
    (c) Does $v = $ constant represent a great circle?

Spherical coordinates for the unit sphere

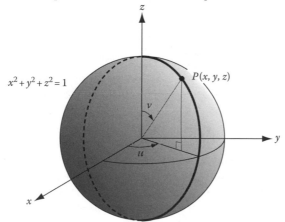

Figure P.7

9.9 Hyperbolic Parallelism and Bolyai's Ideal Points

We now delve a little deeper into the theory of parallels for hyperbolic
geometry, extending the introduction in Chapter 5 to more advanced
concepts. The first item on the agenda is deriving the formula for the
angle of parallelism defined in Chapter 5 (Section **5.7**). Recall that if
A is any point, l is a line not passing through it, and B is the foot of the
perpendicular from A, the line BX for $X \in l$ on one side of B tends to the

asymptotic parallel on that si
measure of $\angle BAX$ approaches
in Figure 9.31. The cosine rel ϵ
(Section **6.12**) is given by

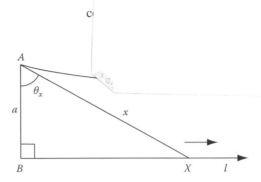

Figure 9.31 Proving Bolyai's formula for the angle of parallelism.

We want to take the limit as $BX \to \infty$ and $x \to \infty$, so that $\theta_x \to A$, the angle of parallelism. First, we show that $\lim_{x \to \infty} \tanh \kappa x = 1$. To that end, observe that with $u = \kappa x$,

$$\tanh u \equiv \frac{e^u - e^{-u}}{e^u + e^{-u}} = \frac{1 - \dfrac{1}{e^{2u}}}{1 + \dfrac{1}{e^{2u}}} \to \frac{1 - 0}{1 + 0} = 1 \quad \text{as} \quad u \,(\text{or } \kappa x) \to \infty$$

Hence, we obtain $\cos A = \lim_{x \to \infty} \cos \theta_x = (\tanh \kappa a)/1 = \tanh \kappa a$; that is,

(1) $$\cos \gamma = \tanh \kappa a$$

The following manipulations will convert **(1)** to a more convenient form. Using the definition for $\tanh u$, write $\cos \gamma = (e^{2\kappa a} - 1)/(e^{2\kappa a} + 1)$ or with $z = e^{2\kappa a}$, $\cos \gamma = (z - 1)/(z + 1)$. Now solve this equation for z in terms of $\cos \gamma$ to obtain $z = (1 + \cos \gamma)/(1 - \cos \gamma) = \cot^2 \tfrac{1}{2}\gamma$, or $e^{2\kappa a} = \cot^2 \tfrac{1}{2}\gamma$. The final result is then

(2) $$e^{-\kappa a} = \tan \tfrac{1}{2}\gamma$$

The notation introduced in Chapter 5 for the angle of parallelism $p(a)$ with respect to the distance a now produces the remarkable formula discovered independently by both Bolyai and Lobachevski (Section 5.7): solving for γ in **(2)** one obtains

(3) $$\gamma = p(a) = 2 \tan^{-1} e^{-\kappa a}$$

The inverse of $p(a) = A$ [using the equivalent form **(2)**] is $\kappa a = \ln \cot \tfrac{1}{2} A$, so that a given angle of parallelism A has a unique "distance of parallelism"

(Figure 9.32), mentioned in Section 5.7. Figure 9.33 illustrates an incredible situation that is possible in hyperbolic geometry, in stark contrast to Euclidean geometry; if the angle of parallelism at A is $1°$ and $m\angle BAC = 2°$, then line l' is divergently parallel to line m, and *the two lines l' and m have a common perpendicular*! Moreover, the distance d between the two lines (as shown) as point X moves away from B, *increases to infinity*!

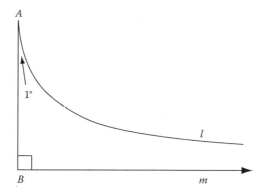

Figure 9.32 Angle of parallelism of measure 1°.

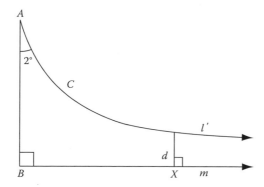

Figure 9.33 Hyperbolic parallels.

Another concept, which Bolyai used in his development of hyperbolic trigonometry, amplifies what is already known about the hyperbolic plane. Due to properties of parallelism, three noncollinear points do not always determine a circle as they do in Euclidean geometry. Suppose, as illustrated in Figure 9.34, that rays h and k are perpendicular to line t at A and B, respectively. With $W \in k$, let a circle passing through W and tangent to t at A be constructed (its center is the intersection C of h and the perpendicular bisector w of AW). Now consider what happens as W approaches B and as the radius of the circle AC increases without bound. In Euclidean geometry, the limiting position of the circle will be the tangent line t itself. This is because A, W, and its reflection W' in h always determine a unique circle. But in hyperbolic geometry, this is not true; the limiting position of the circle centered at C is some intermediate curve different from line l—the subject of some interesting analysis to follow.

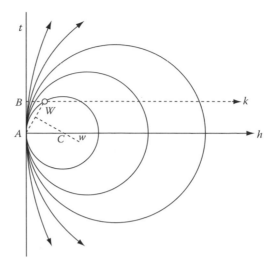

Figure 9.34 Circles in Euclidean geometry.

To facilitate the development, let the hyperbolic plane be augmented by the addition of certain *points at infinity*, called *ideal points*. To that end, consider a family $\mathcal{F}$ of asymptotic parallels in the same direction, where we take for granted the properties of symmetry and transitivity of such asymptotic parallels (see Section 5.8). We agree to add an **ideal point** F^* to the plane corresponding to the family $\mathcal{F}$ where, by convention, each line in the family is said to be **incident** to F^* (or, if you will, **passes through** F^*), as illustrated in Figure 9.35. Different families $\mathcal{F}_1, \mathcal{F}_2, \mathcal{F}_3, \ldots$ of asymptotic parallels correspond to different ideal points $F_1^*, F_2^*, F_3^*, \ldots$ Then, it becomes meaningful to state *two asymptotic parallels intersect* (at an ideal point). This is merely a way of saying (nomenclature) that such asymptotic parallels *belong to the same family $\mathcal{F}$.*

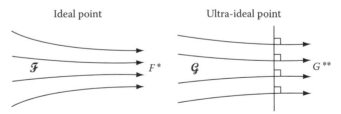

Figure 9.35

Similarly, consider a family $\mathcal{G}$ of divergent parallels, each perpendicular to the same line. We add an **ultra-ideal** point G^{**} to the plane corresponding to the family $\mathcal{G}$ where, by convention, each line in the family formally **passes through** or is **incident to** G^{**}. Again, we say that two lines that are divergently parallel *intersect* at an ultra-ideal point. The original points $P, Q, \ldots$ of the hyperbolic plane will be called **ordinary** points. The hyperbolic plane with the addition of the ideal and ultra-ideal points will be called the **extended hyperbolic plane**. We use the notation Ω for a generic point, either ordinary, ideal, or ultra-ideal ($\Omega = P, F^*,$ or G^{**}).

Definition 1   Three or more lines are said to be **concurrent** in the extended hyperbolic plane iff they pass through an ordinary, ideal, or ultra-ideal point Ω.

The first result will establish a key property to be used from now on.

Theorem 1   Given any family of lines concurrent at Ω and point A lying on one of those lines, there exists a unique point A' lying on any other such line in the family, such that $\angle AA'\Omega \cong \angle A'A\Omega$.

Proof   Since this is obvious if Ω is an ordinary point P, let Ω be either an ideal or ultra-ideal point. If Ω is an ideal point, use the construction for an isosceles asymptotic triangle (Figure 5.45, Section **5.8**), and if Ω is an ultra-ideal point, construct a Saccheri quadrilateral. The details are left as a problem for the reader (see Problem 10). ◣

The points A and A' established in Theorem 1 are called **corresponding points** with respect to Ω (Figure 9.36). As line m varies through Ω, the locus of points A' on m corresponding to A presumably defines a certain curve in the hyperbolic plane. This locus, together with point A itself, will be denoted by the symbol $\mathscr{C}\,[A, \Omega]$. It is now our purpose to describe such curves in more detail. (You should have no difficulty proving that $\mathscr{C}\,[A, \Omega]$ is a circle if Ω is an ordinary point.)

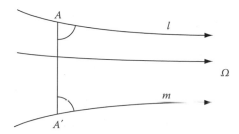

Figure 9.36 Corresponding points.

Definition 2   The set of points $\mathscr{C}\,[A, \Omega]$ just defined is called a **limit curve** (traditionally, a **horocycle**) if Ω is an ideal point, and an **equidistant locus** if Ω is an ultra-ideal point (Figure 9.37). (The generic term **cycle** is used if Ω is not specified as to type.)

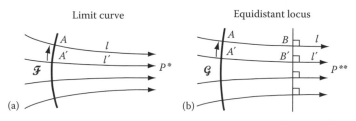

Figure 9.37 Cycles in hyperbolic geometry.

The actual proof of the following result is beyond the scope of this book (not needed in any of the results to follow). Interpreted intuitively, however, it helps us to get a mental image concerning these curves.

Theorem 2   As with the case of an ordinary circle orthogonal to its radii, the curve $\mathscr{C}\,[A, \Omega]$ is an orthogonal trajectory of the family of lines passing through Ω. That is, the curve $\mathscr{C}\,[A, \Omega]$ makes right angles with each line l passing through Ω.

Intuitive proof   Let A' approach A as limit, $A \in l$ and $A' \in l'$.

1. If Ω is an ideal point (Figure 9.35a), then the angle of parallelism of l with respect to AA' tends to $\pi/2$ as the distance $AA' \rightarrow 0$.
2. If Ω is an ultra-ideal point (Figure 9.37b), then l and l' have a common perpendicular BB' and $\lozenge ABB'A'$ is a Saccheri quadrilateral with acute angle BAA' whose limit is a right angle as $A' \rightarrow A$. ◥

A key result in obtaining information about the loci $\mathscr{C}\,[A, \Omega]$ has familiar overtones. The proof was already outlined in a previous problem (Problem 20, Section **5.8**), but, for convenience, we present a complete proof here.

Theorem 3   In the extended hyperbolic plane, the perpendicular bisectors of the sides of any triangle are concurrent.

Proof   Since the theorem is clear if any two of the perpendicular bisectors meet at an ordinary point, assume that two of them, say l and m (the perpendicular bisectors of sides BC and AC), meet at an ultra-ideal point G^{**} (Case 1).

1. Let l and m have a common perpendicular RS, and let X, Y, and Z be the feet on line RS of the perpendiculars from A, B, and C, respectively, as in Figure 9.38. Since $BL = LC$, $LS = LS$, and the angles at L and S are right angles, $\lozenge BLSY \cong \lozenge CLSZ$ by SASAA, hence $BY = CZ$. Similarly, $CZ = AX$. Thus, $BY = AX$ and $\lozenge XYBA$ is a Saccheri quadrilateral with base XY. Therefore, the perpendicular bisector n of segment AB (the base) is also perpendicular to line RS (at the midpoint of segment XY), and by definition of ultra-ideal points, n passes through G^{**}.

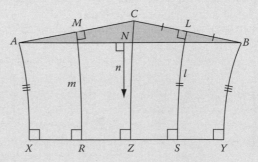

Figure 9.38 Perpendicular bisectors concurrent at Ω.

2. Suppose l and m meet at an ideal point F^*. Consider the perpendicular bisector n of side AB. Then by Case 1 just proven, n cannot meet l at an ultra-ideal point G^{**} or else m must be divergently parallel to l and must have, with l, a common perpendicular $R'S'$ (Hilbert's theorem, Problem 22, Section **5.8**); by part (1), m must also be perpendicular to $R'S'$ hence is not limit parallel to l, a contradiction. Also, n cannot meet l at an ordinary point or else m would pass through that point, contradicting $l \parallel m$. Hence n is asymptotically parallel to l and n passes through F^*. ⬊

Theorem 4 The perpendicular bisector of any chord AB of a cycle $\mathscr{C}\,[D, \Omega]$ passes through Ω. (See Figure 9.39.)

Proof If Ω is an ordinary point, then $\mathscr{C}\,[D, \Omega]$ is a circle centered at Ω and the conclusion follows from the familiar theorem concerning the chord of a circle. If Ω is an ideal point, the result was already observed earlier in Problem 20, Section **5.8** for asymptotic triangle $AB\Omega$. (See Figure 9.39a.) We can assume for the final case that lines $A\Omega$ and $B\Omega$ are divergent parallels having a common perpendicular UV, and that the perpendicular bisector n of AB passes through a point W of segment UV (Figure 3.39b). Then, by SASAA, $\lozenge WMAU \cong \lozenge WMBV$ and $\angle MWU \cong \angle MWV$, hence $n \perp$ line UV (n passes through Ω). ⬊

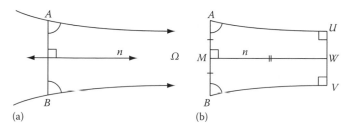

Figure 9.39 Proof of Theorem 4.

The above result is the hyperbolic generalization of the familiar theorem involving ordinary chords of circles. We conclude with another result of this sort.

Theorem 5 Given any three ordinary non-collinear points A, B, and C in the extended hyperbolic plane, there exists a unique locus $\mathscr{C}\,[D, \Omega]$ (a circle or a cycle) passing through them.

Proof Consider the perpendicular bisectors of the sides of $\triangle ABC$, which are concurrent at some point Ω. A candidate for a cycle containing A,B, and C is therefore $\mathscr{C}\,[A, \Omega]$; we proceed to show that $B, C \in \mathscr{C}\,[A, \Omega]$. By examining Figure 9.38, note that $\angle BA\Omega \cong \angle AB\Omega$ when Ω is an ultra-ideal point, since these angles are the summit angles of Saccheri quadrilateral $XYBA$. Similarly, $\angle CA\Omega \cong \angle AC\Omega$ using Saccheri quadrilateral

XZCA. If Ω is an ideal point, $\angle BA\Omega \cong \angle AB\Omega$ since angles of parallelism corresponding to equal distances are equal. Similarly, $\angle CA\Omega \cong \angle AC\Omega$. Finally, if $\Omega = P$, an ordinary point, one uses the isosceles triangle theorem. Thus, in all cases, $\angle BA\Omega \cong \angle AB\Omega$ and $\angle CA\Omega \cong \angle AC\Omega$, so that $B \in \mathscr{C}[A, \Omega]$ and $C \in \mathscr{C}[A, \Omega]$. The uniqueness will be left to the reader; it depends on the exterior angle inequality for ordinary and asymptotic triangles. ◣

Thus, we can see that limit curves and equidistant loci behave formally like ordinary circles, the chief distinction being that their centers need not be ordinary points, and their radii may not be defined. The situation implicit in the last corollary may be summed up dynamically in a manner analogous to that of circles in Figure 9.34. Let rays h and k be perpendicular to line t at A and B, with $W \in k$ and w the perpendicular bisector of segment AW, as shown in Figure 9.40. Finally, let Ω be the point of intersection of w and h (an ordinary, ideal, or ultra-ideal point). The set $\mathscr{C}[A, \Omega]$ is a circle when $\angle WAP$ is small enough (when w intersects h). But as $m\angle WAP$ increases, w becomes a nonintersector of line h. When w is limit parallel to h and thus Ω is an ideal point, $\mathscr{C}[A, \Omega]$ is a limit curve. As $m\angle WAP$ continues to increase, w is hyperparallel to h and $\mathscr{C}[A, \Omega]$ becomes an equidistant loci, finally approaching line t as a final limiting case.

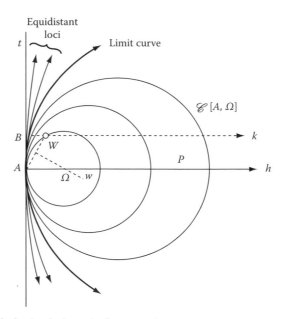

Figure 9.40 Cycles in hyperbolic geometry.

The following problems 5–10 will guide the reader in visualizing the preceding analysis in the Poincaré model, thereby demonstrating the great perfection with which the model emulates any development in abstract reasoning in hyperbolic geometry. It also helps to make the ideal points introduced previously more tangible.

Problems (Section 9.9)

Group A

1. With $\kappa = 1$, find the length $a = AB$ corresponding to an angle of parallelism of $\theta = 1°$ (as in Problem 9, Section **5.8**), by setting $p(a) \approx 0.0174$ rad in **(3)**. What is the length corresponding to 1 millionth of a degree?

2. As a tends to zero as limit, what effect does this have on the angle of parallelism $p(a)$. Use **(3)** to prove your answer.

3. Show that **(2)** is equivalent to $\sin \gamma = \operatorname{sech} \kappa a$.

Group B

4.* Using the result of Problem 3 above and Problem 17, Section **6.12**, show that in Figure P.4, θ is the angle of parallelism for the distance a, as constructed in the directions given below (due to Bolyai and modified by Beltrami):

Start with the given distance $a = AB$ and construct lines l and m perpendicular to line AB at A and B. Choosing C on l close enough to A so that the perpendicular CD at C to l meets m at some point D, swing an arc centered at A having radius $c = BD$, cutting segment CD at E. Then $m\angle EAB \equiv \theta$ is the desired angle of parallelism.

[**Hint:** $\sin \theta = \cos \varphi = \tanh x / \tanh c$; use **(17)** derived in Problem 17, Section **6.12**, and apply some algebra to the resulting expression.]

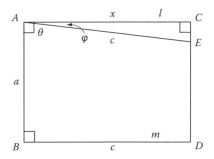

Figure P.4

5. Show by inversion that the Euclidean circles inside the Poincaré disk are also h-circles for S.

6. Consider the three families of concurrent h-lines indicated in Figure P.6. Which of the three have an ordinary point of concurrency? An ideal point? An ultra-ideal point? Where would you place the points at infinity in the figure? Note that the ultra-ideal points are not as well defined; they could be identified with a *pair* of points on **C** or with the common perpendicular of the lines in $\mathcal{G}$.

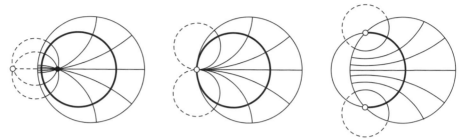

Figure P.6

7. Using the fact that the cycles $\mathscr{C}\ [A,\ \Omega]$ are orthogonal to the lines through Ω, identify a circle, a limit curve, and an equidistant locus in the Poincaré model, observing the three types of concurrent h-lines shown in Figure P.6 for Problem 6.

8. Show that the phenomenon illustrated in Figure 9.40 can be represented in the Poincarè model as shown in Figure P.8.

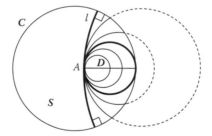

Figure P.8

9. Show that any Euclidean chord AB of **C** that is not a diameter is an equidistant locus (Figure P.9).

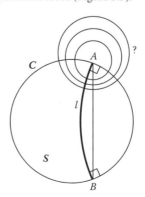

Figure P.9

Group C

10. Prove the existence of a unique point A' on a given line m of a family of concurrent lines corresponding to a given point A on line l of that family, as shown in Figure P.10, if the point of concurrency is

  (a) An ordinary point P.

  (b) An ideal point $P*$ [**Hint:** Consider the angle bisectors at $A \in l$ and any point $B \in m$; lay off $A'E = DA$ and draw AA' and FI with F the midpoint of segment AA' and use SASSS.]

  (c) An ultra-ideal point $P**$.

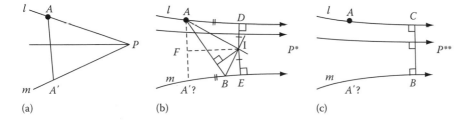

(a)   (b)   (c)

Figure P.10

Appendix A: Sketchpad Experiments

The passages below contain specific instructions for carrying out drawing experiments in the Euclidean plane (often having applications to non-Euclidean geometry). In order to successfully conduct such experiments, the reader needs to know the basics of *The Geometer's Sketchpad*, Version 4.0, Fall 2001, Key Curriculum Press. For your convenience, the most frequently used terms in these experiments will be defined below, with instructions as to their use.

Special Terms

SELECT   To activate the selection of an object already on the screen, place cursor over the Arrow panel in the toolbar at top left of your screen, left click, then move cursor to item to be selected, and click again. More than one object can be selected by holding down the shift key.

CHOOSE   This term instructs you to choose one of the icon panels in the toolbox at top left (such as the Arrow, Point, Circle, Segment, etc.) or various commands at top of screen (such as EDIT, DISPLAY, CONSTRUCT, etc.). These commands include a menu that prompts the selection of desired item to be made.

CONSTRUCT   A special textbook term meaning to create a figure (i.e., to place an object on your screen) or to actually construct an object by using the CONSTRUCT command at top of screen. The tools in the vertical toolbar at the top left of the screen should be selected by left clicking, moving cursor to screen, and activating (by holding left mouse key down and dragging by moving the cursor).

SELECT POINT OF INTERSECTION   If two objects appear to intersect on the screen, either point to the intersection and left click or select the two objects, choose {CONSTRUCT, Intersection}, and left click.

HIDE   In order to make lessons more effective, you should always hide preliminary objects of construction. Hold down shift key, select objects to by erased, then choose DISPLAY, and left click on Hide menu item. Note: *Sketchpad* provides a safeguard to help prevent hiding of unintended items by displaying such terms as "Hide Points" (which means you have selected more than one point to be hidden and does not include any lines or segments) or "Hide midpoint" for a specific item. The term "Hide Objects" means you have selected a combination of lines, segments, points, etc. to be hidden. If you accidentally delete object(s) you did not mean to hide, simply choose EDIT, Undo Hide, and object(s) will appear again.

LABELING   It has been found extremely difficult to keep a labeling sequence A, B, C,... intact while performing multistep constructions, so we do not use the labeling provided by *Sketchpad* to refer to points or lines on the screen. If the automatic labeling is already activated and you want to eliminate it, choose EDIT, Preferences, Text, and under the title Show Labels Automatically, click out arrows in the boxes below.

Geometric Constructions Commonly Used

TRIANGLES, POLYGONS   To draw a polygon choose SEGMENT in Toolbar at left, left click, move cursor to screen, hold down left mouse key and move cursor to desired point, left click, repeat, move cursor in another direction, left click, and continue until you have the number of sides desired, ending with starting point, then choose the select tool (arrow) to end construction. An option is to locate a number of points on the screen, to select the points in order, and then to choose CONSTRUCT, Segments. If you want to move part of a figure, select point or segment, hold down left mouse key, and move cursor (referred to as "dragging"). In order to move an entire figure, you must select either all vertices or all sides of a polygon before dragging to another location. Another choice from CONSTRUCT appears in the item "Interior of Polygon," which produces the interior points of the polygon whose vertices you have selected; you can then select and drag the interior.

PERPENDICULAR/PARALLEL LINES   With line (or segment) AB on the screen, for example, and point C either on or not on the line, select both line and point, then choose CONSTRUCT, Perpendicular/Parallel Line, left click. The line through C perpendicular (or parallel) to line AB will appear on the screen.

ISOSCELES TRIANGLE   With a segment AB selected choose CONSTRUCT, Midpoint; while still selected, hold down shift key and select segment AB, and then choose CONSTRUCT, Perpendicular Line. Locate point C on the perpendicular and, with both A and B selected, choose CONSTRUCT, Segment, and then hide perpendicular and midpoint. Notice that by dragging point C, it remains on the hidden perpendicular, and triangle ABC remains isosceles.

CIRCLE   Circles can be constructed in any one of the three ways: (1) By choosing the circle panel from the toolbar at left, moving cursor to screen and dragging; (2) by selecting two points on the screen and choosing CONSTRUCT, Circle By Center + Point; and (3) by selecting a point and segment anywhere on the screen and then choosing CONSTRUCT, Circle By Center + Radius. You will learn by experience which of the three to use in order to best accomplish the task at hand.

ARC OF CIRCLE   To construct an arc on a circle already on the screen, select two points, say A and B on the circle, and choose CONSTRUCT, Arc On Circle. Circular arc AB will appear in the counterclockwise order. To construct an arc determined by three points, say, A, B, and C, select A, B, and C in that order, choose CONSTRUCT, Arc Through 3 Points, and the arc ABC will appear.

Special note   In all multistep constructions, you must left click and/or choose the select tool repeatedly in order to discontinue drawing items previously selected.

Sketchpad Experiments

1. **Desargues' Theorem**   Choose Ray tool and construct three concurrent rays from point P, then select three points A, B, and C on those three rays (see Figure A.1). Join the points A, B, and C by segments, forming triangle ABC. Repeat with three other points A', B', and C' on those same three rays. Hide the three rays, point P, and the other anchor points for the rays. You now have two triangles in perspective from a point. (If desired, construct the interiors of the two triangles colored with your favorite color [select their vertices and choose CONSTRUCT, Triangle Interior, then select shaded interior and choose DISPLAY, Color].) Using the ray tool, construct corresponding rays AB and $A'B'$ and select their point of intersection L; similarly with rays AC and $A'C'$, intersecting at M. (You may have to adjust the vertices of the triangles in order for the points L and M to appear.) Draw line LM and choose DISPLAY, Line Width, Thick. Finally, draw ray BC, any ray $B'X$ ($X \neq C'$), choose DISPLAY, Line Width, Thin, and select the point N of intersection of rays BC, $B'X$. You are now ready for the experiment. Drag X toward point C' and watch the movement of point N. Does anything happen? With $X = C'$, drag the vertices of the two triangles. What do you observe?

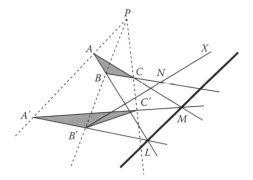

Figure A.1 Desargues' theorem.

2. **Median/Cevian Inequality**   Construct a triangle ABC, then select segment BC and choose CONSTRUCT, Midpoint to locate the midpoint M of side BC. Draw cevian AM. Choosing MEASURE, find the lengths of segments AB, AC (after selecting them) and compute $\frac{1}{2}(AB + AC) \equiv S$. Then find the length of segment AM and compare with S. What is your conclusion? Drag A, B, and C to see if your conclusion persists. For further analysis concerning cevians, let M be any point on BC and compute $p = BM/BC$, $q = MC/BC$, $pAC + qAB$, and compare with $d = AM$ (Figure A.2).

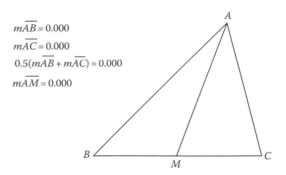

$m\overline{AB} = 0.000$

$m\overline{AC} = 0.000$

$0.5(m\overline{AB} + m\overline{AC}) = 0.000$

$m\overline{AM} = 0.000$

Figure A.2 Median and cevian inequality.

3. **Angle Interior**   Construct line *AB* and semicircle *S* having *AB* as radius (take circle with center at *A* and radius *AB* intersecting line *AB* at *W*, select *B*, *W*, and circle, and choose CONSTRUCT, Arc On Circle, then hide circle.) Construct ray *AC* and intersect with *S* at *P*, then construct the bisector of angle *CAB*, intersecting *S* at *Q*. (Select points *C*, *A*, *B* in that order and choose CONSTRUCT, Angle Bisector.) Construct arc *BQP*, then choose CONSTRUCT, Arc Interior, Arc Sector, using your choice of color. Double click on line *AB* to make it a line of reflection, select *S*, and choose TRANSFORM, Reflect, to create semicircle *S′* opposite *S*. Move ray *AC* so it intersects *S′* and repeat instructions to obtain arc *BQ′P′* and its interior. Hide ray *AQ* and point *Q*. Finally, drag point *B* off screen so that interiors of arcs extend to beyond edge of screen. Now drag point *C* to observe Int ∠*CAB* (Figure A.3).

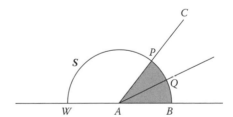

Figure A.3 Displaying the interior of an angle.

4. **Intersections of Three Half-Planes**   Construct line *AB*, points *C* and *D* on that line, the circle *CD* and its intersection *W* with line *AB* and circle *AB*, intersecting line at *B* and *C*. Select *D*, *W*, circle and choose CONSTRUCT, Arc On Circle, CONSTRUCT, Arc Interior, Arc Sector, and color it yellow (under DISPLAY). Hide circle, points *C*, *D*, and *W*, and line *AB*, leaving only anchor points *A* and *B*. Drag *D* off screen so that shading resembles a half-plane region H_1. Select interior of semicircle, then EDIT, Copy, Paste (twice) in order to produce two more half-planes H_2 and H_3. Make a change of colors to turquoise (H_2) and magenta/pink (H_3). Change positions by dragging their respective anchor points, turning one of the half-planes 180°, and arranging other two half-planes to check

following colors of intersections: $H_1 \cap H_2$ (yellow + turquoise) = green, $H_2 \cap H_3$ (turquoise + magenta) = dark blue, and $H_1 \cap H_3$ (yellow + magenta) = red. The overlap of all three (the region to be studied) is black. Now drag half-planes to various positions to see what the black region can become (Figure A.4).

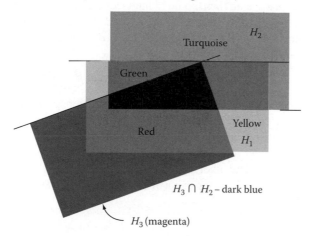

Figure A.4 Intersections of three half-planes.

5. **Bisectors of Angles in a Linear Pair**   Draw any line AB and locate point C such that (ACB). At C draw ray CD. Select points D, C, and B, then choose CONSTRUCT, Angle Bisector. Repeat to obtain the bisector of angle DCA. Measure the angle formed by the bisectors (choose MEASURE, Angle after locating and selecting points E and F on those bisectors and selecting E, C, and F). Use thick for linewidth on rays CE and CF (the bisectors), then drag point D and observe the effect. For a more dynamic feature, draw a circle centered at C and reconstruct rays AD and the angle bisectors, with D on the circle. Hide the circle. Select point D and choose EDIT, Action Buttons, Animation, and choose Once Only and Speed: Slow, OK. A panel will appear entitled Animate Point. Point to this panel and left click. Point D will then rotate once about the hidden circle, counterclockwise (Figure A.5).

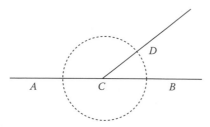

Figure A.5 Angle bisectors experiment.

6. **Euler Line**   This experiment will not only verify the collinearity of the points H, G, and O, but will generate from animation an interesting pattern that the Euler line makes as one vertex of the triangle traverses a circle. Construct ray AP, locate point W on that

ray, and draw a circle centered at *W* with radius *WA*. Draw segment *AB* for the base of the triangle, and locate *C* on the circle. In triangle *ABC*, construct two altitudes and their intersection *H*, the midpoints of two sides, the medians to those sides, and their intersection *G*. Finally, construct perpendiculars at the midpoints and locate their point of intersection *O*. Draw line *HG* to verify collinearity with *O*. Using the MEASURE menu find the distances *HG* and *GO*. Drag vertices *A*, *B*, *C* changing Δ*ABC* to various shapes. What did you find?

For the animation part, hide all objects except the segment *AB*, the points *W*, *C*, and the Euler line *HG*. Select line *HG* and choose DISPLAY, Trace Line. Then select *C* and choose EDIT, Action Buttons, Animation, Speed: other (enter 5). Now activate animation and experiment with the pattern created by the Euler line with different positions of *W* on ray *AB* relative to segment *AB*. Among others, you should obtain a curve that resembles a cardioid. Before starting each experiment, you need to erase the trace tool under DISPLAY (Figure A.6).

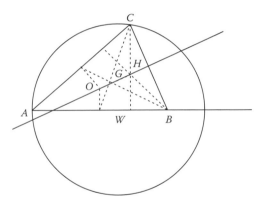

Figure A.6 Euler line of Δ*ABC*.

7. Three Locus Problems Involving Circles:

(a)   A circle is tangent to one side of an angle at point *F* and the diametrically opposite point *G* lies on the other side. Find the locus of the center, as well as any point having a fixed position on that circle (say a point *P* rotated counterclockwise 60° from *F*). See Figure A.7a.

Although the proof is rather involved, the answer is immediate using *Sketchpad*. Draw angle *BAC* and take any point *F* on side *AB*. Construct the perpendicular to line *AB* at *F*, find its intersection *G* with side *AC*, and construct segment *FG*. The midpoint *M* of the perpendicular segment *GF* is the center of the circle. Select *M* and choose DISPLAY, Trace Point. Double click on *M* making it the center of rotation, select *F* and choose TRANSFORM, Rotate, By 60° to locate the second point *P* whose locus we seek. Select *P* and trace its movement from DISPLAY. Select *F* and choose EDIT, Action Buttons, Animation, Speed: Enter 5. The animation demonstrates (but does not prove) the two loci in the problem (Figure A.7a).

(b)   A fixed point A on a circle is one end of a segment whose other end B varies on that circle. Find the locus of the midpoint P of segment AB. Use the animation tool in *Sketchpad* to discover the answer (instructions should not be necessary) (Figure A.7b).

(c)   The top of a ladder leaning against a wall and resting on the horizontal ground is sliding downward at a certain velocity. Find the locus of the midpoint of the ladder. (In calculus, one is supposed to find the velocity of the foot of the ladder at a certain instant; here we are more interested in the geometry involved.) (Figure A.7c).

The animation provided by *Sketchpad* not only shows the answer, but provides the pattern made by the ladder itself (which turns out to be the tangents to the curve $x^{2/3} + y^{2/3} = a^{2/3}$, the hypocycloid of four cusps). In order to simulate the sliding ladder having fixed length, use the following construction on *Sketchpad*: Construct horizontal and vertical segments AB and AC, with $AC < AB$, and locate point P on AC. With P as center and AC as radius (select P, then segment AC, and choose CONSTRUCT, Circle By Center And Radius), intersect the circle with AB at Q. Then draw segment PQ. Hide circle. As P moves on AC, PQ remains constant length, which represents the ladder. Construct the midpoint M of PQ and trace its movement with P animated at slow speed. You may also trace segment PQ to demonstrate its movement.

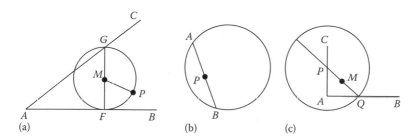

(a)   (b)   (c)

Figure A.7 Three locus problems.

8. **When a Secant Approaches Tangency**   Let O be the center of a circle and AB a secant line. (Locate A and B on the circle and draw line AB. With the MEASURE tool, display the measure of angle OAB on your screen. As you drag point B toward point A on the circle, watch the change in $m\angle OAB$ (Figure A.8).

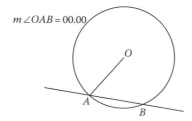

Figure A.8 Secants and tangents.

9. **Ratio Phenomenon**   Draw any triangle ABC. Locate D on BC and E on AB. Using CONSTRUCT, Parallel Line, construct segment EF parallel to BC, as in Figure A.9. Locate the intersection G of cevian AD and segment EF. The question is: if D is the midpoint of BC, then is G the midpoint of EF? Locate the midpoints M and N of BC and EF, respectively. Drag D until it coincides with M; does G coincide with N? More generally, examine the ratios EG/GF and ED/DC, using the MEASURE, Distance tool (select E and G, choose MEASURE, Distance, to find BD, and similarly for DC, EG, and GF); then choose MEASURE, Calculate to find the above ratios. Drag D to any position on BC and observe the ratios. Write down your observations; can you prove your conclusions?

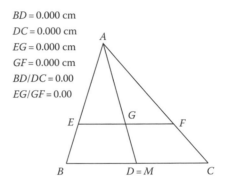

Figure A.9 The ratio *EG/GF*.

10. **The Identity $h^2 = c_1c_2$**   Construct a triangle ABC having acute angles at B and C, and construct the altitude (segment) CD to side AB. Using the MEASURE menu, place the length of CD squared on the screen, along with the product $AD \cdot DB$. While watching these two quantities (CD^2 and $AD \cdot DB$), drag vertices A and C to various positions, including one that makes triangle ABC a right triangle. With CD, AD, and DB playing the roles of h, c_1 and c_2, respectively, what is your conclusion based on this experiment? (Figure A.10)

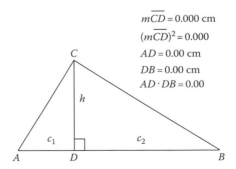

Figure A.10 Geometric mean of two segments.

11. **Angles in Six-Pointed Star**   In order to have some control over what constitutes a six-pointed star, assume that the six points form a convex hexagon and that each side of the star is formed by two diagonals (and not by any side of the hexagon). In the case of a regular hexagon, the angle sum can be obtained by properties of inscribed angles, which is calculated to be 240 (see Figure A.11). Can this value be decreased to, possibly, zero or increased beyond 240 to perhaps 720? Investigate, by using the MEASURE, Angle, and Calculate tools on *Sketchpad* starting with the regular hexagon, placing the angle sum of the six angles on the screen, and dragging the vertices off the circle. At times, the sum remains constant as certain vertices are dragged; find an explanation of this phenomenon, if you can. Write down any conjectures you might make. At least find the smallest and largest values that seem to be possible for this angle sum. (Can you prove these bounds?)

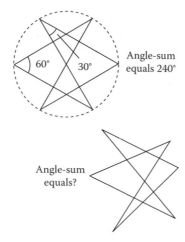

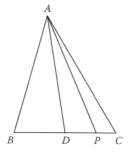

Figure A.11 Six pointed stars.

12. **Angle Bisector and Ratio of Included Sides**   Construct triangle *ABC*, and construct the bisector *AD* of $\angle BAC$ (selecting points *B*, *A*, and *C*, then choosing CONSTRUCT, Angle Bisector, etc.). Locate another point *P* on side *BC* and place the ratios *AB/AC* and *BP/PC* on the screen using the MEASURE menu. Now drag *P* on *BC* and see if you can draw a conclusion from this experiment (Figure A.12).

Figure A.12 Ratio relation involving an angle bisector of a triangle.

13. **Pappus' Theorem on Area** Construct triangle *ABC* and parallelograms I, II, and III on the sides as described in Problem 24, Section **5.12**. For example, to construct I on side *AC*, locate a point *U* and draw segment *UA*, select *U* and segment *AC*, then choose CONSTRUCT, Parallel Line. At *C*, construct line through *C* parallel to segment *UA* in the same manner; select the point of intersection *V*, then construct segments *UV* and *VC*. To find the area, select in order the points *A*, *C*, *V*, *U* and choose CONSTRUCT, Quadrilateral Interior, then while still selected, MEASURE, Area. Do this for parallelogram II on side *BC*, with $\overleftrightarrow{BR} \parallel \overleftrightarrow{CS}$ and $\overleftrightarrow{RS} \parallel \overleftrightarrow{BC}$ (Parallelograms I and II are arbitrary; their choice determines parallelogram III uniquely.) As stated in Problem 24, draw the lines *UV* and *RS* and obtain the point of intersection *P*. Translate segment *CP* to *AQ* as follows: select *P*, then *A*, and choose TRANSFORM, Mark Vector to define the desired translation. Select segment *PC* and point *C*, then choose TRANSFORM, Translate, by Marked Vector. This produces point *Q* such that segments *PC* and *AQ* are parallel and congruent. Complete the parallelogram III = ◊*AQTB*. Construct the interior and area of parallelogram III. Now verify the relation Area I + Area II = Area III. Drag various vertices to change the shape of the triangle and parallelograms (e.g., using vertex *U*) and watch the change in measurements. In addition, drag vertex C to make ∠*ACB* a right angle, and vertices *U* and *R* to make I and II squares. What do you observe? (Figure A.13)

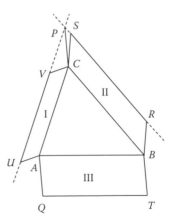

Figure A.13 Does Area I + Area II equal Area III?

14. **Harmonic Construction** Draw a line and locate variable points *A*, *B*, and *C* on that line, with *C* on segment *AB* ≠ midpoint of *AB*. Use *Sketchpad* to create the harmonic construction, as shown, determining point *D*, the harmonic conjugate of *C* with respect to *A* and *B*; use lines to gain more generality. Drag points *P* and *Q* on line *PB* and observe the effect. Does *D* move? What is your conclusion? For further analysis, locate the midpoint *O* of segment *AB* and place the measurements OB^2 and $OC \cdot OD$ on the screen. What result do you observe? Also check the relation $1/AB = \frac{1}{2}(1/AC + 1/AD)$ using the MEASURE tool (Figure A.14).

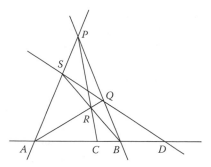

Figure A.14 Harmonic conjugate of *C* with respect to *A* and *B*.

15. **Steiner's Theorem**   Draw a circle and locate the vertices *A*, *B*, and
    C of a triangle on that circle, and draw its sides. Locate point *P*,
    a variable point on the circle. Drop perpendiculars to each of the
    sides, obtaining points *D*, *E*, and *F*. Verify Simpson's theorem by
    constructing line *DE*. Hide points *D*, *E*, and *F* and the perpendicu-
    lars you used to find them. Put a trace on line *DE*. Select point *P*
    and choose EDIT, Action Buttons, Animate, and select Once Only,
    Speed: 5. Hide all lines of construction. Then click on the animation
    panel to observe the motion of the Simpson line as *P* rotates about
    the circumcircle. Further features mentioned in Steiner's theorem
    can be verified with the *Sketchpad* construction of the nine-point
    circle and/or the Morley triangle (shown in the figure). Drag the ver-
    tices of Δ*ABC* to see if Steiner's theorem seems to be valid for all
    triangles, including an obtuse triangle in particular (Figure A.15).

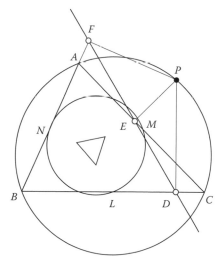

Figure A.15 Rotating the Simson line.

16. **Orthogonal Systems of Circles**   Draw a horizontal segment
    UV about half way across the screen, construct its midpoint *M*
    under CONSTRUCT, then hide the endpoints *U* and *V*. Double
    click on the midpoint to make it a center of rotation, then select
    segment *UV*, and choose TRANSFORM, Rotate, 90°. Locate

point A on UV and a variable point P at one end of the vertical segment. With P selected, select A and choose CONSTRUCT, Circle By Center + Point, to draw circle PA, with center at P. Locate Q at one end of the horizontal segment, and construct segment QP and its midpoint N. With N already selected, select point P and construct circle NP, intersecting circle PA at R. Finally, select Q and R and construct circle QR, which will then be orthogonal to circle PA. Hide all objects, keeping the two segments UV and $U'V'$, the points P and Q, and the two orthogonal circles. Put a trace on the two circles (under DISPLAY). Animate the points P and Q simultaneously as follows: select P and Q, choose EDIT, Action Buttons, Animate. You are prompted to select one of two items listed. Select the first item, and under "bidirectional," select forward, check the Once Only panel, then Speed: 5. Before clicking OK, select the second item and repeat these instructions. Click on the animate panel to generate the families $\mathcal{F}$ and $\mathcal{G}$. Erase traces (under DISPLAY) and drag point A to observe whether the phenomenon changes (Figure A.16).

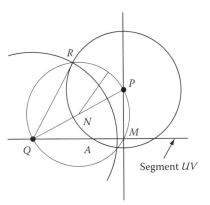

Figure A.16 Generating the families $\mathcal{F}$ and $\mathcal{G}$.

Appendix B: Intuitive Spherical Geometry

A sphere is among the easiest surfaces in three-dimensional geometry to define. It is simply the set of points in space at a fixed distance from some fixed point. (The fixed distance is its *radius*, r, and the fixed point its *center*, O.) A sphere is thus a circular shell, with zero thickness, as would be any object we would call a *surface*. In three-dimensional geometry, a plane that meets a sphere is either tangent to the sphere (having only one point in common with the sphere) or intersects it in a circle. If the plane passes through the center, the circle is of maximal size, called a *great* circle. Suppose for some real number α, the sphere has radius $r = \alpha/\pi$. Using the formula for the circumference of a circle, a great circle would also have radius α/π and its circumference would be given by $2\pi r = 2\pi(\alpha/\pi) = 2\alpha$. Thus, the real number α would have the distinction of being one-half the distance around the sphere, which would presumably be the maximal shortest arc distance measurable on that sphere.

Now imagine a piece of string that passes through two fixed points P and Q and lying on a sphere S (Figure B.1). If the string is gradually pulled tight, it begins to assume the position of what appears to be a great circle arc PAQ. This experiment lends support to (is not a proof of) the principle that the shortest possible arc joining two points on S is the (minor) great circle arc joining them. Thus, in spherical geometry the great circles are the analogs of straight lines. The distance between two points on S is then defined as the *minimal length among the lengths of all curved arcs lying on S having those two points as end points*. The number α mentioned previously is therefore the *maximum of all distances between pairs of point on S*.

Figure B.1 The distance from P to Q on the sphere.

For points P and Q on S, the case $PQ = \alpha$ is exceptional. This happens only when P and Q are the end points of some diameter (are opposite poles), and in this case infinitely many great circles pass through P and Q (since there are infinitely many planes containing line PQ). If on the other hand $PQ < \alpha$, then P and Q are not poles and are therefore not collinear with the center O. In this case the plane through P, Q, and O is unique and determines a unique great circle passing through P and Q. If the term "line" is used in place of "great circle," the preceding analysis may be summarized as follows: *Each pair of points P and Q on S lie on at least one line, and if $PQ < \alpha$, that line is unique.* This is the embodiment of our **Axioms 1**, **2**, and **4**.

Aspects from spherical geometry are often used in geography, and this book makes use of a few terms in this area in order to convey various ideas and examples in spherical geometry more easily. Since not all readers will be familiar with this, we go over the basics here. The earth's surface is a nearly perfect sphere, so we can speak freely of (imaginary) great and small circles and diametrically opposite poles on the surface of the earth. Let's begin with the North and South poles. These two points are determined by the earth's approximate axis of rotation. The longitude lines (called *meridians*) are the great circles passing through the North and South poles. East and west are divided by the unique meridian passing through Greenwich, England (as indicated in Figure B.2), called the *prime* meridian, east being toward the Asian continent, west being toward the American continent. Each meridian is assigned a degree measure, which is the number of degrees in the angle between the plane of the given meridian and that of the prime meridian, as shown in Figure B.3. These angle measures range from $0°$ to $180°$, east or west. The *equator* is the great circle determined by the plane through O perpendicular to the axis. The lines of *latitude* are the circles lying in planes parallel to the equator, ranging from zero (at the equator) up to $90°$ at either pole (north or south). These lines are referred to simply as *parallels*; the equator being the only parallel that is a great circle. The degree measure of each parallel is that of the angle made by the latitude line and the equator, having vertex at O, the center of the earth (as illustrated in Figure B.3). North and south latitudes correspond to Great Britain in the northern hemisphere and Australia in the southern hemisphere.

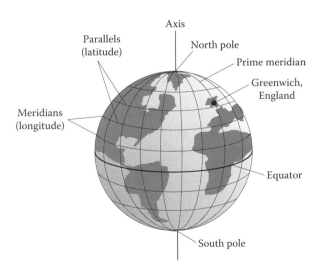

Figure B.2 Terms from geography.

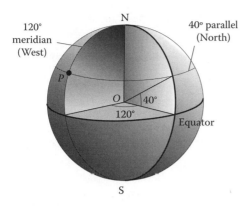

120°
meridian
(West)

40° parallel
(North)

Figure B.3 Coordinates for points.

Every point on the surface of the earth may then be uniquely located using longitude and latitude lines for its coordinates. For example, the position 120°W, 40°N, indicates that point P lying on the intersection of the west longitude 120° and the north latitude 40° (shown in Figure B.3). An interesting formula (based on those proved in Chapter 6) yields the great circle distance d in miles between two points in the same hemisphere and west of the prime meridian. If the coordinates of those two points are (θ_1, φ_1) and (θ_2, φ_2), the great circle distance d between them is given by

$$d = b\cos^{-1}[\sin\varphi_1 \sin\varphi_2 + \cos\varphi_1 \cos\varphi_2 \cos(\theta_1 - \theta_2)]$$

where $b = 69.1$ (the approximate number of miles for each degree on a great circle). For example, the great circle distance from New York City (74°W, 41°N) to Los Angeles (118°W, 34°N) is approximately 2434 mi using this formula (the reader might like to perform this computation). The official atlas entry for this mileage is 2467, the small error being attributed to the value of b used, and the fact that the earth's surface is not a perfect sphere.

Appendix C: Proof in Geometry

The object of any proof in mathematics in general is to provide an explanation for why a certain property is true (or why it is not true) and to write it as clearly and concisely as possible. There is, of course, an art to this process and it is a highly creative endeavor, which makes it difficult to analyze quantitatively. Much of it must simply be learned by practice. But there are a few guidelines that can be mentioned.

In all fields of mathematics, as in geometry, the development takes place one step at a time, and at any given point in the development, all previous results can be used for a proof. This is what makes it difficult for anyone desiring to avoid active participation in such a development. For this reason, in this book we constantly encourage you to think about what is taking place, and why, and to take an active part in the process. Thus, the following ingredients are basic in order to be successful in proof writing:

- Recalling and understanding previous results
- Fully understanding the problem you are working on
- Putting the two together—observing what is relevant among the previous results

One can learn how to prove theorems by tackling problems that are found interesting and/or challenging. This is why there are so many problems in this book—problems of all kinds, practical as well as theoretical. The experiments in *Sketchpad* are also important (the computer software used in this book to illustrate geometric principles). At certain points in our development, we bring up the theory of logic, how to use it in proving theorems and how it places natural limits on mathematics. Although this is very interesting, it is not our main focus, so we do not give a detailed analysis on logic. Such matters are best left to books on the subject; they will not help you much in writing proofs.

Mathematical developments normally begin with accepted statements—*axioms*—which are not proven but are used in order to prove other results—the *theorems*. For example, suppose one of the axioms is: *Each two points lie on exactly one line.* A mere twist of words, or so it seems, is as follows: *If two lines meet, they do so in only one point.* This statement is taken as a theorem to be proved from the above axiom. (Can you provide a coherent explanation?)

Other statements not requiring proof are the definitions. Since they merely establish the terminology to be used, there is nothing to prove: a phenomenon can obviously be given a name without justification. However, very often a definition is the reason for a certain statement one makes and needs to be mentioned in order to clarify the basis for it. For example, consider the two familiar definitions: (1) *A **right angle** is an angle having measure 90°* and (2) *two angles are said to be **congruent** if they have the same measure.* A theorem involving these definitions is the following: *All right angles are congruent.* Your proof might look something like this in two-column, T-proof fashion:

Statements	Reasons
1. Angle A and angle B are right angles.	1. Given
2. The measure of angles A and B each equal 90.	2. Definition of right angle
3. Therefore, angle A is congruent to angle B.	3. Definition of congruent angles

Does this look familiar? If you are used to such outline proofs, that is good, because this means you have some experience in proving theorems, and it is a simple matter to convert any outline proof to paragraph form, the kind exclusively used in this book. One has only to copy the steps, paraphrase where necessary or desirable, and blend the reasons into a clear explanation of the result. Using the paragraph format makes mathematics read like a passage from a novel and renders it less formal. This format has thus become the standard in mathematics, and in particular, in geometry.

We end this brief discussion with an example, which shows how one might reason out the proof of a problem and how two different visualizations of the same problem can lead to different proofs. Suppose we assume the usual angle-angle-side criterion for congruent triangles, that the base angles of an isosceles triangle are equal or congruent and that congruent triangles are those triangles having corresponding sides and angles congruent (definition). The problem involves the configuration *ABPCD* as stated in Figure C.1.

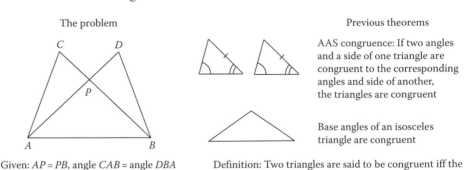

The problem

Given: $AP = PB$, angle CAB = angle DBA
To prove: $AC = DB$

Previous theorems

AAS congruence: If two angles and a side of one triangle are congruent to the corresponding angles and side of another, the triangles are congruent

Base angles of an isosceles triangle are congruent

Definition: Two triangles are said to be congruent iff the corresponding sides and angles are congruent

Figure C.1 A typical geometry problem.

The first obvious thing to look for is whether triangle *CPA* is congruent to triangle *DPB*, and how to justify this; this would directly imply, by definition of congruent triangles, $AC = DB$. With a little thought, however, we can see that angle CAP = angle DBP since they are the differences of two angles that are congruent (here we are using the isosceles triangle property for triangle *PAB* that the base angles are equal [angle PAB = angle PBA] and $AP = PB$ [given]). But now we seem to be stuck. What about angles *CPA* and *DPB*? Is there any reason why they are equal? Insight dawns, and we seem to remember something about vertical angles being congruent. This provides the key, and thus by AAS congruence, triangle

CPA = triangle *DPB* and *AC* = *DB*. Once these thoughts come to light we are ready to write an explanation (proof), going through the above analysis as if we were explaining this to another person. (There is definitely a social aspect to mathematics; one always has the hope that ones efforts will be appreciated by someone else, and a little competition makes it interesting.)

Although the preceding proof was valid, there is a different visualization of the same problem that leads to a shorter (but perhaps not better) proof. The above was based on (1) the image provided by triangles *CPA* and *DPB* (see Figure C.2), but one might observe (2) the overlapping triangles *ACB* and *ABD* as *separate triangles*, which leads to a different proof. Here, note the congruent angles indicated in the figure, which come directly from the given and the theorem about isosceles triangles. Since the triangles have a common side included by the congruent angles, ASA applies to show that triangle *ACB* is already congruent to triangle *ABD*, and therefore *AC* = *DB*.

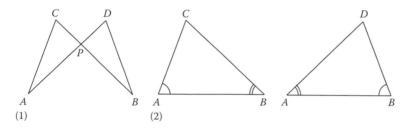

Figure C.2 Two ways to prove it.

This example should help you get started, and to recall some geometry you have had from your high school course. Since calculus is a prerequisite, it is assumed that you have already had considerable work in geometry and trigonometry, which should contribute to your success here.

Appendix D: The Real Numbers and Least Upper Bound

The real number system and the least upper bound concept are used throughout the axiomatic development of geometry, beginning with the coordinatization axiom for rays (**Axiom 6**). It is thus appropriate to outline briefly what constitutes the real numbers. To build the real number system, one begins with the *natural numbers* 1, 2, 3, ..., together with the operations of addition, multiplication, and the usual order properties (for example, $a < a + b$ for all a and b and if $a < b$ and $b < c$, then $a < c$, etc.). This development may be made rigorous using mathematical induction. Next, the *integers* are introduced as equivalence classes of pairs of natural numbers (m, n) representing the difference $m - n$. Thus, one assumes that $(m, n) = (p, q)$ iff $m + q = n + p$ (meaningful in the natural number system). Addition, multiplication, and order are introduced via these pairs, where $0 = (m, m)$, $(m, n) > 0$ iff $m > n$ and $(m, n) < 0$ iff $n > m$. The *rationals* are introduced as ordered pairs of integers: $(m, n) \equiv m/n$, with addition, multiplication, and order being defined in the usual manner. At this point, the Archimedean principle can be proven: *If a and b are any two positive rational numbers, there exists an integer n such that na > b.*

If we mark the integers on an imaginary line, spaced equally apart (a *discrete* system), we find that the rationals begin to "fill up" the number line. But there remain certain "holes," which are occupied by the so-called irrational numbers yet to be constructed (see Figure D.1). One can "complete" the real number system by introducing equivalence classes of *Cauchy sequences* of rational numbers, essentially creating the numbers $\sqrt{2}$, $\sqrt{3}$, π, for example, and all the remaining real numbers. (A **Cauchy sequence** is any sequence $\{a_m\}$ having the property that for each $\varepsilon > 0$ there exists an integer N such that if $n, m > N$ then $|a_n - a_m| < \varepsilon$.) At this point the field and order properties can be shown, and it can also be shown that a *Cauchy sequence of real numbers converges* (i.e., the *limit exists*). It follows that no further numbers can be added, since there is a theorem that states: *there exists only one complete, ordered field*, namely, the real numbers.

The rational field

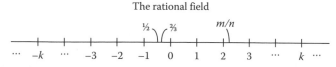

Figure D.1 Filling up the real number line.

There are other forms of completeness that can be established. For example, it can be shown that *every sequence of nonempty, nested intervals contains at least one point.* Pertinent to our development here, the *least upper bound property* may be established. Consider any nonempty set of real numbers that is *bounded*, that is, there exist numbers M and N such that for all x in the set

$$N < x < M$$

and x lies on the interval $[N, M]$. It is apparently a simple observation to conclude that there is a smallest such interval $[a, b]$ such that for all $x \in S$

$$a \le x \le b$$

It might be well to consider some numerical examples. Suppose

(a) $S = \left\{ \dfrac{1}{n} : n = 1, 2, 3, \ldots \right\}$

Here, the smallest closed interval containing S is $[0, 1]$. Note that the right end point, 1, belongs to S, while the left end point, 0, does not. Next, consider the set

(b) $S = \left\{ \dfrac{1}{2x} - \dfrac{1}{x^2} : x > 1 \right\}$

In order to find the bounds on S in this case, some calculus is needed. The derivative of the function $1/2x - 1/x^2$ is $\frac{1}{2}(1/x^2 - 4/x^3)$, which is positive for $x < 4$, zero at $x = 4$, and negative for $x > 4$. Hence, this function has an absolute maximum at $x = 4$. As $x \to 1$, the values of the function $\to -\frac{1}{2}$. We conclude that $[-\frac{1}{2}, \frac{1}{6}]$ is the smallest interval containing S and neither end points belong to S. Note that in these two examples, the right end point b of the smallest interval containing S satisfies the following two properties:

(1) For each element $x \in S$, $x \le b$. That is, b is an **upper bound** of S.
(2) If M is any upper bound for S, then $b \le M$. That is, b is the **least** among all upper bounds for S.

> **Definition**   A number b satisfying properties (1) and (2) above is called the **least upper bound** of S, denoted LUB S.

Note: A similar analysis is made for the left end point a of the least interval $[a, b]$ containing S, called the **greatest lower bound** of S, denoted GLB S. ✎

Of course, the *existence* of the least upper bound (or greatest lower bound) of a bounded set S is important. An outline of the proof goes something like this: Since S is bounded above, let M be any upper bound, and set $b_1 = M$. If b_1 is a least upper bound we are finished. Otherwise, there exists an upper bound b_2 of S less than b_1, and since S is nonempty, $a \le b_2 < b_1$ for any particular element a in S. If b_2 is the least upper bound, we are finished. Otherwise, there exists an upper bound b_3 of S less than b_2 (and greater than or equal to a). Continuing inductively, either at some point we have found a least upper bound b_n or a strictly decreasing sequence $b_1 > b_2 > b_3 > \cdots > b_n > \cdots > a$ of upper bounds of S exists. The remainder of the proof is to show that $\{b_n\}$ is Cauchy and thus has limit b, and then to show that this limit satisfies the properties (1) and (2) above. (An analogous procedure establishes the existence of the greatest lower bound of a set S bounded below (having a *lower bound N*).

We close by showing that the system of rational numbers does not have the least upper bound property. First, note that $\sqrt{2}$ is not of the form m/n for any two positive integers m and n, hence does not exist in the rational numbers. Define the following sequence of rationals (given in terms of decimals)

$a_1 = 1.4$
$a_2 = 1.41$
$a_3 = 1.414$
$a_4 = 1.4142$
$a_n = 1.414\, x_4 \ldots x_n$

where x_n is chosen such that $a_n^2 < 2$. The set $S = \{a_n : n = 1, 2, 3, \ldots\}$ is bounded (by 1.5 for instance) but *has no least upper bound* in the set of rationals.

Appendix E: Floating Triangles/Quadrilaterals

The results needed in the proof of the floating triangle/quadrilateral theorems will be established here. These results can be regarded as mere technicalities, but they must be proved in order to make the development valid. Recall that for the floating triangle theorem, we want to prove that for any ray $AC = h$ and B lying on one side of h such that $m\angle CAB = \theta \to \varphi$ as $B \to A$, then

$$\lim_{\substack{B \to A \\ \theta \to \varphi}} \frac{AC}{AB} = c(\varphi)$$

where $\angle C$ is a right angle. This will be proven first for $\varphi = 0$. A preliminary result will be needed for this purpose.

In any right triangle AFG, right angle at F, with sides less than $\alpha/2$ in length, suppose $\angle A$ is divided into n equal parts by rays AF_k, $F_k \in \overline{FG}$ as shown in Figure E.1. Then

(1) $$FF_1 < FG/n$$

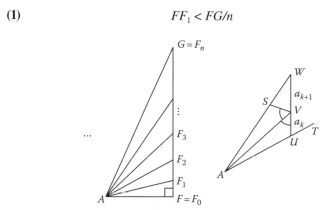

Figure E.1 Proof of preliminary result.

Proof For convenience, write $F_{k-1}F_k = a_k$ $(1 \le k \le n)$, where $F_0 = F$. It will be shown that $a_1 < a_2 < a_3 < \cdots < a_n$. Observing the diagram at right in Figure E.1, with $F_{k-1} = U$, $F_k = V$, and $F_{k+1} = W$, since $\angle AVU$ is acute, $\angle SVA$ can be constructed congruent to $\angle UVA$ in $\angle AVW$ making $\triangle SVA \cong \triangle UVA$ by ASA. By the exterior angle inequality in $\triangle AWU$, $m\angle SWV < m\angle VUT = 180 - m\angle AUV = 180 - m\angle ASV = m\angle WSV$. Hence by the scalene inequality, $a_k = VU = VS < VW = a_{k+1}$ $(1 \le k \le n - 1)$. Thus,

$$nFF_1 = na_1 = a_1 + a_1 + a_1 + \cdots + a_1 < a_1 + a_2 + a_3 + \cdots + a_n = FG$$

which proves **(1)**. ◣

581

Theorem E.1 Let triangle ABC have right angle at C, with $\theta = m\angle A$. Then

$$\lim_{\substack{B \to A \\ \theta \to 0}} \frac{BC}{AB} = 0$$

Proof (See Figure E.2.) Construct $\angle GAF$ having measure $\pi/4$, and drop the perpendicular GF to side AC, $F \in$ ray AC. Let n be any positive integer, and divide $\angle A$ into n equal angles as before. Then by **(1)**, $FF_1 < FG/n$. Since $\theta \to 0$ and $AB \to 0$, we can assume that ray AB falls between rays AC and AF_1 and that (ACF) as shown in Figure E.2. Let ray CB meet AF_1 at D and AG at E. In hyperbolic geometry, by **(3)** Section **6.3**, we have

$$\frac{BC}{AB} < \frac{BC}{AC} < \frac{CD}{AC} < \frac{FF_1}{AF} < \frac{FG/n}{AF} = \frac{1}{n} \cdot \frac{FG}{AF}$$

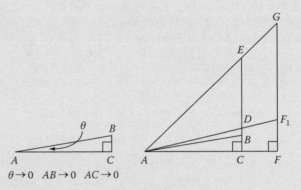

$\theta \to 0 \quad AB \to 0 \quad AC \to 0$

Figure E.2 Proof of Theorem **E.1**.

Since FG/AF is fixed and n is arbitrary (letting $n \to \infty$), this proves $BC/AB \to 0$. In elliptic geometry, note that $m\angle AEC > \pi/4 = m\angle EAC$ by the angle-sum theorem. Hence, $AC > EC$ and, by **(1)**, $CD < EC/n$. Thus

$$\frac{BC}{AB} < \frac{BC}{AC} < \frac{CD}{AC} < \frac{CD}{EC} < \frac{1}{n}$$

and the desired conclusion follows for both geometries. ⬛

Theorem E.2 With the same hypothesis as Theorem E.1

$$\lim_{\substack{B \to A \\ \theta \to 0}} \frac{AC}{AB} = 1$$

Proof By the triangle inequality we have

$$1 - \frac{BC}{AB} = \frac{AB - BC}{AB} < \frac{AC}{AB} < \frac{AB + BC}{AB} = 1 + \frac{BC}{AB}$$

Since $BC/AB \to 0$, the pinching theorem on limits gives us $AC/AB \to 1$. ⬛

One last result will be needed before we prove the main theorem on floating triangles. It involves an isosceles triangle ABD, with legs AB and AD. The midpoint M of base BD is the foot of the perpendicular from A to side BD, so if $AB \to 0$ and $m\angle A \to 0$, then by Theorem E.1, $BD/AB = 2MB/AB \to 0$. Thus

Corollary If the vertices of the base BD of an isosceles triangle ABD converge to the vertex A and if the measure of the vertex angle converges to zero, then the ratio of the base to either leg converges to zero (i.e., $BD/AB \to 0$).

Theorem E.3 Let h be a ray from point A, with $\triangle ACB$ a right triangle having the vertex C of the right angle on h. Then if $m\angle BAC \equiv \theta \to \varphi$ as $B \to A$,

$$\lim_{\substack{B \to A \\ \theta \to \varphi}} \frac{AC}{AB} = c(\varphi)$$

Proof As shown in Figure E.3, construct $\angle DAC$ such that $m\angle DAC = \varphi$, with $AD = AB$, and E the foot of D on h. By betweenness properties, we have $(\overrightarrow{AB}\ \overrightarrow{AD}\ \overrightarrow{AE})$ if $\theta > \varphi$ and $(\overrightarrow{AD}\ \overrightarrow{AB}\ \overrightarrow{AE})$ if $\theta < \varphi$ and either (ACE) or (AEC). With F the foot of D on line BC, we have in either case (since $AB = AD$)

$$\frac{AC}{AB} = \frac{AE \pm CE}{AB} = \frac{AE}{AD} \pm \frac{CE}{FD} \cdot \frac{FD}{BD} \cdot \frac{BD}{AB}$$

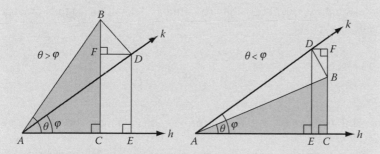

Figure E.3 Proof of floating triangle theorem.

By the corollary above, $\lim BD/AB = 0$ as $AB \to 0$, and since $CE/FD < 2$ (Problem 14, Section **6.4**) and $FD/BD < 1$, it follows that the product of the above three ratios $\to 0$ and that $\lim AC/AB = \lim AE/AD$ as $AC \to 0$. Therefore

$$\lim_{AB \to 0} \frac{AC}{AB} = \lim_{AD \to 0} \frac{AE}{AD} = c(\varphi)$$

as desired. ❧

Note: Because of Theorems E.1 and E.2, the preceding theorem is valid when $\varphi = 0$ or $\varphi = \pi/2$. For example, if $\varphi = 0$ then Theorem E.2 yields $AC/AB = b/c \to 1 = c(0)$ as $AB \to 0$. ⬙

One can use Theorem E.3 to prove the continuity of $c(\theta)$. It will be necessary to use the $\varepsilon{-}\delta$ definition of continuity: Given $\varepsilon > 0$, there exists $\delta > 0$ such that if $0 < |\theta - \theta_0| < \delta$, then $|c(\theta) - c(\theta_0)| < \varepsilon$. Here, we shall use φ for θ_0. Let $\varepsilon > 0$ be given, and consider in Figure E.4, $\triangle ABC$ as B and C approach A and θ approaches φ as limit. By Theorem E.3, $AC/AB \to c(\varphi)$, so for AB sufficiently small and θ sufficiently close to φ (i.e., for some $\delta > 0$, $AB < \delta$ and $|\theta - \varphi| < \delta$)

(*)
$$\left| \frac{AC}{AB} - c(\varphi) \right| < \frac{\varepsilon}{2}$$

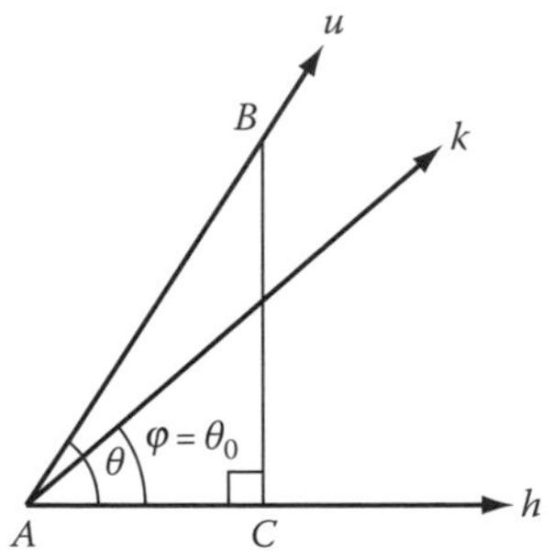

Figure E.4 Continuity of $c(\theta)$.

Now let θ be any particular value such that $0 < |\theta - \varphi| < \delta$ (θ is now fixed, but arbitary, as shown in Figure E.4, with B on a fixed ray u). If $AB < \delta$, the inequality (*) will be valid. The limit as $AB \to 0$ in (*) then produces the desired result

$$|c(\theta) - c(\varphi)| \leq \frac{\varepsilon}{2} < \varepsilon$$

for all θ such that $0 < |\theta - \varphi| < \delta$, proving the continuity of $c(\varphi)$ at $\varphi = \theta_0$ for $0 \leq \varphi \leq \pi/2$. ⬙

The next task is to prove the corresponding results for the function $C(u)$ and the property that produces the floating quadrilateral theorem.

Theorem E.4 Let $\lozenge ABCD$ be a Lambert quadrilateral with right angles at A, B, and C. Then, if A and D converge to P and B and C converge to Q, the limit

$$\lim \frac{AD}{BC} = C(u)$$

exists, where $u = PQ$ (see Figure E.5).

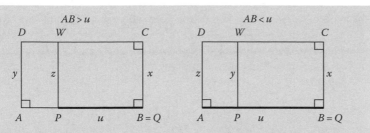

Figure E.5 Problem: Prove that $\lim y/x = \lim z/x$.

Proof By constructing quadrilaterals congruent to $\Diamond ABCD$, it can be assumed that $B = Q$. The case $u > 0$, $AB > u$ will be considered first. Let a fixed Lambert quadrilateral $\Diamond TBRS$ be constructed, with right angles at T, B, and R (Figure E.6), such that $TB > u$, and let P be a point such that (TPQ) and $PQ = u$. Again by the construction of certain congruent quadrilaterals, we may assume without loss of generality that A lies on ray BP and that C and D remain on one side of line PB as AP, CB, and $DA \to 0$. Erect the perpendicular to line PB at P, meeting line CD at W. Since (APB) in this case, it follows that $(UDWC)$. By **(9)** Section **6.2**, which applies to normal trapezoids, we obtain

$$AD < pUT + qWP \quad \text{where } p = AP/TP, q = TA/TP$$

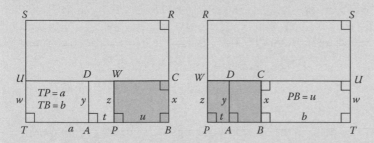

Figure E.6 Proof of floating quadrilateral theorem.

In hyperbolic geometry, the angles at U and W are, respectively, acute and obtuse—reversed in elliptic geometry, so the normal trapezoid $\Diamond TPWU$ has a different base in the two geometries; this does not affect **(9)** as it appears above. Using the notation in the Figure E.6

$$y < pw + qz = pw + (1 - p)z \quad \text{where } p = t/a$$

which can be put in the form

$$\frac{y}{z} - 1 < p\left(\frac{w}{z} - 1\right)$$

In hyperbolic geometry, $w > y > z > x$, so that $y/z > 1$ and $w/z > 1$, and in elliptic geometry, the inequalities are reversed, with $y/z < 1$ and $w/z < 1$. Hence, it follows that in either case

(*) $$\left|\frac{y}{z} - 1\right| < p\left|\frac{w}{z} - 1\right|$$

In hyperbolic geometry, $\lim w/x = C(b)$ where $b = TQ$ and $\lim z/x = C(u) \geq 1$ so that $w/z \to C(a)/C(u)$. In elliptic geometry, $w/z < 1$. Thus, in either case w/z is bounded. Since $t = AP \to 0$ as $A \to P$, then $p = t/a \to 0$ and $x \to 0$ in (∗), thus proving that $|y/z - 1| \to 0$, hence $y/z \to 1$. Therefore

$$\frac{AD}{BC} = \frac{y}{x} = \frac{y}{z} \cdot \frac{z}{x} \to 1 \cdot C(u) = C(u)$$

as desired. In the case $AB < u$, construct $\Diamond TPRS$, as in Figure E.6 (right diagram), and take B on side PT, with (PAB) and (WDC). On examination, the above analysis is seen to be virtually the same, with the same conclusion. Finally, if $u = 0$, the case (PAB) cannot occur, and the diagram on the right in Figure E.6 applies, with $P = B$, $t = AB$, and $z = x$. Again the above analysis applies, where (∗) implies $\lim AD/BC = \lim y/z = 1 = C(0)$. ◿

In order to prove the continuity of $C(u)$ from Theorem E.4, continuity will be proved in the following form: Given $\varepsilon > 0$, there exists $\delta > 0$ such that if $0 < t < \delta$, then $|C(u + t) - C(u)| < \varepsilon$.

Corollary   The function $C(u)$ is continuous for $0 \leq u < \alpha/2$.

Proof   It was shown in the proof of Theorem E.4 that as $x \to 0$ and $t \to 0$, $y/z \to 1$ (Figure E.7). Let $\varepsilon > 0$ be given and set $\varepsilon' = \frac{1}{2} \varepsilon/C(u)$ if $C(u) \neq 0$ (unnecessary if $C(u) = 0$). Then there exists $\delta > 0$ such that for all $x < \delta$ and $t < \delta$

$$\left| \frac{y}{z} - 1 \right| < \varepsilon'$$

At this point in the argument, we regard t as fixed, but arbitrary, such that $t < \delta$. The preceding inequality can be written in the equivalent form (since both x and z are positive)

$$\left| \frac{y}{x} - \frac{z}{x} \right| < \varepsilon' \cdot \frac{z}{x}$$

Take the limit as $x \to 0$; the result is, for all positive $t < \delta$,

$$|C(u+t) - C(u)| \leq \varepsilon' C(u) < \frac{\varepsilon}{C(u)} \cdot C(u) = \varepsilon$$ ◿

Appendix F: Axiom Systems for Geometry

It is interesting to compare the axiom system for unified geometry with other systems. We begin with the system used in this book.

Postulates for Unified Geometry

Undefined terms: point, line, with elementary set theory.

Axiom 1   Each line is a set of points having at least two members.

Axiom 2   There exist two points, and each two points belong to some line.

Axiom 3 [Metric Axiom]   To each pair of points (A, B), distinct or not, there corresponds a real number AB, called *distance*, or the *distance from A to B*, which satisfies the properties

(a) $AB \geq 0$ with equality if and only if $A = B$,
(b) $AB = BA$

Axiom 4   If the distance between two points is less than α, the least upper bound of distance, they lie on exactly one line.

[Notation (ABC) defined.]

Axiom 5   If A, B, C, and D are any four distinct collinear points such that (ABC), then either (DAB), (ADB), (BDC), or (BCD).

Axiom 6 [Ruler Postulate for Rays]   There is a one-to-one correspondence between the points of any ray h and the set of non-negative real numbers x, $0 \leq x \leq \alpha$, called *coordinates*, such that

(a) The coordinate of the endpoint O of h is zero (termed the *origin* of the coordinate system), and all other points have positive coordinates
(b) If $A[a]$ and $B[b]$ are any two points of h as indicated by their coordinates, then $AB = |a - b|$

Axiom 7   To each angle $\overline{hk}$ there corresponds a real number hk, called its *measure*, satisfying the properties

(a) $hk \geq 0$, with equality if and only if $h = k$,
(b) $hk = kh$

Axiom 8   The measure of any angle is less than or equal to the measure of a straight angle.

Axiom 9 [Plane Separation Postulate]   There corresponds to each line l in the plane two regions (sets) H_1 and H_2 having the properties:

  (a) Each point in the plane belongs to one and only one of the three sets l, H_1, and H_2,
  (b) H_1 and H_2 are each non-empty sets,
  (c) If $A \in H_1$ and $B \in H_2$, any segment AB joining A and B intersects l at an interior point on that segment.

Axiom 10 [Angle Addition Postulate]   If point D lies in the interior of $\angle ABC$ or on its sides, then $m\angle ABD + m\angle DBC = m\angle ABC$.

Axiom 11 [Angle Construction Postulate]   Given any line AB and half-plane H determined by that line, for every real number r between 0 and 180, there is exactly one ray AP in H such the $m\angle PAB = r$.

Axiom 12 [SAS Postulate]   If under some correspondence between vertices, two sides and the included angle of one triangle are congruent, respectively, to two sides and the included angle of the other, the triangles are congruent.

Moise's Postulate System for Absolute Geometry*

Undefined terms: point, line, plane, and S the universal set.

Incidence Axioms:

I-0. All lines are subset of S.

I-1. Each pair of distinct points belong to one and only one line.

I-5. Every line contains at least two points and S contains at least three noncollinear points.

Distance Function:

D-0. For each two points A and B, there exists a real number AB associated with those two points.

D-1. $AB \geq 0$ for all A and B.

D-2. $AB = 0$ if and only if $A = B$.

D-3. $AB = BA$ for all A and B.

D-4. **[Ruler Postulate].** Every line l has a coordinate system $f: l \to R$ such that f is one-to-one, onto and for all A, B in l, $|F(A) - F(B)| = AB$.

Plane Separation:

PS-1. The *Plane-Separation Postulate.*   Given any line, the set of all points in the plane not lying on the line is the union of two disjoint sets such that (1) each of the sets is convex and (2) if A belongs to one of the sets and B belongs to the other, then the segment $\overline{AB}$ intersects the line.

---

* Axiom statements have been paraphrased to match language and notation used in this book. See Moise (1990).

Postulates on Angles:

M-1. Each angle $\angle A$ is associated with a real number, denoted by $m\angle A$.

M-2. $m\angle A$ lies between 0 and 180 for any angle $\angle A$.

M-3. **[Angle-Construction Postulate].** Given a half-plane H of any line $\overrightarrow{AB}$, for every number r between 0 and 180, there is exactly one ray $\overrightarrow{AC}$ with P in H, such that $m\angle PAB = r$.

M-4. **[Angle Addition Postulate].** If D is an interior point of $\angle BAC$, then $m\angle BAC = m\angle BAD + m\angle DAC$.

M-5. **[Supplement Postulate].** A linear pair of angles are supplementary.

Congruence Postulate:

SAS. If under some correspondence between two triangles, two sides and the included angle of the first triangle are congruent to the corresponding parts of the second triangle, then the triangles are congruent.

Note: In a former edition, the conclusion in the SAS axiom was, instead of the triangles being congruent, that the remaining corresponding angles be congruent—slightly weaker. It remains an interesting exercise to then prove that the remaining pair of corresponding sides are congruent, providing the desired conclusion. ⬧

Hilbert's Postulates for the Euclidean Plane

Primitive terms: point, line, on, between, congruent

Group I—Connection:

I-1. There is one and only one line passing through any two given distinct points.

I-2. Every line contains at least two points, and for any given line, there is at least one point not on the line.

Group II—Order:

Note: For convenience in stating these axioms, author has taken the liberty of using (ABC) in place of the phrase "B is between A and C."

II-1. If (ACB), then A, B, C are all on the same line, and (BCA) but not (CBA) nor (CAB).

II-2. For any two distinct points A and B there is always a point C such that (ACB) and a point D such that (ABD).

II-3. If A, B, C are any three distinct points on the same line, then one of the points is between the other two.

Group III—Congruence ($\cong$):

III-1. If A and B are distinct points and if A' is a point on line m, then there are two and only two distinct points B' and C' on m such that $(A', B') \cong (A, B)$ and $(A', C') \cong (A, B)$; moreover, A' is between B' and C'.

III-2. If two pairs of points are congruent to the same pair of points, then they are congruent to each other.

III-3. If (ABC) and $(A'B'C')$, and if $(A, B) \cong (A', B')$ and $(B, C) \cong (B', C')$, then $(A, C) \cong (A', C')$.

Group IV—Parallel Postulate:

IV-1. [Playfair's Postulate]. Through a given point A not on a given line L there passes at most one line which does not intersect it.

Group V—Postulate of Continuity:

Definition of order on a segment: Consider a segment AB. Let us call one endpoint, say A, the *origin* of the segment and the other point, B, the *extremity* of the segment. Given two distinct points M and N of AB, we define the relation $M < N$ (or $N > M$) iff either M coincides with A or (AMN). A segment AB, ordered in this way, is called an *ordered segment*.

V-1. [Dedekind's Postulate]. If the points of an ordered segment AB such that $A < B$ are separated into two disjoint sets L and U in such a way that:

(1) Each point of AB belongs to either L or U
(2) The point A belongs to L and point B belongs to U
(3) $L < U$ (each point of L precedes each point of U)

Then there exists a point C on AB such that every pair $P < C$ on AB belongs to L and every point $Q > C$ on AB belongs to U.

Note: Hilbert's first edition had two postulates of continuity: the Archimedean property stated for segments and a peculiar axiom of completeness that implied the usual idea of completeness of a line segment (equivalent to the axiom of completeness for real numbers). Later editions and translations included the version found here. If applied to the ordered field of rationals, it is well known that the Dedekind postulate can be used to construct the real numbers, which, in turn implies the least upper bound property. It can be shown without difficulty that the least upper bound property implies the Archimedean property, making it unnecessary to take the latter as an axiom. ◤

Solutions to Selected Problems

Note: For the convenience of the instructor in selecting possible test problems or other purposes, solutions are given for selected odd-numbered problems. With rare exceptions when deemed helpful by the author even-numbered problems are noted here, and in the text, by an asterisk following the problem number. Asterisk for even-numbered problems indicate answers that appear at the back. The symbol $\Rightarrow$ will be used for logical implication throughout. ⬟

Section 1.2

1. Axiom 0 is violated.
3. $l_{10} = \{D, E\}$; $\frac{1}{2}n(n - 1)$.
8.* The model is: Points—$\{A, B, C, D, E\}$, Lines—$l_1 = \{A, B\}$, $l_2 = \{C, D, E\}$; if a sixth point is F, then there are two models having points A, B, C, D, E, and F: Lines—$l_1 = \{A, B\}$, $l_2 = \{C, D, E, F\}$ and Lines—$l_1 = \{A, B, C\}$, $l_2 = \{D, E, F\}$.

Section 1.3

3. See Problem 5, Section **1.2**.
9. **(b)** K **(c)** A (points are collinear).
15. In any finite affine plane, let families of parallel lines be $F_1, F_2, ..., F_k$, and take these families as "points" to be added; for convenience, call these points the *points at infinity*, and adopt the convention that two parallel lines belonging to the same family F_i *meet at the point at infinity* F_i $(1 \le i \le k)$; result is a projective plane.

Section 1.5

9. Suppose l and m are the $\perp$-bisectors of $\overleftrightarrow{AB}$ and $\overleftrightarrow{BC}$, which meet at some point P; P is equidistant from A and B and is equidistant from B and C; hence it is equidistant from A and C and thus lies on the $\perp$-bisector of $\overleftrightarrow{BC}$. (Fallacy: *in this case, P lies on the $\perp$-bisector of $\overleftrightarrow{BC}$ but not true in general, and cannot be used in the argument*).
11. Every circle (of which there are only two) contains 12 points (see Problem 13).

Section 1.7

3. By definition, $(ABCD) \Rightarrow (ABD)$ and $(BCD) \Rightarrow AD = AB + BD = AB + (BC + CD)$.

5. **(a)** Only (ABC) and (BCD) **(b)** Yes   **(c)** No.

7. Segments $\overline{BC}, \overline{AD}, \overline{DC}$; rays $\overrightarrow{BA}, \overrightarrow{BD}, \overrightarrow{BC}, \overrightarrow{AD}, \overrightarrow{AC}, \overrightarrow{DC}, \overrightarrow{CD}, \overrightarrow{CA}, \overrightarrow{CB}$.

9. **(a)** 9.5 cm.

15. $AC = AB + BC = (p/q + 1)BC = [(p + q)/q]BC \Rightarrow BC = qAC/(p + q)$; $AB = (p/q) \cdot qAC/(p + q) = p/AC(p + q)$.

17. $\angle C \cong \angle DCB \cong \angle ACD$ and $DC/CB = AB/BC = AC/AB = AC/CD \Rightarrow$ $\triangle DBC \sim \triangle ADC \Rightarrow \angle BDC \cong \angle DAC \cong \angle ADB$; let $\theta = m\angle C \Rightarrow m\angle ADB = m\angle ADC - m\angle BDC = \theta - (180 - 2\theta) = 3\theta - 180 = m\angle A = 180 - 2\theta \Rightarrow$ $5\theta = 360$.

19. No.

23. **(b)** $[2.5, 3] \cup [3.5, 4] \cup [4.5, 5] \cup \{2, 5.5\}$ **(c)** $\{2.5, 3.5, 4.5, 5.5\}$
    (d) Let $\lceil x \rceil = n$ be least integer such that $x \le n$. Basic properties: $\lceil x + y \rceil \le \lceil x \rceil + \lceil y \rceil$ and for $x < y, \lceil x \rceil \le \lceil y \rceil$. By definition, $d(x, y) = \lceil |x - y| \rceil = \lceil |(x - z) + (z - y)| \rceil \le \lceil |(x - z)| + |(z - y)| \rceil \le \lceil |(x - z)| \rceil + \lceil |(z - y)| \rceil = d(x, z) + d(z, y)$.

25. Yes: place A, B, C, and D on the x-axis having coordinates 0, 2, 3, and 5.

Section 1.8

3. **(a)** 7   **(b)** 7   **(c)** 3.

5. $AB = 11, BC = 8, AC = 5$; because $AB + BC = 19 > \alpha$.

7. With $A[0]$, $B[x]$, $C[x+3]$, $D[16]$ representing the points in figure for this problem, by given $x + 3 = 16 - x$ or $2x = 13$ and $x = 6.5$ and length = $x + 3 = 9.5$ cm.

9. If (ABC), then $\alpha = AB < AB + BC = AC$ which violates $AC \le \alpha$.

11. **(a)** Segment   **(b)** union of a ray and segment.

13. **(a)** -8 and -16 **(b)** -8 and 14.

15. $BD = 12$ with $\alpha \ge 15$.

17. 15, 19, 27, or 61, depending on order.

Section 1.10

3. Let $X \in \overline{AB}$. If $X = A$ then $X \in \overline{AC}$; $X = B \Rightarrow (AXC) \Rightarrow X \in \overline{AC}$. Otherwise (AXB) with $(ABC) \Rightarrow (AXBC) \Rightarrow (ABC)$ and $X \in \overline{AC} \Rightarrow$ $\overline{AB} \subseteq \overline{AC}$. If $\overline{AB} = \overline{AC}$, choose X such that $(ABXC) \Rightarrow X \notin \overline{AB} \Rightarrow$ $\overline{AB} \subset \overline{AC}$.

5. Let $\overline{PQ}$ be any segment with length $c \equiv AB/3$ (ray coordinatization axiom); construct $\overline{AC} \cong \overline{PQ}$ on $\overrightarrow{AB}$ and $\overline{CD} \cong \overline{PQ}$ on $\overrightarrow{BC}$ (segment construction theorem). Then $(ACDB)$ with $AC = CD = c$, (ADB) and $(ACD) \Rightarrow DB = AB - AD = AB - (AC + CD) \Rightarrow DB = 3c - 2c = c$.

9. Suppose $S = \overline{AB}$, $T = \overline{CD}$, and $A = C$ ($A \ne B$ by hypothesis). $S = T \Rightarrow$ $\overline{AB} = \overline{AD}$; if $B \ne D$ then $(ABD) \Rightarrow \overline{AB} \subset \overline{AD}$ (Problem 3), a contradiction.

11. In $\triangle ACM$, $d \le a + \frac{1}{2}c$ and in $\triangle CMB$ $d \le b + \frac{1}{2}c$. Sum to obtain $2d \le a + b + c \le a + b + (a + b)$ by the triangle inequality $\Rightarrow 2d \le 2a + 2b$ or $d \le a + b$.

15. The induction step (from order k to order $k + 1$) is obtained by letting $r = m/2^{k+1}$ for m odd $(=2n + 1)$, and by assuming the inequality has been proven for any dyadic rational p of order k). Since D_r is the midpoint of segment $\overline{EF}$ where $AE = (m - 1)/2^{k+1}AB \equiv pAB$ and $AF = (m + 1)/2^{k+1}AB \equiv p'AB$, and p and p' are dyadic rationals of order k, by **(2)**, $d_r < \frac{1}{2}(d_p + d_{p'}) < \frac{1}{2}[p_k a + (1 - p_k)b] + \frac{1}{2}[p'_k a + (1 - p'_k)b] = \frac{1}{2}[(p_k + p'_k)a + (2 - p_k - p'_k)b] = ra + (1 - r)b$.

Section 2.1

3. 165.

5. If $hu = 70$.

7. 105.

9. Assuming the scale on the protractor is perfectly uniform, each number on the protractor is $180/179 \approx 1.0056$ too large and an angle-sum of $180°$ is measured as $180° \times 1.0056 = 181.008° \Rightarrow$ error $= 1.008°$.

10.* $179.7°$ (reverse steps in solution to Problem 9).

11. **(a)** $90° + 150° = 240°$ **(b)** $270° - 30° = 240°$ **(c)** $m\angle AOB = 90°$, $m\angle BOC = 150°$; $(m\angle AOC = 120°)$.

13. $m\angle AOC' + m\angle C'OB = 60° + 30° = 90° = m\angle AOB$.

15. The remaining betweeness relation is (ukv): $uv = hv - hu = (hk \equiv kv) - (hk - uk) = kv + uk$.

Section 2.3

1. **(a)** Convex   **(b)** Not convex.

3. **(a)** Not convex   **(b)** Convex.

5. If $\angle ABC$ is convex and (ADC) then $D \in \overline{AC} \subseteq \angle ABC \equiv \overrightarrow{BA} \cup \overrightarrow{BC} \Rightarrow D$ is an interior point of $\overrightarrow{BA}$ or $\overrightarrow{BC}$, or both. If both, then by Theorem 3, Section **1.8**, $\overrightarrow{BA} = \overrightarrow{BC}$ and $\angle ABC$ is degenerate. Otherwise, with A and D on line AB, then C and D lie on line AB = line BC; $D \notin \overrightarrow{BC} \Rightarrow D$ lies on the opposite ray $\overrightarrow{BC'} \Rightarrow \overrightarrow{BA} = \overrightarrow{BD} = \overrightarrow{BC'}$ and $\angle ABC$ is a straight angle.

7. If A and C lie on opposite sides of line l there exists X on l such that $(AXC) \Rightarrow l$ meets AB by Postulate of Pasch, a contradiction.

9. Reflexive/symmetry laws routine; for transitive law, suppose $A \approx B$ and $B \approx C$, to prove $A \approx C$. If $A \not\approx C$, then l contains X such that (AXC) by definition of $\approx$ $\Rightarrow l$ passes through Y such that (AYB) (Postulate of Pasch) contradicting $A \approx B$. Let H_1 be equivalence class containing any point $A \notin l$ (using **Axiom 0**) $\Rightarrow$ if $E \in \overline{CD}$ for $C, D \in H_1$ then $C \approx A \approx D \Rightarrow E \approx A$ (if not, a contradiction results again using Pasch) and $E \in H_1$ proving H_1 is convex. Taking any point F on l and G such that (AFG); then $G \not\approx A$ and let H_2 = equivalence class containing G, which is convex by the same argument as above. If $P \in H_1$ and $Q \in H_2$ then $P \not\approx Q$ and by definition, there exists $R \in l$ such that (PRQ).

Section 2.4

5. (a) $144°$ (b) $141°$, $147°$.

7. $44°$.

9. (a) 126   (b) 128.

11. $(h'uk) \Rightarrow h'u + uk = h'k$; by linear pair theorem, $h'u = 180 - hu$ and $h'k = 180 - hk \Rightarrow$ (by substitution) $(180 - hu) + uk = (180 - hk)$, or $-hu + uk = -hk \Rightarrow$ either $u = h$ or (ukh).

13. If either of the angles hk, hu, or uk is a straight angle, we already have either (huk), (hku), or (hku). Otherwise, $W \notin (h, k, h', k')$. Let H_1 and H_2 be half-planes determined by $h \cup h'$, and H_3 and H_4 that by $k \cup k'$, where notation is chosen so that H_1 contains k and H_4 contains h. Then, by the plane separation postulate, the whole plane equals either h, k, h', k' or $\text{Int}\angle hk$, $\text{Int}\angle h'k$, $\text{Int}\angle h'k'$, or $\text{Int}\angle k'h$, hence $W \in$ either $\text{Int}\angle hk$, $\text{Int}\angle h'k$, $\text{Int}\angle h'k'$, or $\text{Int}\angle k'h \Rightarrow$ (by angle addition postulate) either (huk), $(h'uk)$, $(h'uk')$, or $(k'uh) \Rightarrow$ either (huk), (ukh) (Problem 11), $(h'uk')$, or (uhk) (Problem 11). Case $(h'uk')$ remains $\Rightarrow h'u + uk' = h'k'$ or $(180 - hu) + (180 - uk) = hk$ (linear pair and vertical pair theorems). Hence, by the linear pair theorem, $hu' + u'k = hk$ and $(hu'k)$.

14.* For example, if (hku) then $hk + ku = hu$ or $hu + uk = hk + ku + uk > hk$.

17. $A = C = 45$, $B = 135$.

19. $A = 60$, $B = 120$, $C = 30$.

23. By Theorem 6, $(\overrightarrow{ABADAC}) \Rightarrow D \in \text{Int}\angle BAC$ (converse of Axiom 10) $\Rightarrow \overrightarrow{AD}$ meets $\overline{BE}$ at a point F such that (BFE) (crossbar theorem). It remains to prove (AFD). The same argument proves ray $\overrightarrow{BE}$ meets $\overline{AD}$ at F' such that $(AF'D)$, and we now have lines AD and BE meeting at points F' and $F \Rightarrow F' = F$ or $FF' = \alpha$. If (AFD) does not hold, then (ADF) does (since $F \in \overrightarrow{AD}$). Then, with $(AF'D)$, one obtains $(AF'DF)$ or $(AF'F)$, which implies $F'F < \alpha \Rightarrow F' = F$ and (AFD).

Section 2.6

1. 95.

3. (a) $360 - (150 + 110) = 100$ (b) Ray ST' lies between rays SR and SW by applying Theorem 7, Section 2.4 $\Rightarrow m\angle WST = 180 - m\angle WST' = 180 - (m\angle WSR - m\angle T'SR) = 180 - 110 - (180 - 150) = 100$.

5. (a) 71   (b) 88   (c) 99.5.

7. Construct point C' such that $(C'BC)$; thus C' lies on the D-side of line $AB \Rightarrow m\angle ABC' = 180 - m\angle ABC = m\angle ABD$ (linear pair theorem). According to Axiom 11 there can be only one such ray BD on the D-side of line AB and rays BD and $\overrightarrow{BC'}$ coincide.

9. Construct $\angle HEF \cong \angle ABC$ such that $(\overrightarrow{EGEHEF}) \Rightarrow$ ray EH meets segment DF at G at an interior point G of ray EH (crossbar) $\Rightarrow \angle GEF = \angle HEF \cong \angle ABC$.

11. Let H be either half-plane determined by line l and containing points A and B, let $C \in l$. If $AB = \alpha$ then A and B are extreme points, and by

Corollary **C**, Section **2.6**, (*ACB*). But this puts *A* and *B* on opposite sides of line *l*, a contradiction.

15. By definition of triangle interior, $P \in \text{Int}\angle BAC \Rightarrow$ ray *AP* meets side *BC* at an interior point *Q* (crossbar); if *l* passes through *A*, *B*, or *C* we are finished by preceding remark applied to either ray *AP*, *BP*, or *CP*. Hence, assume *l* does not pass through *A*, *B*, or *C* (Figure S.1). $P \in \text{Int}\angle ABC \Rightarrow (\overrightarrow{BA}\overrightarrow{BP}\overrightarrow{BC})$ (**Axiom 10**) $\Rightarrow$ (*APQ*) (Theorem **6**, Section **2.4**). Since *l* passes through *P* and does not pass through *Q*, *A* or *C*, by Postulate of Pasch applied to $\triangle AQC$, *l* meets segment *AC* or *QC*, thus side *AC* or side *BC* of $\triangle ABC$, from which rest of theorem follows by postulate of Pasch.

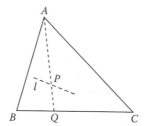

Figure S.1 Problem 15.

17. By result of Problem 13, Section **2.4**, either (*huk*), (*hku*), (*khu*), or (*hu'k*). If (*huk*) then *u* passes through an interior point *W* of $\angle hk$ and thus a point *B* such that (*ABC*) (converse of **Axiom 10** and crossbar), where *h*, *u*, and *k* are the rays *OA*, *OB*, and *OC*, respectively (see Figure S.2). If (*ADBC*) then *D* is an interior point of both $\angle hu$ and $\angle hk$, by definition. But this is a denial that $\angle hu$ and $\angle hk$ are adjacent.

By similar reasoning, one cannot have (*hku*) nor (*khu*) $\Rightarrow$ (*hu'k*) $\Rightarrow hk = hu' + u'k = 180 - hu + 180 - uk \Rightarrow hk + hu + hk = 360$.

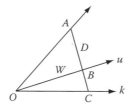

Figure S.2 Problem 17.

Section 3.2

1. (**a**) $AB = 2, BC = AC = 1 + \sqrt{3}$ (**b**) 60 for all three angles   (**c**) Isosceles.

3. (**b**) $a = 16, b = 12, c = 20$ (**c**) $a^2 + b^2 = 256 + 144 = 400 = c^2$.

5. (**a**) $R(4, 3)$ (**b**) 4 blocks.

Section 3.5

3. $x = 90$, $y = 9$.

5. $x = 3$, $y = 8$, $AB = 16$, $m\angle B = 21$.

7. By vertical pair theorem, $\angle AMB \cong \angle DMC$, so by SAS $\triangle AMB \cong \triangle DMC \Rightarrow$ (by CPCPC) $m\angle MDC = m\angle MAB = 90$ and $CD = BA = 9$.

9. By isosceles triangle theorem, $AE = EB$; by vertical pair theorem, $\angle DEA \cong \angle CEB \Rightarrow$ (by ASA) $\triangle DEA \cong \triangle CEB \Rightarrow$ (by CPCPC) $x = DA = CB = 5$ and $y = CE = DE = 7$.

11. Since (PAS) and (PBQ), $PA = PB \Rightarrow AS = BQ$. Since $SR = QR$ and $\angle S \cong \angle Q$, $\triangle ASR \cong \triangle BQR \Rightarrow AR = BR$. By hypothesis, $AB = AR = BR$.

13. Assume from diagram (ABE) and $(ACF) \Rightarrow AE = AB + BE = AC + CF = AF$, $\angle EAC \cong \angle FAB$, $AC \cong AB$. By SAS $\triangle AEC \cong \triangle AFB \Rightarrow$ (by CPCPC) $BF = CE$ and $\angle E \cong \angle F$, together with $\overline{BC} \cong \overline{CB} \Rightarrow$ (by SAS) $\triangle BEC \cong \triangle CFB$ and $\angle EBC \cong \angle FCB$. By linear pair theorem, $m\angle ABC = 180 - m\angle EBC = 180 - m\angle FCB = m\angle ACB$.

15. Side $\overline{BD}$ is not included side of $\angle A$, $\angle ADB$; proof valid only if $AD = BD$.

17. (Use Figure 3.23.) For $X \in l$, let (PQS) with $PQ = QS = \alpha/2$. Assume first that $QX < \alpha$; $PX \neq \alpha$ or else l passes through P since $X \in l$, contrary to hypothesis. Then in triangles PQX and SQX, $PQ = QS$, $\angle PQX \cong \angle SQX$ and $QX = QX \Rightarrow$ (by SAS) $\triangle PQX \cong \triangle SQX \Rightarrow m\angle QXP \cong \angle QXS$. By the criterion for perpendicularity, Section **2.4**, line $PX \perp l$. Then $PX = \alpha/2$ by the theorem. (If $QX = \alpha$ let Y be the midpoint of segment QX on l, and apply the argument to PY, then to PX.)

19. Let $\triangle ABC$ have right angle at C and legs CA, CB of length $\alpha/2$. By Corollary B, every point X on line BC has length $\alpha/2$ and line $AX \perp BC$. Thus $AB = \alpha/2$ and the angle at B is a right angle. The same argument applies to side AC to show the angle at A is a right angle.

21. **(a)–(c)** $\angle 1 = \angle 2 = \angle 3 = \angle 4$ **(d)** $\angle 2 = \angle 3 = 90 = \angle 1 \Rightarrow$ lines AP and $AQ \perp$ line PQ, contradiction.

Section 3.7

3. $z \leq x < y$.

5. $\overline{CD}$ is smallest, $\overline{AD}$ is largest.

7. $\angle UTV$.

9. By scalene inequality, $IJ < JK < IK$ and $IJ < JK < KL$. Thus the construction shown in the figure is valid with $\triangle IJK \cong \triangle WKJ$ by SAS $\Rightarrow IK = WJ$ with $\angle JWL$ obtuse $\Rightarrow JW < JL$ and the maximum segment length is JL.

11. 249, 251.

15. $m\angle A = 45$ (use tri-rectangular triangles.)

16.* $m\angle A = 180p$. (For example, in the solution to Problem 15, where in tri-rectangular triangle ABP, where $BC = \frac{1}{2} BP = \alpha/4$, divide segment BP into four congruent segments forming four congruent triangles, with four congruent angles at A. It follows that the cases $p = \frac{1}{4}$, $p = \frac{1}{2}$, and $p = \frac{3}{4}$ lead to the respective angle values of $\frac{1}{4} m\angle BAP \equiv \frac{1}{4}$ (90), $\frac{1}{2}$ (90), and $\frac{3}{4}$ (90). Generalize, using mathematical induction.)

17. By the exterior-angle inequality in $\triangle ABD$, $\angle 4 > \angle 1 = \angle 2 \Rightarrow$ (by sca-lene inequality) $BC > DC$; $\angle 3 > \angle 2 = \angle 1 \Rightarrow AB > AD$. Therefore, $AB + BC > AD + DC = AC$. [**Note:** The standard proof of the scalene inequality uses the isosceles triangle theorem and exterior-angle inequality for $\alpha = \infty$, and thus is proved without triangle inequality. Above proof of triangle inequality then becomes valid in such treatments.]

19. We are to prove that if (ABC) does not hold for collinear points A, B, and C, then $AB + BC > AC$. By Theorem **2**, Section **1.8**, either (ACB), (BAC), or $(AB*C) \Rightarrow$ (respectively) $AB + BC = (AC + CB) + BC > AC$, $AB + BC = AB + (BA + AC) > AC$, or $AB + BC = \alpha - AB* + \alpha - B*C = 2\alpha - AC > \alpha > AC$.

21. To show $AC > AF$, by exterior-angle inequality $m\angle AFC > m\angle B = m\angle ACF \Rightarrow$ (scalene inequality) $AC > AF$; $m\angle FAC < m\angle BAC = m\angle ACF \Rightarrow FC < AF$.

22.* By postulate of Pasch ray WP must meet either side YZ or XZ; former possibility violates exterior-angle inequality $\Rightarrow WP$ contains interior point Q such that $(XQZ) \Rightarrow \triangle XWQ \cong \triangle ABC$ and $AC = XQ < XZ$. To prove $BC < YZ$ use similar argument, constructing $\triangle YW'Q' \cong \triangle BAC$ with W' on side XY.

Section 3.8

3. Use HA.

5. $\triangle YZW \cong \triangle ABC$ (SAS), $YW = AB = XY \Rightarrow \angle X \cong \angle W \cong \angle A \Rightarrow$ (AAS) $\triangle ABC \cong \triangle XYZ$.

9. With $\angle ECD \cong \angle CDF$ (acute angles) in Figure for this problem, let M be midpoint of CD and A and B the feet of M on lines EC and DF (which will fall on rays CE and DB by Theorem **2**, Section **3.6**). Now $MC = MD < \alpha/2$ so by Corollary **B**, Section **3.5** $MA < \alpha/2$ and $MB < \alpha/2$; by the corollary in Section **3.7** AC and $BD < \alpha/2 \Rightarrow$ HA applies $\Rightarrow \triangle ACM \cong \triangle BDM \Rightarrow \angle AMC \cong \angle BMD$ and, by converse of vertical-angle theorem (Problem 8, Section **2.6**), (AMB) holds and line AB is common perpendicular.

17. In triangles BFC and CEB, $\angle FBC \cong \angle ECB$, $BC = CB \Rightarrow$ (by HA) $\triangle FBC \cong \triangle ECB \Rightarrow \angle GBC \cong \angle GCB \Rightarrow BG = GC$ (isosceles triangle theorem); $AB = AC$, $AG = AG \Rightarrow$ (by SSS) $\triangle BAG \cong \triangle CAG \Rightarrow \angle BAG \cong \angle GAC$; (BGE) follows from (BFA) and $(\overrightarrow{CBCFCA}) \Rightarrow (\overrightarrow{ABAGAC})$ and $\overrightarrow{AG}$ bisects $\angle BAC$.

Section 3.9

5. $y + z = a, x + z = b, x + y = c \Rightarrow$ (adding in last two equations) $2x + y + z = b + c \Rightarrow 2x + a = b + c \Rightarrow 2x = b + c - a = 2s - 2a \Rightarrow x = s - a$; rest follows in similar manner.

7. $AE = 5$.

13. Suppose perpendicular bisectors of sides BC, AC at L and M meet at $O \Rightarrow O$ equidistant from B and C, and from A and $C \Rightarrow OB = OC = OA \Rightarrow O$ equidistant from A and $B \Rightarrow$ (by perpendicular bisector theorem) O lies on third perpendicular bisector of side AB at midpoint N.

Section 4.3

5. If $AB = XY$, $\angle B \cong \angle Y$, $BC = YZ$, $\angle C \cong \angle Z$ and $\angle D \cong \angle W$, then $\Diamond ABCD \cong \Diamond XYZW$.

7. By SAS $\triangle ABD \cong \triangle BAC \Rightarrow AC = BD$ (CPCPC).

9. If $AD = XW$, by SASAS $\Diamond ABCD \cong \Diamond XYZW$; if $AD \neq XY$, it follows that $\angle A$ and $\angle X$ are supplementary.

10.* By Theorem **1**, Section **4.1**, $(\overrightarrow{ABAC}AD) \Rightarrow \overrightarrow{AC}$ passes through interior of $\angle BAD \Rightarrow$ (crossbar) $\overrightarrow{AC}$ meets $\overline{BD}$ at interior point E; $(\overrightarrow{CDCACD}) \Rightarrow$ ray $\overrightarrow{CA}$ meets $\overline{BD}$ at $E' \Rightarrow E = E'$ since both lie on segment AC (intersection of rays AC and CA) $\Rightarrow E$ lies on both diagonals.

13. By hypothesis for convex quadrilaterals $\Diamond ABCD$ and $\Diamond XYZW$, $\angle A \cong \angle X$, $AB = XY$, $\angle B \cong \angle Y$, $BC = YZ$, and $\angle C \cong \angle Z$. (See Figure S.3.) Draw diagonals AC and $XZ \Rightarrow \triangle ABC \cong \triangle XYZ$ (SAS) $\Rightarrow AC = XZ$, $\angle CAB \cong \angle ZXY$, and $\angle ACB \cong \angle XZY$. Since $\angle DAB \cong \angle WXY$ and $\angle DCB \cong \angle WZY$, betweenness relations relative to convex quadrilaterals $\Rightarrow \angle DAC \cong \angle WXZ$ and $\angle DCA \cong \angle WZX \Rightarrow$ (by ASA) $\triangle DAC \cong \triangle WXZ \Rightarrow AD = XW$, $\angle D \cong \angle W$, and $DC = WZ$.

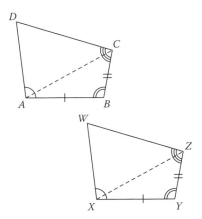

Figure S.3 Problem 13.

19. $\angle 1 + \angle 2 = 90 < \angle 1 + \angle 3 \Rightarrow \angle 2 < \angle 3$. Thus in triangles ADB and CBD, the SAS inequality implies $CD < AB$.

Section 4.4

5. $A = 179.28°$, $s = 0.0125\,662\,879$, $a = 0.9999\,803\,261$, $P = 6.2831\,439\,5$, $K = 3.1415\,029\,971$.

7. 3, 5, 15, 17, 51, 85 and products with powers of $2 \leq 100$.

11. 20; the number of diagonals equals the number of segments joining n points (n objects taken 2 at a time, or $\frac{1}{2}n(n-1)$ minus the number of sides: $\frac{1}{2}n(n-1) - n$.

12.* There is no problem with the ⊥-bisectors of two consecutive sides intersecting in elliptic geometry; for Euclidean geometry, one must prove that the ⊥-bisector of P_1P_2 passes through the point directly opposite, P_k, where $k = \frac{1}{2}(n + 3)$ when n is odd, and passes through the midpoint M_k of the side opposite P_kP_{k+1} for $k = \frac{1}{2}(n+2)$ when n is even (intuitively clear). This is intuitively clear mainly because the ⊥-bisector of P_1P_2 is the line of symmetry l for the vertices of the polygon on opposite sides of l (i.e., opposite vertices are equidistant from l). The details will be omitted, but this idea leads to a fairly straightforward proof for unified geometry. It follows that the ⊥-bisectors of consecutive sides AB and BC of a regular polygon $ABCDE...$ meet at some point O. The rest is simply using the properties of regular polygons to prove that O is equidistant from the vertices A, B, C, D, E,..., and it suffices to prove this only for only the first four vertices A, B, C, and D by change of notation.

Section 4.5

7. $KM = KN$ and $TM = TN \Rightarrow K$ and T are equidistant from M and N, and must lie on the ⊥-bisector of segment MN.

11. If l were not tangent, it intersects C at a second point $Q \Rightarrow OP < \alpha/2$, $OQ < \alpha/2$; by isosceles triangle theorem, $\angle OQP \cong \angle OPQ = $ right angle $\Rightarrow OP = OQ = \alpha/2$. (Theorem **2**, Section **3.5**).

13. By the theorem of Problem 12, $\overline{OA} \perp \overline{PA}$, $\overline{OB} \perp \overline{PB}$. If $PO < \alpha/2$, all three sides of right triangles OPA and OPB are of length $< \alpha/2$ (Corollary **C**, Section **3.7**) and HL applies $\Rightarrow \Delta PAO \cong \Delta PBO \Rightarrow PA = PB$. This is the standard argument for $PO < \alpha/2$; if $PO \geq \alpha/2$, then construct $\angle POB \cong \angle PAO$ on the opposite side of line PO as A with $OB = OA$, and draw PB. Then $\Delta POB \cong \Delta POA$ by SAS and $\angle PBO \cong \angle PAO$ $\Rightarrow \overline{OB} \perp \overline{PB}$ and line PB is tangent to circle O, with $PA = PB$.

15. By theorem of Problem 13, $MB = MA = MB$; since circles C and D are on opposite sides of line l, so are points B and $C \Rightarrow (BMC)$ and $M = $ midpoint of segment BC.

17. Letting z and w be distances from points of contact of circle with sides BC and DA to the vertices, then $x + y = a, y + z = b, z + w = c$ and $w + x = d \Rightarrow a + c = x + y + z + w = b + d; a + b + c + d = 2s = 2(a + c) \Rightarrow s = a + c$.

19. Draw UR passing through center O; then in polygons $URWV$ and $URST$, $VU = TU$, $\angle VUO \cong \angle TUO$ (in Problem 13 $\Delta PAO \cong \Delta PBO$ making $\angle OPA \cong \angle OPB$), $UR = UR$, $\angle ORW \cong \angle ORS$, and $RW = RS \Rightarrow \Diamond URWV \cong \Diamond URST$ (by SASAS) $\Rightarrow VW = TS$.

Section 5.4

1. (a) Parallel lines $\Rightarrow$ angles at A, B are supplementary from alternate-angle theorem (Euclidean geometry).   (b) 9,550 mi.

3. Square: 91; pentagon: 109; triangle: 160, 11, 11.

5. (a) 137   (b) 12.

6.* **(a)** Given: angle-sum of $\triangle ABC = 180$; summing angles in figure,

$$m\angle A + m\angle B + m\angle C = \angle 1 + (\angle 3 + \angle 4) + \angle 6 =$$
$$= (\angle 1 + \angle 2 + \angle 3) + (\angle 4 + \angle 5 + \angle 6) - \angle 2 - \angle 5 =$$
$$= \text{Angles}(\triangle ABD) + \text{Angles}(\triangle ADC) - 180$$

The above equals 180, and if angle-sum of either right triangle < 180, the angle-sum of the other >180, contradicting Legendre's first theorem, contradiction. Hence Angle-sum ($\triangle ADC$) = 180. **(b)** Construct segment $BE \perp$ segment DB, with $BE = AD$; $90 = \angle 1 + \angle 7 = \angle 1 + \angle 3$ $\Rightarrow \angle 7 = \angle 3 \Rightarrow \triangle ADB \cong \triangle BEA$ by SAS $\Rightarrow \angle 8 = \angle 1$, $\angle E \cong \angle D \Rightarrow$ all four angles of $\lozenge ADCE$ are right angles.

7. By construction, let rectangle $EBFG \cong$ rectangle $ADBE$ have side EB in common (Figure S.4) $\Rightarrow \lozenge ADFG$ a quadrilateral having four right angles; repeating this procedure in all directions produces rectangles having arbitrarily large sides.

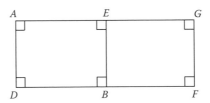

Figure S.4 Problem 7.

8.* By result of Problem 7, one can construct a rectangle $QBCD$ with sides containing legs of $\triangle PQR$ (Figure S.5); draw lines shown and show, in stages, angle-sum of triangles PQB and PQR are each 180. For example, suppose it has been established that angle-sum ($\triangle PQB$) = 180. Then Angles ($\triangle PQR$) + Angles ($\triangle PRB$) = Angles ($\triangle PQB$) + $\angle 8 + \angle 7 = 360$; by Legendre's Second Theorem it follows that Angles ($\triangle PQR$) = Angles ($\triangle PRB$) = 180.

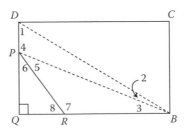

Figure S.5 Problem 8.

9. All right triangles have angle-sum 180, so in Figure P.6a for Problem 6, let $\triangle ABC$ be given with angle-sum ≤ 180 and acute angles at B and C, and let D be foot of A on side BC; with angles as marked in figure,

$$\text{Angles}(\Delta ABC) = \angle 1 + (\angle 3 + \angle 4) + \angle 6 =$$

$$= (\angle 1 + \angle 3 + \angle 2) + (\angle 4 + \angle 5 + \angle 6) - \angle 2 - \angle 5 =$$

$$= 180 + 180 - (\angle 2 + \angle 5)$$

$$= 360 - 180 = 180.$$

11. 1.5708, 1.5708, and 0.5236; due to the existence of similar triangles and triangles of different sizes having angles 88°, 88°, and 4°, unique triangle does not exist.

13. The value π may be replaced by α without changing the proof; thus **(3)** becomes $m\angle BAC = 180 \cdot BC/\alpha$.

15. **(b)** In middle diagram of figure for problem, hemisphere = K + I + II + III $\Rightarrow$ (1) K + I + II + III = 2π; in diagram far right shows three lunes: K + I = $\pi A/90$, II + K = $\pi C/90$, K + III = $\pi B/90$. Sum produces (2) $3K$ + I + II + III = $(\pi/90)$ $(A + B + C)$; substitute $2\pi - K$ for I + II + III $\Rightarrow 3K + 2\pi - K = (\pi/90)$ $(A + B + C) \Rightarrow$ $K = (\pi/180)$ $(A + B + C) - \pi = (\pi/180)$ $(A + B + C - 180)$.

Section 5.8

1. 15.

3. **(a)** $12\theta - 1800 = 12 \Rightarrow \theta = 151$.

5. Since $\alpha = \infty$, a row of congruent octagons can be placed side by side without overlapping (in fact, the entire plane can be tiled with these octagons) $\Rightarrow$ area of plane $> nK$, where K is the area of one of these octagons and n = any positive integer.

7. In Figure 5.30, B is interior to angle formed by two hyperparallels, with line l passing through $B \Rightarrow$ if line through B meets one side, it cannot meet other side; explicit example: $\angle EAE'$ and point B in Figure 5.30.

9. Bolyai's formula is $\tan \frac{1}{2}A = e^{-a}$ so we must solve $\tan 0.5° = e^{-a}$ or $.008727 = e^{-a} \Rightarrow a = 4.74$.

10.* If l lies in the half-plane opposite H, then any point A on l could be joined to any point B on n by a segment AB which must meet line m, contrary to hypothesis.

11. 25°; $a = 9.48$ produces $A \approx 1°$.

13. π.

Section 5.12

1. $x = y = z = 60$ ($\Delta SPU \cong \Delta QPV$ by LA $\Rightarrow UP = VP$.

3. **(a)** $\angle ABC \cong \angle DBA$; by angle-sum theorem, $\angle BAC \cong \angle BDA \Rightarrow$ (by AA similarity) $\Delta ABC \sim \Delta DBA$. **(b)** corresponding sides proportional $\Rightarrow \angle ADC \cong \angle BCD \cong \angle DBC$ and $\angle ACD \cong \angle BCD$ by the isosceles triangle theorem $\Rightarrow$ by AA similarity, $\Delta ADC \sim \Delta DBC$. **(b)** BD/BC = $DA/DC = AC/BD \Rightarrow BD^2 = AC \cdot BC$. **(c)** By angle-sum theorem

$m\angle BDC = 180 - 2\theta = m\angle A = m\angle ADB$; by Euclidean exterior angle theorem, $\theta = m\angle DBC = 2m\angle A = 2(180 - 2\theta) \Rightarrow \theta = 360 - 4\theta \Rightarrow \theta = 72$. **(c)** $\theta = m\angle ABC = m\angle ACB = 2m\angle D$; $m\angle D = 180 - 2\theta \Rightarrow$ (by substitution) $\theta = 2(180 - 2\theta) = 360 - 4\theta \Rightarrow 5\theta = 360$ or $\theta = 72°$.

5. $37.5°$.

7. $x = 40$, $y = 15$.

9.   Let l and m be $\perp$ bisectors of sides BC and AC of $\triangle ABC$. If $l \parallel m$, then line BC is $\perp$ to m (a line $\perp$ to one of two parallel lines is $\perp$ to the other also). But $\overleftrightarrow{AC} \perp m \Rightarrow \overleftrightarrow{BC} \parallel \overleftrightarrow{AC}$ and A, B, and C are collinear (lines $\perp$ to the same line are parallel). Therefore, l meets m at some point O, and by familiar argument O is equidistant from A, B, and C $\Rightarrow$ a circle centered at O passes through vertices.

12.*   (For SSS) (See Figure S.6.) If $k > 1$, construct $\overline{AD} \cong \overline{XY}$ on $\overline{AB}$ and $\overline{DE} \parallel \overline{BC}$, $(AEC) \Rightarrow AC/AE = AB/AD = AB/XY = k = AC/YZ \Rightarrow AE = \cancel{Y}Z$.

Similarly, $BC/DE = AC/AE = k = BC/YZ \Rightarrow DE = YZ \Rightarrow \triangle ADE \cong \triangle XYS$ by SSS $\Rightarrow \angle A \cong \angle X$ and $\angle B \cong \angle ADE \cong \angle Y$ (corresponding angles) $\Rightarrow \angle C \cong \angle Z \Rightarrow \triangle ABC \sim \triangle XYZ$ by definition.

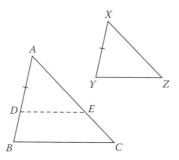

Figure S.6 Problem 12.

15. $ab = a(ph) = (pa)h = ch$.

17.   If $P \in \text{Int}\angle ABC$, let the line through P parallel to side AB meet the other side BC at Q; show that any line PR for R such that (BQR) meets both rays BC and BA.

23.   **(a)** If $\lozenge ABCD$ is parallelogram, diagonal creates congruent triangles ABC and $CDA \Rightarrow AB = CD$, $BC = DA$. **(b)** From $\triangle ABC \cong \triangle CDA$, $\angle A \cong \angle C$; use Corollary **A**, Section **5.9**, for the rest. **(c)** If diagonals AC and BD meet at P, $\triangle ABP \cong \triangle CDP$ by AAS $\Rightarrow AP = PC$, $BP = PD$.

25.   **(a)** If $m\angle C < 90 = m\angle C'$, $AB < A'B'$ (SAS inequality) $\Rightarrow c^2 < d^2 = a^2 + b^2$. **(b)** is the Pythagorean theorem, and **(c)** is logically equivalent to **(a)**. For corollary, use trichotomy on $m\angle C$ versus 90.

27.   Euler line e of $\triangle ABC$ is line of collinearity of O, H, and G; Euler line e_k of Δ_k is line of collinearity of O_k, H_k, and G_k, that is, line $O_k H_k = H_k G_k$ for each k. By **(a)** of Problem 26, $G_k = G$ and by **(b)**, $O_k = H_{k+1}$ $\Rightarrow e = OG = H_1 G = e_1 = O_1 G = H_2 G = e_2 = O_2 G = H_3 G = e_3$, and so on. Thus, the circumcenters O_k all lie on e ($k = 1, 2, 3, \ldots$).

29. (See Figure S.7.) Draw line *KR* where *R* is midpoint of *LM*, inter-secting *WM* at *S*; construct equilateral triangle *TLM* on *LM* as base (*T* on segment *KR*) $\Rightarrow m\angle TMS = 10 = m\angle WKS$; $m\angle SMK = 10 = m\angle SKM \Rightarrow \Delta SMK$ is isosceles $\Rightarrow SM = SK$, $\angle TSM \cong \angle WSK$ (vertical angles) $\Rightarrow$ (by AAS) $\Delta TMS \cong \Delta WKS \Rightarrow WK = TM = LM$.

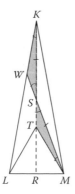

Figure S.7 Problem 29.

35. Polar coordinates for *C* in the two figures are, respectively, (b, A) and $(a, \pi - B) \Rightarrow b \sin A = y = y' = a \sin (\pi - B) = a \sin B \Rightarrow b/\sin B = a/\sin A$.

36.* Distance formula applied to coordinates of points $B(c, 0)$ and $C(x, y)$ [having polar coordinates (b, A)], produces $a^2 = BC^2 = (x - c)^2 + (y - 0)^2 = (b \cos A - c)^2 + (b \sin A)^2 = b^2\cos^2A - 2bc \cos A + c^2 + b^2\sin^2A = b^2(\cos^2A + \sin^2A) + c^2 - 2bc \cos A = b^2 + c^2 - 2bc \cos A$.

Section 6.4

1. By **(8)** Section **6.2** (elliptic geometry) $PV/PS > QW/QR \Rightarrow x/21 > 5/15 \Rightarrow x > 7$.

3. Use **(3)** Section **6.2** for lower bounds for both *x* and *y*; for upper bounds apply **(9)** Section **6.2** $\Rightarrow x < \frac{1}{3} \cdot 20 + \frac{2}{3} \cdot 17 = 18$; $y < \frac{2}{3} \cdot 20 + \frac{1}{3} \cdot 17 = 19$.

5. $21 < x < 30$.

7. $x = 23.4$.

9. Upper bound follows from Corollary **D**, Section **6.1**: $RS < WV = 12$; use **(11)** Section **6.2** for lower bound: $RS > (8/12) WV + (4/12) TU = \frac{2}{3} \cdot 12 + \frac{2}{3} \cdot 9 = 11$.

11. Euclidean geometry: $x = 4$, $y = 16.8$; Hyperbolic geometry: $3 < x < 4$, $16 < y < 16.8$; elliptic geometry: $4 < x < \alpha$, $16.8 < y < 20$.

13. Euclidean geometry: $x = 3.1$, $y = 18.5$; hyperbolic geometry: $3 < x < 3.1$, $18 < y < 18.5$; elliptic geometry: $3.1 < x < 9.3$, $18.5 < y < 18$.

15. In Figure P.15, as $x \to 0$, $XY \to 0$; by **(9)** $LM < (q - x) DE (2q - x) + qXY (2q - x) \Rightarrow$ (by limits) $LM \leq \frac{1}{2}DE$. **(b)** By **(9)**, $DE < aBC/(a + b) + bLM/(a + b) \leq aBC/(a + b) + \frac{1}{2} bDE/(a + b)$; multiply by $2(a + b)$: $2(a + b)DE < 2aBC + bDE \Rightarrow (2a + b)DE < 2aBC \Rightarrow DE/BC < 2a/(2a + b) = AE/AC$.

17. **(8)** Section **6.2** give first two inequalities directly, in addition to (*) $(x + y)/z < (a + b)/c$. To obtain $x/z < a/c$ use $1 + y/x > 1 + b/a$ or $x/(x + y) < a/(a + b)$; multiply by (*) to complete.

Section 6.5

1. $\angle ADC$ acute $\Rightarrow \angle ADB$ obtuse $\Rightarrow AB > AD \Rightarrow AC/AD > AC/AB \Rightarrow$
   $c(\theta_1) = \lim AC/AD \geq \lim AC/AB = c(\theta_2)$.
3. $s(\theta) = \lim a/c$ and $c(\theta) = \lim b/c \Rightarrow$ (quotient theorem on limits)
   $t(\theta) = (\lim a/c)/(\lim b/c) = \lim (a/c)/(b/c) = \lim a/b$.
5. Let $\triangle ABC$ be isosceles right triangle, right angle at $C \Rightarrow$ as $B, C \to A$,
   angle-sum $\triangle ABC \to \pi$; if $\lambda =$ either non-right angle of $\triangle ABC$, then $\lim$
   $\lambda = \pi/4$. By floating triangles, with $a = b$, $\lim a/b = 1 = t(\lim \lambda) = t(\pi/4)$.
7. **(a)** $\theta - \pi/2 \equiv \varphi$ is less than $\pi/2$ so $s(\varphi) = c(\pi/2 - \varphi) = c(\pi/2 - (\theta - \pi 2))$
   $= c(\pi - \theta)$. Therefore, by definition $s(\theta - \pi/2) = s(\varphi) = c(\pi - \theta) = -c(\theta)$.
   (b) $c(\varphi) = s(\pi/2 - \varphi) = s(\pi/2 - (\theta - \pi/2)) = s(\pi - \theta) = s(\theta)$.

Section 6.6

1. $s^2(\theta) = 1 - c^2(\theta) = \frac{1}{2}[2 - 2c^2(\theta)]$ and $2c^2(\theta) = c(2\theta) + 1$ (from **(4)** Section
   6.6); substitution produces $s^2(\theta) = \frac{1}{2}[2 - (c(2\theta) + 1)] = \frac{1}{2}[1 - c(2\theta)]$.
   Replacing θ by $\frac{1}{2}\theta$, one obtains $s^2(\theta/2) = \frac{1}{2}[1 - c(\theta)] \Rightarrow$ half-angle
   identity $s(\theta/2) = \sqrt{\dfrac{1 - c(\theta)}{2}}$.
3. Start with **(4)** Section **6.6**: $c(2\theta) = 2c^2(\theta) - 1 = 2c^2(\theta) - [s^2(\theta) + c^2(\theta)]$
   $= c^2(\theta) - s^2(\theta)$.
5. $s(2\theta)/[c(2\theta) + 1] = 2s(\theta) c(\theta)/[2c^2(\theta) - 1 + 1] = s(\theta)/c(\theta) = t(\theta)$.
7. Right triangles ACD and BED have $\angle D$ in common, so $m\angle EBD =$
   $m\angle CAD = \theta$ and $\triangle EBD \sim \triangle CAD$:

$$\sin\theta\cos\varphi + \cos\theta\sin\varphi =$$

$$= \frac{ED}{BD}\cdot\frac{AC}{AB} + \frac{BE}{BD}\cdot\frac{BC}{AB}$$

$$= \frac{BE}{AB}\left(\frac{ED}{BD}\cdot\frac{AC}{BE} + \frac{BC}{BD}\right) = \frac{BE}{AB}\left(\frac{ED}{BD}\cdot\frac{CD}{ED} + \frac{BC}{BD}\right)$$

$$= \frac{BE}{AB}\left(\frac{CD}{BD} + \frac{BC}{BD}\right) = \frac{BE}{AB}\cdot\frac{BD}{BD}$$

$$= \frac{BE}{AB} = \sin(\theta + \varphi)$$

Section 6.9

3. Identity **(3)**, Section **6.9** $\Rightarrow C(2a) = 2C^2(a) - 1 = 2\cdot 4^2 - 1 = 31$; $C(4a)$
   $= 2\cdot 31^2 - 1 = 1921$.
5. Squaring both sides: $S^2(x) = [1 - C^2(x)]\,\sigma$. Multiply both sides by σ to
   obtain the desired form (note that $\sigma^2 = 1$).
7. $S(2a) = \sqrt{5}/2$; $S(4a) = 3\sqrt{5}/2$

Section 6.12

1. Elliptic geometry: $c = 1.798$; hyperbolic geometry: $c = 2.444$.
3. $A = 0.8505$ rad. $\approx 48.7°$.

5. $\cosh a \cosh b = \cosh c \Rightarrow \tan A \tan B = \dfrac{\tanh a}{\sinh b} \cdot \dfrac{\tanh b}{\sinh a} = \dfrac{\tanh a}{\sinh a} \cdot$

$\dfrac{\tanh b}{\sinh b} = \dfrac{1}{\cosh a} \cdot \dfrac{1}{\cosh b} = \dfrac{1}{\cosh c} = \operatorname{sech} c.$

9. Using Figure S.8, $\cos A = \tanh \tfrac{1}{2}a / \tanh a$. Apply the half-angle identity from hyperbolic trigonometry (analogous to ordinary trigonometry): $\tanh \tfrac{1}{2}x = \sinh x / (1 + \cosh x)$.

$$\cos A = \left. \dfrac{\sinh a}{1 + \cosh a} \middle/ \tanh a = \dfrac{\sinh a}{1 + \cosh a} \cdot \dfrac{\cosh a}{\sinh a} = \dfrac{\cosh a}{1 + \cosh a}\right.$$

$\Rightarrow \sec A = (1 + \cosh a)/\cosh a = \operatorname{sech} a + 1$. To obtain **(14)**, merely solve **(13)** for $\cosh a$: $\cosh a = 1/(\sec A - 1) = \cos A/(1 - \cos A)$.

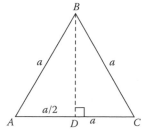

Figure S.8 Problem 9.

11. $\tan A = \sin A / \cos A = \dfrac{S(a)/S(c)}{T(b)/T(c)} = \dfrac{S(a)}{S(c)} \cdot \dfrac{T(c)}{T(b)} =$

$$\dfrac{S(a)}{C(c)T(b)} = \dfrac{S(a)}{C(a)C(b)T(b)} = \dfrac{T(a)}{S(b)} \cdot$$

13. **(a)** $c = 9195$ mi. $(a = 3010 \to 0.760101, \; b = 8120 \to 2.050505)$.
(b) $B = 49.1°$.

15. From table, $\cot A = \sinh \kappa a / \tanh \kappa a = \cosh \kappa a$; as $a \to 0$, $\cosh \kappa a \to 1 \Rightarrow \cot A \to 1$ and $A \to 45°$.

16.* Series expansion of $\cos c = \cos a \cos b$ is $\left(1 - \dfrac{c^2}{2!} + \dfrac{c^4}{4!} - \dfrac{c^6}{6!} + \cdots\right) =$

$\left(1 - \dfrac{a^2}{2!} + \dfrac{a^4}{4!} - \dfrac{a^6}{6!} + \cdots\right)\left(1 - \dfrac{b^2}{2!} + \dfrac{b^4}{4!} - \dfrac{b^6}{6!} + \cdots\right)$; terms of degree 4 or higher are then dropped without significant error to give $1 - \tfrac{1}{2}c^2 = (1 - \tfrac{1}{2}a^2)(1 - \tfrac{1}{2}b^2)$ or $1 - \tfrac{1}{2}c^2 = 1 - \tfrac{1}{2}a^2 - \tfrac{1}{2}b^2 + \tfrac{1}{4}a^2b^2 \Rightarrow c^2 = a^2 + b^2 - \tfrac{1}{2}a^2b^2 \Rightarrow c^2 = a^2 + b^2.$

17. Hint produces relation $\sinh c / \sinh w = \tanh a / \tanh w \Rightarrow \sinh c \tanh w = \tanh a \sinh w \Rightarrow \sinh c = \tanh a \cosh w \Rightarrow \sinh c = \tanh a \cdot (\cosh a \cosh b) = \sinh a \cosh b$, which is **(16)**. To obtain **(17)**, use $\triangle ADC$ to obtain $\tanh d / \tanh w = \cos \varphi = \sin \theta = \sinh b / \sinh w \Rightarrow \tanh d = \sinh b \tanh w / \sinh w = \sinh b / \cosh w = \sinh b / \cosh b \cosh a.$

18.* $\tanh a \cosh d = \tanh a \cosh w / \cosh c = \tanh a \cosh a \cosh b / \cosh c = \sinh a \cosh b / \cosh c$; now use **(16)** to complete by elementary substitution. Use $\tanh x / x \to 1$ as $x \to 0$ for the limit relation.

Section 7.2

3. By distributive law: $[AB \cdot CE] + (AC \cdot DE) + [AD \cdot EC] + AE \cdot BD + (BE \cdot CA) = (AB + DA)CE + AC(DE + EB) + AE \cdot BD = DB \cdot CE + AC \cdot DB + AE \cdot BD = DB(CE + AC + EA) = DB(0) = 0$.

7. $20^2 = x \cdot 12^2 + (1 - x) \cdot 13^2 - x(1 - x) \cdot 5^2 \Rightarrow 25x^2 - 50x - 231 = 0$ with solution $x = 21/5$ ($BD = 21$ with (BCD)), or $-11/5$ ($BD = 11$ with DBC)).

11. 2, 3, and 4.

13. Since we want to show that $a < b$ implies $t_a > t_b$, consider the inequality (from (2)): $2\sqrt{bcs(s-a)}/(2s-a) > 2\sqrt{acs(s-b)}/(2s-b)$ or $b(s-a)(2s-b)^2 > a(s-b)(2s-a)^2$; since $a < b$, it suffices to prove $(s-a) \cdot (2s-b)^2 > (s-b)(2s-a)$. To that end, expanding and simplifying (judicial factoring helps), we obtain the equivalent inequality $(as + bs - ab) \cdot (b - a) > 0$, or simply $as + bs - ab > 0$ since $b - a > 0$; but $as + bs = (a + b)s = \frac{1}{2}(a + b)(a + b + c) > (a + b)a > ab$.

15. (a) $5m_c^2 - m_a^2 - m_b^2 = 5(\frac{1}{2}a^2 + \frac{1}{2}b^2 - \frac{1}{4}c^2) - (c^2 + \frac{1}{4}a^2 + \frac{1}{4}b^2) \Rightarrow \frac{9}{4}(a^2 + b^2 - c^2) = 0$ iff $\angle C = $ right angle. (b) $2K/h_a = a^2$, $2K/h_b = b^2$, $2K/h_c = c^2 \Rightarrow 2K(1/h_a^2 + 1/h_b^2 - 1/h_c^2) = a^2 + b^2 - c^2 = 0$ iff $\angle C = $ right angle.

16.* $PG^2 = \frac{1}{3} PA^2 + \frac{2}{3} PL^2 - \frac{2}{9} AL^2 = \frac{2}{3}(\frac{1}{2} PB^2 + \frac{1}{2} PC^2 - \frac{1}{4} a^2) - \frac{2}{9}(\frac{1}{2} b^2 + \frac{1}{2} c^2 - \frac{1}{4} a^2) = \frac{1}{3}(PA^2 + PB^2 + PC^2) - \frac{1}{9}(a^2 + b^2 + c^2)$.

17. $OA = OB = OC = R \Rightarrow$ [from (6)] $OG^2 = \frac{1}{3}(3/R^2) - \frac{1}{9}(a^2 + b^2 + c^2)$.

19. Triangle having three medians as sides has area ¾ that of $\triangle ABC$ [see Figure S.9] $\lozenge BLC'M$ and $\lozenge ANCC'$ are parallelograms $\Rightarrow K' = $ Area $\triangle ALC' = 2$(Area $\triangle AWL$) $= 2$(Area $\triangle ALC - $ Area $\triangle LWC$) $= 2$(Area $\triangle ALC - \frac{1}{4}$ Area $\triangle BMC$) $= 2(\frac{1}{2}K - \frac{1}{8}K) = \frac{3}{4}K \Rightarrow$ (by (4) applied to $\triangle ALC'$) $K = \frac{4}{3}K' = \frac{4}{3}\sqrt{m_s(m_s - m_a)(m_s - m_b)(m_s - m_c)}$

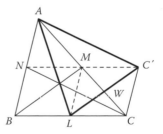

Figure S.9 Problem 19.

21. Substitute $2ah_a^2$ for $4K$ in (11).

Section 7.4

1. $x = 6$ in (a) and (b).

3. $LW = 12$, $WM = 4$.

4.* $\sqrt{365}$

5. $\frac{1}{2}\sqrt{305}$.

7. Using the hint given, Figure S.10 shows diameter $CD \perp$ line CP, and AD drawn. With angles indicated as in the figure, by the inscribed angle theorem, $m\angle ABC = \angle 1 = 90 - \angle 3 = \angle 2$.

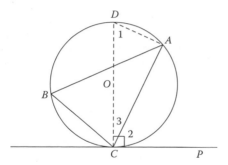

Figure S.10

9. **(c)** (1) With A denoting given minor arc, choose C on line OB, as in figure (c) for problem; $\triangle AOC$ isosceles, exterior angle theorem $\Rightarrow$ $m\angle AOB = 2\theta = mA$ (definition, arc measure). (2) If A = semicircle, for any point C on circle C not on A, $m\angle ACB = 90 \Rightarrow mA = 2 \cdot 90 = 180$. (3) (See Figure S.11.) If A = major arc, let C, C' be any two points on arcs A, A', respectively, lying on opposite sides of line $AB \Rightarrow$ (by (1) and definition of arc measure) $mA = 2m\angle AC'B = 2(180 - m\angle ACB) = 360 - 2m\angle ACB = 360 - mA' = 360 - m\angle AOB$.

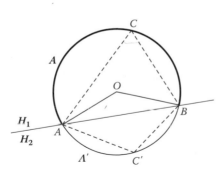

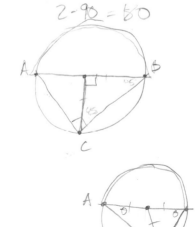

Figure S.11 Problem 9.

11. $x = \frac{1}{2}(80) = 40$; $y = 180 - (x + 25) = 115 \Rightarrow 115 = \frac{1}{2}(z + 60) \Rightarrow$ $z = 2(115) - 60 = 170$.
13. $10\sqrt{2}$.
15. $4\sqrt{3}\sin 20°$.
16.* **(a)** $m(\text{arc } WE) = m(\text{arc } WEU) - m(\text{arc } EU) = 180 - (\varphi + \omega) \Rightarrow$ $m\angle UCD = \frac{1}{2}(180 - \varphi - \omega + 3\omega) = 90 - \frac{1}{2}\varphi + \omega$. **(b)** $90 - \frac{1}{2}\varphi + \omega = 30 + \theta = 30 + \frac{1}{2}(\varphi + 2\omega) \Rightarrow \varphi = 60$. **(c)** $\angle ECO$ is vertical to $\angle UCD$ of measure $90 - \frac{1}{2}(60) + \omega = 60 + \omega =$ measure of central angle $\angle EOU \Rightarrow EC = EO =$ radius.
19. Let $\lozenge ABCDE$ = regular pentagon $\Rightarrow \lozenge ABCD$ is cyclic $\Rightarrow$ (Ptolemy) $AB \cdot CD + BC \cdot AD = AC \cdot BD \Rightarrow a^2 + ab = b^2 \Rightarrow a/b + 1 = b/a$.
21. By result in Example 1 in Section **7.3**, Power P (circle C_1) $= PT^2 =$ Power P (circle C_2).

23. (a) $\angle ACL'$ is inscribed in semicircle $\Rightarrow$ lines $L'C$, AC are $\perp$, as are lines BH, AC $\Rightarrow$ lines $L'C$, BH are parallel; similarly, lines HC, BL' are parallel $\Rightarrow \lozenge HCL'B$ is a parallelogram. **(b)** Diagonals BC, HL' bisect each other $\Rightarrow L$ is midpoint of both BC and HL'.

25. $16K^2 = 4(ab + cd)^2 - (a^2 + b^2 - c^2 - d^2)^2 = [2(ab + cd) + (a^2 + b^2 - c^2 - d^2)][2(ab + cd) - (a^2 + b^2 - c^2 - d^2)] = [(a^2 + 2ab + b^2) - (c^2 - 2cd + d^2)] [(c^2 + 2cd + d^2) - (a^2 - 2ab + b^2)] = [(a + b) + (c - d)][(a + b) - (c - d)] \cdot [(c + d) + (a - b)][(c + d) - (a - b)] = (a + b + c - d)(a + b - c + d) (a - b + c + d)(-a + b + c + d) = (2s - 2d)(2s - 2c)(2s - 2b) (2s - 2a) = 16(s - a)(s - b)(s - c)(s - d)$.

Section 7.5

1. $[KW, TQ] = -8$.

3. $x = -8$.

4.* $xy + 21 = 5x + 5y$

5. $x + 21/y = 5x/y + 5$ or $x \to 5$ as $y \to \infty$, C approaches midpoint of $\overline{AB}$ as limit.

7. (b) It must be shown that if μ is any of the six values given in Theorem 1, $1/\mu$ and $1 - \mu$ will produce another value in the list; trivial if $\mu = \lambda$ or $\mu = 1 - \lambda$, so let $\mu = 1/\lambda \Rightarrow 1/\mu = \lambda$, $1 - \mu = 1 - 1/\lambda$; $\mu = 1 - 1/\lambda \Rightarrow 1/\mu = 1/(1 - 1/\lambda) = \lambda/(\lambda - 1)$, $1 - \mu = 1/\lambda$; $\mu = 1/(1 - \lambda)$ $\Rightarrow 1/\mu = 1 - \lambda$, $1 - \mu = 1 - 1/(1 - \lambda) = \lambda/(\lambda - 1)$; $\mu = \lambda/(\lambda - 1) \Rightarrow 1/\mu = (\lambda - 1)/\lambda = 1 - 1/\lambda$, $1 - \mu = 1 - \lambda/(\lambda - 1) = 1/(1 - \lambda)$.

8.* In view of Problem 7, it suffices to set $\lambda =$ any one of the five other values in Theorem 1; $\lambda = 1/\lambda \Rightarrow \lambda^2 = 1 \Rightarrow \lambda = \pm 1 \Rightarrow \lambda = -1$ and other two values generated by operations (1), (2) in Problem 7(b); $\lambda = 1 - \lambda \Rightarrow \lambda = \frac{1}{2}$ and other two values generated by (1), (2); λ cannot coincide with either $1 - \lambda$ or with $1/(1 - \lambda)$ for real λ; finally, $\lambda = \lambda/(\lambda - 1) \Rightarrow \lambda = 2$.

9. By construction, $EO = OC$, $OF = OB$, and $OF^2 = OE \cdot OD \Rightarrow OB^2 = OC \cdot OD$.

Section 7.6

3. Linearity no. $= \dfrac{12}{5} \cdot \dfrac{3}{12} \cdot \dfrac{5}{3} = 1$.

5. D, E, F collinear $\Rightarrow$ linearity no. $= -1 \Rightarrow \dfrac{1}{4} \cdot \dfrac{BD}{-(4 - BD)} \cdot \dfrac{2}{1} = 1 \Rightarrow$ $BD = 8$.

7. (a) From (1), $K'/K = [(\frac{1}{3})^3 - (\frac{2}{3})^3]^2/(\frac{1}{3} - \frac{4}{9})^3 = (7/27)^2/(7/9)^3 = \frac{1}{7}$.
(b) 4/13, 3/7. **(c)** $K'/K = [(1/p^3 - (p - 1)^3/p^3]^2/[1/p + (p - 1)^2/p^2]^3$x $= [(1 - (p - 1)^3]^2/[p + (p - 1)^2]^3 = (p^3 - 3p^2 + 3p - 2)^2/(p^2 - p + 1)^3 = (p - 2)^2 (p^2 - p + 1)^2/(p^2 - p + 1)^3 = (p - 2)^2/(p^2 - p + 1)$.

9. $AX/XB = AF/FB \Rightarrow (AB + BX)/XB = (AB + BF)/FB \Rightarrow AB/XB - 1 = AB/FB - 1 \Rightarrow XB = FB \Rightarrow X = F$.

11. $\triangle ABD \sim \triangle CBF \Rightarrow BD/FB = AB/CB = c/a$; similarly, $\triangle BCE \sim \triangle ACD$ $\Rightarrow CE/DC = BC/AC = a/b$, $\triangle CAF \sim \triangle ACF \Rightarrow AF/EA = CA/BA = b/c$ $\Rightarrow \begin{bmatrix} ABF \\ CDF \end{bmatrix} = \dfrac{c}{a} \cdot \dfrac{a}{b} \cdot \dfrac{b}{c} = 1 \Rightarrow$ altitudes AD, BE, CF concurrent. [These ratios also apply if $\triangle ABC$ has obtuse angle, but extra work required.]

13. In $\triangle PAB$, menelaus points N, B', A' collinear, hence $\begin{bmatrix} PAB \\ NB'A' \end{bmatrix} = -1$;

similarly, $\begin{bmatrix} PBC \\ LC'B' \end{bmatrix} = -1$ and $\begin{bmatrix} PCA \\ MA'C' \end{bmatrix} = -1$; product of these three

produces ratio product $\dfrac{PA'}{A'A} \cdot \dfrac{AN}{NB} \cdot \dfrac{BB'}{B'P} \cdot \dfrac{PB'}{B'B} \cdot \dfrac{BL}{LC} \cdot \dfrac{CC'}{C'P} \cdot \dfrac{PC'}{C'C} \cdot \dfrac{CM}{MA}$.

$\dfrac{AA'}{A'P} = -1$ which reduces to $\dfrac{AN}{NB} \cdot \dfrac{BL}{LC} \cdot \dfrac{CM}{MA} = \begin{bmatrix} ABC \\ LMN \end{bmatrix} = -1$ and by

Menelaus' theorem, L, M, N are collinear.

15. (Use figure for problem.) Cevians AD, BE, CF concurrent (given); to show AP, BQ, and CR concurrent. By Ceva's theorem,

$\begin{bmatrix} ABC \\ DEF \end{bmatrix} = \dfrac{AF}{FB} \cdot \dfrac{BD}{DC} \cdot \dfrac{CE}{EA} = 1$; secant theorem produces $AF \cdot AR =$

$AQ \cdot AE$ or $AF/AE = AQ/AR$; similarly, $BD/BF = BR/BP$, $CE/CD =$

$CP/CQ \Rightarrow$ product yields $\dfrac{AF}{AE} \cdot \dfrac{BD}{BF} \cdot \dfrac{CE}{CD} = \dfrac{AQ}{AR} \cdot \dfrac{BR}{BP} \cdot \dfrac{CP}{CQ} \Rightarrow$

$\dfrac{AQ}{QC} \cdot \dfrac{CP}{PB} \cdot \dfrac{BR}{RA} = \begin{bmatrix} ACB \\ PRQ \end{bmatrix} = 1 \Rightarrow$ lines AP, CR, BQ concurrent.

Section 7.7

1. When radii of circles are equal.

3. $O(h, k)$ = center, r = radius $\Rightarrow h = -a/2$, $k = -b/2$, $r^2 = h^2 + k^2 - c$;
Power $P(x_0, y_0) = PO^2 - r^2 = (x - h)^2 + (y - k)^2 - (h^2 + k^2 - c) = x_0^2 + y_0^2 - 2hx_0 - 2ky_0 + c \equiv x_0^2 + y_0^2 + ax_0 + by_0 + c$.

4.* If $P(x_0, y_0)$ is any point on radical axis of two circles, by Problem 3,
$x_0^2 + y_0^2 + a_1 x_0 + b_1 y_0 + c_1 = x_0^2 + y_0^2 + a_2 x_0 + b_2 y_0 + c_2$ or $(a_1 - a_2) x_0 + (b_1 - b_2) y_0 + (c_1 - c_2) = 0$; conversely, if $(a_1 - a_2)x + (b_1 - b_2)y + (c_1 - c_2) = 0$, then powers of $P(x, y)$ with respect to circles are equal $\Rightarrow$ above is equation of radical axis.

5. Using congruent triangles angle between tangents to two intersecting circles at one point of intersection is congruent to angle at other point. (See Figure S.12.)

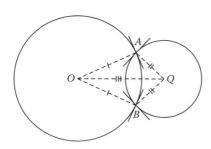

Figure S.12 Problem 5.

7. Let radical axes of C_1, C_2 and C_1, C_3 meet at $P \Rightarrow$ Power $P(C_1) =$ Power $P(C_2)$, Power $P(C_1) =$ Power $P(C_3) \Rightarrow$ Power $P(C_2) =$ Power $P(C_3)$ and P lies on third radical axis.

9. Given construction produces secant lines passing through intersections $C \cap E$, $D \cap E$, which are radical axes of (C, E), $(D, E) \Rightarrow$ perpendicular is desired radical axis since these lines are concurrent at P (Problem 5).

10.* Use construction of Problem 9 to obtain radical axis l of C_1, C_2, which intersects line of centers $m \equiv \overleftrightarrow{OQ}$ at point W; construct tangents WT and WT' from W to C_1, C_2 with points of contact T and T' (only one of these is needed in construction required); since $WT^2 =$ Power $W(C_1) =$ Power $W(C_2) = WT'^2$, the circle C centered at W and passing through T and T' will be orthogonal to C_1, $C_2 \Rightarrow C$ belongs to the family $\mathcal{F}$ (where C_1, C_2 belong to $\mathcal{G}$); let C meet m at A and $B \Rightarrow$ every circle through A and B belongs to $\mathcal{F}$ and is orthogonal to C_1, C_2, including circle through A, B, and P, which can now be constructed.

11. $x^2 + y^2 + \lambda x - (2\lambda + 5)y = 0$ (by substitution of coordinates of C and D into general equation and setting $a = \lambda$).

13. (a) $x^2 + y^2 + ax + by + 1 = 0$. (b) (by substitution of coordinates) $x^2 + y^2 + ax - (2a + 6)y + 1 = 0$, for real a. (c) $B(\tfrac{2}{5}, \tfrac{1}{5})$.

15. Radical axis is that between circle $\lambda = 0$ and any other: $x^2 + y^2 = a^2 + b^2$ and $x^2 + y^2 + \lambda(ax + by) = a^2 + b^2 \Rightarrow \lambda(ax + by) = 0$, $\lambda \neq 0 \Rightarrow ax + by = 0$.

16.* (a) By result of Problem 15, radical axis given by $[a(t) - a(0)]x + [b(t) - b(0)]y + c(t) - c(0) = 0$; since this must be a fixed line $ax + by + c = 0$ for certain constants a, b, c, coefficients must be proportional to a, b, $c \Rightarrow a(t) - a(0) = at'$, $b(t) - b(0) = bt'$, $c(t) - c(0) = ct' \Rightarrow a(t) = at' + a'$, $b(t) = bt' + b'$, $c(t) = ct' + c'$ (with obvious definitions for a', b', c').

17. Slope of ellipse ($\lambda < b^2 < a^2$) given by implicit differentiation of $x^2/(a^2 - \lambda) + y^2/(b^2 - \lambda) = 1 \Rightarrow 2x/(a^2 - \lambda) + 2yy'/(b^2 - \lambda) = 0 \Rightarrow y' = -x (b^2 - \lambda)/y(a^2 - \lambda)$; slope of hyperbola ($b^2 < \mu < a^2$) given same way: $y' = -x (b^2 - \mu)/y(a^2 - \mu)$; product of slopes $= (x^2/y^2) \cdot (b^2 - \lambda)(b^2 - \mu)/[(a^2 - \lambda)(a^2 - \mu)]$ at any point (x, y) of intersection. Coordinates of such points given by solving system of given equations for x^2, y^2 (Cramer's rule): $x^2 = 1/(b^2 - \mu) - 1/(b^2 - \lambda)]/k = (\mu - \lambda)/[k(b^2 - \lambda)(b^2 - \mu)]$ where $k =$ determinant of coefficients; similarly, $y^2 = (\lambda - \mu)/[k(a^2 - \lambda)(a^2 - \mu)] \Rightarrow x^2/y^2 = -(a^2 - \lambda)(a^2 - \mu)/[(b^2 - \lambda)(b^2 - \mu)]$. When substituted for x^2/y^2 in above product of slopes, result $= -1$ proving orthogonality.

Section 8.1

1. (a) $x = \dfrac{x'}{y' - 1}$, $y = \dfrac{y'}{x' - 1}$ (b) $v\colon y = 1$ (c) $l \cap m = P(3, 1)$; $l'\colon y' = \tfrac{1}{3}$ $x' + \tfrac{5}{6}$, $m'\colon y' = \tfrac{1}{3}x' + \tfrac{4}{3}$.

3. $fg\colon x'' = \dfrac{x + y}{y}$, $y'' = \dfrac{y + 2}{y}$; $gf\colon x'' = \dfrac{x + y}{x + y + 6}$, $y'' = \dfrac{y + 2}{x + y + 6}$.

5. Conditions $x' = x$, $y' = y$ lead to cubic equation $x^3 - 3x^2 - 4x + 12 = 0$ which may be solved by (group) factoring; fixed points: $(\pm 2, 4)$, $(3, 9)$.

9. $1 = (a + 1)(1)/(a + b + 1)$, $1 = (b + 1)(1)/(a + b + 1) \Rightarrow a = 0, b = 0$.

10.* **(a)** $x'^2 + \dfrac{(y' + 1)^2}{2} = 1$ (ellipse).   **(b)** $y' = -\frac{1}{2}x' + \frac{1}{2}$ (parabola).

(c) $-\dfrac{x'^2}{2} + \dfrac{(y' - 2)^2}{2} = 1$ (hyperbola).

11. Asymptotes for general hyperbola $\varepsilon(x^2/a^2 - y^2/b^2) = 1$ ($\varepsilon = \pm 1$) given by $x/a = \pm y/b$; for the hyperbola of Problem 10(c), asymptotes are $y' = \pm x' + 2$; these are images of tangent lines $y = \pm x + 2$.

15. Entire plane maps to $y' = -2x' + 1$.

17. **(b)** (substitute coordinates of points into general equation and solve for a, b, c in terms of d): $10x - 3y - 7z = 13$.

19. $(P_1 P_2)^2 = (P_0 P_1)^2 + (P_0 P_2)^2 \Rightarrow (x_1 - x_2)^2 + (y_1 - y_2)^2 + (z_1 - z_2)^2 = (x_0 - x_1)^2 + (y_0 - y_1)^2 + (z_0 - z_1)^2 (x_0 - x_2)^2 + (y_0 - y_2)^2 + (z_0 - z_2)^2$; expand using binomial square formula, and simplify. Focusing on just the x coordinates (which are independent of the y and z coordinates): $x_1^2 - 2x_1 x_2 + x_2^2 = x_0^2 - 2x_0 x_1 + x_1^2 + x_0^2 - 2x_0 x_2 + x_2^2 \Rightarrow -2x_1 x_2 = 2x_0^2 - 2x_0 x_1 - 2x_0 x_2 \Rightarrow x_0 x_1 - x_1 x_2 - x_0^2 + x_0 x_2 = 0 \Rightarrow x_1(x_0 - x_2) - x_0(x_0 - x_2) = 0 \Rightarrow (x_1 - x_0)(x_0 - x_2) = 0$.

21. Solve equations of planes simultaneously for x, y in terms of z: $x = z + 1$, $y = -2z - 1$; if $z = 2t - 1$, $x = 2t$, $y = -4t + 1$ (real t); conversely, such a point will satisfy both equations and lie in both planes (check).

23. **(b)** $x = 2t + 3$, $y = 3t + 1$, $z = -4t$. **(c)** Equations for general t and s lead to $2t + 3 = x = 3s - 2$, $-t + 1 = y = 4s - 2$, $t + 4 = z = -2s + 5 \Rightarrow t = -1$, $s = 1 \Rightarrow (x, y, z) = (1, 2, 3)$, point of intersection; direction nos. are, respectively 2: -1: 1 and 3: 4: $-2 \Rightarrow aa' + bb' + cc' = 2 \cdot 3 + (-1) \cdot 4 + 1 \cdot (-2) = 0 \Rightarrow$ lines are perpendicular.

Section 8.2

4.* **(a)** $(t, -2)$ for real t (a line of fixed points, $y = -2$).   **(b)** $(-2, 2)$.

5. No, first two determine entire line of fixed points (Corollary C).

7. Map given $\triangle ABC$ with midpoints M, N of sides AC, AB to right $\triangle PQR$, right angle at R and M and N to midpoints S (midpoint of hypotenuse PQ) and T (midpoint of leg PR); theorem obvious for TS by use of similar triangles QST and QPR. Properties of parallelism and ratios of parallel segments $= \frac{1}{2}$ are invariant under affine transformations, thus theorem for $\triangle ABC$.

11. From **(2)**, $m = (c + dm)/(a + bm) \Rightarrow bm^2 + (a - d)m = c$ for all real m; take values 0, 1, 2 for $m \Rightarrow b = c = 0$, $a = d$; result is $x' = ax$, $y' = ay$.

14.* By definition of mappings, for any fixed A as reference point in $\mathcal{P}$ (see Figure S.13), lines PP_1 and AA_1 are parallel to e, and $P_1 P'$ and $A_1 A'$ are parallel to $f \Rightarrow PP_1$ and AA_1 are parallel to each other, as are $P_1 P'$ and $A_1 A' \Rightarrow$ plane of $PP_1 P' \parallel$ plane of $AA_1 A' \Rightarrow \overleftrightarrow{PP'} \parallel \overleftrightarrow{AA'} \equiv m$ (fixed line), proving first part of affine reflection. For ratios, in right triangles PQP_1 and $PP_1 Q$ where Q is point of intersection of line PP' with $\mathcal{P} \cap \mathcal{P}_1$ we have $\tan \theta = P_1 Q/PQ$, $\tan \varphi = P_1 Q/P'Q \Rightarrow P'Q/PQ = \tan \theta/\tan \varphi = k$, where θ and φ are constant angles which lines e and f make with plane $\mathcal{P}$.

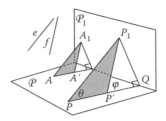

Figure S.13

15. If $c = 0$, then by (1) $ad \neq 0$ and (3) $\Rightarrow ab = 0 \Rightarrow a \neq 0$ and $b = 0$ ($c = 0$ implies $b = 0$ in same manner); (b) implies $a = d$ or $a = \pm d \Rightarrow$ conclusion in problem. Otherwise, $b \neq 0$ and $c \neq 0 \Rightarrow a = -cd/b$ by (c). Substituting into (b) $\Rightarrow c^2 d^2/b^2 + c^2 = c^2(d^2/b^2 + 1) = c^2(d^2 + b^2)/b^2 = b^2 + d^2 \Rightarrow c^2/b^2 = 1 \Rightarrow c = \pm b$, $a/c = \mp d/c$ or $a = \mp d$.

Section 8.3

1. (a) General affine.   (b) Dilation.   (c) Similitude.
6.* (a) 13   (b) 1.
7. Draw (construct) any circle D tangent to sides of given angle A (vertex) and draw line AP, intersecting D at Q; dilation with factor $k = AP/AQ$ maps D to circle tangent to sides of angle and passing through P.
9. (a) With P any point not on line AB, similitude maps $\triangle ABP$ to $\triangle A'B'P'$ with dilation factor $k = A'B'/AB = 1$ (by hypothesis) $\Rightarrow$ similitude = isometry. (b) With A = origin, let similitude be given by $x' = ax - \varepsilon by + x_0$, $y' = bx + \varepsilon ay + y_0$; A and B map to (x_0, y_0) and $(ax - \varepsilon by + x_0, bx + \varepsilon ay + y_0) \Rightarrow$ (by distance formula) $A'B'^2 = (ax - \varepsilon by)^2 + (bx + \varepsilon ay)^2 = (a^2 + b^2)(x^2 + y^2)$ (by algebra); by hypothesis, $AB = A'B' \Rightarrow x^2 + y^2 = (a^2 + b^2)(x^2 + y^2) \Rightarrow a^2 + b^2 = 1 = k \Rightarrow$ mapping is isometry.
11. No: dilation is similitude with only one fixed point.
13. By similar triangles P' stays in line with O and $P \Rightarrow P' \in$ ray OP; line AP parallel to line $A'P' \Rightarrow k = OP'/OP = OA'/OA$ by parallel projection $\Rightarrow$ mapping $P \to P'$ is dilation $\mathcal{D}[O, k]$. (One doubles image by setting A = midpoint of $\overline{OA}$.)
15. $A \to D \Rightarrow -3a + 2b + x_0 = 10$, $-3a + 2b + y_0 = -9$. First equations in $B \to E$ and $C \to F$ yield $3a + 4b + x_0 = 28$, $a - 10b + x_0 = -18$ which, with $-3a + 2b + x_0 = 10$ yields unique values for a, b, x_0; in same manner second equations give c, d, y_0. Solution produces mapping $x' = 2x + 3y + 10$, $y' = 3x - 2y + 4$ which is form for similitude.

Section 8.4

5. In figure for this problem, θ and φ are equal to acute angles of right triangle $QRO \Rightarrow \varphi = 90 - \theta \Rightarrow$ angles between rays and transversal QR supplementary (to prove) $\Rightarrow$ two rays are parallel.

9. Incenter of $\triangle DEF$ = orthocenter H of $\triangle ABC$ (which can be easily constructed) $\Rightarrow$ one side of $\triangle ABC$ is line perpendicular to HD at D.

10.* It was shown in Example 3 that ST (base of isosceles $\triangle AST$) = p = perimeter of orthic triangle, that $AS = AT = AD = h_a$, and that $m\angle SAT = 2A \equiv 2m\angle BAC \Rightarrow \frac{1}{2} ST/SA = \cos(90 - \frac{1}{2}A) = \sin\frac{1}{2}A \Rightarrow p = ST = 2SA \sin\frac{1}{2}A = 2h_a\sin\frac{1}{2}A$.

11. $P(5.375, 0)$

17. (a) Let Q be any point on t; $\angle WPG \cong \angle QPF_2$ (vertical angles), $\angle QPF_2 \cong \angle F_1PG$ (from given) $\Rightarrow \angle WPG \cong \angle F_1PG \Rightarrow$ (from $PF_1 = PW$) line $PG = t$ is perpendicular bisector of segment $WF_1 \Rightarrow Q$ equidistant from F_1, W; $QF_1 + QF_2 = WQ + QF_2 > WF_2 = WP + PF_2 = PF_1 + PF_2 \Rightarrow Q \notin E \Rightarrow t$ meets E at only point $P \Rightarrow t$ tangent to E at P.

Section 8.5

3. A nontrivial isometry is a product of either one, two, or three reflections; since reflections are opposite, a product of odd number is opposite $\Rightarrow$ a direct isometry must be product of exactly two reflections $\Rightarrow$ it is either a translation or rotation

5. If O is a fixed point, (1) Section **8.3** reduces to $x' = ax - \varepsilon by$, $y' = bx + \varepsilon ay$; fixed point (x, y) exists iff $x = ax - \varepsilon by$, $y = bx + \varepsilon ay \Rightarrow (a - 1)x - \varepsilon by = 0$ and $bx + (\varepsilon a - 1)y = 0$, with $a^2 + b^2 = 1$; if $\varepsilon = -1$ there are multiple fixed points given by $y = bx/(a + 1) \Rightarrow \varepsilon = 1$, by hypothesis and mapping becomes $x' = ax - by$, $y' = bx + ay$, a rotation.

7. $22.6°$.

11. If P is a point on one side, rotate square S about P through angle of $60°$; rotated square S' will intersect S at some point Q', whose inverse image Q lies on $S \Rightarrow$ both Q, $Q' \in S$, $PQ = PQ'$, and $m\angle QPQ' = 60 \Rightarrow \triangle QPQ'$ is equilateral.

13. Draw line through center of D perpendicular to l, then translate C in direction of l so that its center is translated onto perpendicular; if translated circle C' intersects D at A' and B', then line $A'B'$ is desired line ($A'B' = AB$).

15. Perform half-turn on C; its image will intersect D at points A and B', the image of point B whose reverse image lies on C, with $AB = AB' \Rightarrow$ line AB' is desired line.

Section 8.6

3. $A' = (2, 2)$, $B' = (2, 0)$ (fixed point), $C' = (2, -2)$; circle through A, B, C passes through O and maps to line through $B' = B$ perpendicular to line OB ($x = 2$); points A', B', C' lie on this line, in agreement.

4.* (See Figure S.14.)

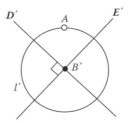

Figure S.14 Inverse image of diagram for Problem 4.

5. (See Figure S.15.)

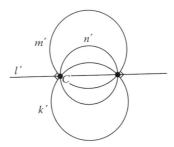

Figure S.15 Inverse image of diagram for Problem 5.

7. By definition of inversion, $(OP')^2 = r^4/(OP)^2 \Rightarrow$ (by distance formula) $(\lambda x)^2 + (\lambda y)^2 = r^4/(x^2 + y^2) \Rightarrow \lambda^2 = r^4(x^2 + y^2)^{-2} \Rightarrow \lambda = r^2(x^2 + y^2)^{-1} \Rightarrow$ $x' = \lambda x = r^2x(x^2 + y^2)^{-1}$, $y' = r^2y(x^2 + y^2)^{-1}$.

9. Circle D orthogonal to circle of inversion $x^2 + y^2 = r^2$ iff its equation is of form $x^2 + y^2 + ax + by + r^2 = 0$. Since inversion is self-inverse, for convenience, let given equation be written in primes and use direct substitution of **(9)** to find image D': $[r^2x/(x^2 + y^2)]^2 + [r^2y/(x^2 + y^2)]^2 +$ $a[r^2x/(x^2 + y^2)] + b[r^2y/(x^2 + y^2)] + r^2 = 0 \Rightarrow r^4x^2/(x^2 + y^2)^2 + r^4y^2/(x^2 +$ $y^2)^2 + ar^2x/(x^2 + y^2) + br^2x/(x^2 + y^2) + r^2 = 0$; multiply both sides by $(x^2$ $+ y^2)/r^2$: $r^2x^2/(x^2 + y^2) + r^2y^2/(x^2 + y^2) + ax + by + (x^2 + y^2) = 0 \Rightarrow r^2(x^2 +$ $y^2)/(x^2 + y^2) + ax + by + x^2 + y^2 = 0$; original equation results, proving $D' = D$.

11. In Figure S.16, let D be given circle; draw line l through center of inversion O and center of D cutting D at A and B, forming diameter AB of D. Let P be any point on D and P' its inverse (case (OAB) assumed, although there are other cases to consider). As P varies on D, $m\angle APB = 90 \Rightarrow$ (exterior angle equality) $m\angle OAP = 90 + m\angle OBP \Rightarrow$ (by **(4)**) $m\angle OP'A' = 90 + m\angle OP'B' \Rightarrow m\angle A'P'B' = m\angle OP'A' - m\angle OP'B' = 90 \Rightarrow P'$ varies on circle having diameter $A'B'$. Statement about tangents is true because lines through O and tangents to curves are invariant under inversion and lines.

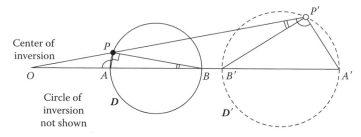

Figure S.16 Problem 11.

12.* By results in Example 2, circumcircle $[O, R]$ maps to $[U, \tfrac{1}{2} r]$ via incircle, whose distance is d from center of $[O, R]$ to center of circle of inversion I; relation between radii $a \equiv R$ and $b \equiv \tfrac{1}{2} r$ of circle and image circle with respect to radius of circle of inversion $c \equiv r$ and distance d is **(7)**, which becomes $\tfrac{1}{2} r = Rr^2/|R^2 - d| \Rightarrow 1 = 2Rr/(R^2 - d)$, equivalent to **(10)**.

13. Mechanism produces inverse points P and P': let W be foot of O on AB $\Rightarrow OP \cdot OP' = (OW - PW)(OW + PW) = OW^2 - PW^2 = OA^2 - AW^2 - (AP^2 - AW^2) = OA^2 - AP^2 = \text{constant} \Rightarrow P$ and P' are inverse points $\Rightarrow$ circle through O traced by P maps to line traced by P'.

Section 9.1

1. $c = 0.604$, $A = 0.644$ rad $\approx 36.9°$, $B = 1.595$ rad $\approx 146.8°$.

3. $A = C = 90°$, $c = \pi/2$.

5. $\cos c = \dfrac{2 - \sqrt{3}}{\sqrt{3}} = 0.1547005 \Rightarrow c = 1.415$, $a = 0.607$, $b = 1.726$.

9. In $\triangle ABD$: $\cosh b = \cosh d \cosh c_1 - \sinh d \cdot \sinh c_1 \cos (\pi - \theta) = \cosh d \cosh c_1 + \sinh d \sinh c_1 \cos \theta$; in $\triangle ACD$: $\cosh a = \cosh d \cosh c_2 - \sinh d \sinh c_2 \cos \theta$. Divide two equations respectively by $\sinh c_1$ and $\sinh c_2$, then sum to eliminate term $\sinh d \cos \theta$: $\cosh a/\sinh c_2 + \cosh b/\sinh c_1 = \cosh d(\cosh c_2/\sinh c_2 + \cosh c_1/\sinh c_1) = \cosh d/\sinh c$; by solving equation for $\cosh d$, desired relation **(3)** results.

11. From last step in hint: $(\sin a \sin b)Q = -\tan c_1 \tan c_2 \cos a \cos b + \sin^2 h \cos c = -\sin c_1 \sin c_2 (\cos a/\cos c_2)(\cos b/\cos c_1) + (1 - \cos^2 h) \cdot \cos (c_1 + c_2) = -\sin c_1 \sin c_2 \cos^2 h + \cos c - \cos^2 h (\cos c_1 \cos c_2 - \sin c_1 \sin c_2) = \cos c - \cos^2 h \cos c_1 \cos c_2 = \cos c - \cos a \cos b = \sin a \cdot \sin b \cos C \Rightarrow Q = \cos C$.

Section 9.4

1. With $a = b = c$, show that $s - b = s - c = \tfrac{1}{2} a$ and substitute into **(4)**. Section **9.3**.

3. Elliptic case: $\cos A = -0.25002 \Rightarrow 2^2 < 1^2 + 2^2 - 4 \cos A \Rightarrow 4 < 6.0007$; hyperbolic case: $\cos A = 0.29973 \Rightarrow 2^2 > 1^2 + 2^2 - 4 \cos A \Rightarrow 4 > 3.08256$.

5. $c^2 > d^2 + a_1^2 - 2da_1\cos(\pi - \theta) = d^2 + a_1^2 + 2da_1\cos \theta$; $b^2 > d^2 + a_2^2 - 2da_2\cos \theta \Rightarrow c^2/a_1 + b^2/a_2 > d^2/a_2 + d^2/a_1 + a_1 + a_2 \Rightarrow c^2/a_1 + b^2/a_2 - a > d^2(a/a_1a_2)$; solving for d^2 produces relation **(7)**.

7. In $\triangle ABC$, $f^2 < \frac{2}{3}a^2 + \frac{1}{3}b^2 - \frac{2}{9}c^2$; in $\triangle ACD$, $d^2 < \frac{2}{3}f^2 + \frac{1}{3}AD^2 - \frac{2}{9}b^2 < \frac{2}{3}(\frac{2}{3}a^2 + \frac{1}{3}b^2 - \frac{2}{9}c^2) + \frac{1}{3}(\frac{2}{3}c)^2 - \frac{2}{9}b^2 = \frac{4}{9}a^2 + \frac{2}{9}b^2 - \frac{4}{27}c^2 + \frac{4}{27}c^2 - \frac{2}{9}b^2 = \frac{4}{9}a^2 \Rightarrow d < \frac{2}{3}a$.

9. $\tan \frac{1}{2}B = \tan r/\sin x = \tan r/\sin(s - b) \Rightarrow \tan r = \sin(s - b) \cdot$

$[\sin\frac{1}{2}B/\cos\frac{1}{2}B] = \sin(s-b) \cdot \sqrt{\dfrac{\sin(s-a)\sin(s-c)}{\sin s \sin(s-b)}}$ which simplifies

to desired identity.

Section 9.5

7. $\tan \frac{1}{2}K' = \tan\frac{1}{2}(A + B + C - \pi) = \dfrac{\sin\frac{1}{2}(A + B - \pi/2)}{\cos\frac{1}{2}(A + B - \pi/2)} =$

$\dfrac{\frac{\sqrt{2}}{2}\sin\frac{1}{2}(A + B) - \frac{\sqrt{2}}{2}\cos\frac{1}{2}(A + B)}{\frac{\sqrt{2}}{2}\cos\frac{1}{2}(A + B) + \frac{\sqrt{2}}{2}\sin\frac{1}{2}(A + B)} = \dfrac{\sin\frac{1}{2}(A + B) - \cos\frac{1}{2}(A + B)}{\sin\frac{1}{2}(A + B) + \cos\frac{1}{2}(A + B)} =$

$\dfrac{\cos\frac{1}{2}C\cos\frac{1}{2}(a' - b') - \sin\frac{1}{2}C\cos\frac{1}{2}(a' + b')}{\cos\frac{1}{2}C\cos\frac{1}{2}(a' - b') + \sin\frac{1}{2}C\cos\frac{1}{2}(a' + b')} =$

$\dfrac{\cos\frac{1}{2}(a' - b') - \cos\frac{1}{2}(a' + b')}{\cos\frac{1}{2}(a' - b') + \cos\frac{1}{2}(a' + b')} =$

$\dfrac{-2\sin\frac{1}{2}a'\sin(-\frac{1}{2}b')}{2\cos\frac{1}{2}a'\cos(-\frac{1}{2}b')} = \tan\frac{1}{2}a'\tan\frac{1}{2}b'$, where, throughout, $a' = \kappa a$,

$b' = \kappa b$.

9. Perimeter: In $\triangle OAM$, $\sin \pi/n = \sin \kappa AM/\sin \kappa r \Rightarrow \sin \kappa P/2n = \sin \kappa r \cdot \sin \pi/n$. Area (using **(7)** for $\triangle OAM$ and $K/k = 2n \cdot$ Area $\triangle OAM/k$): $\tan K'/4n = \tan\frac{1}{2}\kappa a \tan\frac{1}{2}\kappa AM = \tan\frac{1}{2}\kappa a \tan\frac{1}{2}\kappa P/4n$.

10.* Circumference: Let $n \to \infty$ in following equations after multiplying

by n, and use $P_n \to C$; $\dfrac{\kappa P_n}{2n} \cdot \dfrac{\sin \kappa P_n/2n}{\kappa P_n/2n} = \sin \kappa r \cdot \dfrac{\pi}{n} \cdot \dfrac{\sin \pi/n}{\pi/n}$; using the

limit $\sin x/x \to 1$ as $x \to 0$, this becomes, in the limit, $\dfrac{\kappa C}{2} = \sin \kappa r \cdot \pi \Rightarrow$

$\kappa C = 2\pi \sin \kappa r$. Area: $\dfrac{K'_n}{4n} \cdot \dfrac{\tan K'_n/4n}{K'_n/4n} = \tan \kappa a_n/2 \cdot \dfrac{\kappa P_n}{4n} \cdot \dfrac{\tan \kappa P_n/4n}{\kappa P_n/4n}$;

$(n \to \infty)$, $P_n \to C$, $K_n \to A$, $a_n \to r$, and $\tan x/x \to 1$ as $x \to 0 \Rightarrow$

$\dfrac{A/k}{4} = \tan\frac{1}{2}\kappa r \cdot \dfrac{\kappa C}{4} = \tan\frac{1}{2}\kappa r \cdot \dfrac{2\pi \sin \kappa r}{4} \quad \Rightarrow \quad A = 2k\pi\tan\frac{1}{2}\kappa r \cdot$

$(2 \sin \frac{1}{2} \kappa r \cos \frac{1}{2} \kappa r) = 4k\pi \sin^2 \frac{1}{2} \kappa r$.

Section 9.6

3. If a and b are coordinates of A and B on line l, by coordinatization theorem for lines, $|AB| = |b - a| = AB$ if $|b - a| \leq \pi$ (when $|AB| \leq \pi$), but if $|AB| = |b - a| > \pi$, then $AB = 2\pi - |b - a| = 2\pi - |AB|$.

4.* (Unified proof). Since D is collinear with B and C, Stewart's theorem is $S(AB)C(CD) + S(BD)C(CA) + S(DA)C(CB) = 0 \Rightarrow S(AB)\ C(d) = S(DB)C(b) + S(AD)C(a) \Rightarrow C(d) = [S(DB)/S(AB)]C(b) = [S(AD)/S(AB)]\ C(a) = pC(b) + qC(a)$.

5. $\mathcal{L}\begin{bmatrix} ABC \\ DEF \end{bmatrix} = \dfrac{sAF}{sFB} \cdot \dfrac{sBD}{sDC} \cdot \dfrac{sCE}{sEA}$ and $\mathcal{L}\begin{bmatrix} ACB \\ DFE \end{bmatrix} = \dfrac{sAE}{sEC} \cdot \dfrac{sCD}{sDB} \cdot \dfrac{sBF}{sFA}$

which is observed to be reciprocal of former.

7. $\mathcal{R}[AB, CD] = \dfrac{sAC \cdot sBD}{sAD \cdot sCD} = \mathcal{R}[AB, CX] = \dfrac{sAC \cdot sBX}{sAX \cdot sBC} \Rightarrow \dfrac{sBD}{sCD} = \dfrac{sBX}{sBC} \Rightarrow$

$X = D$ or $X = D^*$ (Lemma 1).

9. With (BAD), in $\triangle AA'D$ and $\triangle BB'D$ we have $\dfrac{S(a)}{S(AD)} = \sin\theta = \dfrac{S(b)}{S(BD)}$

which yields the desired result; if (BDA), angles at D are vertical

angles $\Rightarrow \dfrac{S(a)}{S(AD)} = \sin\theta = \dfrac{S(b)}{S(DB)}$; (the two results are different only

if signed distances are used).

11. If D, E, F are collinear, by Problem 10 linearity number $= -1$; con-

versely, suppose $\mathcal{L}\begin{bmatrix} ACB \\ DFE \end{bmatrix} = -1$; by Lemma 2 line determined by two

menelaus' points, say D and E, intersects third side (in this case, AC) at

some point X. By first part, $-1 = \mathcal{L}\begin{bmatrix} ABC \\ DEX \end{bmatrix} = \mathcal{L}\begin{bmatrix} ABC \\ DEF \end{bmatrix} \Rightarrow$ (by property

(7)) $X = F$ or $F^* \Rightarrow$ line DE passes through F.

13. In triangles DCE and ACD, $\tan C = \dfrac{T(e)}{S(CE)} = \dfrac{T(d)}{S(DC)} \Rightarrow \dfrac{sCE}{sDC} = \dfrac{T(e)}{T(d)}$;

similarly for other two ratios in linearity number.

15. Observing Figure S.17, in $\triangle PAB$ cevians PC, AQ, BS are concurrent at

$R \Rightarrow \mathcal{L}\begin{bmatrix} PAB \\ CQS \end{bmatrix} = 1$; in $\triangle PBA$ menelaus' points S, Q, D are collinear $\Rightarrow$

$\mathcal{L}\begin{bmatrix} PBA \\ DSQ \end{bmatrix} = -1$; product gives $\dfrac{sPS}{sSA} \cdot \dfrac{sAC}{sCB} \cdot \dfrac{sBQ}{sQP} \cdot \dfrac{sPQ}{sQB} \cdot \dfrac{sBD}{sDA} \cdot \dfrac{sAS}{sSP} = -1$

which (by cancellation) reduces to $\dfrac{sACsBD}{sCBsDA} = \mathcal{R}[AB, CD] = -1$.

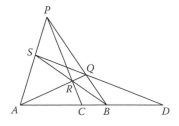

Figure S.17 Problem 15.

17. But for notation, same as Problem 14, Section **7.6**.

Section 9.7

3. (See Figure S.18.)

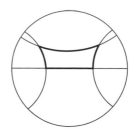

Figure S.18 Problem 3.

7. (AOB) $\Rightarrow$ $(AB) = (AO) - (OB)$ $\Rightarrow$ $\sinh(AB) = \sinh(OA) \cdot$
$\cosh(OB) + \cosh(OA)\sinh(OB) = \dfrac{2a}{1-a^2} \cdot \dfrac{1+b^2}{1-b^2} + \dfrac{1+a^2}{1-a^2} \cdot \dfrac{2b}{1-b^2} =$
$\dfrac{2(a+b+ab^2+a^2b)}{(1-a^2)(1-b^2)} = \dfrac{2(a+b)(1+ab)}{(1-a^2)(1-b^2)}.$

9. Take $D[Q, s]$ as circle of inversion, which defines reflection in line m (Q is center of circular arc m). One need only prove A maps to O in order that $(MA) = (MO)$; in right triangle OUQ, UA is altitude to hypotenuse $\Rightarrow UA^2 = AQ \cdot OA$, with $s = QU$. Observe: $AQ \cdot OQ = AQ$ $(OA + AQ) = UA^2 + AQ^2 = QU^2 = s^2$, proving that A and O are inverse points.

Section 9.8

1. $(AB) = 10$, $(AC) = \sqrt{218} \approx 14.76$, $(AD) = \sqrt{212} \approx 14.56$ (see Figure S.19).

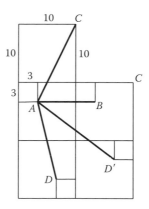

Figure S.19 Problem 1.

7. **(a)** This circle is intersection of plane $y = x$ and sphere $\Rightarrow u = \pi/4 \Rightarrow x = (\sqrt{2}/2)\sin v$, $y = (\sqrt{2}/2)\sin v$, $z = \cos v$ (v = parameter). **(b)** $z = x$, $x^2 + y^2 + z^2 = 1$. **(c)** Not unless $v = \pi/2$.

Section 9.9

1. 18.6 for angle of one millionth degree [use (**2**)].

3. $\tan \frac{1}{2} \gamma = e^{-\kappa a} \Rightarrow \dfrac{1-\cos\gamma}{\sin\gamma} = \dfrac{\sin\gamma}{1+\cos\gamma} \Rightarrow \cosh \kappa a = \dfrac{e^{\kappa a} + e^{-\kappa a}}{2} =$

    $\frac{1}{2}\left(\dfrac{1+\cos\gamma}{\sin\gamma} + \dfrac{1-\cos\gamma}{\sin\gamma} \right) = \dfrac{1}{\sin\gamma} \Rightarrow \sin\gamma = \operatorname{sech} \kappa a.$

4.* In right triangle ACE, $\cos \varphi = \dfrac{\tanh \kappa x}{\tanh \kappa c} = \sin\theta \Rightarrow$ (by (**15**) Section **6.12**)

    $\cosh \kappa a = \tanh \kappa c/\tanh \kappa x \Rightarrow \sin\theta = \operatorname{sech} \kappa a \Rightarrow \theta = p(a).$

7. Arcs are, in order mentioned: ordinary circle in S, circle in S tangent to C, and arc of circle (or limiting case) in S that intersects C at two distinct points.

9. Recall Problem 9, Section **9.7**; it was shown that chord UV was reflected in h-line m to circular arc passing through $O \Rightarrow$ chord is reflection of equidistant locus so is itself an equidistant locus. (If points of line n are equidistant from a set of points E, reflection in h-line preserves equidistant behavior, and maps n to another line $\Rightarrow$ image set E' is also equidistant locus with respect to line n'.

Bibliography

Books

Bonola, R. 1955. *Non-Euclidean Geometry*. New York: Dover Publications.

Burton, D.M. 1998. *History of Mathematics: An Introduction*, 2nd edn. New York: McGraw Hill Publishers.

Coxeter, H.S.M. 1992. *The Real Projective Plane*, 3rd edn. New York: Springer.

Coxeter, H.S.M. 2001. *Introduction to Geometry*, 2nd edn. (paperback). New York: Wiley.

Coxford, A., Z. Usiskin, and D. Hirschorn. 1991. *The University of Chicago Study Mathematics Project, Geometry*. Glenview: Scott, Foresman and Company.

Eves, H. 1963. *A Survey of Geometry*, Vol. 1. Boston, MA: Allyn and Bacon.

Faber, R.L. 1983. *Foundations of Euclidean and Non-Euclidean Geometry*. New York: Marcel Dekker, Inc.

Forder, H.G. 1953. *Coordinates in Geometry*, Bulletin No. 41. New Zealand: Auckland University College.

Greenberg, M.J. 2009. *Euclidean and Non-Euclidean Geometries*, 4th edn. New York: W.H. Freeman and Company.

Heath, T.L. 1956. *The Thirteen Books of Euclid's Elements*, Vol. 1. New York: Dover Publications.

Heilbron, J.L. 1998. *Geometry Civilized: History, Culture, and Technique*. Oxford, U.K.: Clarendon Press.

Hilbert, D. 1902. *The Foundations of Geometry*. Chicago, IL: The Open Court Publishing Company.

Jacobs, H. 2003. *Geometry*. New York: W.H. Freeman and Company.

Moise, E.E. 1990. *Elementary Geometry from an Advanced Standpoint*, 3rd edn. Reading, MA: Addison Wesley Longman.

Palmer, C.I. and C.W. Leigh. 1914. *Plane and Spherical Trigonometry*. New York: McGraw Hill Book Company.

Proclus. 1992. *A Commentary on the First Book of Euclid's Elements*, reprinted. trans. G.R. Morrow, with notes and introduction. Princeton, NJ: Princeton University Press.

Struik, D.J. 1950. *Lectures on Classical Differential Geometry*. Reading, MA: Addison Wesley Publishing Company, Inc.

Shultz, J.E., K.A. Hollowell, W. Ellis Jr., and P.A. Kennedy. 1953. *Geometry*. Austin, TX: Holt, Rinehart and Winston.

Wolfe, H.E. 1945. *Introduction to Non-Euclidean Geometry*. New York: Holt Rinehart and Winston.

Journal Articles

Lockwood, E.H. 1953. Simson's line and its envelope. *Mathematical Gazette* **37**: 124–125.

Milnor, J. 1996. Hyperbolic geometry: The first 150 years. *The Mathematical Heritage of Henri Poincaré, Proceedings of Symposia in Pure Mathematics of the American Mathematical Society* **39**: 25–40.

Index

Geometric Symbols

AB (distance), 19
α (maximum distance), 25
$\overleftrightarrow{AB}$ (line), 26
$\overrightarrow{AB}$ (ray), 29
AB, segment, 29
A^* (polar, extremal opposite), 37, 88
$\angle ABC$ (angle), 53
hk (angle), 53
$m\angle ABC$ (angle measure), 53
hk (angle measure), 53
β (maximum angle measure), 55
$\perp$ (perpendicular), 74
ΔABC (triangle), 95
$\cong$ (congruent), 97
$\Diamond ABCD$ (quadrilateral), 162
γ (angle of parallelism), 231
$p(a)$ (angle of parallelism function), 235
$\parallel$ (parallel), 251
$\sim$ (similar), 262
$<$ (inequality for unified
              geometry), 281
$c(\theta)$ (cosine function), 300–301
$s(\theta)$ (sine function), 307
$C(u)$ (cosine, hyperbolic cosine), 317
σ (sign indicator), 322

Theorems and Identities

(Note: Numbers in brackets indicate
theorem number)

Section 1.6
    Theorem, 26
    (1), 25

Section 1.7
    Theorem 1, 28
    Theorem 2, 29
    Theorem 3, 31

Section 1.8
    Betweenness for 3 collinear points
        [2], 37
    Coordination of lines [5], 41
    Corollaries A, B, 38
    Theorem 3, 38
    Theorem 4, 40
    (1)–(10), 36–37
    (11)–(12), 38
    (13)–(14), 39–40

Section 1.10
    Extension construction, 48
    Midpoint construction, 48
    Opposite ray theorem, 49
    Segment construction, 47
    (1)–(2), 50–51
    (3), 52

Section 2.3
    Corollary, 64
    Postulate of Pasch, 66
    Theorems 1–2, 64
    Theorems 3–4, 65

Section 2.4
    Converse of Axiom 10, 73
    Criterion for perpendicularity, 74
    Crossbar theorem, 74
    Linear pair theorem, 71
    Theorem 2, 72
    Theorem 3, 73
    Theorem 6, 77
    Theorem (Problem 15), 81
    (1)–(3), 69

Section 2.5
    Angle bisection, 84
    Angle construction, 83
    Circular protractor, 85

Section 2.6
    Existence of triangles, 89
    Theorem 1, 88
    Corollaries A–C, 88
    Corollary D, 89

Section 3.2
    (1), 100

Section 3.3
    ASA criterion, 110
    Isosceles triangle theorem, 112
    SSS congruence theorem, 112

Section 3.4
    SAS congruence theorem, 115

Section 3.5
    Perpendicular bisector theorem
        [1], 117
    Theorem 2, 119
    Theorem 3, 122
    Corollary A, 120
    Corollaries B–D, 121

Section 3.6
    Exterior angle inequality [1], 130
    Theorem 2, 132
    Corollaries A–B, 131
    Corollaries C–E, 132
    (1), 129

Section 3.7
    SAS inequality, 137
    Steiner–Lehmus theorem, 143
    Scalene inequality, 135
    Triangle inequality, 133
    Theorem 1, 135
    Theorem 2, 136
    Theorem 3 (Problem 24), 143
    Corollary, 137
    (1), 133
    (2)–(3), 134–135

Section 3.8
    AAS congruence criterion, 144
    Angle-bisector, 149
    HA theorem, 148
    HL theorem, 147
    LA theorem, 147
    SSA theorem, 146

Section 3.9
    Theorem 1, 155
    Theorem 2, 156

Section 4.1
    Theorem 1, 163
    Theorem 2, 164
    Corollary, 164

Section 4.2
    ASASA congruence, 165
    SASAS congruence, 165

Section 4.3
    Saccheri's theorem, 172
    Theorem 1, 166
    Theorem 2, 167
    Theorem 3, 168
    Theorems 4–5, 171
    Corollary A, 172
    Corollary B, 173

Section 4.4
    Existence of regular polygon, 180
    Theorem of Gauss, 178
    Theorem (Problem 13), 185
    Corollary, 181

Section 4.5
    Secant theorem [6], 193
    Two-circle theorem
        (Problem 25), 201
    Theorem 1, 190
    Theorem 2, 192
    Theorem 4 (Problem 12), 197

Theorem 5 (Problem 13), 197
    Corollaries A–B, 191
    Corollary C, 192
    Corollary D, 194
    (1), (2), 189

Section 5.1
    Legendre's First Theorem, 208
    Legendre's Second Theorem, 210
    Theorem 1, 205
    Theorem 2, 206
    Theorem 3, 207
    Corollaries A–B, 206

Section 5.2
    AAA congruence, 213
    Additivity of excess, 212
    Angle-sum theorem, 210
    Theorem 1, 210
    Theorem 6, 213

Section 5.3
    Theorem 1, 215
    Theorems 2–3, 216
    Corollary, 215

Section 5.4
    Distance-angle theorem, 219

Section 5.5
    Angle-sum theorem, 228
    Exterior angle inequality, 228
    Corollary B, 228

Section 5.6
    AAA congruence, 231
    Theorem 1, 229
    Theorem 2, 231

Section 5.7
    Theorem 1, 233
    Theorem 2, 235

Section 5.8
    AA, BA theorems, 244
    Hilbert's theorem, 251
    Symmetry of parallels, 243
    Transitivity of parallels, 243
    Theorem (Problem 15), 248
    Corollary, 244

Section 5.9
    Alternate interior angles [1], 252
    Angle-sum theorem, 253
    Corresponding angles, 252
    Exterior angle theorem, 253
    (1), 255

Section 5.10
    Parallel projection, 259–260

Section 5.11
    AA similarity criterion, 262
    Orthic triangle theorem, 266

SAS similarity criterion, 263
Side-splitting theorem, 260
SSS similarity criterion, 263

Section 5.12
Morley's theorem, 275
Theorem (Problem 25), 273
Theorem (Problem 26), 274
(1)–(7), 266–267

Section 6.1
Angle-sum theorem, 282

Section 6.2
Midline bisector theorem, 286
Side-splitter theorems, 289, 294
(1)–(5), 285

Section 6.3
Side-splitter theorem, 295

Section 6.4
Inequality result
        (Problem 14), 304

Section 6.5
(1)–(5), 307–308
(6)–(7), 309

Section 6.6
(1)–(5), 312–313

Section 6.7
(1), 314
(2)–(6), 315–316

Section 6.8
(1)–(2′), 317–318
(6)–(7), 319–320

Section 6.9
(1)–(4), 321

Section 6.10
(1)–(2), 322
(3)–(7), 323
(8)–(10), 324
(11)–(12), 325

Section 6.11
Pythagorean relation, 327

Section 6.12
(1)–(3), 328
(4)–(10), 329–330
(11), 334
(12)–(15), 336–337
(16)–(17), 338–339

Section 7.1
Cevian formula, 345
Euler's identity (1), 343
Stewart's theorem (2), 343
Theorem, 342
Corollary, 342

Section 7.2
Altitude formulas (3), 350
Angle bisector formulas (2), 348
Heron's formula (4), 350
Median formulas (1), 347
(5)–(7), 352
(8)–(12), 353

Section 7.3
Inscribed angle theorem, 357
Power theorem, 355
Secant-tangent theorem, 356
Two-chord theorem, 356
Two-secant theorem, 356
Corollaries A–B, 358
Corollaries C–D, 359

Section 7.4
Nine-point circle theorem
        [3], 362
Ptolemy's theorem [2], 360
Theorem 1, 360
Theorem 4, 364

Section 7.5
Theorem 1, 375
Theorem 2, 379
Theorem 3, 379
Corollary, 375
(1)–(4), 377

Section 7.6
Ceva's theorem, 386
Menelaus' theorem, 385
Theorem of Pappus, 398
(1)–(2), 382
(3)–(5), 383

Section 7.7
Theorem 1, 401
Theorem 2, 403
Theorem 3, 405
Corollary, 405
(1), 399
(2)–(4), 404–405
(5)–(7), 406
(8)–(9), 409–410

Section 8.1
(1), 414

Section 8.2
Fundamental theorem, 431
Theorems 1–2, 427
Theorem 3, 430
Theorem (Problem 15), 436
Corollaries A–B, 427
Corollaries C–D, 428
Corollary E, 431
(1), 432
(2), 435

Section 8.3
    Theorem, 443
    (1)–(2), 439
    (3), 443

Section 8.4
    Fagnano's theorem [1], 454
    Fundamental theorems, 456
    Normal equation of line, 462
    Corollaries A–B, 456
    (1)–(2), 455–456
    (3)–(4), 457–458

Section 8.5
    Theorem, 465
    (1), 466
    (2)–(3), 468

Section 8.6
    Theorems 1–2, 480
    Theorem 3, 481
    (1)–(6), 478–479
    (7)–(8), 484–485
    (9)–(10), 487

Section 9.1
    Cevian formula, 495
    Law of cosines (2), 493
    Law of sines (1), 491

Section 9.2
    (1)–(8), 496
    (9)–(10), 497

Section 9.3
    (1), 498
    (2), 499
    (3)–(5), 500

Section 9.4
    Cosine inequality (3), 503
    Median inequality (4), 503
    Theorem 1, 501
    Theorem 2, 503
    (5), 504

Section 9.5
    Equations of Gauss (3), 508
    (1)–(2), 508
    (5)–(6), 510
    (7)–(9), 513–514

Section 9.6
    Ceva's theorem, 522
    Menelaus' theorem, 522
    Corollary, 522
    (1)–(5), 515
    (6)–(7), 519

Section 9.7
    Theorem 1, 535
    Corollary, 535

Section 9.8
    Theorem, 544

General

Absolute geometry, 95
Acute angle
    definition, 71
    hypothesis of, 169
Addition identities for sine, cosine, 314
Addition identity, *see* Trigonometric
        identities
Additivity of
    arc measure, 368
    defect, 229
    excess, 212
Adjacent angles, 71
Affine plane
    9-point model, 13–14
    postulates for, 13
Affine reflection, 429
Affine transformation, 425–426
    coordinate form, 432
    fundamental theorem, 431
    invariants, 434
Aliquot triangle, *see* Cevian triangle
Alligator, hinge theorem, *see* SAS
        inequality
Alternate interior angles, 151,
        205–206, 252
Altitudes
    concurrence of, 265, 525
    formula for, 350
    normal trapezoid, 285
    triangle, 22, 154, 274
Angle, 53
    acute, 71
    addition postulate, 71
    bisector, 84
    central, 187, 356
    construction postulate, 83
    degenerate, 53
    doubling theorem, 92
    of incidence, of reflection,
        448–449
    inscribed, in a circle, 187, 356
    inscribed in a semi-circle,
        187–188, 358
    measure of, 54
    obtuse, 72
    of parallelism, 232–236
    of a polygon, 177
    related to distance, 217, 221
    right, 71
    straight, 53–54
    of a triangle, 96
    trisectors, 91, 184
Angles about a point, 93
Angle-sum, 131, 171
    elliptic geometry, 210
    Euclidean geometry, 253
    hyperbolic geometry, 224, 228
    Saccheri's theorem, 172

Apothem, 182, 513
Archimedean principle, 208, 218, 261, 577
Archimedes (217–212 BC), 217–218
Arc measure
    additivity, 356, 368
    definition, 367
Arc of a circle, 187, 192, 357, 368
Area
    circle, 514
    concept, elliptic geometry, 212
    concept, hyperbolic geometry,
        228–229
    cyclic quadrilateral, 359–360, 373–374
    regular polygon, 182, 513
    triangle, Euclidean, 350, 353, 499,
    triangle, non-Euclidean, 510, 513
Area-preserving transformation, 437
Astroid, see Hypocycloid of 4 cusps
Asymptotic
    parallels, 232, 237, 249, 548
    triangle, 236–244
Axiomatic spherical geometry, 87
Axiom M, 115–116, 455
Axioms, axiomatic systems, 2, 30
    categorical, 5
    consistency of, 5
    Hilbert's postulates, 589–590
    independence of, 5
    Moise's system, 588–589
    unified geometry, 587–588
Axis, 114, 429
Base, base angles, of
    asymptotic triangle, 237–239
    isosceles triangle, 111
    normal trapezoid, 285
    Saccheri quadrilateral, 166
    trapezoid, 256
Belongs to, 3
Beltrami, Eugenio (1835–1900), 526, 539
Beltrami–Klein model, 526
Beltrami–Poincaré half-plane model,
        539, 542
Beltrami's pseudosphere, 542
Betweenness
    collinear points, 27, 36
    concurrent rays, 55
    transformations and, 415, 425–426
Bisector of angle, 84
Bisector, perpendicular, 21–22, 117
Bisectors of a triangle, 154
    concurrence of, 156
    formula for, 348
Bolyai, János (1802–1860), 95, 226, 236,
        306, 526
Bolyai–Lobachevski formula, 546
Born, Max, 102
Brahmagupta (c. 600), 373
Carom, 460
Cartesian coordinates, see Rectangular
        coordinates

Categorical system, 5
Cauchy sequence, 577
Center of
    circle, 186
    inversion, 477
    projection, 412
    rotation, 464
Central angle of a circle, 187, 356
Central projection, 412
Central reflection (half-turn), 475
Centroid of a triangle, 23, 264, 274
Ceva, Giovanni (1647–1736), 344, 381
Ceva's theorem, 386, 522, 524
Cevian
    definition, 344
    formula, 345, 495, 517, 523
    inequality, 51–52, 129, 506
    triangle, 382
Chord of a circle, 187
Circle, 103–104, 186–187
    area, 514
    circumference, 514
    elementary properties, 187–188
    equation for, 404
    exterior of, 189, 191
    inscribed angle, 192
    interior of, 189, 191
    power with respect to, 353,
        364–365
    regular polygons, and, 179–181
Circle theorems
    inscribed angle, 357
    secant, 193
    secant-tangent, 356
    tangent, 197
    two-chord, 356
    two-secant, 356
Circular arc, see Arc of a circle
Circular functions, 277
Circular inversion, see Inversion
Circular protractor theorem, 85, 93
Circumcenter, triangle, 24, 156, 264
Circumcircle
    regular polygon, 181
    triangle, 156
Circumference of a circle, 25, 39,
        514, 569
Classical forms for
    $c(\theta)$, 314–316
    $C(x)$, 322, 325
    $s(\theta)$, 316
Closed half-plane, 63
Closed interior of
    angle, 71
    circle, 189
Closed polygonal path, 162
Collinear, noncollinear points, 20, 27,
        37–38
Combinatorial mathematics, 8,
Common notions, in the *Elements*, 1

Compass, straight-edge construction,
  see Euclidean construction
Complementary angles, 71
Completeness, property of, 577
Composition (product), 414
Concentric circles, 25
Concurrence of
    altitudes, 22, 155, 274, 522
    angle bisectors, 156
    medians, 22–23, 263
    perpendicular bisectors, 24–25, 156,
        249, 550
Concurrence of parallel lines, 11, 549
Concurrent rays, 55
Concyclic points, 359
Cone, 543
Confocal conics, 399
Conformal, 437, 480
Congruence, congruent
    angles, 72, 97
    equivalence relation, 96
    quadrilaterals, 165
    segments, 47, 97
    triangles, 97
Congruence criteria for quadrilaterals
    ASASA, 165
    SASAA, 166, 175
    SASAS, 165
    SSAAA, 166, 176
    SSASA, 174
    SSSSA, 166
Congruence criteria for right triangles
    HA, 148
    HL, 147
    LA, 147
Congruence criteria for triangles
    AA, 244
    AAA, 213, 231
    AAS, 144
    ASA, 110
    BA, 244
    SAS (Axiom 12), 110
    SsA, 153
    SSA, 146
    SSS, 112, 127
Conjugate, see Harmonic
Consecutive vertices
    polygon, 162, 176
    polygonal path, 162
    quadrilateral, 162
Consistent
    axioms, 5
    hyperbolic geometry, 535
Constant of proportionality, 262;
        see also Similar triangles
Construction, see Euclidean
        construction
Continuity of
    $c(\theta)$, 584
    $C(x)$, 586

Continuum of points, 46–47
Contrapositive, 63
Converse of
    Axiom 10, 73
    Pythagorean theorem, 273
    triangle inequality, 201
    vertical pair theorem, 91
Convex
    polygon, 177
    quadrilateral, 163–164, 166, 360
Convex set, 60, 189
Coordinate equations; see also
        Equations of Coordinate
        forms for transformations
    affine, 432
    dilation (homothetic), 443
    inversion, 487
    isometry, 439
    projectivity, 414
    reflection in a line, 457–458
    rotation, 467–468
    similitude, 439
    translation, 466
Coordinates for
    concurrent rays, 85
    points on a line, 39, 41
    E^2, rectangular, 276
    E^2, polar, 278
    E^3, 422
    sphere, 540, 571
Correspondence, see Transformations
Corresponding angles, 205–206, 252
Corresponding parts in triangles, 97–98
Corresponding points (cycles), 549
Cosine function, 277
Cosines, law of, 279
Countable set, 45
CPCPC, 98
Crossbar theorem, 74
Cross ratio, 374–375, 517; see also
        Generalized cross ratio
Cycle, 549
Cyclic
    group, 476–477
    quadrilateral, 359–360
Cylinder, 542
Dart, 127
Defect, 228–229
Degenerate angle, 53
Degree measure, 54
Deltoid, 393
Dense, 45
Desargues, theorem of, 16, 397, 413, 525
Determinant, 415, 421
Diagonals of a quadrilateral, 162
Diameter of a circle, 187
Dilation transformation, 442
Directed distance, 341, 515
    cevian formula, 345
    Euler's identity, 343

properties of, 342
 Stewart's theorem, 343
Direction numbers, 423–425
Direct, indirect reasoning, 63
Direct, opposite transformation, 437
Discrete
 metric, 20
 plane, 45–46
Disk, 189
Distance, 18–19, 25–26; *see also* metric
 angle, related to, 221
 directed, 341
 of parallelism, 234
 in Poincare's model, 528
 point to line, 121, 463
 point to ray, 148
 on a sphere, 18, 26, 569
Distance-preserving, *see* Isometry
Divergent parallels, 234, 237, 548
Dot product, *see* Inner product
Double elliptic geometry, 62, 216
Duality, 55–56, 77, 81, 85
Dyadic rationals, 52, 290–291,
      315, 325
Einstein, Albert (1879–1955), 102, 540
Elements, see Euclid's *Elements*
Ellipse, 103, 105, 463
Elliptic geometry, 87, 204, 210–221
 angle related to distance, 217–221
 angle-sum theorem, 210
 law of cosines, 493
 law of sines, 491
 orthogonal (parallel) projection, 298
 pole-polar theory, 214–217
 trigonometry of right triangle, 330
Empty set, 3
End-point(s), 29
Equal, 96
Equation(s) of, *see also* Coordinate
      forms
 circle, 404
 confocal conics, 399
 line, 424–425, 458, 463
 orthogonal circles, 400, 405–406
 plane, 424
Equations of Gauss, 507–508
Equator, 570
Equidistance, 117, 148–149, 159
Equidistant curve (locus), 108, 549, 552
Equivalence, relation, 96
Euclid (c. 300 BC), 3, 27, 111, 128–129,
      178, 526
Euclidean construction for
 angle trisectors, 91, 184
 arithmetic mean, 272
 equilateral triangle, 180, 183
 geometric mean, 33, 272, 367
 golden ratio, 33
 harmonic conjugate, 389
 regular pentagon, 33, 269

regular polygon, *see* Gauss, theorem of
 tangent to circle, 361, 381
 tangent to ellipse, 463
Euclidean exterior angle theorem, 253
Euclidean geometry, 2, 25, 204, 251
 angle-sum theorem, 253
 law of cosines, 279
 law of sines, 278–279
 parallel projection, 259–260
 Pythagorean theorem, 266–267
 similar triangles, 260, 262
 trigonometry, 268
Euclid's *Elements*, 1–2, 27–28, 96, 113,
      170, 183
Euclid's Fifth Postulate, 1, 169, 525–526
Euler line, 155, 264–265, 444
Euler's
 formula, 487
 identity, 343, 523
Eves, Howard, 21
Excess, 210
Excircle, 363
Exhaustion, method of, 217; *see also*
      Archimedes
Existence of regular polygon, 180
Extended, extension of
 Euclidean plane, 11
 hyperbolic plane, 548
 segment, 48
Exterior angle of a triangle, inequalities
 definition, 129
 elliptic geometry, 224
 Euclidean geometry, 253
 hyperbolic geometry, 228
 unified geometry, 241
Exterior point of
 circle, 189, 191
 polygon, 186
Extremal opposites, 88; *see also* Poles,
      Polar opposites
Fagnano, J.F., 453
Fagnano's theorem, 454
Fano's geometry, 10
Fejr, L., 453
Fermat, Pierre de, (1601–1665), 179
Fermat
 point (of a triangle), 391, 470
 prime (numbers), 178–179
Feuerbach's theorem, 363
Finite compactness, 46
Finite geometry, 6, 8
 9-point affine plane, 13–15
 7-point projective plane, 10
 13-point projective plane, 16
Fixed point of a transformation, 414, 417
Floating quadrilaterals, 318, 584–586
Floating triangles, 308, 581–584
Foot (of perpendicular), 121
Forder, H.G. (1889–1981), 306
Formula for geographical distance, 571

Fundamental theorem of
  affine transformation, 431
  isometry, 456
  projective transformations, 421
  similitudes, 456
γ (angle of parallelism), 231, 236,
  244, 546
Gardner, Martin, 124
Gauss, Carl Fredrich (1777–1855), 178,
  526, 539
  equations of, 507–508
  theorem of, 178
Generalized cross ratio, 481, 528
General linear group, *see* Linear group
Geodesic, 540
Geometer's Sketchpad (see also
    Sketchpad Experiments)
  applied in geometry, 370, 376,
    381–382, 470, 573
  instructions for, 557–559
Geometrically faithful model, 5
Geometric mean, 33, 367
Gergonne, J.D. (1771–1859), 397
Gergonne point, 397
GLB, *see* Greatest lower bound
GL(2), *see* linear group
Glukemeyer, 541
Golden ratio, 33
Great circle on a sphere, 25, 30, 39–40,
  60, 569–570
Greatest lower bound
  definition, 578
  used in limits, 300, 317
Group, subgroup, 476; *see also* Linear
    group
Half-angle identities, 313, 321, 324,
  498–501
Half plane, 63
Half-turn, 475
Hampton Court Palace, 195
Harmonic conjugate, 17, 378, 389,
  421, 525
Harmonic mean, 379
Hemisphere, 59, 70, 570
Heron's formula, proof of, via
  cevian formula, 352
  equations of Gauss (unified),
    507–510
  trigonometry (Euclidean), 499,
    507–510
Hilbert, D. (1862–1943), 27–28, 46, 541
Hilbert's theorem, 251
Hinge (alligator) theorem, *see* SAS
    inequality
Historical notes
  Archimedes, 218
  Bolyai, 226
  Euclid, 3
  Gauss, 178
  Hilbert, 28

Lambert, 169
Lobachevski, 226
Minkowski, 102
Pasch, 66
Saccheri, 169
h-object, in Poincare's model, 528–529
Hofmann, J E., 470
Homothetic mapping, *see* Dilation
Horocycle, 549
Hyperbola, 103, 109, 420, 464
Hyperbolic geometry, 204
  angle of parallelism, 231–232
  angle-sum theorem, 228
  asymptotic parallelism, 232
  asymptotic triangles, 237
  divergent parallelism, 234
  equidistant loci, 250, 549, 552, 555
  law of cosines, 493
  law of sines, 491
  orthogonal (parallel) projection, 298
  trigonometry of right triangle, 330
Hyperbolic
  parallel postulate, 204
  parallels (*see* Divergent parallel)
  sine, cosine, 323
  trigonometry, 323–324; *see also*
    Trigonometry for unified
    geometry
Hypocycloid
  3-cusps, 393
  4-cusps, 563
Hypotenuse of a right triangle, 132, 135
Hypotheses of Saccheri, 169, 175
Ideal point, *see* Point at infinity
Identities, *see* Trigonometric identities
Identity, in a group, 476
Identity mapping, 428
Image, 411
Incenter of a triangle, 156, 255, 264,
  483, 487
Incircle of a triangle, 156, 363, 397, 483
Inclination, *see* Slope
Independence of axioms, 5
Indirect proof, 63
Inequalities
  cevian, 51, 129, 140, 506
  cosine, 503
  exterior angle, 130, 241
  Lambert quadrilateral,
    involving a, 304
  median, 50–51, 503
  midline, 290–291
  polygonal, 134
  ratios (midlines), 285, 288–289
  SAS, 137
  scalene, 135
  triangle, 19, 128; *see also* Triangle
    inequality, proof of
Inequality concept for unified
    geometry, 281

Inner product, 423
Inradius, formula for, 353, 506
Inscribed angle of
    circle, 187, 192, 356
    semicircle, 187–188, 358
Inscribed angle theorem, 357
Interior angles along a transversal, 205
Interior of
    angle, 69–70
    circle, 189, 191
    convex quadrilateral, 176
    polygon, 186
    ray, 29
    segment, 29
    triangle, 92
Intrinsic geometry of surfaces, 541
Invariants of
    affine transformations, 434
    inversions, 486
    isometries, 445
    projective transformations, 418
    similitudes, 445
Inverse of a
    group element, 476
    transformation, 415, 426, 478
Inversion, 477
    invariants, 486
    involutory property, 478
    properties, 478–479
Isometry, 114, 119
Isomorphic models, 5
Isosceles
    trapezoid, 256
    triangle, 111, 132
    triangle theorem, 112
Jordan, C. (1828–1922), 195
Jordan closed curve theorem, 195
Kant, Immanuel (1724–1804), 206, 526
Kite, 127
Klein, Felix, (1849–1925), 204, 489,
        526, 539
Klein's models, 489, 526, 539
Lambert, Johann H. (1728–1777), 167
Lambert quadrilateral, 167–168, 283,
        294, 316
    formulas, 338–339
    on a sphere, 167
Lattice points, 45
Lattitude, 570
Law of sines (limiting case), 309
Laws of sines and cosines
    Euclidean geometry, 278–279
    unified geometry, 490–493, 497
Least upper bound, 578
    angle measure, 55
    angle of parallelism, 231
    distance, 25
    used in limits, 300, 317
Legendre Adrien Marie (1752–1833),
        207, 209, 526

Legendre's
    first Theorem, 208
    second Theorem, 210, 222, 226
Legs of
    isosceles triangle, 111
    right triangle, 132
    Saccheri quadrilateral, 166
    trapezoid, 256
Length of a segment, 47
L'Huillier's formula, 510
L'Huillier, Simon (1760–1810), 510
Limit curve, see Equidistant locus
Limit parallel, see Asymptotic parallel
Linear group, 476
Linearity number, 383, 519
Linear pair theorem, 71
Linear transformation, see
        Transformations
Line coordinates, see Ruler postulate
Line of fixed points, see Axis
Line of symmetry, 118
Line reflection, 114, 119, 429, 448
Line(s)
    asymptotically parallel, 243
    axioms of alignment, 3, 26
    coordinate equation(s), 277, 424
    determined by two points, 4, 26
    direction numbers, 423, 425
    divergently parallel, 243
    inclination, slope, 277
    as intersection of two planes, 412
    as subset of plane, 3, 412
    as undefined term, 2–3
Lobachevskian geometry, 204
Lobachevski, Nicolai (1793–1856), 204,
        226, 236, 306, 526
Locally Euclidean, 543
Locus, 116–117
Locus of points equidistant from
    endpoints of a segment, 117
    line (hyperbolic geometry),
        549, 552
    sides of an angle, 148–149
    two sets, 159–160
Logic, see Proof
Longitude, 570
LUB, see Least upper bound
Lükemeyer, G., 541
Lune, 70, 224
Major, minor arc of circle, 367–368
Manhatten metric, 101; see also Taxicab
        geometry
Matrices, Matrix theory, 437–438
    representation of linear
        transformations, 476
    subgroups, 476
Mean
    arithmetic, 367
    geometric, 33, 367
    harmonic, 379

Measure of
    angle, 54
    central angle, 368
    inscribed angle, 368
    segment (length), 47
Median inequality, 50–51, 503
Median of a trapezoid, 256
Median(s) of a triangle, 22–23, 154
    concurrence of, 155, 263, 525
    formula for, 347
Menelaus of Alexandria (c. 100), 381
Menelaus' theorem, 385, 522, 524
Meridian, 570
Metric, *see also* Distance
    axioms for, 19
    characterization of parallelism,
        234–235
Metric betweenness, 103; *see also*
        Betweenness
Metric space, 19–20, 46
Midline, 284
Midline bisector theorem, 286; *see also*
        Side-splitter theorems
Midline inequality, 290
Midpoint connector theorem, 258
Midpoint construction theorem, 48
Midpoint, of a segment, 48
Minimum
    distance from point to line, 122
    distance from point to set, 148, 160
    perimeter of inscribed triangle,
        see Fagnano's theorem
    sum of distances, *see* Fermat point
Minkowski, Hermann
        (1864–1902), 102
Models for axiomatic systems, 4–5, 30;
        see also Surface theory
    elliptic geometry, 116
    hyperbolic geometry, 535
Moise, Edwin, 2, 47
Morley, Frank (1860–1937), 275
Morley triangle, 275
Motion, 114
Mutually orthogonal
    circles, 402
    conics, 399
    coordinate form, 402, 406
Nagel, C.H., 390
Nagel point, 390
Napolean (Bonaparte)
        (1769–1821), 472
Napolean's theorem, 472
n-con
    quadrilateral, 166
    triangle, 100
Nine-point circle, 362
    Feuerbach's theorem, 363
    properties of, 362–363, 444–445
Non-Euclidean geometry, *see* Elliptic,
        Hyperbolic geometry

Normal
    equation for lines, 458, 462
    to plane, 424
    trapezoid, 285
Obtuse angle
    definition, 72
    hypothesis of, 169, 175
Open half plane, 63
Open segment, ray, *see* Interior of
        Opposite
    interior angle, 129
    poles (sphere), 87, 570
    rays, 31, 49
    sides (angles) of quadrilateral, 162
    transformation, 437
Order of
    affine plane, 14
    polygonal path, 176
    projective plane, 11
Order of points on a line, *see*
        Betweenness
Orientation of points in the plane, 437
Origin of coordinate system, 35
Origin of ray, 29
Orthic triangle theorem, 266
Orthocenter, 264
Orthogonal, *see* Mutually orthogonal;
        Perpendicular
Orthogonal projection, 297, 305
Overlapping triangles, 99, 575
Pappus of Alexandria (c. 300), 273
Pappus, theorem involving
    area, 273
    collinearity, 15, 398, 420, 525
Parabola, 103, 464
Parabolic geometry; *see* Euclidean
        geometry
Paragraph proof, 574
Parallelism in Euclidean geometry
    projection, 259–260
    properties F and Z, 252
    symmetry law, 259
    transitivity law, 259
Parallelism in hyperbolic geometry,
        232–234
    angle of, 231, 233, 235
    asymptotic, 232, 234, 237, 548
    distance of, 236
    divergent, 234, 548
    symmetry, 243
    transitivity, 243
Parallelism in unified geometry,
        203–204, 206
Parallel lines, 11, 13, 203–206,
        231–236, 427; *see also*
        Parallelism
Parallelograms
    definition, 256
    elementary properties, 256
    translations, 466

Parallel postulate, for; *see also* Euclid's
    Fifth Postulate
  affine geometry, 13
  elliptic geometry, 204
  hyperbolic geometry, 204
  parabolic geometry, 204
Parallel projection
  in Euclidean geometry, 259–260,
      270, 414
  in unified geometry, 297–298
Parametric equations for
  line, 424–425
  orthogonal circles, 406
Pasch, Moritz (1843–1930), 65–66
Pasch, postulate of, 66, 68
Peano, Giuseppe (1858–1932), 66
Peaucellier, A. (1832–1932), 488
Peaucellier's cell, 488
Pentagon, regular, 33, 178, 269, 371
Perimeter
  orthic triangle, 454
  regular polygon, 513–514
Perpendicular, 21–22, 71, 73–74
  between two lines, 151, 168, 206,
      234, 241, 252
  bisector, 21–22, 24, 117, 154
  criterion for, 74, 423
  distance, 121
Perpendicular-bisector theorem, 117
Perpendicular, foot of, 121
Pi, 182
Plane
  equation in E^3, 424
  properties in E^3, 412
  as undefined term, 3
Plane separation postulate, 63
Playfair, John (1746–1819), 203
Playfair's postulate, 203, 590
Poincaré, Henri (1854–1912), 526
Poincaré's disk model, 525–537,
      554–555
  angle measure, 530
  distance, 528
  parallelism, 535, 537
Point
  at infinity (ideal point), 14, 237, 412, 548
  reflection, *see* Central reflection
  of tangency, contact, 187
  ultra-ideal, 548
  as undefined term, 2–3
Polar opposites (on a line), 37, 87
Pole-polar theory, 214–217
Poles, geographical, 570; *see also*
      Extremal, Polar opposites
Polygon
  angles, sides of, 176
  convex, 177
  definition of, 176
  diagonals of, 162
  regular, 177–182

Polygonal inequality, 134
Polygonal path, 162, 176
  closed, 162
  simple, 162
Pons asinorum, 111, 124
Positive definite property
  of angle measure, 54
  of distance, 19
Positive orientation, 437
Postulate(s)
  of continuity, 590
  in the *Elements*, 1
  of Pasch, 65, 250
Power of a point, 353, 364–365
Power theorems, 356
Pre-image, 426–427
Prime numbers, 11, 178–179
Proclus (c. 410–485), 170
Proclus' argument, 227
Product (compostition)
  closure, in a group, 476
  of two transformations, 414
Product-to-sum identity for, *see also*
      Trigonometric identities
  $c(\theta)$, 312–313
  $C(x)$, 321, 507
  $s(\theta)$, 312–313
  $S(x)$, 507
  cos x, 315
  cosh x, 324
  sinh x, 500
Projection, *see* Projective
      transformation
Projective plane
  infinite, 11, 62
  7-point model for, 10
  13-point model for, 16
  postulates for, 10
Projective transformation, 414
  fixed points, 417
  fundamental theorem, 421
  invariants, 418
Proof in geometry, 63, 573–575
  axioms and definitions, 573
  outline (T-proof), 574
Proportional, 262
Protractor, 57–58; *see also* Circular
      protractor theorem
Pseudosphere, 542
Ptolemy, Claudius (c. 85–165), 360
Ptolemy's theorem, 360
Pythagorean theorem in
  Euclidean geometry, 266–267
  unified geometry, 325–327
Quadrilateral, 162
  angles and sides of, 162
  convex, 163, 166
  cyclic, 359–360
  diagonals of, 162, 164
  Saccheri, 166–168

Quadrilaterals, congruence of, *See* Congruence criteria
Radian measure, 55
Radical axis, 365, 403, 482
Radius of
    circle, 187
    inversion, 477
    sphere, 569
Ratio inequalities, *see* Inequalities
Ray, 29
Real number line, 35
Real number system, 577–579
Rectangle in unified geometry, 167, 171, 256
Rectangular (Cartesian) coordinates (E^2)
    2-dimensional (E^2), 276–277
    3-dimensional (E^3), 422–425
Reflection, *see also* Line reflections
    in a line, 448, 457
    in a point, *see* Half-turns
Reflexive law of congruence, 96
Region, 412
Regular polygon, 177–182
    area of, 182, 513
    circumcircle of, 181
    definition of, 177
    Euclidean construction of, 33, 178–179
    existence of, 180
    perimeter of, 182, 513
Relative consistency, 535
Right angle
    definition, 71
    hypothesis of, 169
Right asymptotic triangle, 238
Right triangle
    definition, 132
    congruence of, *see* Congruence criteria
    properties in unified geometry, 135–137
    Pythagorean theorem, 266–268, 327
    trigonometry of, *see* Trigonometry
Rigid motion, *see* Motion
Round-up metric, 34
Rotation
    coordinate form of, 468
    definition, 464
Ruler Postulate, 35
Russell, Bertrand (1872–1970), 2
Saccheri quadrilateral
    on a sphere, 167
    three hypotheses for, 169
    in unified geometry, 166, 282
Saccheri, Girolamo (1667–1773), 166, 169
Saccheri's theorem, 172
Saddle surface, 225

SAS hypothesis, 99
SAS postulate, 110, 113–116
    proved in Poincare's model, 534
    verified on the sphere, 116
Scale factor, *see* Similar triangles
Scalene inequality, 135
Scheiner, Christolph (1575–1650), 447
Scheiner's pantograph, 447
Secant of a circle, 187
Secant-tangent theorem, 356
Secant theorem, 193
Segment
    construction theorems, 47–48
    definition, 29
Semicircle
    definition of, 187
    inscribed angle of, 188
Semiperimeter, 157, 348–349
Set theory, 3
Shape of
    circle, 189–192
    triangle, 501–502
Side of a line, 63
Sides of
    angle, 53
    asymptotic triangle, 237
    polygon, 176
    polygonal path, 176
    quadrilateral, 162
    triangle, 96
Side-splitter theorem
    Euclidean geometry, 260
    Lambert quadrilaterals, 294
    normal trapezoids, 289
    right triangles, 295
Similar triangles
    in Euclidean geometry, 260–263
    non-existence of, 230
    in unified geometry, 299–300
Similarity criteria
    AA, 262
    SAS, 263
    SSS, 263
Similitude, 438–441, 456
Simple
    closed curve, 194
    polygonal path, 162
Simson line, 392–393
Simson, Robert (1687–1768), 392
Sine function, 277
Sines, law of, 278–279
Single elliptic geometry, 62, 216
Sketchpad Experiments, 16–17, 51, 70, 78, 81, 158, 196, 270–273, 393, 559–568; see also *Geometer's Sketchpad*
Slope, for lines
    defined in E^2, 27↔↔7
    under transformation, 435

Sphere (spherical geometry), 25, 59–61,
        543, 569–571
    angle measure, 60
    distance, 18, 26, 60, 337, 569
    great circle, 26, 60, 569
    hemisphere, 59, 70, 570
    lune, 70
    as model for elliptic geometry,
        60–61, 116, 210
Spherical coordinates, 540, 571
SSA hypothesis, 145
Steiner, Jacob (1796–1863), 393
Steiner–Lehmus theorem, 143, 351–352
Steiner's
    curve, 393
    porism, 488
    theorem, 393
Straight angle, 53–55, 71
Subgroup, 476
Subset, 3
Summit, summit angles, 166–167
Sum of angles, see Angle-sum
Sum-to-product identity, 507; see also
        Product-to-sum
Superposition, 96, 113
Supplementary angles, 71
Surfaces in E^3, intrinsic geometry
        of, 540
    cone, 543
    cylinder, 543
    sphere, 543–544
    tractroid, 542
Symmetry in geometry, 118
Symmetry law (property) of
    angle measure, 54
    asymptotic parallelism, 243
    congruence relation, 96
    distance, 19
    divergent parallelism, 243
    Euclidean parallelism, 259
Synthetic, 1, 27, 47, 352
Tangent function, 277
Tangent theorem, 197
Tangent to
    circle, 187, 191, 197
    ellipse, 463
    sphere, 60–61
Taxicab geometry, 103–104
Tiling, hyperbolic plane, 538
T-proof, 574
Tractrix curve, 542
Tractroid, 542
Transformation group $GL(2)$, 476
Transformation proof of
    centroid, existence of, 431
    Fagnano's theorem, 454
    Fermat's point, existence of, 470
    Napoleon's thoerem, 472
    Nine-point circle, properties of,
        444–445

Transformations, see also Projective
        transformations
    affine, 425
    affine reflection, 429
    half-turn, 475
    inversion, 477
    isometry, 438
    linear, 414
    line reflection, 448
    rotation, 464
    similitude, 438
    translation, 464
Transitive law (property) of
    asymptotic parallelism, 243
    congruence, 96
    Euclidean parallelism, 259
Translation, 464, 466
Transversal of
    angles, 76–77
    lines, 205–206, 246, 252
Trapezoid, 256
Triangle in Euclidean geometry
    area, Heron's formula, 352, 499
    Euler line, 155, 264
    Euler's formula, 487
    Fagnano's theorem, 454
    Fermat point, 391, 470
    Gergonne's point, 397
    law of cosines, 279
    law of sines, 278–279
    Morley triangle, 275, 393
    Nagel point, 390
    nine point circle, 362–363, 393
    orthic triangle, 266
    Simson line, 392–393
    trigonometry, for right triangles, 268
Triangle inequality, 19
    concurrent rays, 80
    proofs, 128, 133, 141
Triangle in unified geometry, 89, 95
    altitudes of, 22, 154, 274, 522
    angle sum, 282
    area, 510, 513
    asymptotic, 236–245
    circumcenter, 156, 264
    circumcircle, 156
    incenter, incircle, 156, 255
    law of cosines, 493, 497
    law of sines, 491, 497
    right, 129, 132, 135–137
    and Saccheri quadrilateral, 171–172
    trigonometry, for right
        triangles, 330
    tri-rectangular, 129, 136
Trichotomy property, 63
Trigonometric identities, 311–315, 496,
        498–499
Trigonometric formulas
    equations of Gauss, 508
    half-angle, 498, 500

law of cosines, 493, 497
law of sines, 491, 497
Trigonometry of right triangle in
  elliptic geometry, 330
  Euclidean geometry, 268
  hyperbolic geometry, 330
Trisecting an angle, 91, 184
Two-circle theorem, 201
Uncountable, 45–46
Undefined terms, 2
Undergraduate research projects, 35,
  109, 159–160, 186
Unified geometry, 95, 203
  angle-sum of a triangle, 282
  area of a triangle, 510, 513
  inequality concept, 281
  "in the small", 213, 231

Lambert quadrilaterals, 283
law of cosines, 497
law of sines, 497
parallel projection, 298; *see also*
  Orthogonal projection
Saccheri quadrilaterals, 283
similar triangles, 299
trigonometry for, 496
Universal set, 3
Upper half-plane model, 539
Vertex angle, isosceles triangle, 111
Vertex of an angle, 53
Vertical angles (vertical pair), 59, 72
Vertical pair theorem, 72
Vertices of a triangle, polygon, 96, 162, 176
Wallace, W. (1768–1873), 392
Warehouse problem, 471